Genome Analysis in Eukaryotes
Developmental and Evolutionary Aspects

Genome Analysis in Eukaryotes
Developmental and Evolutionary Aspects

R.N. Chatterjee
L. Sánchez

Springer-Verlag

Narosa Publishing House

EDITORS
Dr. R.N. Chatterjee
Department of Zoology, University of Calcutta
Calcutta-700 073, India

Dr. Lucas Sanchez
Cenrtro Dr Investigaciones Biologicas (CSIC)
Velaquez, Madrid, Spain

ISBN 3-540-63524-6 Berlin Heidelberg New York
ISBN 0-387-63524-6 New York Berlin Heidelberg
ISBN 81-7319-164-6 Narosa Publishing House, New Delhi

Printed in India.

Preface

The early development of different animals is extremely diverse with great variability in cell numbers and size, in rates of development, in cell lineage patterns etc. Differences and similarities in developmental strategies among organisms are stably transmitted in generation after generation. This is so because of the genetic control of developmental processes. Thus, it is the genes and their interactions that we should look for the underlying rules that govern development. These rules started to be understood through the genetic and molecular biology analyses of developmental processes in different organisms. Consequently, a comprehensive view about the mechanisms of how organisms are designed and built is emerging in recent years. In spite of the variability of developmental processes, some common genetic rules controlling these processes appear to be in action. In addition, the relationship between the processes of development in different types of embryos is a topic of great relevance to evolutionary genetics.

The present volume focuses on an wide spectrum of topics ranging from cell cycle regulation to development and evolution. This book starts with an article of control of cell cycle proliferation during animal evolution and development by N.G. Brink (Chapter 1) showing how cell cycle proteins interact with the genes in different animals and the evolution of the cell cycle system in a multicellular organisms. Differentiation in *Athalia rosae* and *Bombyx mori* has surprising about to tell us about developmental patterns in the organism other than *Drosophila*. We also received very useful reviews of chapters from Laurie Tompkins, Pedro Santamaria, Roland Rosset, Pedro P. López, Begoña Granadino, Michèle Thomas-Delaage, Neel B. Randsholt to include genetics of development, behaviour and evolution in *Drosophila*. The sex determination mechanism in *Sciara* has been reviewed by A.L.P. Perondini (Chapter 7). An evolutionary link between replication and the establishment of repressive chromatin structures has been reviewed elegantly by Francesco De Rubertis and Pierre Spierer (Chapter 10). Thus, we have been fortunate to have extraordinary cooperation from the research community. We appreciate the time and effort that these authors have invested in this book.

Our hope is that this book can guide beginners all the way through a developmental aspect and evolution in animals and at the same time provide established investigators with new ideas and thoughts.

Finally our appreciation are for M/s Narosa Publishing House, New Delhi, for their efforts in bringing out this volume in the present form.

R.N. CHATTERJEE
LUCAS SÁNCHEZ

Contributors

N.G. Brink
School of Biological Sciences. The Flinders University of South Australia, Adelaide, Australia.

R.N. Chatterjee
Department of Zoology, University of Calcutta, 35 Ballygunge Circular Road, Calcutta - 700019, India.

Francesco De Rubertis
Department of Zoology and Animal Biology, University of Geneva, 30 quai Ernest-Ansermet, CH-1211 Geneva 4, Switzerland.

K.P. Gopinathan
Department of Microbiology and Cell Biology, Indian Institute of Science, Bangalore - 560012, India.

Begoña Granadino
Centro de Investigaciones Biológicas, Velázquez 144, 28006 Madrid, Spain.

Masatsugu Hatakeyama
Department of Biology, Faculty of Science, Kobe University, Nada, Kobe 657, Japan.

Omana Joy
Department of Microbiology and Cell Biology, Indian Institute of Science, Bangalore - 560012, India.

Pedro P. López
Centro de Investigaciones Biológicas, Velázquez 144, 28006 Madrid, Spain.

Kugao Oishi
Department of Biology, Faculty of Science, Kobe University, Nada, Kobe 657, Japan.

A.L.P. Perondini
Departamento de Biologia, Instituto de Biociéncias, Universidade de São Paulo, C. Postal 11461, CEP 05422-970, São Paulo, Brazil.

Neel B. Randsholt
Centre de Genetique Moleculaire du C.N.R.S., F-91198 Gif sur Yvette Cedex, France.

Roland Rosset
Laboratoire de Génétique et Physiologie du Dévelopement IBDM, Pare Scientifique de Luminy, CNRS Case 907, 13288 Marseille, Cedex 9, France.

Lucas Sánchez
Centro de Investigaciones Biológicas, Velázquez 144, 28006 Madrid, Spain.

Pedro Santamaria
Centre de Génétique Moléculaire du C.N.R.S., F-91198 Gif sur Yvette Cedex, France.

Masami Sawa
Department of Biology, Aichi University of Education, Kariya, Aichi 448, Japan.

Amit Singh
Department of Microbiology and Cell Biology, Indian Institute of Science, Bangalore - 560 012, India.

Pierre Spiere
Department of Zoology and Animal Biology, University of Geneva, 30, quai Ernest-Ansermet, CH-1211, Geneva 4, Switzerland.

Michèle Thomas-Delaage
Laboratoire de Génétique et Physiologie du Dévelopement IBDM, Pare Scientifique de Luminy, CNRS Case 907, 13288 Marseille Cedex 9, France.

Laurie Tompkins
Department of Biology, Temple University, Phiadelphia, Pennsylvania 19122, U.S.A.

Contents

Genome Analysis in Eukaryotes: Developmental and evolutionary aspects
R.N. Chatterjee and L. Sánchez (Eds)
Copyright © 1998 Narosa Publishing House, New Delhi, India

1. Control of Cell Proliferation During Development and Animal Evolution

N.G Brink

School of Biological Sciences, The Flinders University of South Australia
Adelaide, Australia

I. Introduction

It is not the purpose of this chapter to review the cell cycle and the role of cell division within this cycle since the task would be gigantic and in addition most aspects of this complex cellular process have been extensively reviewed by a number of authors in recent years. However, I will briefly summarize the main molecular genetic controls which operate during the cell cycle and accompanying cell division and then discuss how far these control mechanisms have been conserved during the evolution of multicellular animals. I will review the variations in this control process which have occurred in a range of multicellular animals from simple marine invertebrates to complex vertebrate systems, and will only make reference to the extensive yeast literature (reviewed in Nurse, 1990; Forsburg and Nurse, 1991) where it is relevant to this review. I will not consider plants as there are some differences in cell division which probably arose early in evolution and do not easily fit into a discussion of cell proliferation control in animals.

All multicellular animals commence development as a single celled zygote following fertilization. This zygote divides many times to produce a mature adult which may have as few as 959 cells as in *Caenorhabditis elegans*, to as many as 10^{15} cells in most mammals. Although the origin of multicellular animals is still a matter of debate, it is likely that they either arose following cellularization of a multinucleate cytoplasm in a free swimming ciliate or the association of free swimming flagellates into a colony followed by subsequent diversification of these cells (see Hanson, 1977 for review). Animals with two cell layers (diploblasts) arose from a single celled ancestor whilst animals with three cell layers (triploblasts) either arose from these diploblasts or independently from a single celled ancestor. Based on molecular evidence, Christen et al (1991) suggest the latter alternative is more likely. During late pre-Cambrian (about 680 million years ago) both diploblastic and triploblastic forms probably co-existed side by side. They may have separated from a common ancestor some 800 million years ago (reviewed by Conway Morris, 1993) although based on the evidence from the Burgess Shales there may have been an explosion in diversity which generated all major body plans around 570 million years ago (see Levinton, 1992; Gould, 1994 for discussion). Whatever, the temporal interrelationships between the diploblastic and triploblastic animals, cell division and cell differentiation are common to both. If these two groups arose from a common ancestor and there was a sudden explosion in body forms giving rise to all the present day Phyla, then cell proliferation mechanisms, which are highly ordered genetically controlled processes essential for survival, are likely to be shared by all animals indicating a high degree of evolutionary conservation. The evidence supporting this will be reviewed in this chapter.

Dillon (1960) reported that nuclear organization during cell division was established in unicellular ciliates except for the formation of astral rays which are confined to metazoans. Kubai (1975) and Health (1980) have extensively reviewed mitosis in single celled eukaryotes and find the basic process of cell division is conserved although there is some diversity which probably arose during the evolution of these cells from prokaryote ancestors. In metazoan animals, cell division is initiated in the oocyte following fertilization and many division cycles subsequently take place during the cleavage stages of embryogenesis. Because cleavage cycles are rapid and highly ordered, both spatially and temporally, they are excellent systems for examining genetic control of the cell cycle. Several animal species ranging across the phylogenetic tree have served as model systems in these studies. In mammals, on the other hand, as it is difficult to isolate and manipulate the early embryos, studies of the cell cycle usually involve the use of cultured cells or cycling in the haemopoietic cell lineage (see McConnell, 1991 for review). A brief description of embryonic cleavage cycles in several animals is given in the following section.

II.   Cleavage Cycles

Amongst multicellular animals in which cleavage patterns have been studied in any detail, it appears that marine invertebrates (e.g. *Arbacia, Strongylocentrotus, Asterina, Ilyanassa* and some other molluscs) as well as nematodes (e.g. *Caenorhabditis*), have highly predictable and invariant cleavage which results in groups of cells having specified fates by virtue of their position (see Davidson, 1986). Davidson further describes cleavage in vertebrate embryos (e.g. *Xenopus*) as variable due to specification of particular blastomeres depending on their position within the embryo. This also applies to *Drosophila* and other insects where cleavage is syncitial.

The detailed mechanism of sea urchin cleavage has been discussed elsewhere (see Horstadius, 1973; Cameron and Davidson, 1991 for reviews). Cleavage is described as radial. The first two cleavage divisions are meridional and at right angles to each other with the third division being equatorial thereby producing eight equal sized blastomeres, an upper tier and a lower tier each containing four blastomeres. During the fourth cleavage the animal quartet of cells (upper tier) each undergo a meridional division producing a single layer of eight equal sized mesomeres. On the other hand, in the vegetal quartet of cells, the division spindle forms off centre because of the asymmetrically placed nucleus producing four larger macromeres on top of four micromeres. This division is equatorial. The three cell types have different developmental fates (Davidson, 1986). During the fifth cleavage cycle all cells undergo an equatorial division. The mesomeres divide to produce two tiers of eight cells and the macromeres yield two tiers of four equal sized cells. However, in the micromeres the spindle is placed off centre producing four large micromeres and four small micromeres with these two groups of cells having different fates (Khaner and Wilt, 1991). This division occurs slightly later than the other divisions. The seventh cleavage is meridional producing an embryo with 128 cells. All division cycles up to the tenth, when gastrulation begins, are synchronous, with S (DNA synthesis) alternating with M (mitotic division) and with no distinct gap periods (G_1 or G_2) separating them. Since spindles always form in this manner it suggests that the cytoskeleton must respond either to some internal or external cues to regulate spindle orientation. However, this orientation is not necessary for normal development since application of pressure which alters spindle orientations does not prevent normal pluteus larvae being formed (see Horstadius, 1973).

In the starfish, *Asterina*, as well as the sea cucumber *Synapta*, the pattern of cleavage is slightly different (Dan-Sohkawa, 1976; Dan-Sohkawa and Satoh, 1978; Mita, 1983). The first three cleavages are similar to that in sea urchins, the fourth division is meridional, the fifth equatorial with the plane of subsequent divisions alternating up until the tenth when gastrulation begins. All cells are of equal size. The ten cycles are synchronous and lack gap periods. The cell cycle time is about 35 minutes.

In echinoderms, it has been proposed that cleavages are regulated by a timing mechanism which is controlled by the nucleo-cytoplasmic ratio since half embryos begin asynchronous cleavages one cycle earlier (Mita, 1983).

Spiral cleavage is observed in case of molluscs. Here successive tiers of cells in an embryo do not have their axis of division perpendicular or parallel to the existing layer of cells, rather are acute angles generating tiers of cells in spirals. This cleavage pattern is also common amongst annelids. The limpet *Patella vulgata* illustrates the basic pattern of cleavage (van den Bigelaar, 1977). After an initial lag period of about two hours following fertilization, the egg divides to produce two equal sized blastomeres. Subsequent divisions occur every 30 minutes until the fifth cleavage after which cycles lengthen and become asynchronous because the group of cells begin to differentiate. The third cleavage is asymmetrical yielding four micromeres at the animal pole and four macromeres at the vegetal pole. The third and subsequent cleavage divisions are radial. In the freshwater mollusc, *Limnea,* cleavage is similar to that in the limpet (Hess, 1971). Centrifugation does not alter the pattern of cleavage suggesting that there may be some genetic control of spindle formation and orientation. This correlates with the genetic determination of dextral and sinistral coiling in snails (Hess, 1971). Some molluscs (e.g. *Dentalium* and *Ilyanassa)* extrude a bulb of cytoplasm from the fertilized egg just prior to first cleavage. This produces two unequal sized blastomeres (AB and CD), which at the four cell stage produces three equal sized blastomeres (A, B and C) and a large fourth blastomere (D). For the next four cleavage cycles one of the derivatives of this D blastomere continues to divide unequally and also slightly later than the other blastomeres (Clement, 1962). The D blastomere localizes information essential for specifying the fate of derived blastomeres as well as enabling certain cell interactions to take place (Render, 1989). It also probably contains information for the timing of its successive divisions and the orientation of the spindle during these divisions.

In the nematode, *C. elegans,* the embryonic cleavage divisions have been well documented (Sulstion et al, 1983). The first division in the cylindrical egg is unequal producing a larger AB cell and a smaller P_1 cell. This unequal first division is dependent on an organized microfilament network. The second division is asynchronous with the AB cell dividing slightly earlier than the P_1 cell. The spindle in the AB cell is oriented transversely and produces two equal size cells. All subsequent divisions within this lineage remain equal and synchronous and occur every 15 minutes (approx). The spindle in the P_1 cell, on the other hand, is oriented longitudinally and results in an EMS cell and a smaller P_2 cell. The EMS cell undergoes an equational division to produce two cells of equal size before the P_2 cell divides unequally on a longitudinal spindle to produce a C cell and a smaller P_3 cell. The P_3 cell divides to produce a D cell and a smaller P_4 cell. The orientation of the spindle appears to be dependent on the function of at least four genes (the *par* genes) which was experimentally revealed by application of cytochalasin B, a microfilament inhibitor (see review Strome, 1989). All the major cell lineages are now established and subsequent cell divisions are synchronous within any lineage but asynchronous with respect to each other. Not all products of subsequent divisions are viable with one of every two products undergoing programmed cell death (Sulston and Horvitz, 1977).

In *Drosophila,* embryonic cleavages occur within a syncitium (Rabinowitz, 1941; Sonnenblick, 1950; Zalokar and Erk, 1976). Following fertilization, the two pronuclei divide and the male and female sets of chromosomes subsequently intermingle at the spindle poles before fusing. The resultant zygotic nuclei divide synchronously about every 8 minutes for the first nine cleavage cycles (Zalokar and Erk, 1976; Foe and Alberts, 1983; Karr and Alberts, 1986). The first seven cycles take place within the interior of the egg, cycles 8 and 9 occur during migration of the nuclei to the egg periphery. Cycle 10 takes place at the periphery of the embryo. Cycles 10 to 13 are longer, with cycle 13 taking 21 minutes but all remain synchronous (Edgar and Schubiger, 1986). As in the case of sea urchins, the nucleo-cytoplasmic ratio appears to be important in specifying the number of cycles of division since in the presence of haploid nuclei there is an extra division cycle (Edgar et al, 1986). In wild-type embryos, the first ten cycles have alternating periods of S and M, whilst there is a very short G_2 period in the remaining cycles. During the fourteenth cleavage cycle, cell walls form around the 6000 syncitial nuclei and cell division occurs within 27 domains which are asynchronous with respect to each other, but synchronous within any one domain (Foe, 1989). Cells dividing within each of these domains seem to have G_2 periods of varying length. Cells in most of these domains undergo two further rounds of cycling before differentiating. Only in certain tissues (nervous system, imaginal discs and germ line) are there further rounds of cycling. However, in those cells which differentiate, endoreplication occurs giving rise to polyteny. Here the cell cycle has periods of G and S with polytenization occurring in a defined temporal and spatial sequence beginning at germ band retraction during tissue differentiation (Smith and Orr-Weaver, 1991).

Following fertilization in *Xenopus* there are twelve synchronous divisions (Signoret and Lefresne, 1971). However, because of the yolk mass the cells in the vegetal hemisphere divide slightly later than those in the animal hemisphere. The first cycle takes about 90 minutes and the subsequent 11 cycles occur every 35 minutes (Newport and Kirschner, 1982). Cycle 12 takes about 42 minutes. There is no G_1 or G_2 during these cycles. After the twelfth division, the cell cycle becomes longer and there is an increasing asynchrony between groups of cells with distinct gap phases in the cell cycle (Newport and Kirschner, 1982). Based on experimental manipulations it has been suggested that the mid blastula transition (when G_1 and G_2 phases first appear) is determined by the nucleo-cytoplasmic ratio.

The cell cycle events which take place in the early mouse embryo have been described by McConnell (1991). The first cleavage is of a longer duration (about 20 hours) and has a long G_1 period and is dependent on maternal information for its completion. Subsequent cycles only take about 12 hours and are fully regulated by the zygotic genome. It has been reported that they lack or have a very short G_1 phase (Gamow and Prescott, 1970). All cycles after the first one have a short G_2 period.

III. Genetic Regulation of the Cell Cycle

The cleavage cycles described in the previous section conform to the same basic pattern. Differences are confined to the length of the cycle and whether the Gap 1 and Gap 2 phases are present or not. The orientation of the spindle in any particular cycle of division also does not influence the cleavage cycle. In this section I will review the genetic control of the cell division cycle.

Masui and Markert (1971) injected an extract from unfertilized *Xenopus* eggs into stage 6 oocytes which triggered the entry of cells into meiosis by initiating germinal vesicle breakdown. When the same factor was injected into fertilized embryos where protein synthesis was arrested, mitotic cycling was restored (Newport and Kirschner, 1984). This factor was eventually purified (Dunphy et al, 1988; Gautier et al, 1988; Lohka et al, 1988; Draetta et al, 1989; Labbe et al, 1989a; b) and is called MPF (maturation promoting factor or mitosis promoting factor). It is present at high concentration at metaphase and low concentration at interphase.

At the same time attention was directed to the molecular components which constitute MPF. The basic component is a kinase (see Lohka, 1989; Lewin, 1990; Norbury and Nurse, 1992 for reviews) and is referred to as $p34^{cdc2}$ kinase, after an homologous protein found in *Schizosaccharomyces pombe* and the gene which codes for this kinase. It is now referred to as CDK1. There appear to be a number of possible substrates for this kinase including H1 histone, lamins, vimentin, caldesmon, cyclin B, nucleolin, myosin light chain, and $p60^{src}$ and others (see Lewin 1990; Norbury and Nurse, 1992; Nigg, 1995, for reviews). Some of these proteins may have a direct role in the cell cycle, whilst others may initiate further cascades in the control. The other component of MPF is referred to as cyclin. As the cell cycle progresses, $p34^{cdc2}$ undergoes cycles of activation and inactivation which is correlated with cycles of dephosphorylation and phosphorylation (see reviews Hunt, 1989; Lewin, 1990; Pines, 1991; Norbury and Nurse, 1992). Briefly during interphase $p34^{cdc2}$ associates with cyclin and this complex is phosphorylated during G_2. Entry into M results in the dephosphorylation of tyrosine residues on $p34^{cdc2}$ by tyrosine phosphatase which is a homologue of yeast $p80^{cdc25}$ (see Norbury and Nurse, 1992 for review) and the exit from M requires the degradation of cyclin at the metaphase-anaphase transition. The activity of $p80^{cdc25}$ also decreases upon exit from M phase. Cyclin was initially isolated from marine invertebrates (Evans et al, 1983) and is found in cycling cells from *Xenopus, Drosophila*, humans, starfish, sea urchins, clams, limpets and a number of other cell types. Evans and his colleagues initially isolated two cyclins (A and B). Both these cyclins are present in all dividing animal cells with cyclin B appearing later and also persisting later than cyclin A. Cyclin B appears to be a cyclin degraded at the metaphase-anaphase transition. Several other cyclins as well as a number of cyclin dependent kinases which appear to have a role in regulating events at the G_1/S boundary have recently been reported (see Pines, 1993; Nigg, 1995 for reviews). Their role during the cell cycle as well as the possible evolutionary significance for this diversity is discussed in Section IV.

In addition to driving the cell cycle, MPF also phosphorylates a number of other proteins including: histones, (leading to chromosome condensation), lamins (associated with nuclear envelope breakdown), intermediate filaments and a number of other non-histone proteins (see Jacobs, 1992; Ghosh and Paweletz, 1993 for reviews). These appear to be independently controlled events and not part of a developmental cascade. Therefore part of the cell division control process requires coordination of different cell cycle events.

The cell cycle can only proceed through M phase if there is a properly organized and functional division spindle permitting normal segregation of replicated chromosomes. The synthesis and organization of these spindles has been extensively reviewed (see Dustin, 1984; Mazia, 1987; Mitchison, 1988; Balczon and Brinkley 1990; McIntosh and Koonce, 1989; McIntosh, 1991; Fuller and Wilson, 1992; Kimble and Kuriyama, 1993; Wadsworth, 1993). Spindles are dynamic structures composed of microtubules polymerized from α- and β-tubulin heterodimers. They are organized by duplicated microtubule organizing centres (MTOC's) which separate during prometaphase because of their association with the spindle poles (centrosomes). MTOC's initiate spindle formation within electron dense material associated with each centriole and this is mediated at least in part by gamma tubulin (Oakley et al, 1990; Zheng et al, 1991; Joshi et al, 1992; Raff et al, 1993; Stearns and Kirschner, 1994). Microtubules have their slower growing minus ends embedded within the MTOC with their fast growing plus ends adding tubulin subunits as they move either into the peripheral cytoplasm (astral microtubules) or towards the opposite spindle pole becoming interzonal microtubules (Belmont et al, 1990). These microtubules rapidly continue to assemble and disassemble, but eventually some are captured by one of kinetochores of a chromatid pair (Hayden et al, 1990). The captured chromosome then undergoes a series of rapid poleward and antipoleward movements along the captured microtubule until both chromatids are captured resulting in the chromosome moving towards the central region of the spindle (Rieder and Alexander, 1990; Rieder and Salmon, 1984). The remaining interzonal microtubules from each pole also become stabilized as their ends interact with each other or a non-microtubule matrix surrounding the spindle and which may be composed of intermediate filaments (Cande et al, 1977; Rebhun and Palazzo, 1988). Microtubules appear to become stabilized as cells pass from prometaphase into metaphase. Cyclin B which is nuclear, localised during mitosis (Pines and Hunter, 1991a, b) may be instrumental in this stablization by interacting with the microtubule-associated protein MAP4 (Ookata et al, 1993). P34^{cdc2} kinase appears to initiate microtubule destabilization during late metaphase (Verde et al, 1990; Faruki et al, 1992) by phosphorylating centrosomal proteins which decreases the steady state length of microtubules and increases the rate of microtubule nucleation, but this is not correlated with phosphorylation of MAP2. It may however be associated with cyclin B proteolysis at the metaphase-anaphase transition. Centrosomes can initiate the formation of both microtubule and microfilament networks independently of nuclear presence (Karsenti et al, 1984; Raff and Glover, 1988) suggesting that they cycle independently

(Prescott, 1987; Yasuda et al, 1991), but respond to cell cycle cues to ensure integration of nuclear and cytoskeletal processes during mitosis.

Precise chromosome segregation during mitosis and meiosis depends on the kinetochore and associated centromeric regions. In animal cells, there is an increase in the number of microtubule binding sites in this region compared to yeast and this may be related to a higher efficiency for capturing microtubules in these cells (Zinkowski et al, 1991; Bloom, 1993). The kinetochores are complex structures composed of multiple copies of repetitive contromere specific DNA interspersed with a DNA linker region and associated with a number of centromere binding proteins (CENP's) which probably have some role in kinetochore assembly or the attachment to spindle microtubules (Saitoh et al, 1992; Bloom, 1993; Tomkiel et al 1994; Brown et al, 1994). In addition, there are a number of proteins required for chromosome movement on the mitotic spindle (see Gorbsky, 1992; Wadsworth, 1993; Sawin and Endow, 1993; Earnshaw and Pluta, 1994; Rieder and Salmon, 1994 for reviews).

It is now clear that chromosome movement towards the poles is not a "reeling in" of chromosomes attached to spindle fibres, but a "motoring" of chromosomes along spindle fibres (Mitchison, et al, 1986; Gorbsky et al, 1988; Nicklas, 1989; Wadsworth, 1993). There are apparently two kinetochore based motors of opposite polarity which direct chromosome movement along the spindle fibres (Hyman and Mitchison 1991a). One of them is an ATP-dependent minus-end directed motor represented by a family of molecules known as dyneins which may facilitate movement of chromosomes towards the poles during anaphase A. Another family of molecules, kinesins, direct plus end movement of chromosomes and seem to be located at poles and may be partly responsible for some of the rapid random movement of chromosomes during prometaphase as well as the separation of poles during prometaphase and also during anaphase B (see McIntosh and Pfarr, 1991; Gorbsky, 1992, for reviews), although their exact function remains unclear. These motors will be discussed more fully in Section IV.

A group of inner centromere proteins (INCENP's) have been isolated from mammalian cells. These proteins also termed "passenger proteins" alternate between residing on chromosomes where they may determine the plane of cell division and subsequent chromatid separation (Earnshwa and Bernat, 1990), and being localised in the midbody region where their exact function remains unknown although they may play a role in cytokinesis (see Bloom, 1993; Earnshaw and Pluta 1994, for reviews). Rappoport (1986) has reviewed the mechanisms of cytokinesis. He has examined geometric rearrangements of the mitotic apparatus within cells of various shapes and finds mitosis can be delayed by experimentally inducing a change in shape in these cells. A number of proteins seem to be associated with the contractile ring which forms between the two late anaphase nuclei. These proteins include actin, α-actinin, myosin II, filamen, INCENP's and others (see Satterwhite and Pollard, 1992, for review).

Exit from mitosis is presumably coupled with nuclear envelope reassembly and cytokinesis. This apparently requires calcium release and the binding of membrane vesicles and lamins to chromosomes. The sequence of these events is not clear as discussed in the review of Hutchison et al (1994). They have proposed three models-(i) lamins associate with chromosomes followed by nuclear envelope proteins, (ii) the nuclear envelope precursors assemble followed by the organization of lamin and (iii) simultaneous and cooperative binding of the nuclear envelope components. Of the three models, the authors are in favour of third model which allows rearrangement of the nuclear cytoskeleton along with associated chromosomes after each division cycle.

Clearly mitosis is a highly complex but integrated process. However, the analysis of the genes involved in this process remains a major challenge. Based on studies in yeast, Hartwell and Weinert, (1989) have suggested that the kinase-cyclin complex which drives the cell cycle is subjected to regulation either in G_1 or G_2. They have used the term "check point control" to describe this regulation. Thus M will not proceed if S is not completed and entry into S does not occur if mitosis or cytokinesis remain incomplete. There have been a number of recent studies in animal cells to analyse these control points and these will be discussed in Section IV. A diagram depicting the major cytoskeletal changes whose function needs to be integrated with various cell cycle proteins during various stages is shown in Figure 1.

IV. Evolution of Cell Cycle Control Systems

A variety of invertebrates and vertebrates have been used to investigate the mechanism and control of cell division in animals. Initially, a number of workers used various organisms as models for microscopic studies coupled with experimental evidences. For example, scientists have used a ciliated protozoa, such as *Tetrahymena Sp.* (see Kubai, 1975; Health, 1980 for reviews) and marine invertebrates, notably the sea urchin (where the intact mitotic apparatus was successfully isolated from oocytes in division) for investigating the structure and organisation of the mitotic spindle (Mazia and Dan, 1952). Subsequently, starfish, polychaete, annelids, some molluscs as well as insect neuroblasts or male germ line cells were successfully used to analyse the mitotic spindle (see Inoue, 1981 for review) because the cells were large, divided frequently and were suitable for histochemical staining as well as for microscopic analysis. Amongst vertebrates, Kirschner and his colleagues made extensive use of *Xenopus* egg extracts to investigate microtubule dynamics during interphase of the first cleavage cycle (Gard and Kirschner, 1987a, b; Stearns and Kirschner, 1994). This system was also found suitable for the purification of MPF (Lohka et al, 1988) and the synthesis and degradation of the cyclin component of MPF during the S and M phases of the cell cycle (Murray and Kirschner, 1989). With the advent of immunohistochemistry and confocal imaging a number of studies

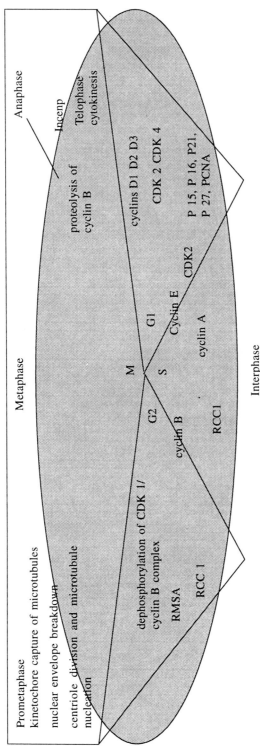

A Typical Cell Cycle
(Most Components Evolutionary Conserved)

Fig. 1.   Schematic diagram of a cell cycle of a multicellular animal. The phases of the cycle—M, G1, S and G2 are noted in each division near centre. The relative size of each division does not reflect cycle timing. Cyclins D1, D2 and D3 interact with CDK2 and CDK4 in the early to mid phases of G1, and cyclin E plus CDK2 regulates entry into S. p15, p16, p21, p27 are some of the cyclin kinase inhibitors which, along with PCNA and other molecules regulate the events in G1. Cyclin A combines with CDK1 during S and cyclin B with CDK1 during early G2 and these complexes are activated at the G2-M boundary. During cell division the cell cycle oscillator must be coupled with changes in the cytoskeleton some of which are shown in the left hand corner of the diagram. Some of the cytoskeletal changes are mediated by proteins such as RMSA, RCC1 and others. The metaphase-anaphase transition followed by telophase is a critical stage of the cycle and is characterised by the proteolysis of cyclin B and movement of INCENP proteins from nucleus to cytoplasm.

have used cell culture systems. The diversity of experimental systems used in cell division and cell cycle studies also stems from the general belief that the mitotic process is ubiquitous and the basic features have been conserved throughout evolution. The evidence supporting this view will be reviewed in this section.

Cell cycle regulation

Studies on the genetic regulation of the cell cycle had their origins in budding and subsequently fission yeast (Hartwell et al, 1974; Nurse et al; 1976: Beach et al, 1982). Genetic studies also have been carried out in *Drsophila* because it is a well documented experimental model in classical genetics and also suitable for molecular analysis. Earlier studies on genetic regulation of the cell cycle in *Drosophila* (Edgar and O'Farrell, 1989; Lehner and O'Farreell, 1989, 1990; O'Farrell et al, 1989; Jimenez et al, 1990) established a homology between yeast and *Drosophila* cyclins A and B, cdc2 kinase as well as *cdc25* and the *Drosophila* gene *string,* the gene which dephosphorylates MPF permitting an entry into metosis in both organisms. In addition, many genetic mutants have been isolated in *Drosophila* which have enabled the genetical dissections of the control of the cell division cycle (see Gonzalez et al, 1994, for review).

In other metazoan animals, components of the cell cycle oscillator other than the cell division kinase, cyclins A, B, C, D, E (Koff et al, 1991; Lew et al, 1991; Xiong, et al, 1991) and F (Chang et al, 1994) have been reported from mammalian cells. A number of forms of cyclin D have also been reported (see Sherr, 1993 for review). Cyclin C (Leopold and O'Farrell, 1991) and cyclin E (Richardson et al, 1993) are present in *Drosophila,* suggesting that, cyclins A, B, C and E are present in both invertebrates as well as vertebrates. In mammals, at least three cyclins (A, D and E) function during the G_1 and/or S phase of the cell cycle or at the G_1/S boundary and appear to have some homology with CLN gene products from *Saccharomyces cerevisiae* (see reviews Sherr 1993; 1994). Lehner and O'Farrell (1989) and Minshull et al (1990) have analysed the amino acid sequence of cyclin A and B from a number of organisms including sea urchin, clam, human, Xenopus. *S. pombe* and *Drosophila.* From these observations they concluded that some highly conserved common sequences are present in cyclin A. However, only about 30% homology has been found between cyclins A and B. An extensive study of the amino acid sequences of all cyclins by O'Farrell and Leopold (1991) show that when cyclins A, B, D and E are compared, homology is restricted to four domains containing about 157 amino acids. This region has been referred to as the cyclin box. When cyclin C and the *Saccharomyces crevisiae CLN* genes are included in the abovementioned group then the homology is restricted to two domains each containing 15 and 24 amino acids respectively. Only two amino acids are completely conserved in all cyclins. O'Farrell and Leopold (1991) also suggested that the two amino acids based on similar homology with the *ras* oncogene protein are important for interaction with other proteins (namely cdc2 or related kinases). Furthermore, the spatial and temporal distribution and tissue specificity of these cyclins may be dependent on regions C-terminal to cyclin box.

In addition to the structural homology, there also appears to be a functional homology as a cyclin protein from one organism can overcome a cell cycle deficiency caused by the absence of this protein in an entirely unrelated organism. Swensen et al (1986) demonstrated the functional homology of G_2 cyclin by the injection of clam cyclin A protein into *Xenopus* oocytes which induced entry into meiotic M-phase. Cyclins C and E in *Drosophila* (Leopold and O'Farrell, 1991; Richardson et al, 1993) and the various putative G_1 cyclins in human HeLa cells also provide Cln (G_1 cyclin) function in yeast (Lew et al, 1991). However, these authors also found that human cyclin B rescued *cln*- yeast mutants which was not the case when yeast G_2 cyclin was used in the assay. This may suggest some looseness in the binding of human cyclins which may be related to secondary structure around the cyclin box. Pines and Hunter (1994) have shown that a 42 amino acid region located N-terminally outside the cyclin box is necessary for cytoplasmic location of cyclin B and nuclear localization of cyclin A during interphase.

A similar homology exists for the cell cycle kinases. This was first demonstrated by Lee and Nurse (1987) when they found a completely conserved sequence of 16 amino acids (PSTAIRE sequence) in *Schizosaccharomyces pombe*, *Saccharomyces cerevisiae* and starfish M phase kinases. The same sequence was subsequently found in *Xenopus* (Fang and Newport, 1991) who reported two cdc2 kinases with different molecular weights, one controlling the G_1-S transition and the other affecting the S-G_2 transition. In *Drosophila*, two related kinases were also found with the product, one of them (Dmcdc2) complementing the corresponding mutant (cdc2) in yeast. The other (Dmcdc2c) probably has homology to CDK2 reported in human cells (Hunter and Pines, 1991). Both products share 56% homology including the PSTAIRE sequence (Lehner and O'Farrell, 1990; Jimenez et al. 1990). In humans a whole family of cdc2 related proteins have been reported (Meyerson et al, 1992). Five of them share about 44%–65% homology and a conserved PSTAIRE sequence with a couple of exceptions. They bind different G_1 cyclins. Seven of them appear to be unique. The various cyclins may target different CDK'S to different locations in cells or tissues or promote the binding of other regulatory molecules to the CDK/cyclin complex. These and other possible functions are discussed by Pines (1993). What is the reason for the increased number of CDK's and cyclins in higher organisms? It is possible that these extra molecules have evolved in vertebrates as a consequence of the duplication of the basic *cdc2*, cyclin A and cyclin B genes present in simple marine invertebrates.

These duplicated genes would progressively evolve new, but partially overlapping functions. The associated redundancy in the control system would increase chances of survival if one molecule is knocked out as a consequence of gene mutation. Cyclin F may be one such example as this cyclin appears to be present throughout the cell cycle (Chang et al, 1994).

Structure and function of the mitotic apparatus

The mitotic apparatus has been conserved in all muticellular animals, although there are differences present in the structure when compared with single celled eukaryotes (see reviews Kubai, 1975; Heath, 1980). Here the organization and function of the basic components of the mitotic spindle apparatus in metazoan animals, particularly in *Drosophila* and vertebrate cells the two most widely studied systems, have been discussed. Microtubules, centrosomes and associated proteins, microtubule associated proteins (MAP's) centromeric associated proteins (CENP's), microfilaments and their affiliated proteins have all been identified along with the spindle and therefore are likely to have some role in its function. However, the functions of many of these proteins is not yet understood. In addition it is likely that there are other MAP's and CENP's which have not yet been isolated.

Microtubules

In vertebrates there are 6 α-tubulin and 6 β-tubulin genes specifying at least 5 isotypes in each case (Cleveland, 1987). In *Drosophila* four isoforms have been identified (see Gonzalez et al 1994, for review). Cleveland (1987) has outlined two alternative hypotheses to explain this diversity, where each isoform either performs its own unique function or is multifunctional, with different regulatory sequences accounting for the differing temporal and spatial specificities. The products of every gene are generally conserved except that in the carboxy-terminal regions show variations. It is likely that the different isoforms arose from gene duplications with some subsequent evolutionary divergence in function. The variability in the carboxy-teminal region may permit association with different proteins allowing specificity in microtubule function, along with some redundancy. Vertebrates with more isoforms may display more redundancy. Some of these binding proteins may be microtubule associated proteins (MAP's).

Microtubule associated Proteins (MAP's)

Kellogg et al (1989) have undertaken a systematic search for MAP's in *Drosophila*, and have raised antibodies to 24 proteins which have specific locations to centrosomes, asters, spindles or kinetochores. This study indicates that there are numerous proteins required for the function of the mitotic spindle, and many of the vertebrate homologues have not yet been isolated. Some of the MAP's described by Kellogg are "motor proteins". There are two families of motor proteins. They are kinesins which control movement from minus to plus (i.e. they are "plus" end directed motors) and dyneins, which are "minus" end directed motors (see Carpenter, 1991; Hyman and Mitchison, 1991a; McIntosh and Pfarr, 1991; Sawin and Endow, 1993 for reviews).

Genes coding for mitotic spindle associated kinesin motors, were first identified in fungi and homologous genes have been described in *Drosophila* (Stewart et al, 1991; Heck et al, 1993) and *Xenopus* (Le Guellec et al, 1991). Duplicated spindle poles fail to separate during prophase in mutants with a consequent failure of chromosome segregation. Furthermore, mRNA products only appear to be present in proliferating cells of *Drosophila* and *Xenopus* (see Goldstein, 1993, for review). The motor domain as well as the tail domain are highly conserved between these mitotic kinesins compared to other membrers of the family suggesting an interaction with other components of the mitotic apparatus to facilitate their function (Goldstein, 1993)

Dynein has been located within the kinetochores and along parts of the spindle (Pfarr et al, 1990), and together with kinesins may be partly responsible for the rapid movement of chromosomes during prometaphase (Rieder and Alexander, 1990). However, the underlying mechanisms explaining these rapid poleward-antipoleward movements is not understood and may involve a combination of microtubule motors and general microtubule instability (Rieder and Salmon, 1994). Nevertheless, as more microtubules attach to kinetochores there is a stabilization of prometaphase movement (Hyman and Mitchison, 1991b). However, Hyman and Mitchison found no evidence to confirm that dynein is a motor controlling poleward movement of chromosomes at anaphase even when all kinetochores are attached to microtubules.

Two mutants in *Drosophila, nod* (Carpenter, 1973; Zhang et al, 1990) and *ncd* (Endow et al, 1990) which cause meiotic chromosomal non-disjunction code for kinesin-like proteins. nod^+ codes for a typical plus-directed kinesin, however, ncd^+ has its kinesin motor at the carboxy terminal end and is therefore a minus-directed motor (Walker et al, 1990; McDonald et al, 1990). However, it is not clearly understood which molecules are functioning as minus-end directed motors and how they direct chromosome movement at anaphase. Motor proteins are associated with a number of other cellular activities in addition to chromosome segregation at mitosis or meiosis (Goldstein, 1993; Rasmusson et al, 1994). Because of this complexity, coupled with the lack of knowledge, it is not possible to conclude that evolution has led to an increase in the number of these molecules in vertebrates to ensure the integrity of the mitotic process in these organisms.

Actin

Cande et al, (1977) reported that both actin and tubulin are components of the mitotic spindle but are distributed differently. Subsequently, it was suggested that actin has a role in chromosome condensation and mitosis (Maupin and Pollard, 1986). However, the precise function of actin during division remains unclear. In both invertebrates and vertebrates actin is coded by a family of six or more genes (see Reece et al, 1992 for review). A number of animal species have been examined including sea urchin, scallop, *Drosophila, Xenopous,* chicken along with several mammalian species and it has been found that the different isoforms of actin are about 90% conserved with the greatest divergence occurring in the 20 amino terminal residues. Four of these residues vary characteristically between isoforms and may indicate specific binding sites to other actin molecules or actin binding proteins (Reece et al, 1992; Miller et al, 1989). For example, the actin related molecules of centractin in mammalian cells (Clark and Meyer, 1992) and Arp1 in *Drosophila* (Fyrberg *et al.* 1994) are components of dynactin which is probably necessary for the action of dynein *in vivo* (Muhua et al, 1994). On the other hand, in *Drosophila*, Miller et al (1989) reported that most of the actin binding proteins were found in furrows at metaphase and are probably involved in cytokinesis. Mutations in genes coding for myosin light and heavy chains in *Drosophila* (Karess et al, 1991; Young et al, 1993) also disrupt cytokinesis suggesting that they may interact with actin and associated proteins to promote cytokinesis at the end of telophase (see Satterwhite and Pollard, 1992 for review). How actin is involved in the cell division process is far from clear at the present time.

Checkpoint Control Genes
One of the major challenges in recent times is to unravel how the numerous gene controlled events of the cell cycle are integrated and regulated in a way which safeguards the integrity of the process and consequently has been the basis of many recent investigations. This problem was initially identified by Hartwell and Weinert (1989) and the hypothesis further elaborated by Murray (1992). Genes which regulate passage through S and M are referred to as checkpoint controls. There are three places where these controls will operate-(i) entry into mitosis, (ii) exit from mitosis and (iii) a 'decision' point where the cell becomes committed to cycle again (Murray, 1992).

Several examples of mutants in checkpoint genes in yeast which either control entry into M (Hoyt et al, 1991; Li and Murray, 1991; Murray, 1992; Weinert and Hartwell, 1993) or S (Moreno and Nurse, 1994) have been described and this has served as an impetus to screen for similar controls in animal cells. In mammalian cells various CDK/cyclin complexes form in a particular temporal sequence during G_1 in response to mitogenic signals and therefore reflect the point within the cell cycle which any particular cell has reached (see Pines, 1993; Sherr, 1994 for reviews). In these cells there are also several regulatory proteins which appear to inhibit the function of these cyclin/kinases (see Sherr, 1994; Peter and Herskowitz, 1994, for reviews). These proteins are referred to as CKI's (cyclin dependent kinase inhibitor) proteins. For example, factor p21 which is induced by the protein from the tumour suppressor gene p53 (El-Deiry et al, 1993) is a universal inhibitor of cyclin kinases (Xiong et al, 1993). The anti-mitogenic factor, TGF-β, has been found to arrest cells in G_1. These cells have a factor (p27) which is an inhibitor of CDK2 which in turn reduces the amount of cyclin E/CDK2 and thereby arrests cells in G_1 (Polyak et al, 1994). p27 also interacts with cyclin D and CDK4. Factor p16 (Serrano et al, 1993) has also been reported to arrest cells in G_1 by inhibiting cyclin D/CDK4. This protein has four ankyrin motifs which may be involved in down regulation of CDK4. These same ankyrin repeats have been found associated with genes which regulate S-phase entry in *Drosophila* (Axton et al, 1994), suggesting that insects also have CKI's. However, whether the number of genes are similar to that in mammalian cells remains to be seen as all possible genes have probably not been isolated and characterized.

Proliferating cell nuclear antigen (PCNA) is a co-factor of DNA polymerase, and it associates with CDK/cyclin/CKI complexes at G_1/S to facilitate cell proliferation (Baserga, 1991; Sherr, 1994). This factor is also present in *Drosophila* during embryonic nuclear division cycles (Yamaguchi et al, 1991).

In embryos which undergo rapid cleavage without accompanying cell growth, checkpoint controls do not commence with the first division, but may function before the mid blastula transition when the zygotic genome begins to transcribe. For example, in *Xenopus,* blocking DNA synthesis after the 6th cleavage division prevents cells from subsequently entering mitosis (Dasso and Newport, 1990) suggesting a likely genetic control point at the G_2/M transition. There may be two reasons for this. If a proportion of the DNA is unreplicated then ERK2 (a mitogen MAP kinase) is activated causing mitotic arrest by blocking the degradation of cyclin B and a consequent failure of spindle assembly (Minshull et al, 1994). Alternatively, RCC1 (Regulator of Chromosome Condensation) is a chromatin associated protein (Ohtsubo et al, 1989) which may detect incomplete DNA replication and therefore regulate entry into M. The original mutation, *tsBN2,* was isolated from BHK21/13 hamster cells and is temperature sensitive. It failed to synthesize any DNA when cultured at the non-permissive temperature and also failed to enter mitosis indicating a dependency on S phase completion (Nishimoto et al, 1978). Human RCC1 cDNA has been isolated. It codes for a protein of 421 amino acids (Ohtsubo et al, 1987). This protein has been found to rescue RCC1 mutants in yeast, *Xenopus*, and hamster cells (see Dasso, 1993).

Passage into M depends on the presence of the RMSA-1 protein which acts as a substrate for cdc2 kinase. This protein which binds to chromosomes is present in insect and mamalian cells and is probably essential for assembly of the mitotic spindle. There are likely to be other proteins with complimentary or similar functions which have not yet been isolated or adequately characterised. In *Xenopus* and mammalian cells, metaphase to anaphase transition and exit from mitosis is not only dependent on inactivation and/or degradation of the cyclin cofactor, but also on loss or proteolysis of some unidentified protein(s) (Holloway et al, 1993).

One such protein may be CENP-E which initially binds to kinetochores before subsequently localizing to the midbody before being lost (degraded?) at the end of mitosis (Brown et al, 1994)

Therefore, it appears that many of the proteins which regulate entry into M from G_2 and exit from M in vertebrate cells are also present in insects as well as yeast and may indicate that this control point has been conserved through evolution. However, it is very likely that many proteins of this type will be isolated in vertebrate systems which may again reflect duplication or redundancy of these control points.

V. Conclusions and Future Directions

It is evident from the foregoing discussion that the basic structure and organization of the mitotic spindle apparatus has remained unchanged throughout evolution of multicellular animals. This also applies to motor proteins, as strong homologies in structure and function have been found between vertebrates and invertebrates. The basic cyclin/CDK cell cycle regulatory mechanism has also been conserved in all eukaryotes. During the explosion of metazoan animal evolution which purportedly occurred during the Cambrian period about 570 million years ago (Levinto, 1992; Gould, 1994), the common ancestor(s) presumably shared the same mechanism for cell division, although minor differences have occurred in this mechanism amongst unicellular eukaryotes since both Heath (1980) and Kubai (1975) described variations amongst present day descendants of these unicellular forms. However, it must be said that information on the cell cycle in diploblastic animals (e.g. Cnidarians) is fairly sketchy at present.

Have any variations arisen in the control of the cell cycle and cell division throughout animal evolution? Probably yes, but these controls seem to be associated with modulation of the basic cell division pattern. Checkpoint or feedback controls which integrate the various steps in mitosis and preserve the integrity of the process appear to be more complex in vertebrate cells. At present much effort is being directed to understand the role of actin and intermediate filaments in cell division as well as elucidating the role of the various MAP's CENP's, INCENP' (passenger proteins) and centrosome associated proteins which have been or are still to be isolated. The isolation of conditional mutants coupled with homology and complementation studies which were successful in dissecting cell cycle control, will play an important part in these studies.

Most components of the cell cycle oscillator in vertebrate systems have probably now been isolated although it is possible that others will still be discovered (see Nigg, 1995 for a review). However it is clear that the exact role and function of these elements is still far from being understood. The numerous cyclins and CDK proteins isolated from vertebrate cells reflect some duplication and associated redundancy in the system which ensures that cells divide normally. This redundancy is further modulated by a number of CKI proteins which probably interact with a whole set of other modulators during G_1 of the cell cycle. Consequently, there appear to be more of these controls than in *Drosophila*, which cannot merely be explained as a lack of research activity utilizing this organism, but is likely to reflect a system of 'fine tuning' which has evolved in more complex vertebrate cells to ensure that cells divide or fail to divide in response to particular developmental cues. In *Drosophila* some evidence has also been cited regarding the different developmental expression of components of the cell cycle oscillator (see Edgar and O'Farrell, 1989; Richardson et al, 1993; Knoblich et al, 1994) which may be related to 3'untranslated sequences (Dalby and Glover, 1992) or alternatively to upstream regulatory sequences. These integrative events with early development has generated interest and in a few years some interesting results, are expected.

The vast array of MAP's CENP's, INCENP's, motor proteins as well as other proteins associated with the mitotic spindle probably have a role in coupling the chromosome replication and segregation cycle with the centrosome replication, segregation and function cycle. *Drosophila* research in the next few years should result in the characterization of several genes discussed in the review by Gonzalez et al (1994) which are likely to fall into this class. For example, the separation of spindle poles or the maintenance of this separation may be under the control of *merry-go-round* (Gonzalez et al, 1994) and *urchin* genes (Wilson et al, 1992) along with some other genes. Glover and his colleagues (Gonzalez et al. 1994) also reported that the abnormal spindle structure and organization are affected by the genes *haywire* and *polo* (whose products are found associated with the spindle at some stage of the cell cycle) and are likely to be coding for either MAP's, CENP's or centrosome associated proteins. There may also be several other genes in this class. However, as discussed above, for the control of the mitotic oscillator we may expect the number of gene products required for monitoring entry into M from G_2, ensuring the cohesion of chromatids at metaphase and their subsequent segregation at anaphase coupled with the proteolysis of cyclins to be much greater in vertebrates when compared with invertebrates. In any case an answer should be forthcoming within the next few years as there is much interest in the control of these processes (Sigrist et al, 1995; King et al, 1995; Tugendreich et al, 1995).

One area which remains to be investigated is the orientation of the division spindle during cleavage divisions in embryos. A simple explanation which is generally put forward is that the orientation of the spindle within the cell is controlled by physical factors and orientations during subsequent divisions depend on the orientation during the previous division. There are several factors antagonisting this view. Firstly, in the case of echinoderms it has been found that no micromeres are produced in starfish although they occur in sea urchins, and in addition to this successive divisions do not always alternate between equational and meridional. Furthermore, maloriented chromosomes delay metaphase-anaphase transition in sea urchins (Sluder et al, 1994). Secondly, as cited in *Caenorhabditis*, molluscs and *Drosophila* the orientations of spindles is under genetic control as it is disrupted by mutations. In *Caenorhabditis*, spindle orientation are disrupted by *par* mutation in specific ways (Kempheus et al, 1988; Cheng et al, 1995), and this one appears to be similar to the case of treating embryos with cytochalasin B (Hill and Strome, 1988) suggesting that microfilaments are essential to preserve the normal orientations of spindles. In *Limnea*, a simple mutation converts sinistral coiling into dextral coiling of the shell (Hess, 1971) and this is a consequence of different spindle orientation. In *Drosophila*, two mutants *spno* and *dos* (Webster et al, 1992; Craig and Brink, 1996) show delay in cleavage and also depict spindles which are maloriented and sometimes fused or linked in chains. The latter mutant also has spindles without asters. Similar phenotypes are produced following treatment with cytochalasin (Callaini et al, 1992), which points to a possible role for actin in spindle organization in syncitial embryos.

The next ten years should be exciting in cell division cycle research as they should help to clarify the functional significance of the additional proteins required to regulate the cell cycle in higher eukaryotic cells.

REFERENCES

Axton, J.M., Shamanski, F.L., Young. L.M., Henderson, D.S., Boyd, J.B. and Orr-Weaver, T.L. (1990). The inhibitor of DNA replication encoded by the *Drosophila* gene *plutonium* is a small, ankyrin repeat protein. *EMBO J.* **13**, 462–470.

Balczon, R.D. and Brinkley, B.R. (1990). The kinetochore and is its roles during cell division. In *"Chromosomes: Eukaryotic, Prokaryotic and Viral"*, (K.W. Adolph ed.) Vol I, pp 167–189 CRC Press.

Baserga, R. (1991). Growth regulation of the PCNA gene *J. Cell Sci.* **98**, 433–436.

Beach, D., Durkacz, B. and Nurse, P. (1982). Functionally homologous cell cycle control genes in budding and fission yeast. *Nature* **300**, 706–709.

Bloom, K. (1993). The centromere frontier: Kinetochore components, microtubule based motility, and the CEN-value paradox. *Cell* **73**, 621–624.

Belmont, L.D., Hyman, A.A., Sawin, K.E., and Mitchison, T.J. (1990). Real-time visualization of cell cycle-dependent changes in microtubule dynamics in cytoplasmic extracts. *Cell* **62**, 579–589.

Brown, K.D., Coulson, R.M.R., Yen, T.J. and Cleveland, D.W. (1994). Cyclin-like accumulation and loss of the putative kinetochore motor CENP-E results from coupling continuous synthesis with specific degradation at the end of mitosis. *J. Cell Biol.* **125**, 1303–1312.

Callaini, G., Dallai, R., and Riparbelli, M.G. (1992). Cytochalasin induces spindle fusion in the syncitial blastoderm of the early *Drosophila* embryo. *Biol Cell* **74**, 249–254.

Cameron, R.A. and Davidson, E.H. (1991). Cell type specification during sea urchin development. Trends Genet. **7**, 212–218.

Cande, W.Z., Lizarides, E. and McIntosh, J.R. (1977). A comparison of the distribution of actin and tubulin in the mammalian mitotic spindle as seen by indirect immunofluorescence. *J. Cell Biol* **72**, 552–567.

Carpenter, A.T.C. (1973). A mutant defective in distributive disjunction in *Drosophila melanogaster. Genetics* **73**, 393–428.

Carpenter, A.T.C. (1991). Distributive segregation: Motors in the polar wind? *Cell* **64**, 885–890

Chang, B., Richman, R. and Ellidge, S. (1994). Human cyclin F. *EMBO J.* **13**, 6087–6098.

Cheng, N.N., Kirby, C.M., and Kempheus, K.J. (1995). Control of cleavage spindle orientations in *Caenorhabditis elegans:* The role of the genes *par-2* and *par-3. Genetics* **139**, 549–559.

Christen, R., Ratto, A., Broine, Perasso, R., Grell, K.G. and Adoutti, A. (1991). An analysis of the origin of metazoans using comparison of partial sequence of the 28S RNA reveals an early emergence of triploblasts. *EMBO J.* **10**, 499–503.

Clark, S.W. and Meyer, D.I. (1992). Centractin is an actin homologue associated with the centrosome. *Nature* **359**, 246–250.

Cleveland, D.W. (1987). The multitubulin hypothesis revisited: what have we learned? *J. Cell Biol.* **104**, 381–383.

Clement, A.C. (1962). Development of *Ilyanassa* following removal of the D macromere at successive cleavage stages. *J. Exp. Zool.* **149**, 193–215.

Conway Morris, S. (1993). The fossil record and the early evolution of the Metazoa. *Nature* **361**, 219–225.

Craig, S and Brink, N. (1996). A mutation *dosach* in *Drosophila* which effects aster formation and nuclear migration during cleavage. *Biol Cell* **87**, 45–54.

Dalby, B and Glover, D.M. (1992). 3'non-translated sequences in *Drosophila* cyclin B transcripts direct posterior pole accumulation late in oogenesis and peri-nuclear association in syncytial embryos. *Development* **115**, 989–997.

Dan-Sohkawa, M. (1976). 'Normal' development of denuded eggs of the starfish *Asterina pectinifera. Dev. Growth Differ.* **18**, 439–445.

Dan-Sohkawa, M. and Satoh, N. (1978). Studies on dwarf larvae developed from isolated blastomeres of the starfish *Asterina pectinifera. J. Embryol. Exp. Morphol.* **46**, 171–185.

Dasso, M. (1993). RCC1 in the cell cycle: the regulator of chromosome condensation takes on new roles. Trends Bio. Sci. **18**, 96–101.

Dasso, M. and Newport, J.W. (1990). Completion of DNA replication is monitored by a feedback system that controls the initiation of mitosis *in vitro*: studies in *Xenopus. Cell* **61**, 811–823.

Davidson, E.H. (1986). Gene activity in early development. Academic Press, New York.

Dillon, L.S. (1960). Comparative cytology and the evolution of life. *Evolution* **16**, 107.

Draetta, G., Luca, F., Westendorf, J., Brizuela, L., Ruderman, J. and Beach, D. (1989). cdc2 protein kinase is complexed with both cyclin A and cyclin B: evidence for proteolytic inactivation of MPF. *Cell* **56**, 829–838.

Dunphy, W.G., Brizuela, L., Beach, D. and Newport, J.W. (1988). The *Xenopus cdc2* protein is a component of MPF, a cytoplasmic regulator of mitosis. *Cell* **54**, 423–431.

Dustin, P. (1984). Microtubules and Mitosis. In Microtubules. Springer Verlag, Berlin.

Earnshaw, W.C. and Pluta, A.F. (1994). Mitosis. *Bioessays* **16**, 639–643.

Earnshaw, W.C. and Bernat, R.L. (1990). Chromosomal passengers: Towards an integrated view of mitosis. *Chromosoma* (Berl) **100**, 139–146.

Edgar, B.A. and O'Farrell, P.H. (1989). Genetic control of cell division patterns in the *Drosophila embryo string. Cell* **57**, 177–187.

Edgar, B.A., Kiehle, C.P. and Schubiger, G. (1986). Cell cycle control by the nucleo-cytoplasmic ratio in early *Drosophila* development. *Cell* **44**, 365–372.

Edgar, B.A. and Schubiger, G. (1986). Parameters controlling transcriptional activation during early *Drosophila* development. *Cell* **44**, 871–877.

El-Deiry, W.S., Tokino, T., Velculescu, V.E., Levy, D.B., Parsons, R., Trent, J. M., Lin, D., Mercer, E., Kinzler, K.W. and Vogelstein, B. (1993). WAF1, a potential mediator of p53 tumor suppression. *Cell* **75**, 817–825.

Endow, S.A., Henikoff, S. and Soler-Niedziela, L. (1990). Mediation of meiotic and early mitotic chromosome segregation in *Drosophila* by a protein related to kinesin. *Nature* **345**, 81–83.

Evans, T., Rosenthal, E., Youngbloom, J., Distel, D. and Hunt, T. (1983). Cyclin: a protein specified by maternal mRNA in sea urchin eggs that is destroyed at each cleavage division. *Cell* **33**, 389–396.

Fang, F. and Newport, J.W. (991). Evidence that the G_1–S and G_2–M transitions are controlled by different cdc2 proteins in higher eukaryotes. *Cell* **66**, 731–742.

Faruki, S., Doree, M. and Karsenti, E. (1992). cdc-2 kinase-induced destablization of MAP-2-coated microtubules in *Xenopus* egg extracts. *J. Cell Sci.* **101**, 69–78.

Foe, V, E. (1989). Mitotic domains reveal early commitment of cells in *Drosophila* embryos. *Development* **107** 1–22.

Foe, V.E. and Alberts, B.M. (1983). Studies of nuclear and cytoplasmic behaviour during the five mitotic cycles that precede gastrulation in *Drosophila* embryogenesis. *J. Cell Sci.* **61**, 31–70.

Forsburg, S.L. and Nurse, P. (1991). Cell cycle regulation in the yeasts *Saccharomyces cerevisiae* and *Schizosaccharomyces pombe. Annu. Rev. Cell Biol.* **7**, 227–256.

Fuller, M.T. and Wilson, P.G. (1992). Force and counterforce in the mitotic spindle. *Cell* **71**, 547–550.

Fyrberg, C., Ryan, L., Kenton, M. and Fyrberg, E. (1994). Genes encoding actin related proteins in *Drosophila. J. Mol. Biol.* **241**, 498–503.

Gamow, E. and Prescott, D.M. (1970). The cell life cycle during early embryogenesis of the mouse. *Exp. Cell Res.* **59**, 117–123.

Gard, D.L. and Kirschner, M.W. (1987a). Microtubule assembly in cytoplasmic extracts of *Xenopus* oocytes and eggs. *J. Cell Biol.***105**, 2191–2201.

Gard, D.L. and Kirschner, M.W. (1987b). A microtubule associated protein from *Xenopus* eggs which specifically promotes assembly at the plus end. *J. Cell Biol.* **105**, 2203–2215.

Gautier, J. Norbury. C. Lohka, M., Nurse, P., and Maller, J. (1988). Purified maturation-promoting factor contains the product of a *Xenopus* homolog of the fission yeast cell cycle control gene $cdc2^+$. *Cell* **54**, 433–439.

Ghosh, S. and Paweletz, N. (1993). Mitosis: Dissociability of its events. *Int. Rev.* Cytol. **144**, 217–258.

Goldstein, L.S.B. (1993). With apologies to Scheherazade: Tails of 1001 kinesin motors. *Annu Rev. Genet.* **27**, 319–351.

Gonzalez, C., Alphey, L. and Glover, D.M. (1994). Cell cycle genes of *Drosophila. Adv. Genet.* **31**, 79–139.

Gorbsky, G.J., Sammak. P.J. and Borisy, G.G. (1988). Microtubule dynamics and chromosome motion visualized in living anaphase cells. *J. Cell Biol.* **106**, 1185–1192.

Gorbsky, G.J. (1992). Chromosome motion in mitosis *Bioessays* **14**, 73–80.

Gould, S.J. (1994). Evolution of life on earth. *Sci. Amer.* **271** (4), 63–69.

Hanson, E.D. (1977). The origin and early evolution of animals. Wesleyan Pitman, London.

Hartwell, L.H. and Weinert, T.A. (1989). Checkpoints: controls that ensure the order of cell cycle events. *Science* **246**, 629–634.

Hartwell, L.H., Culotti, J., Pringle, J., and Reid, B. (1974). Genetic control of the cell division cycle in yeast. *Science* **183**, 46–51.

Hayden, J.H., Bowser, S.S. and Rieder, C.L. (1990). Kinetochores capture astral microtubules during chromosome attachment to the mitotic spindle: direct visualization in live newt lung cells. *J. Cell Biol.* **111**, 1039–1045.

Heath, B. (1980). Variant mitoses in lower eukaryotes: Indicators of the evolution of mitosis. *Int. Rev. Cytol.* **64**, 1–80.

Heck, M.M.S., Pereira, A., Pesavento, P, Yannoni, Y., Spradling, A.C. and Goldstein, L.S.B. (1993). The kinesin-like protein KLP2 is essential for mitosis in *Drosophila*. *J. Cell Biol.* **123**, 665–679.

Hess, O. (1971). Freshwater gastropods. In *Experimental embryology of marine and freshwater invertebrates.* (G. Reverberi. ed.). North Hollond Publishing Company.

Hill, D.P. and Strome, S. (1988). An analysis of the role of microfilaments in the establishment and maintenance of asymmetry in *Caenorhabditis elegans* embryos. *Dev. Biol.* **125**, 75–84.

Holloway, S., Glotzer, M., King, R.W. and Murray, A.W. (1993). Anaphase is initiated by proteolysis rather than by the inactivation of maturation-promoting factor. *Cell* **73**, 1393–1402.

Horstadius, S. (1973). *Experimental Embryology of Echinoderms.* Oxford Univ. Press (Clarendon), London.

Hoyt, M.A., Totis, L. and Roberts, B.T. (1991). *S. cerevisiae* genes required for cell cycle arrest is response to loss of microtubule function. *Cell* **66**, 507–517.

Hunt, T. (1989). Maturation-promoting factor, cyclin and the control of M-phase. *Curr. Opin. Cell Biol.* **1**, 268–274.

Hunter, T. and Pines, J. (1991). Cyclins and cancer. *Cell* **66**, 1071–1074.

Hutchison, C.J., Bridger, J.M., Cox, L.S. and Kill, I.R. (1994). Weaving a pattern from disparate threads: lamin function in nuclear assembly and DNA replication. *J. Cell Sci.*, **107**, 3259–3269.

Hyman, A.A. and Mitchison, T.J. (1991a). Two different microtubule-based motor activities with opposite polarities in Kinetochores. *Nature* **351**, 206–211.

Hyman, A.A. and Mitchison, T.J. (1991b). Regulation of the direction of chromosome movement. *Cold Spr. Harb. Symp. Quant. Biol.* **56**, 745–750.

Inoue, S. (1981). Cell division and the mitotic spindle. *J. Cell Biol.* **91**, 131s–147s.

Jacobs, T. (1992). Control of the cell cycle. *Dev. Biol.* **153**, 1–15.

Jimenez, J., Alphey, L., Nurse, P., and Glover, D. (1990). Complementation of fission yeast *cdc2*^ts and *cdc25*^ts mutants identifies two cell cycle genes from *Drosophila:* a *cdc2* homologue and *string. EMBO J.* **9**, 3565–3571.

Joshi, H.C., Palacios, M.J., McNamara, L.R. and Cleveland, D.N. (1992). Gamma-tubulin is a centrosomal protein required for cell cycle dependent microtubule nucleation. *Nature* **356**, 80–83.

Karess, R.E., Chang, X., Edwards, K.A., Kuckarni, S., Aguilera, I. and Kiehart, D.P. (1991). The regulatory light chain of nonmuscle myosin is encoded by *spaghetti squash,* a gene required for cyctokinesis in *Drosophila. Cell* **64**, 49–62.

Karr, T.L. and Alberts, B.M. (1986). Organization of the cytoskeleton in early *Drosophila* embryos. *J. Cell Biol.* **102**, 1494–1509.

Karsenti, E., Newport, J., Hubble, R., and Kirschner, M. (1984). Interconversion of metaphase and interphase microtubule arrays, as studied by the injection of centrosomes and nuclei into *Xenopus* eggs. *J. Cell Biol.* **98**, 1730–1745.

Kellogg, D.R., Field, C.M. and Alberts, B.M. (1989). Identification of microtubule-associated proteins in the centrosome, spindle and kinetochore of the early *Drosophila* embryos. *J. Cell Biol.* **109**, 2977–2991.

Kemphues, K.J., Preiss, J.R., Morton, D.G. and Cheng, N. (1988). Identification of genes required for cytoplasmic localization in early *C. elegans* embryos. *Cell* **52**, 311–320.

Khaner, O. and Wilt, F. (1991). Interactions of different vegetal cells with mesomeres during early stages in sea urchin development. *Development* **112**, 881–890.

Kimble, M. and Kuriyama, R. (1993). Functional components of microtubule-organizing centers. *Int. Rev. Cytol.* **136**, 1–50.

King, R.W., Peters, J.M., Tugendreich, S., Rolfe, M., Hieter, P., and Kirschner, M.W. (1995). A 20S complex containing CDC 27 and CDC 16 catalyses mitosis specific conjugation of ubiquitin to cyclin *B. Cell* **81**, 279–288.

Knoblich, J.A., Sauer, K., Jones, L., Richardson, H., Saint, R. and Lehner, C.F. (1994). Cyclin E controls S phase progression and its down-regulation during *Drosophila* embryogenesis is required for the arrest of cell proliferation. *Cell* **77**, 107–120.

Koff, A., Cross F., Fisher, A., Schumacher, J., Leguellec, K., Philippe, M. and Roberts, J.M. (1991). Human cyclin E, a new cyclin that interacts with two members of the *CDC2* gene family. *Cell* **66**, 1217–1228.

Kubai, D.F. (1975). The evolution of the mitotic spindle. *Int. Rev. Cytol.* **43**, 167–227.

Labbe, J.C., Picard, A., Peaucellier, G., Casadore, J.C., Nurse, P. and Doree, M. (1989a). Purification of MPF from starfish: identification of the H1 histone kinase P34^{cdc2} and a possible mechanism for its periodic activation. *Cell* **57**, 253–263.

Labbe, J.C., Carpony, J.P., Caput, D., Cavadore, J.C., Derancourt, J., Kaghad, M., Lelias, J.M., Picard, A. and Doree, M. (1989b). MPF from starfish oocytes at first meiotic metaphase is a heterodimer containing one molecule of cdc2 and one molecule of cyclin *B. EMBO J.* **8**, 3053–3058.

Lee, M.G. and Nurse, P. (1987). *cdc2* kinase, *CDC28* kinase and starfish M-phase kinase all contain a 16 amino acid conserved sequence which differs from other protein kinases. *Nature* **327**, 31–35.

LeGuellec, R., Paris, J., Couturier, A., Roghi, C and Philippe, M. (1991). Cloning by differential screening of a *Xenopus* cDNA that encodes a kinesin-related protein *Mol. Cell Biol.* **11**, 3395–3398.

Lehner, C.F. and O'Farrell, P.H. (1989). Expression and function of *Drosophila* cyclin A during embryonic cell cycle progression. *Cell* **56**, 957–968.

Lehner, C.F. and O'Farrell, P.H. (1990). *Drosophila cdc2* homologs: a functional homolog is coexpressed with a cognate variant. *EMBO J.* **9**, 3573–3581.

Leopold, P and O'Farrell, P.H. (1991). An evolutionary conserved cyclin homolog from *Drosophila* rescues yeast deficient in G$_1$ cyclins. *Cell* **66**, 1207–1216.

Levinton, S. (1992). The big bang of animal evolution. *Sci Am.* **267**, 52–59.

Lew, D.J., Dulic, V. and Reed, S.I. (1991). Isolation of three novel human cyclins by rescue of G$_1$ cyclin (Cln) function in yeast. *Cell* **66**, 1197–1206.

Lewin, B. (1990). Driving the cell cycle: M phase kinase, its partners and substrates. *Cell* **61**, 743–752.

Li, R. and Murray, A.W, (1991). Feedback control of mitosis in budding yeast. *Cell* **66**, 519–531.

Lohka, M. (1989). Mitotic control by metaphase promoting factor and cdc proteins. *J. Cell Sci.* **92**, 131–135.

Lohka, M.L., Hayes, M.K. and Maller, J.I. (1988). Purification of maturation-promoting factor, an intracellular regulator of early mitotic events. *Proc. Nalt. Acad. Sci.* USA **85**, 3009–3013.

Maupin, P. and Pollard, T.D. (1986). Arrangement of actin filaments and myosin-like filaments in the contractile ring of actin-like filaments in the mitotic spindle of dividing HeLa cells. *J. Ultrastruct. Mol. Struct.* **94**, 92–103.

McConnell, J. (1991). Molecular basis of cell cycle control in early mouse embryos. *Int. Rev. Cytol.* **129**, 75–90.

McDonald, H.B., Stewart, R.J. and Goldstein, L.S.B. (1990). The kinesin-like *ncd* protein of *Drosophila* is a minus end-directed microtubule motor. *Cell* **63**, 1159–1165.

McIntosh, J.R. (1991). Structural and mechanical control of mitotic progression. *Cold Spr. Harb. Quant. Biol.* **56**, 613–619.

McIntosh, J.R. and Koonce, M.P. (1989). Mitosis. *Science* **246**, 622–628.

McIntosh, J.R. and Pfarr, C.M. (1991). Mitotic motors. *J. Cell Biol.* **115**, 577–585.

Mazia, D. (1987). The chromosome cycle and the centrosome cycle in the mitotic cycle. *Int. Rev. Cytol.* **100**, 49–92.

Mazia, D. and Dan, K. (1952). The isolation and biochemical characterization of the mitotic apparatus in dividing cells. *Proc. Natl. Acad. Sci.* USA **38**, 826–838.

Masui, Y. and Markert, C.L. (1971). Cytoplasmic control of nuclear behaviour during meiotic maturation of frog oocytes. *J. Exp. Zool.* **177**, 129–146.

Meyerson, M., Enders, G.H., Wu, C.I, Su, L.K., Gorka, C., Nelson, C., Harlow, E. and Tsai, L.H. (1992). A family of human cdc-2-related protein kinases. *EMBO J.* **11**, 2909–2917.

Miller, K.G., Field, C.M. and Alberts, B.M. (1989). Actin binding proteins from *Drosophila* embryos: a complex network of interacting proteins detected by F-actin affinity chromatography. *J. Cell Biol.* **109**, 2963–2975.

Minshull, J., Golsteyn, R., Hill, C.S. and Hunt, T. (1990). The A and B-type cyclin associated with cdc2 kinases in *Xenopus* turn on and off at different stages of the cell cycle. EMBO J. **9**, 2865–2875.

Minshull, J., Sun, H., Tonks, W.T. and Murray, A.W. (1994). A MAP-kinase dependent spindle assembly checkpoint in *Xenopus* egg extracts. *Cell* **79**, 475–485.

Mita, I. (1983). Studies on factors affecting the timing of early morphogenetic events during starfish embryogenesis. *J. Exp. Zool.* **225**, 293–299.

Mitchison, T.J. (1988). Microtubule dynamics and kinetochore function in mitosis. *Ann. Rev. Cell Biol.* **4**, 527–549.

Mitchison, T.J., Evans, L., Schulze, E. and Kirschner, M.W. (1986). Sites of microtubule assembly and disassembly in the mitotic spindle. *Cell* **45**, 515–527.

Moreno, S. and Nurse, P. (1994). Regulation of progression through the G1 phase of the cell cycle by the *rum1*[+] gene. *Nature* **367**, 236–242.

Muhua, L., Karpova, T.S. and Cooper, J.A. (1994). A yeast actin-related protein homologous to that in vertebrate dynactin complex is important for spindle orientation and nuclear migration. *Cell* **78**, 669–679.

Murray, A.W. (1992). Creative block: cell cycle checkpoints and feedback controls. *Nature* **359**, 599–604.

Murray, A.W. and Kirschner, M.W. (1989). Cyclin synthesis drives the early embryonic cell cycle. *Nature* **339**, 275–286.

Newport, J. and Kirschner, M. (1982). A major developmental transition in early *Xenopus* embryos. I Characterization and timing of cellular changes at midblastula stage. Cell **30**, 675–686.

Newport, J. and Kirschner, M. (1984). Regulation of the cell cycle during early *Xenopus* development. *Cell* **37**, 731–742.

Nicklas, R.B. (1989). The motor for poleward chromosome movement in anaphase is in or near the kinetochore. *J. Cell Biol.* **109**, 2245–2255.

Nigg, E.A. (1995). Cyclin-dependent protein kinases: Key regulators of the eukaryotic cell cycle. *Bioessays,* **17**, 471–480.

Nishimoto, T., Eilen, E. and Basilico, C. (1978). Premature chromosome condensation in a *ts* DNA-mutant of BHK cells. *Cell* **15**, 475–483.

Norbury, C. and Nurse, P. (1992). Animal cell cycles and their control. *Annu. Rev. Biochem.* **61**, 441–470.

Nurse, P. (1990). Universal control mechanism regulating onset of M-phase. *Nature* **344**, 503–508.

Nurse, P., Thuriaux, P. and Nasmyth, K. (1976). Genetic control of the division cycle of the fission yeast *Schizosaccharomyces pombe*. *Mol. Gen. Genet.* **146**, 167–178.

Oakley, B.R., Oakley, C.E., Yoon Y. and Jung, M.K. (1990). Gamma-tubulin is a component of the spindle polebody that is essential for microtubule function in *Aspergillus nidulans*. *Cell* **61**, 1289–1301.

O'Farrell, P.H., Edgar, B.A., Lakich, D. and Lehner, C.F. (1989). Directing cell division during development. *Science* **246**, 635–640.

O'Farrell, P. and Leopold, P. (1991). A consensus of cyclin sequences reveals homology with the *ras* oncogene. *Cold Spr. Harb. Symp. Quant. Biol.* **56**, 83–92.

Ohtsubo, M., Kai, R., Furuno, N., Sekiguchi, T, Sekiguchi, M, Hatashida, H., Kuma, K., Miyata, T., Fukushige, S., Murotsu, T., Matsubara, K. and Nishimoto, T. (1987). Isolation and characterization of the active cDNA of the human cell cycle gene (*RCC1*) involved in the regulation of onset of chromosome condensation. *Genes Dev.* **1**, 585–593.

Ohtsubo, M., Okazaki, H. and Nishimoto, T. (1989). The RCC1 protein, a regulator for the onset of chromosome condensation, locates in the nucleus and binds to DNA. *J. Cell Biol.* **109**, 1389–1397.

Ookata, K., Hisanaga, S., Okumura, E., Kishimoto, T. (1993). Association of P34[cdc2]/ cyclin B complex with microtubules in starfish oocytes. *J. Cell Sci.* **105**, 873–881.

Peter, M. and Herskowitz, I. (1994). Joining the complex: cyclin dependent kinase inhibitory proteins and the cell cycle. *Cell* **79**, 181–184.

Pfarr, C.M., Coue, M., Grissom, P.M., Hays, T.S., Porter, M.E. and McIntosh, J.R. (1990). Cytoplasmic dynein is localized in kinetochores during mitosis. *Nature* **345**, 263–265.

Pines, J. (1991). Cyclins: Wheels within wheels. *Cell Growth Diff* **2**, 305–310.

Pines, J. (1993). Cyclins and cyclin-dependent kinases: take your partners. Trends Biol. Sci. **18**, 195–197.

Pines, J. and Hunter, T. (1991a). Cyclin-dependent kinases: a new cell cycle motif. Trends Cell Biol. **1**, 117–121.

Pines, J. and Hunter, T. (1991b). Human cyclins A and B1 are differentially located in the cell and undergo cell cycle-dependent nuclear transport. *J. Cell Biol.* **115**, 1–17.

Pines, J. and Hunter, T. (1994). The differential localization of human cyclins A and B is due to a cytoplasmic retention signal in cyclin B. *EMBO J.* **13**, 3772–3781.

Polyak, K., Kato, J.Y., Solomon, M.J., Sherr, C.J., Massague, J. Roberts, J.M. and Koff, A. (1994). p27[kip1], a cyclin-cdk inhibitor, links transforming growth factor TGF-β and contact inhibition to cell cycle arrest. *Genes Dev.* **8**, 9–22.

Prescott, D.M. (1987). Cell reproduction. *Int. Rev. Cytol.* **100**, 93–128.

Rabinowitz, M. (1941). Studies on the cytology and early embryology of the egg of *Drosophila melanogaster J. Morphol* **69**, 1–49.

Raff, J.W., and Glover, D.M. (1988). Nuclear and cytoplasmic cycles continue in *Drosophila* embryos in which DNA synthesis is inhibited with aphidicolin. *J. Cell Biol.*, 107, 2009–2019.

Raff, J.W., Kellogg, D.R. and Alberts, B.M. (1993). *Drosophila γ-tubulin is* part of a complex containing two previously identified centrosomal MAPs. *J. Cell Biol.* **121**, 823–825.

Rappoport, R. (1986). Establishment of the mechanism of cytokinesis in animal cells. *Int. Rev. Cytol.,* **105**, 245–281.

Rasmusson, K., Serr M., Gepner, J., Gibbons, I. and Hays, T.S. (1994). A family of dynein genes in *Drosophila melanogaster. Mol. Biol. Cell* **5**, 45–55.

Rebhun, L.I. and Palazzo, R.E. (1988). *In vitro* reactivation of anaphase-B in isolated spindles of the sea urchin egg. *Cell Motil Cytoskeleton* **10**, 197–209.

Reece, K.S., McElroy, D. and Wu, R. (1992). Function and evolution of actins. In Evolutionary Biology, 26; (Hecht, Wallace and MacIntyre eds.) Plenum Press, New York. pp 1–34.

Render, J. (1989). Development of *Ilyanassa obsoleta* embryos after equal distribution of polar lobe material at first cleavage. *Dev. Biol.* **132**, 241–250.

Richardson, H.E., O'Keefe, L.V., Reed, S.I. and Saint, R. (1993). A *Drosophila* G$_1$-specific cyclin E homolog exhibits different modes of expression during embryogenesis. *Development* **119**, 673–690.

Rieder, C.L. and Alexander, S.P. (1990). Kinetochores are transported poleward along a single astral microtubule during chromosome attachment to the spindle in newt lung cells. *J. Cell Biol.* **110**, 81–95.

Rieder, C.L. and Salmon, E.D. (1994). Motile kinetochores and polar ejection forces dictate chromosome position on the vertebrate mitotic spindle. *J. Cell Biol* **124**, 223–233.

Saitoh, H., Tomkeil, J.E., Cook, C.A., Ratrie, H.R., Maurer, M., Rothfield, N.F. and Earnshaw, W.C. (1992). CENP-C, an autoantigen in scleroderma, is a component of the human inner kinetochore plate. *Cell* **70**, 115–125.

Satterwhite, L.L. and Pollard, T.D. (1992). Cytokinesis. *Curr. Opin. Cell Biol.* **4**, 43–52.

Sawin, K.E. and Endow, S.A. (1993). Meiosis, mitosis and microtubule motors. *Bioessays* **15**, 399–407.

Serrano, M., Hannon, G.J. and Beach, D. (1993). A new regulatory motif in cell cycle control causing specific inhibition of cyclin D/CDK4. *Nature,* **366**, 704–707.

Sherr, C.J. (1993). Mammalian G1 cyclins. *Cell* **73**, 1059–1065.

Sherr, C.J. (1994). G1 phase progression: cycling on cue. *Cell* **79**, 551–555.

Signoret, J. and Lefresne, J. (1971). Contribution a l'etude de la segmentation de lioeuf d'axolotl. I. Definition de la transition blastuleenne. *Ann. Embryol. Morphogen.* **4**, 113–123.

Sigrist, S., Jacobs, H., Stratmann, R., and Lehner, C.F. (1995). Exit from mitosis is regulated by *Drosophila fizzy* and the sequential destruction of cyclins A, B and B3. *EMBO J* **14**, 4827–4838.

Smith, A.V. and Orr-Weaver, T.L. (1991). The regulation of the cell cycle during *Drosophila* embrygenesis: the transition to polyteny. *Development* **112**, 997-1008.

Sonnenblick, B. (1950). The early embryology of *Drosophila melanogaster.* In Biology of *Drosophila,* (M. Demerec ed.), John Wiley & Sons, New York, pp 62–167.

Stearns, T. and Kirschner, M. (1994). *In vitro* reconstitution of centrosome assembly and function: The central role of γ-tubulin. *Cell* **76**, 623–637.

Stewart, R.J., Pesavento, P.A., Woerpel, D.N. and Goldstein, L.S.B. (1991). Identification and partial characterization of six members of the kinesin superfamily in *Drosophila. Proc. Natl. Acad. Sci.* USA **88**, 8470–8474.

Strome, S. (1989). Generation of cell diversity during early embryogenesis in the nematode *C. elegans. Int. Rev. Cytol.* **114**, 81–123.

Sulston, J.E. and Horvitz, H.R. (1977). Post-embryonic cell lineages of the nematode *Caenorhabditis elegans. Dev. Biol.* **82**, 110–156.

Sulston, J.E., Schierenberg, E., White, J.G. and Thomson, N. (1983). The embryonic cell lineage of the nematode *Caenorhabditis elegans. Dev. Biol.* **100**, 64–119.

Swenson, K.L., Farrell, K.M. and Ruderman, J.V. (1986). The clam embryo protein cyclin A induces entry into M phase and the resumption of meiosis in *Xenopus* oocytes. *Cell* **47**, 861–870.

Tomkiel, J.E., Cooke, C.A., Saitoh, H., Bernat, R.L. and Earnshaw, W.C. (1994). CENP-C is required for maintaining proper kinetochore size and for a timely transition to anaphase. *J. Cell Biol.* **125**, 531–545.

Tugendreich, S., Tomkiel, J., Earnshaw, W., and Hieter, P. (1995). CDC27 H's co-localizes with CDC16 H's to the centrosome and mitotic spindle and is essential for the metaphase to anaphase transition. *Cell* **81**, 261–268.

Van den Bigelaar, J.A.M. (1977). Development of dorsoventral polarity and mesentoblast determination in *Patella vulgata*. *J. Morphol.* **154**, 157–186.

Verde, F., Labbe, J.C., Doree, M. and Karsenti, E. (1990). Regulation of microtubule dynamics by *cdc2* protein kinase in cell-free extracts of *Xenpus* eggs. *Nature* **343**, 233–238.

Wadsworth, P. (1993). Mitosis: spindle assembly and chromosome motion. *Curr. Opin. Cell Biol.* **5**, 123–128.

Walker, R, A., Salmon, E, D. and Endow, S.A. (1990). The *Drosophila claret* segregation protein is a minus-end directed motor molecule. *Nature* **347**, 780–782.

Webster, M., Moretti, P. and Brink, N. (1992). *Supernova (spno)*, a new maternal mutant producing variable-sized cleavage nuclei in *Drosophila*. *Genet. Res.* **60**, 131–137.

Weinert, T.A. and Hartwell, L.H. (1993). Cell cycle arrest of *cdc* mutants and specificity of the *RAD9* checkpoint. *Genetics* **134**, 63–80.

Wilson, P.G., Heck, M. and Fuller, M.T. (1992). Monastral spindles are generated by mutations in *urchin*, a *bimC* homolog in *Drosophila*. *Mol. Biol. Cell* **3**, 343a.

Xiong, Y., Connolly, T., Futcher, B. and Beach, D. (1991). Human D-type cyclin. *Cell* **65**, 691–699.

Xiong, Y., Hannon, G.J., Zhang, H., Casso, D., Kobayashi, R. and Beach, D. (1993). p21 is a universal inhibitor of cyclin kinases. *Nature,* **366**, 701–704.

Yamaguchi, M., Date, T. and Matsukage, A. (1991). Distribution of PCNA in *Drosophila* embryo during nuclear division cycles. *J. Cell Sci.* **100**, 729–733.

Yasuda, G.K., Baker, J., and Schubiger, G. (1991). Independent roles of centrosome and DNA in organizing the *Drosophila* cytoskeleton. *Development* **111**, 379–391.

Young, P.E., Richman, A.M., Ketchum, A.S. and Kiehart, D.P. (1993). Morphogenesis in *Drosophila* requires nonmuscle myosin heavy chain function. *Genes. Dev.* **7**, 29–41.

Zalokar, M. and Erk, I. (1976). Division and migration of nuclei during early embryogenesis of *Drosophila melangaster*. *J. Microsc. Biol. Cell* **25**, 97–106.

Zhang, P., Knowles, B.A., Goldstein, L.S.B. and Hawley, R.S. (1990). A kinesin-like protein required for distributive chromosome segregation in *Drosophila*. *Cell* **62**, 1053–1062.

Zheng, Y., Jung, M.K. and Oakley, B.R. (1991). Gamma-tubulin is present in *Drosophila melanogaster* and *Homo sapiens* and is associated with the centrosome. *Cell* **65**, 817–823.

Zinkowski, R.P., Meyne, J., and Brinkley, B.R. (1991). A centromere-kinetochore complex: a recent subunit model. *J. Cell Biol.* **113**, 1091–1110.

2. Maternal Information and Genetic Control of Oogenesis in *Drosophila*

Michèle Thomas-Delaage and Roland Rosset

Laboratoire de Génétique et Physiologie du Dévelopement IBDM,
Parc Scientifique de Luminy, CNRS Case 907
13288 Marseille, Cedex 9, France

I. Introduction

In *Drosophila,* by the time the egg is laid, it has received all the genetic information that is required: (i) for the cell divisions that will occur until gastrulation; (ii) for determining the anteroposterior and dorsoventral axes of polarity of the future embryo, and for delimiting the major morphogenetic areas—anterior, posterior and terminal; and (iii) to ensure the perennity of the species (germ cell determinants). This information is accumulated in the oocyte in the course of its development, that is to say during oogenesis, and results from the activity of the maternal genome alone. Zygotic transcription will only begin two hours after the egg is laid, in the pre-blastoderm stage embryo (10th division cycle).

Numerous articles and reviews have dealt with this maternal information. Our contribution here makes no claim as to being an exhaustive study; our aim is rather to undertake a survey of the most recent data and to provide an integrated overview of the subject. This two-fold objective explains the extensive bibliography, which includes, in addition to the reference works to which the reader will frequently be referred (Mahowald and Kambysellis, 1980, Lawrence, 1992, Lasko, 1994), data drawn from the most up-to-date research, which in some cases may require generally accepted ideas to be reconsidered. Particular attention will be paid to problems that are not usually dealt with in this context, such as the genetic and molecular screening used to identify the genes that carry this maternal information, and the pleiotropic expression of certain genes, which may hinder the exploitation of the results of these screenings.

We shall deal with the following points: (a) screening for genes involved in oogenesis; (b) oogenesis: morphology of the ovary and genetic control of oogenesis; (c) acquisition of polarity; and (d) oocyte maturation.

II. Screening for Genes Involved in Oogenesis.

Much of our understanding of oogenesis in *Drosophila* has come from genetical studies. In order to identify the specific function of the genes involved in oogenesis, it is necessary to distinguish between those that play a role in the maintenance of the organism (metabolism, growth, division), or in a mechanism that is involved in oogenesis and also in other stages of development (we will discuss an example later on with *Notch* and *Delta*), and those that fulfill a specific function that is indispensable to oogenesis or to embryonic development (maternal genes *stricto sensu*).

The genetic screening used to identify these genes involved isolating mutants referred to as 'female sterile'. The phenotype of these mutants may then provide an insight into the defective stage or mechanism.

An important observation is the absence of egg deposition. This may correspond to a defect in the germ line or the egg chamber, to early degeneration of the oocyte or of the egg chamber or to blockage of oogenesis. Another group of mutants correspond to a visible alteration in the morphology of the oocyte, evidenced either by an abnormality in the size or appearance of the egg, or by a defect in the establishment of the axes of polarity. Finally, eggs presenting a normal appearance may fail to develop normally, which suggests intervention of genes that are active during oogenesis, but whose function is only necessary during embryonic development.

The mutations that have been involved in oogenesis may be classified in two main groups: (a) those that affect the functions expressed specifically in the ovary and b) those that correspond to genes expressed at different stages of development. During screening, the genes of the first group are generally represented by several alleles, whereas those of the second group are represented by only one or two alleles that are often hypomorphic. Analysis of screenings for chromosome X has shown that the majority of the female sterile mutations are indeed alleles of genes essential for zygotic viability (Perrimon et al., 1986). The first screenings carried out were based on the hypothesis that genes had functions that were either strictly zygotic or strictly maternal. Genetic analysis has gradually undermined this position, and has demonstrated that it was only true for certain genes, most of them having pleiotropic effects.

In order to characterize the zygotic genes whose products are essential during oogenesis, an approach has been made in obtaining mosaic females of wild phenotype but with a mutant germ line. The study of oogenesis or embryonic development in these animals can therefore provide information on the role of these genes. In addition, this type of "chimera" makes it possible to identify the respective roles of the germ line and somatic cells in the development of the oocyte (Frey and Gutzeit, 1986). Several techniques for this have already been developed. The first consists in injecting pole cells taken from mutant embryos into a wild-type embryo (Ilmensee and Mahowald, 1974). Another technique involves using a sterile dominant mutation associated with the germ line, which results in atrophied ovaries in females. In the case of the mutations $ovoD^1$ carried by chromosome X, one can induce in a heterozygotic female (+ ovoD^l/m ovo +)—therefore sterile—germ-line clones of the genotype (m ovo +/m ovo +) which are capable of producing eggs whose phenotype can be examined and whose development can be studied (Perrimon and Gans, 1983). The frequency of these events, after irradiation by X rays, is not very high. Thus a significant improvement is the use of site-specific yeast recombinase (FLP) and its targets (FRT); this means that the number of mosaic individuals can be increased (Chou and Perrimon, 1992).

One of the questions that arise from the various screenings is regarding the saturation of each of the chromosomes in mutations that have a maternal effect. The level of saturation is difficult to determine. Recent studies of more than 800 lethal mutations linked to the X chromosome led to the conclusion that 60% of them had a maternal effect, which confirmed their pleiotropic effects. On the basis of the results obtained for chromosome X, it is suggested that in *Drosophila*, very few genes [between 75 and 250] are used exclusively during oogenesis. They might constitute a group of genes involved in important decisions (Perrimon et al., 1986, Schüpbach and Wieschaus, 1991).

Another way of tackling the problem of the maternal contribution, using a molecular approach, is to seek genes that are only expressed during oogenesis (Ambrosio and Schedl, 1984, Stephenson and Mahowald, 1987, Aït Ahmed et al., 1987). Such genes are characterized for their involvement in embryonic development and in oogenesis. This method, which is complementary to that described above, results in the characterisation of genes whose transcription is limited to the egg chamber and which were not detected by genetic screening (Lantz et al., 1992).

Finally, the enhancer trap method, which is based on the specificity of regulatory elements, should also result in identification of genes with particular specificity of expression in the egg chamber. It is, for example, obvious that the follicle cells play determining roles in the establishment of positional information in the oocyte; insertion of a reporter gene may provide evidence of this (Fasano and Kerridge, 1988).

Using the above mentioned methods one can therefore be able to identify the genes that play a role in oogenesis. Thereafter, by experimental analysis one can specify the stages at which these genes intervene, and the molecular mechanisms in which they are involved. Subsequent analysis therefore can establish the way the maternal information necessary for development in the egg. It is evident that the establishment of this information cannot be dissociated from the oogenesis process itself.

To conclude this section of genetic and molecular screening, we would like to illustrate the complexity of genetic analysis by means of two examples. The *Notch* and *Delta* genes are considered as two neurogenic genes in *Drosophila*: their inactivation leads to the formation of an excess of neuroblasts at the expense of epidermoblasts. The study of these genes has shown that they are, in addition, expressed in the ovary. The use of temperature sensitive alleles has made it possible to show that their inactivation lead to an excess of posterior follicle cells and to a defect in the formation of the follicle stalk. Furthermore, in *notch* mutants, the antero-posterior polarity is altered: *bicoid* mRNA has a double anterior and posterior localization and *oskar* mRNA is no longer detected at the posterior pole (Ruohola et al., 1991). *Brainiac,* which is also a neurogenic gene, plays a role in the establishment of dorsoventral polarity. It alters the morphology of the chorion secreted by the follicle cells, while it is expressed in the oocyte (Goode et al, 1992). Thus, certain genes that are involved in the process of neurogenesis have a discrete function during oogenesis. In both cases, they play a role in a system of intercellular signalling.

III. Oogenesis: Morphology of the Ovary and Genetic Control of Oogenesis

Oogenesis is a stage of development during which the egg is formed and the genes involved in the maternal information are expressed.

It is a process that is highly regulated from the genetic point of view, leading to the formation of an egg ready to be laid with the protective envelopes (vitelline membrane and chorion), along with its polarity and all the maternally expressed genetic information.

A brief outline of the morphology of the genital tract with some insight into its genetic control has been cited here.

A. General architecture of the ovary

The ovary of *Drosophila* is of the polytrophic meroistic type (i.e. nurse cells directly associated with the oocyte), as against the telotrophic type where the nurse cells are grouped at a distance from the oocyte, to which they are attached by a long filament. The anlage of the ovary is formed at the step of embryonic development, when the somatic ridges derived from the lateral mesoderm of fifth abdominal segment are colonised by the germ cells (about 12 elements per anlage) segregated at the posterior pole of the egg at the pre-blastoderm stage (9th to 14th mitotic cycle). Major rearrangements occur during the larval stages and especially during the pupal life; namely, multiplication of the somatic and germ elements, separation of the mass of the gland into 13 to 17 ovarioles, associating the gonads to the other components of the genital tract, ducts and accessory glands developed from the genital discs (King 1970).

Each ovariole is made up of: (a) a **terminal filament**, a single-file array of a few somatic cells (6-9), which connects each ovariole to the other ovarioles of the same ovary; (b) a **germarium** which contains germ cells and somatic elements; it is here where the future egg chambers are formed, "the basic functional units", and that differentiation of the oocyte occurs; and (c) a **vitellarium** which groups egg chambers at different stages of development, starting from the budding stage in the germarium (stage 1), to the stage when the mature egg (stage 14) is liberated into the efferent ducts where it may be fertilized by the spermatozoa stored in the spermatheca during copulation.

1. The terminal filament

The morphogenetic development of the ovarioles begins in the ovary of the 3rd stage larva by the differentiation of the terminal filaments (King, 1970). This process is disturbed by mutations in the *bric-a-brac* gene, which reduces the number of precursors of the cells of the terminal filaments and thus the number of ovarioles that are formed (Sahut-Barnola et al., 1995).

2. The germarium

This part of the ovariole has been subdivided into 3 regions to make it easier to account for the events that take place there: **region 1** contains stem cells I (2 or 3 per germarium according to Wieschaus and Szabad, 1979), stem cells II and cystoblasts, produced by the complete division of a stem cell I, and clusters of 2, 4, or 8 cells issued from incomplete cytokinesis of a cystoblast; **region 2a** groups the clusters (5 to 6 together, on average) of 16 cells formed from the 4 cycles of division of the cystoblasts; the definitive follicle chambers appear in **region 2b**, when mesodermal cells detach from the tunica propria (an acellular membrane which surrounds the germarium) and intercalate between the 16-cell clusters; in **region 3**, a monolayer of follicle cells surrounds the germ-cell cluster leading to the creation of the very first typical egg chamber; some somatic elements are also arranged in follicular stacks separating 2 consecutive chambers. **(Figure 1A)**

In any case in this part of the ovariole and at this initial stage of development two events occur that are of critical importance for the entire oogenetic process: firstly, the harmonious division of the stem cells that results, from one cell and 4 successive incomplete mitoses, in the formation of a cluster of 16 cells, the cystocyte; secondly (and virtually simultaneously), the differentiation, within this cluster, of the presumptive oocyte which remains associated by means of cellular bridges with 15 other elements of the cluster, which later develop into nurse cells.

These two stages are subjected to a system of genetic regulation, the deciphering of which is still open to speculation on many points, although there has been some progress recently.

3. The vitellarium

This is the part of the ovariole where the young egg chamber, formed in region 3 of the germarium, would develop and grow, passing from stage 1 (the final stage in the germarium) to stage 14, which corresponds to the mature oocyte stage (King, 1970). Stages 1 to 7 are known as pre-vitellogenesis

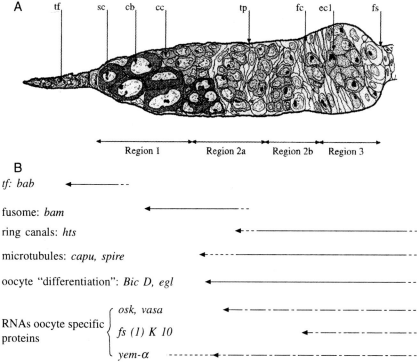

Fig. 1. A—Schematic representation of the wild-type germarium of *D. melanogaster* (original drawing from Koch and King, 1966). Region 1 contains stem cells (sc), cystoblasts (cb) and 2-cell, 4-cell and 8-cell clusters of cystocytes (cc). Region 2a groups recently completed 16–cell clusters. Region 2b contains lens-shaped clusters of 16 cystocytes which become surounded by follicle cells (fc) migrating from the *tunica propria* (tp). Region 3 corresponds to the first typical egg chamber (ec1). Other abbreviations: tf: terminal filament; fs: follicle stack.

B—Some of the genes expressed in the germarium during the very first steps of oogenesis. *bab: bric- a-brac; bam: bag of marbles; Bic D: Bicaudal D; capu: cappuccino; egl: egalitarian; fs (1) K10: female sterile (1) K10; hts: hu-li tai shao; osk: oskar; yem-α: yemanuclein-α.*

and stages 8 to 14 are called vitellogenesis. During the pre-vitellogenetic stages, the nuclei of the nurse cells undergo intensive polytenization (256C at stage 7) while the oocyte remains diploid.

Stage 8 marks the beginning of the vitellogenic phase, during which the oocyte would grow considerably (its volume increases by a factor of 90 000) thanks to the supply of material originating in the nurse cells and the flow of yolk proteins produced by the fat body and the follicle cells.

These cells are also implicated in the synthesis of the egg's envelopes: the vitelline membrane (stage 10) and the chorion (stages 11 to 13). A few cells (6 to 10), also of follicular origin, the border cells, would form the micropylar pore, which would be the passage of entry of the spermatozoa at the time of fertilization.

Every step of the process (cytoplasmic transfer from the nurse cells to the oocyte, migration of the follicle cells, production of the proteins of the yolk and of the egg membranes, etc.) is of course subjected to very precise genetic control that has been studied for a long time and is very well documented in the surveys published by Spradling (1993) and Lasko (1994).

It is during this period, and especially from stages 8 to 10, there appears localization of RNAs and proteins that are involved in the assembly of the anteroposterior and dorsoventral positional information, and the germ cell determinants at the posterior pole.

B. Genetic control of stem cell divisions.

In response to a signal, the nature of which is yet to be elucidated, one of the 3 stem cells in the apex of the germarium undergoes complete mitosis leading to the formation of another stem cell and a cystoblast; the latter undergoes 4 consecutive incomplete mitotic divisions, leading to the formation of a cluster of 16 cells, the cystocyst.

At the end of the first division of the cystoblast, the two cystocytes remain connected to each other; the mitotic spindle is not resorbed: instead it develops into an organelle known as the **fusome** (Lin et al., 1994) which is maintained from one division to the next, thus forming a real network from one cell to another. This is the **polyfusome**, which will serve as a matrix for the construction of the **ring canals** (Robinson et al., 1991).

It is easy to imagine that if the fusome did not form, the division of the cystoblast would then be a complete mitosis which would produce 2 cells of the same type; since this process would be repeated indefinitely, the product would not be a functional unit consisting of 15 nurse cells and 1 oocyte, all interconnected, but a mass of identical cells.

So the first genes that are involved in oogenesis are indeed those that determine the type of division occuring at this stage, within the germarium: their mutation leads to the proliferation of cells of the cystoblast type, and gives rise to the formation of tumours. The *bag of marbles* gene (*bam*) is one of them (Mc Kearin and Spradling, 1990). The *bam* mutation results in the appearance of tumoral egg chambers, that is to say germaria full of stem-like cells; the stem cells here divide indefinitely, each time producing 2 stem cells, or a new stem cell and a cystoblast; the process is blocked at this very early step. It has been established that the *bam* product is one of the proteins that play a key role in the constitution of the **fusome** (Mc Kearin and Ohlstein, 1995).

The adducin-like, that is a product of the *hu-li tai shao* (*hts*) gene (Yue and Spradling, 1992) and the α-spectrin are also components of the **fusome**. These products, along with the product of the *kelch* gene (Xue and Cooley, 1993), are responsible for the maintenance and consolidation of the cytoplasmic bridges, i.e. ring canals, that remain in place between the cells during the 4 mitotic cycles.

Mutation in any of these genes results in the formation of abnormal cystocyts (reduced number of cells) and thus prevents the establishment of functional egg chambers.

C. Genetic control of oocyte differentiation.

In a cluster of 16 cystocytes formed after 4 incomplete divisions of the cystoblast, all the cells are apparently identical; the only difference is the number of cellular bridges that link them viz 2 cells are connected to 4 other cells and thus have 4 bridges, 2 cells have 3 bridges, 4 cells have 2 and 8 cells are connected to only one other cell. However, not all these cells share the same fate: one cell will become the oocyte, the other 15 will become the nurse cells. This differentiation of the oocyte is a major event and the process has been extensively studied.

Many hypotheses have been proposed by King, (1970); Mahowald and Strassheim, (1970); Carpenter, (1979, 1994); and Bohrmann et al., (1992) to explain the control mechanisms of oocyte differentiation in *Drosophila.*

Among the models proposed by Suter and Steward (1991), that refers to the early differentiation by asymmetric distribution of a "factor of differentiation" present in the stem cell is consistent with a number of recent findings, in particular the demonstration by Lin and Spradling (1995) of a single spectrosome in only one of the cells of a 2-cell cluster. This organelle would be the point of anchorage of the fusome-polyfusome machinery which organises a system of one-way transport for the benefit of the presumptive oocyte.

These elements, that are indicative of very early asymmetry, disappear when the mature ring canals are formed. Other mechanisms are now established that continue to ensure this preferential transport to the oocyte: these are the polarized microtubules whose organising centre (MTOC) is localized in the oocyte (Theurkauf et al. 1993, Stebbings et al., 1995); they are also the products of the *Bicaudal D (Bic D)* and *egalitarian* genes that are known to play a role in oocyte differentiation (Driever and Nusslein-Volhard, 1988; Suter et al., 1989; Bohrmann et al., 1992; Ran et al., 1994). Their mutation gives rise to abortive egg chambers (they degenerate at about stage 6), lacking the oocyte: the 16 cells are "pseudo nurse cells". Now the *Bic D* gene product presents a similarity of sequence with the heavy chain of myosin, lamin, desmin, keratin and other proteins of the intermediary filaments, and as a component of the cytoskeleton, its function may be to localize or stabilize products (messenger RNAs and proteins) specifically addressed to the oocyte (Suter et al. 1989). It is also known that the *Bic D* mutation disturbs the reorganization of the microtubule network (Cooley and Theurkauf, 1994; Ran et al. 1994).

Together these recent findings indicated that the differentiation of an oocyte from amongst the 16 cells of the cystocyte is a state that is acheived very early. An indicator of this asymmetry is the presence of the spectrosome as soon as the first division of the cystoblast occurs. A system is then assembled that is made up of the fusome, the microtubules and the ring canals, which directs the transport to one cell, the oocyte, of cellular molecules and organelles originating in the nurse cells; this mechanism remains active throughout oogenesis (**Figure 1B**).

D. Acquisition of polarity.

The asymmetric structure of the egg when it is laid is quite apparent, and makes it possible to identify the anterior extremity by the presence of the micropyle apparatus and the chorionic appendages and also distinguish the flattened dorsal side from the more convex ventral side.

The shape of the egg and the determination of the axes of polarity of the future embryo are established during oogenesis. Thus in the ovariole, which itself is oriented according to the antero-posterior axis of the mother, the egg chambers exhibit polarization as soon as they become distinct. The oocyte is the posterior most cell of the cystocyte, and receives, at the level of what will be its anterior extremity, all the material addressed to it by the 15 nurse cells. The localization of the nucleus, which migrates into the antero-dorsal position from stage 7 of oogenesis, specifies this polarization.

Associated with these morphological elements are events that are directly linked to the expression of maternal genetic information which gives rise to 4 independent systems: a) the **anterior system**, based on the establishment of an antero-posterior gradient of the *bicoid* gene product; b) the **posterior system**, necessary for the formation of the abdomen, and whose effective agent is the *nanos* gene; c) the **terminal system** involved in the delimitation of the head and the tail of the embryo, and which depends on the localized activation of the *torso* gene product, at the ends of the egg; and d) the dorso-ventral system depending on the activation of a receptor, the *Toll* gene product, that is present all along the ventral midline of the egg.

Let us recall that the establishment of the axes of polarity of the oocyte results from interaction between the germ line and the follicle cells which contribute to the structure of the egg chamber. Thus it would appear to be increasingly evident that the polarity of the oocyte does not result from an intrinsic mechanism, but necessitates the proper assembly of the 16 germ cell cluster and of the layer of follicle cells. These cells have their own patterns of expression (Fasano and Kerridge, 1988), which shows that they contain positional information whose establishment results from cell-to-cell interactions between the germ line cells and the somatic cells. The interaction exist between these two components which results in the polarization of the egg and thus of the embryo.

1. Polarization of the oocyte

One of the first observable events in the polarization of the oocyte is the positioning of this cell to the rear of the egg chamber just forming, by a process that has yet to be explained. The microtubule network, organised from the microtubule-organising centre (MTOC) situated at this stage in the posterior part of the oocyte, then allows the selective transport of RNAs and proteins to the oocyte (Theurkauf et al., 1992). This rear positioning is absolutely essential: in *spindle-C* mutants, where it fails to occur, the result is the acquisition by the posterior follicle cells of an anterior identity (Gonzales-Reyes and St Johnston, 1994). In the acquisition of this posterior identity, the signalling system controlled by the *gurken, torpedo* and *cornichon* genes plays a major role. These genes must be expressed in the follicle cells for *torpedo* and in the oocyte for *gurken* and *cornichon* (Roth et al., 1995).

At stage 8 of oogenesis, the polarity of the microtubule network is inverted and the MTOC is positioned in the anterior part (Theurkauf et al., 1992; 1993). This inversion, which is accompanied by the disappearance of the posterior MTOC, involves the protein kinase A in the oocyte (Lane and Calderon, 1994). The integrity of the network is necessary for the localization of mRNAs such as those of *bicoid* or *oskar.* Following this repolarization, the nucleus acquires an anterior position near the cortex which will become the dorsal region of the oocyte. This double mechanism is also dependent on the *torpedo-gurken-cornichon* signalling pathway. The *gurken* mRNA is then localized around the nucleus and the Gurken protein will confer a dorsal identity on the follicle cells near the nucleus. *gurken* encodes a molecule related to TGFα and *torpedo* for a receptor of the EGF family. They are also responsible for the transmission of a signal from the oocyte to the follicle cells; *cornichon* and *brainiac* participate in the emission of the *gurken* signal. Thus the *gurken/torpedo* system plays a role, within different cells, in the establishment of the two axes of polarity (Gonzales-Reyes et al., 1995). Here again we find a situation where a mechanism is used at different stages of development.

2. Anterior-posterior polarity

The anterior- posterior pattern of the embryo is organised by two determinants localized at the two ends of the egg which behave as morphogens (Nusslein-Volhard et al., 1987). They are mRNAs which, once activated, give birth to a gradient of proteins where in one case, *bicoid,* will stimulate the transcription of target genes, and in the other, *nanos,* participate in blocking the translation of a maternal mRNA, *hunchback.*

a: The anterior determinant

The anterior determinant has been identified as the product of *bicoid.* At fertilization, *bicoid* mRNA is localized anteriorly in the oocyte. This localization is detectable from stage 9-10 of oogenesis. It is the result of selective transport from the nurse cells to the oocyte, followed by its stabilisation. The role of the microtubule network in this process has been demonstrated recently to be involved in the maintenance, as the effect of destabilising drugs indicates (Pokrywka and Stephenson, 1991), and also in the transport, which involves motor proteins, for the transfer from the nurse cells to the oocyte or for positioning within the ooplasm. Proteins of the dynein family have been identified to play this role in the oocyte.

Sequences involved in the localization of *bicoid* have been identified in the 3'UTR part of its mRNA which confers the anterior localization. The transport requires the *exuperentia* gene product, and the maintenance of *swallow* (Chao et al., 1991). Subsequently, after activation of the egg, Staufen protein will attach itself to this mRNA. The alterations resulting from their mutations act at different stages of oogenesis: stage 9 for the *exuperentia* mutant, stage 9-10 for *swallow* and 10-12 for *staufen*. This has given rise to the suggestion that *exuperentia* may be involved in transport from the nurse cells to the oocyte, a movement that results in the anterior localization of the *bicoid* mRNA (McDonald et al., 1991, 1993) and *swallow* would intervene to maintain this anterior localization whereby *staufen* would play a role when the egg is laid (Ferrandon et al., 1994). The integrity of the microtubule network is necessary for this localization (Pokrywka and Stephenson, 1991).

b: The posterior determinant

Genetic screening has made it possible to identify a class of maternal genes that have the peculiarity of being involved both in the segmental organization of the posterior abdominal region and in the establishment of the polar plasm that generates the germ cell line. These are the genes of the posterior group: *oskar, vasa, tudor, nanos, pumilio, staufen, valois, mago nashi, germcell less* (Boswell and Mahowald, 1985; Lasko and Ashburner, 1988a, b; Hay et al., 1990; Golumbesky et al., 1991), to which might be added *cappuccino* and *spire*, which, as we shall see later, play a role in the establishment of the two systems, anteroposterior and dorso-ventral.

It should be noted, however, that *oskar* plays a determining role in the formation of the polar plasm (Lehmann and Nusslein-Volhard, 1986; Kim-Ha et al., 1991; Ephrussi and Lehmann, 1992), and that *nanos* is the recognized posterior determinant (Lehmann and Nusslein-Volhard, 1991; Wang and Lehmann, 1991; Wharton and Struhl, 1991). A good illustration of the interdependence of these genes is the fact that it is the correct assembly of the polar plasm that results in the sequestration of the *nanos* mRNA at the posterior pole where its function is required.

It is, howeyer, the products of *nanos* and *pumilio* that are directly involved in the process of posterior polarization. Pumilio is a protein which has the property of fixing directly onto the *hunchback* mRNA (Murata and

Wharton, 1995). The Nanos protein gradient formed after activation of the egg will induce the formation of an inverse gradient of *hunchback* mRNA (Irish et al., 1989). The polarized effect of *nanos* results from the localization of its mRNA at the posterior pole of the egg where it is translated. Probably, interactions with the system of posterior localization then allow its translation (Gavis and Lehmann, 1992, 1994).

c: The germ cell determinants

For the embryologist and the cytologist, they represent the most extraordinary demonstration of the direct continuity of a cell line, the germ line, which is maintained from one generation to the next. In Diptera, using the simplest histological techniques, it is possible to detect, at the posterior end of the newly-laid egg, a small cap of granules; this is the polar plasm, the oosome, the polar or oosomian granules, depending on the author. The name "germ cell determinants" was given to them by Hegner (1911): having altered the posterior polar region of the egg of a Coleoptera (*Calligrapha*) with a heated needle, he noted that the resulting larva was perfectly viable, but sterile. Geigy (1931), experimenting on *D. melanogaster,* was able to prove that destruction of the polar plasm led to the establishment of gonads reduced to their somatic component alone. The polar granules thus confer on the cells that engulf them at the moment of the cellularisation of the blastoderm the properties of germ cells; this has been proved by transplantation experiments (Illmense and Mahowald, 1974; Okada et al., 1974). Mahowald (1971a) had previously demonstrated the continuity of the existence of the polar granules throughout the life of the fly and their accumulation at the posterior end of the oocyte from stage 10 of oogenesis. Mahowald (1971b) also deserves the credit for being the first to detect the presence of RNAs among the oosomian components; specific proteins were subsequently identified (Waring et al., 1978).

The discovery of the genes of the posterior group in general and the study of *oskar* in particular have resulted in significant advances in our understanding of the subject. *oskar* mRNA is in fact a component of the polar plasm and in particular exhibits the characteristics of a determinant: sterility if it is suppressed, induction of germ cells where it is transplanted (Kim-Ha et al., 1993). It is already present in the presumptive oocyte among the 16 cells of the cystocyst, in the germarium. As oogenesis proceeds, *oskar* mRNA is detected in two accumulation areas at the two ends of the oocyte (stage 8). At stage 9, it acquires its typical and definitive localization at the posterior end where it will be translated.

The correct positioning of *oskar* mRNA is under the control of several genes (Mahowald, 1992) from its first appearance in the germarium, where the products of *egalitarian* and *Bicaudal D* are necessary for its accumulation in the cell which will differentiate into an oocyte (Kim-Ha et al., 1991; Ran et al., 1994). Mutations in the *cappuccino* and *spire* genes affect the transitory accumulation of *oskar* mRNA in the anterior region of the oocyte (Manseau and Schüpbach, 1989). The products of *staufen* and *mago nashi* are necessary for the final step of localization, when the *oskar* mRNA moves to the posterior pole (St Johnston et al., 1991, 1992; Newmark and Boswell, 1994). The protein Bruno fixes on the *oskar* mRNA and thus prevents its translation as long as it is not localized at the posterior end of the oocyte (Kim-Ha et al., 1995).

We have noted above, for *oskar,* the functional importance of correct localization of the RNA; this is also the case for the RNAs of *bicoid* (Berleth et al., 1988) and *gurken* (Neuman-Silberberg and Schüpbach, 1994). In all three cases, the localization of the RNA corresponds to that of the protein. Other RNAs may present a transitory concentration in the oocyte, with no direct relation to the site of expression of the protein, and thus to the site where the function of the gene is exercised. This is what Serano and Cohen (1995) refer to as "gratuitous mRNA localization" in the case of the *fs (1) K10* gene (Prost et al., 1988). It is probably the case also for the *yemanucleine*-α gene (Aït Ahmed et al., 1992), whose mRNA presents an expression in a distinct antero-posterior gradient at stages 9 and 10, but whose protein is, in the germarium at the onset then throughout oogenesis, exclusively present in the nucleus of the oocyte. In the *spire* mutant, the *yem*-α mRNA has lost its characteristic gradient distribution pattern; the protein is however still present in the nucleus (Thomas-Delaage, unpublished results) but the nucleus no longer has its usual anterodorsal localization: it appears to float within the cytoplasm, which tends to confirm the hypothesis of the role played by the disorganization of the microtubule network, which has already been cited as a consequence of the *cappuccino* and *spire* mutations (Theurkauf, 1994). Staufen, a RNA-binding protein specific for double-stranded RNAs, would be a likely candidate to mediate of the associations between mRNA and microtubule network (Ferrandon et al., 1994).

3. The terminal system

The determination of the two ends of the embryo is under the control of a group of genes referred to as 'terminal' (Casanova, 1990). Mutations in these genes affect the development of the head and the tail of the embryo simultaneously. Since the phenotype of the various mutants is similar, it has been concluded that these genes play a role in the same process. These genes are transcribed during oogenesis either in the nurse cells (germ line) or in the follicle cells (of somatic origin). Some, for example *torso, torsolike* and *trunk,* are only involved in the establishment of the ends of the embryo; *torso* plays a key role in this process (Klinger et al., 1988). In a *torso* mutant embryo, abnormalities are apparent in the anterior part of the head and structures beyond the 7th abdominal segment. The *torso* gene codes for a tyrosine kinase activity receptor (Sprenger et al., 1989; Sprenger and Nusslein-Volhard, 1992; Doyle and Bishop, 1993). It is ubiquitously present at the surface of the embryo and results from the translation, during embryonic development, of an mRNA stored in the oocyte (Casanova and Struhl, 1993).

The receptor should be activated only at the ends of the egg. The genes involved in this process have been ordered in relation to *torso*. Thus *trunk* and *torso like* are necessary for the activation of *torso*. The product of *torso like*, which is only transcribed in a few follicle cells at the ends of the egg chamber, has been attributed the function of activator ligand of *torso*. This protein does indeed possess the characteristics of a secreted protein (Savant-Bhonsale and Montel, 1993). Recent results have however modified this model. The product of *trunk* would now appear to be this ligand. Like *spätzle*, it codes for a secreted protein capable of being cleaved and presenting a similar carboxy terminal domain. This protein synthesized in the oocyte would be activated locally by cleavage in the terminal parts of the embryo and would become the *torso* ligand (Casanova et al., 1995). Its highly localized activation would thus depend on *torsolike*.

In all, about ten genes have been identified (Perkins et al., 1992; Lu et al., 1993). They are involved in the steps necessary for activation of *torso*, or in the following steps in the case of *corkscrew, son of sevenless, ras, Draf* and *Dsor*. It should be noted that the mutations in these genes have pleiotropic lethal effects, which suggests that they have other functions besides their roles in the formation of the ends of the embryo.

Studies on the mode of transduction of the signal between the tyrosine kinase receptors and the nucleus have brought to light a very high degree of conservation of the molecules in the various organisms. In *Drosophila*, the genes playing a role in this process, therefore situated downstream of *torso*, also play a role in the cascade which is involved in the induction of the photoreceptor cell R7. These are notably *corkscrew, son of sevenless, ras, Draf* and *Dsor*. The parallels between the two mechanisms are underlined by the capacity of a constitutive allele of *torso* to activate the development of R7 in a *sevenless* mutant (Dickson et al., 1992). Downstream of this cascade, the mechanism that leads to the activation of *tailless* and *huckebein* has yet to be determined.

It is important to understand in particular how the same cascade can activate different targets of the MAP kinase (*tailless* and *huckebein* on the one hand, and *yan* and *panlid* on the other). The specificity of the mechanism depends on receptor-ligand couple (here *torso/torsolike)* on the one hand and on the other on the MAP kinase targets.

The maternal terminal system results in the activation in the embryo of the *tailless* and *huckebein* genes, which are considered as the two targets of *torso*, but also have an effect on the activity of the *bicoid* anterior morphogen and on the dorsoventral system. This activity is selectively masked at the ends of the embryo. In both cases, it can be demonstrated that this repression is linked to the activation of the cascade that transmits the signal from *torso* to *ras*, *raf* and the MAP kinase. We thus observe an interaction between the different maternal systems that are responsible for the establishment of the embryonic pattern (Rusch and Levine, 1994; Ronchi et al., 1993).

4. Dorsoventral patterning

This occurs in two steps that are chronologically quite distinct but that both result from interactions between somatic cells: the follicle and germ cells: the oocyte (Schüpbach, 1987). The first step occurs during stages 8 to 10 of oogenesis, and results in the acquisition of a dorsal identity by the follicle cells in response to stimulation emanating from the nearby nucleus; in the second step, the follicle cells situated in the opposite area, therefore presumptive ventral, then produce a ligand that will specifically activate a receptor molecule distributed ubiquitously at the surface of the oocyte, which leads to the zygotic transcription of genes responsible for the acquisition of a ventral identity.

The first visible sign of dorsoventral asymmetry in the egg chamber is the dorsal positioning of the oocyte nucleus at stage 8 of oogenesis. The dorsalizing action of the nucleus has been demonstrated by experiments involving destruction by laser (Montell et el., 1991) which result in ventralization of the egg envelope. It may thus be supposed that the nucleus may be the source of a dorsalizing signal which, because of its limited diffusion, only activates the nearest cells; the products of *gurken* and *torpedo* are involved in this signalling system, *gurken* producing a ligand that will associate with the *torpedo* product that is homologous, in *Drosophila*, with the EGF (epidermal growth factor) receptor.

The localization of the *gurken* transcript, in the anterodorsal corner of the oocyte, in close association with the nucleus, is altered in many mutants such as *cappucino* and *spire*: the *gurken* mRNA remains dispersed all around the anterior end of the oocyte, a pattern that it normally presents in the first stages of oogenesis (stages 1 to 7). These mutants exhibit an exaggerated dorsalized pattern, with the chorion appendages sometimes forming a ring right around the circumference of the egg. One possible interpretation is that the dorsalization signal may have been received by too many cells. The genes *fs (1) K10* (Prost et al., 1988) and *squid* (Kelley, 1993) also play a role in this first stage in the definition of a dorsal identity. The function of *K10* would be to establish a negative regulator of the expression of the dorsalizing signal (Forlani et al., 1993).

Once the follicle cells closest to the nucleus have acquired a dorsal identity, the next step in the dorsoventral polarization process may then take place; this involves the coordinated activity of a set of genes of the 'dorsal group' which includes: *gastrulation defective, dorsal, nudel, tube, pipe, snake, easter, Toll, spätzle, pelle* and *windbeutel*.

The gene *dorsal* is responsible for the dorsoventral asymmetry of the embryo. It intervenes in the ventral and lateral regions. Its activation corresponds to the establishment, in the nuclei of the ventral part, of a Dorsal protein gradient. This protein belongs to the family of transcription regulators: synthesized uniformly in the cytoplasm of the embryo, it is associated to the protein Cactus. The translocation of *Dorsal* within the nuclei of the embryo results from the acquisition by the follicle cells of a ventral identity. They transmit to the embryo a signal that•is the product of the *spätzle* gene after its activation by the *windbeutel, pipe* and *nudel* genes. This ligand thus activated in the perivitelline fluid binds to a receptor coded by the gene *Toll.* The different steps of the process, from the activation of *Toll* to the dissociation of the Dorsal-Cactus complex, are now beginning to be understood (Belvin et al., 1995).

IV. Oocyte Maturation

The nucleus of the oocyte is blocked at metaphase I of meiosis. The progression to anaphase appears to be mechanically impeded by the chiasmas between bivalents (Mc Kun et al., 1993). At the end of oogenesis, the nuclear membrane disappears. The synthesis of the proteins is active in the cytoplasm, less so in the mitochondria (Zalokar, 1976).

The mature oocytes can be stored in virgin females for several days (up to 18 days) before being fertilized and producing viable embryos; conversely, unfertilized oocytes that are laid will degenerate (Wyman, 1979).

Two independent events make possible the development of the egg: its activation and its fertilization. Activation, which can be achieved *in vitro* (Mahowald et al., 1983), allows meiosis to be completed. It is then possible to detect the four haploid products, one of which will fuse with the male pronucleus at the issue of the first mitosis.

While no quantitative difference is apparent in the protein synthesis, one can observe the specific activation of the translation of certain mRNAs, those of the *bicoid, nanos, Toll* and *torso* genes, for example. The activation of the translation of *bicoid, Toll* and *torso* are accompanied by the elongation of the poly A tail (Sallés et al., 1994); this is a situation that is frequently described during maturation of the oocyte. Activation of the translation of the *nanos* mRNA is governed by a different mechanism.

What enables an oocyte to become an egg?

The maternal information present in the oocyte enables it to reach the stage of syncitial blastoderm. The diploid state is not necessary for replication of the DNA, mitosis and transcription to occur ... since haploid embryos, which result from both paternal and maternal mutations, may reach the end of embryogenesis. We tend to suppose that the spermatozoa brings, to the egg activated by laying, the centrosomes which the oocyte lacks.

REFERENCES

Aït-Ahmed, O., Bellon, B., Capri, M., Joblet, C., and Thomas-Delaage, M. (1992). The yemanuclein-α : a new *Drosophila* DNA-binding protein specific for the oocyte nucleus. *Mech. Dev.*, **37**, 69–80.

Aït-Ahmed, O., Thomas-Cavallin, M., and Rosset, R. (1987). Isolation and characterization of a region of the *Drosophila* genome which contains a cluster of differentially expressed maternal genes (*yema* gene region). *Dev. Biol.*, **122**, 153–162.

Ambrosio, L., and Schedl, P. (1984). Gene expression during *Drosophila melanogaster* oogenesis: analysis by *in situ* hybridization to tissue sections. *Dev. Biol.*, **105**, 80–92.

Belvin, M., Jin, Y., and Anderson, V. (1995). Cactus protein degradation mediates *Drosophila* dorso-ventral signaling. *Genes Dev.*, **9** 783–793.

Berleth, T., Burri, M., Thoma, G., Bopp, D., Richstein, S., Frigerio, G., Noll, M., and Nüsslein-Volhard, C. (1988). The role of localization of *bicoïd* RNA in organizing the anterior pattern of the *Drosophila* embryo. *EMBO J.*,**7**, 1749–1756.

Bohrmann, J., Frey, A., and Gutzeit, H.O. (1992). Observations on the polarity of mutant *Drosophila* follicles lacking the oocyte. *Roux's Arch. Dev. Biol.*, **201**, 268–274.

Boswell, R.E., and Mahowald, A.P. (1985). *Tudor,* a gene required for assembly of the germ plasm in *Drosophila melanogaster. Cell,* **43**, 97–104.

Carpenter, A.T.C. (1979). Synaptonemal complex and recombination nodules in wild type *Drosophila melanogaster* females. *Genetics*, **92**, 511–541.

Carpenter, A.T.C. (1994). *Egalitarian* and the choice of cell fates in *Drosophila melanogaster* oogenesis. In: "*Germline development*-Ciba Foundation Symposium, (Wilzy, Chichester), **182**, pp 223–254.

Casanova, J. (1990). Pattern formation under the control of the terminal system in the *Drosophila* embryo. *Development*, **110**, 621–628.

Casanova, J., and Struhl, G. (1993). The *torso* receptor localizes as well as transduces the spatial signal specifying terminal body pattern in *Drosophila*. *Nature*, **362**, 152–155.

Casanova, J., Furriols, M., Mc Cormick, C.A., and Struhl, A. (1995). Similarities between *trunk* and *spätzle*, putative extra cellular ligands specifying body pattern in *Drosophila*. *Genes Dev.*, **9**, 2539–2544.

Chao, Y.C., Donahue, K., and Pokrywka, N. (1991). Sequence of *swallow*, a gene required for the localization of *bicoid* message in *Drosophila* eggs. *Dev. Genet.*, **12**, 333–341.

Chou, J.B. and Perrimon, N. (1992). Use of a yeast site-specific recombinase to produce female germline chimeras in *Drosophila*. *Genetics*, **131**, 643–653.

Cooley, L., and Theurkauf, W. (1994). Cytoskeletal functions during *Drosophila* oogenesis. *Science*, **266**, 590–596.

Dickson, B., Sprenger, F., and Hafen, E. (1992). Prepattern in the developing *Drosophila* eye revealed by an activated torso-sevenless chimeric receptor. *Genes Dev.*, **6**, 2327–2339.

Doyle, H., and Bishop, J. (1993). *Torso* a receptor tyrosine kinase required for embryonic pattern formation, shares substrates with the *sevenless* and EGF–R pathways in *Drosophila. Genes Dev.*, **7**, 633–646.

Driever, W., and Nüsslein-Volhard, C. (1988). A gradient of bicoid protein in *Drosophila* embryo. *Cell.*, **54**, 83–93.

Ephrussi, A., and Lehmann, R. (1992). Induction of germ cell formation by *oskar. Nature*, **358**, 387–392.

Fasano, L., and Kerridge, S. (1988). Monitoring positional information during oogenesis in adult *Drosophila*. *Development*, **104**, 245–253.

Ferrandon, D., Elphick, L., Nüsslein-Volhard, C., and St Johnston, D. (1994). Staufen protein associates with 3'UTR of *bicoid* mRNA to form particles that move in a microtubule-dependant manner. *Cell*, **79**, 1221–1232.

Forlani, S., Ferrandon, D., Saget, O., and Mohier, E. (1993). A regulatory function for *K10* in the establishment of dorso-ventral polarity in the *Drosophila* egg and embryo. *Mech. Dev.*, **41**, 109–120.

Frey, A., and Gutzeit, H. (1986). Follicle cells and germline cells both affect polarity in *dicephalic* chimeric follicles of *Drosophila*. *Roux's Arch. Dev. Biol.*, **195**, 527–532.

Gavis, E., and Lehmann, R. (1992). Localization of *nanos* RNA controls embryonic polarity. *Cell*, **71**, 301–313.

Gavis, E., and Lehmann, R. (1994). Translational regulation of *nanos* by RNA localization. *Nature*, **369**, 315–318.

Geigy, R. (1931). Action de l'ultraviolet sur le pôle germinal dans l'oeuf de *Drosophila*, *Rev. Suisse Zool*, **38**, 187–288.

Golumbeski, G.S., Bardsley, A., and Tax, F. (1991). *Tudor*, a posterior group gene of *Drosophila melanogaster*, encodes a novel protein and a messenger RNA localized during mid-oogenesis. *Genes Dev.*, **5** 2060–2070.

Gonzales-Reyes, A., and St Johnston, D. (1994). Role of oocyte position in establishment of anterior-posterior polarity in *Drosophila*. *Science*, **266**, 639–642.

Gonzales-Reyes, A., Elliot, H., and St Johnston, D. (1995). Polarization of both major body axes in *Drosophila* by *gurken-torpedo* signaling. *Nature*, **375**, 654–658.

Goode, S., Wright, D., and Mahowald, A. (1992). The neurogenic locus *brainiac* cooperates with the *Drosophila* EGF receptor to establish the ovarian follicle and to determine its dorsal-ventral polarity. *Development*, **116, 177–192.**

Hay, B., Jan, J.Y., and Jan, Y.N. (1990). Localization of *vasa,* a component of *Drosophila* polar granules, in maternal effect mutants that alter embryonic anterioposterior polarity. *Development*, **109**, 425–433.

Hegner, R.W. (1911). Germ cell determinants and their significance. *Amer. Nat.*, **45**, 385–397.

Ilmensee, K., and Mahowald, A.P. (1974). Transplantation of posterior polar plasm in *Drosophila* in induction of germ cells at the anterior pole of the egg. *Proc. Natl. Acad. Sci.*, USA, **71** 1016–1020.

Irish, V., Lehmann, R., and Akam, M. (1989). The *Drosophila* posterior-group gene *nanos* functions by repressing *hunchback* activity. *Nature*, **338**, 646–648.

Kelley, R. (1993). Initial organization of the *Drosophila* dorso-ventral axis depends on an RNA-binding protein encoded by the *squid* gene. *Genes Dev.*, **7**, 948–960.

Kim-Ha J., Kerr, K., and Mc Donald, P. (1995). Translational regulation of *oskar* RNA by Bruno, an ovarian RNA-binding protein is essential. *Cell*, **81**, 403–412.

Kim-Ha, J., Smith, J.L., and Mc Donald, P.M. (1991). *Oskar* mRNA is localized to the posterior pole of the *Drosophila* oocyte. *Cell*, **66**, 23–34.

Kim-Ha, J., Webster, P.J., and Smith, J.L. (1993). Multiple RNA regulatory elements mediate distinct steps in localization of *oskar* mRNA. *Development*, **119**, 169–178.

King, R.C. (1970). Ovarian development in *Drosophila melanogaster*. Academic Press, New York and London.

Klinger, M., Erdelyi, M., and Szabad, J. (1988). Function of *torso* in determining the terminal anlagen of the *Drosophila* embryo. Nature, **335**, 275–277.

Koch, E.A., and King, R.C. (1966). The origin and early differentiation of the egg chamber of *Drosophila melanogaster J. Morph.*, **119**, 283–304.

Lane, M.E., and Kalderon, D. (1994). RNA localization along the anteroposterior axis of the *Drosophila* oocyte requires PKA- mediated signal transduction to direct normal microtubule organization. *Genes Dev.*, **8**, 2986–2995.

Lantz, V., Ambrosio, L., Schedl, P; (1992). The *Drosophila orb* gene is predicted to encode sex specific germline RNA-binding proteins and has localised transcripts in ovary and early embryos. *Developement*, **115**, 75–88.

Lasko, P.F. (1994). Molecular genetics of *Drosophila* oogenesis. M.B.I.U. (*Molecular Biolog Intelligence Unit*). R.G. Landes Company, Austin.

Lasko, P.F. and Ashburner, M., (1988a). Posterior localization of *vasa* protein correlates with, but is not sufficient for, pole cell development. *Genes Dev.*, **4**, 905-922.

Lasko, P.F., and Ashburner, M. (1988b). The product of the *Drosophila* gene *vasa* is very similar to eukaryotic initiation factor-4a. *Nature*, **335**, 611–617.

Lawrence, P.E. (1992). "The making of a fly". (Blackwell Scientific Publications). Oxford.

Lehmann, R., and Nüsslein-Volhard, C. (1986). Abdominal segmentation, pole cell formation and embryonic polarity require the localized activity of *oskar* a maternal gene of *Drosophila*. *Cell*, **47**, 141–152.

Lehmann, R., and Nüsslein-Volhard, C. (1991). The maternal gene *nanos* has a central role in posterior pattern formation of the *Drosophila* embryo. *Development*, **112**, 679-692.

Lin, H., and Spradling, A.C. (1995). Fusome asymmetry and oocyte determination in *Drosophila*. *Dev. Genet.*, **16**, 6-12.

Lin, H., Yue, L., and Spradling, A.C. (1994). The *Drosophila* fusome, a germline specific organelle, contains membrane skeletal proteins and functions in cyst formation. *Genes Dev.*, **120**, 977–956.

Lu, X., Chou, T., and Williams, N. (1993). Control of cell fate determination by $p21^{ras}$/ras^1, an essential component of *torso* signalling in *Drosophila*. *Genes Dev.*, **7**, 621–632.

Mahowald, A. (1971a). Polar granules of *Drosophila* III. The continuity of polar granules during the life cycle of *Drosophila*. *J. Exp. Zool*, **176**, 329–344.

Mahowald, A. (1971b). Polar granules of *Drosophila* IV. Cytochemical studies showing loss of RNA from polar granules during early stages of embryogenesis. *J. Exp. Zool*, **176**, 345-352.

Mahowald, A.P. (1992). Germ plasm revisited and illuminated. *Science*, **255**, 1216–1217.

Mahowald, A.P., and Kambysellis, M.P. (1980) Oogenesis. In "Genetics and Biology of *Drosophila*". (M. Ashburner and T.R.F. Wright, eds), **Vol.2d**, 141–224. Academic Press, London.

Mahowald, A.P., and Strassheim, J.M. (1970). Intercellular migration of centrioles in germarium of *Drosophila melanogaster*. *J. Cell Biol.*, **45**, 306–320.

Mahowald, A.P., Goralski, T.J., and Caulton, J.H. (1983). *In vitro* activation of *Drosophila* egg. *Dev. Biol.*, **98**, 437–445.

McDonald, P., Kerr, K., Smith, J.L., and Leask, A. (1993). RNA regulatory element BLE I directs the early step of *bicoid* mRNA localization. *Development*, **118**, 1233–1243.

McDonald, P., Luk, S., and Kilpatrick, M. (1991). Protein encoded by the *exuperentia* gene is concentrated at sites of *bicoid* messenger RNA accumulation in *Drosophila* nurse cells but not in oocytes or embryos. *Genes Dev.*, **5**, 2455–2466.

Manseau, L.J., and Schüpbach, T. (1989). *Cappuccino* and *spire* two unique loci required for both the antero-posterior and dorso ventral patterns of the *Drosophila* embryo. *Genes Dev.*, **3**, 1437-1452.

Mc Kearin, D., and Ohlstein, B. (1995). A role for the *Drosophila bag-of-marbles* protein in the differentiation of cytoblasts from germline stem cells. *Development*, **121**, 2937–2947.

Mc Kearin, D., and Spradling, A. (1990). *bag of marbles*: a *Drosophila* gene required to initiate both male and female gametogenesis. *Genes Dev*, **4**, 2242–2254.

Mc Kun, K., Jang, J., Theurkauf, W., and Hawley, R. (1993). Mechanical basis of meiotic metaphase arrest. *Nature*, **326**, 364–366.

Montell, D., Keshishian, H., and Spradling, A. (1991). Laser ablation studies of the role of the *Drosophila* oocyte nucleus in pattern formation. *Science*, **254**, 290–293.

Murata, Y., and Wharton, R. (1995). Binding of pumilio to maternal *hunchbach* mRNA is required for posterior patterning in *Drosophila* embryos. *Cell* **80**, 747–756.

Neuman–Silberberg, F., and Schüpbach, T. (1994). Dorsoventral axis formation in *Drosophila* depends on the correct dosage of the gene *gurken*. *Development*, **120**, 2457–2463.

Newmark, P., and Boswell, R. (1994). The *magonashi* locus encodes an essential product required for germ plasm assembly in *Drosophila*. *Development*, **120**, 1303–1313.

Nüsslein-Volhard, C., Frohnhöfer, H.G., and Lehmann, R. (1987). Determination of anteroposterior polarity in the *Drosophila* embryo. *Science*, **238**, 1675–1681.

Okada, M., Kleinman, A., and Schneiderman, H.A. (1974). Restoration of fertility in sterilized *Drosophila* eggs by transplantation of polar cytoplasm. *Dev. Biol.*, **37**, 43–54.

Perkins, L., Larsen, I., and Perrimon, N. (1992). *corkscrew* encodes a putative protein tyrosine phosphatase that function to transduce the terminal signal from the receptor tyrosine *torso*. *Cell*, **70**, 225–236.

Perrimon, N., and Gans, M. (1983). Clonal analysis of the tissue specificity of recessive female sterile mutations in *Drosophila melanogaster* using a dominant female sterile mutation *Fs (1) Kl237*. *Dev. Biol.*, **100**, 365–373.

Perrimon, N., Mohler, D., Engstrom, L., and Mahowald, A.P. (1986). X linked female sterile loci in *Drosophila melanogaster*. *Genetics*, **113**, 695–712.

Pokrywka, N., and Stephenson, E. (1991). Microtubules mediate the localization of *bicoid* mRNA during *Drosophila* oogenesis. *Development*, **113**, 55–66.

Prost, E., Deryckere, F., Ross, C., Haenlin, M., Pantesco, V., and Mohier, E. (1988). Role of the oocyte nucleus in determination of the dorso-ventral polarity of *Drosophila* as revealed by molecular analysis of the *K10* gene. *Genes Dev.*, **2**, 891–900.

Ran, B., Bopp, R., and Suter, B. (1994). Null alleles reveal novel requirements for *Bic-D* during *Drosophila* oogenesis and zygotic development. *Development*, **120**, 1233–1242.

Robinson, D.N., Cant, K., and Cooley, L. (1991). Morphogenesis of *Drosophila* ovarian ring canals. *Development*, **120**, 2015–2025.

Ronchi, E., Treisman, J., Dostani, N. Struhl, G., and Desplan, C. (1993). Down-regulation of the *Drosophila* morphogen *bicoid* by the torso-receptor mediated signal transduction. *Cell*, **74**, 347–355.

Roth, S., Neuman-Silberberg, S., Barcelo, G., and Schüpbach, T. (1995). *cornichon* and the EGF receptor signaling process are necessary for both anterior-posterior and dorsal-ventral pattern formation in *Drosophila*. *Cell*, **81**, 967–978.

Ruohola, H., Bremer, K., Baker, D., Swedlow, J., Jan, L., and Jan, Y. (1991). Role of neurogenic genes in establishment of follicle cell fate and oocyte polarity during oogenesis in *Drosophila*. *Cell*, **66**, 433–449.

Rusch, J., and Levine, M. (1994). Regulation of the dorsal morphogen by the *toll* and *torso* signaling pathways: a receptor tyrosine kinase selectively masks transcriptional repression. *Genes Dev.*, **8**, 1247–1257.

Sahut-Barnola, I., Godt, D., Laski, F., and Couderc, J.L. (1995). *Drosophila* ovary morphogenesis: analysis of terminal filament formation and identification of a gene required for this process. *Dev. Biol.*, **170**, 127–135.

Salles, F.J., Lieberfarle, M.E., Wreden, C., Gergen, J.P., Strickland, S. (1994). Coordinate initiation of *Drosophila* development by regulated polyadenylation of maternal messenger RNAs. *Science*, **266**, 1996–1999.

Savant-Bhonsale, S., and Montell, D. (1993). *torso-like* encodes the localized determinant of *Drosophila* terminal pattern formation. *Genes Dev.*, **7**, 2548–2555.

Schüpbach, T. (1987). Germ line and soma cooperate during oogenesis to establish the dorso-ventral pattern of egg shell and embryo in *Drosophila melanogaster*. *Cell*, **49**, 699–707.

Schüpbach, T., Wieschaus, E. (1991). Female sterile mutations on the second chromosome of *Drosophila melanogaster* II. Mutations blocking oogenesis or altering egg morphology. *Genetics*, **129**, 1119–1136.

Serano, T., and Cohen, R. (1995). Gratuitous mRNA localization in the *Drosophila* oocyte. *Development*, **121**, 3013–3021.

Spradling, A.C. (1993). Developmental genetics of oogenesis. In "The Development of *Drosophila melanogaster*" (M. Bate and Martinez-Arias A., eds), vol. **1**, 1–70. Cold Spring Harbor Laboratory Press.

Sprenger, F., and Nüsslein-Volhard, C. (1992). *torso* receptor activity is regulated by a diffusible ligand produced at the extracellular terminal regions of the *Drosophila* egg. *Cell*, **71**, 987–1001.

Sprenger, F., Stevens, L., and Nüsslein-Volhard, C. (1989). The *Drosophila* gene *torso* encodes a putative receptor tyrosine kinase. *Nature*, **338**, 478–483.

St Johnston, D., Beuchle, D., and Nüsslein-Volhard, C. (1991). *staufen* a gene required to localize maternal RNAs in the *Drosophila* egg. *Cell*, **65**, 51–63.

St Johnston, D., Brown, N., and Gall, J. (1992). A conserved double-stranded RNA-binding domain. *Proc. Natl. Acad. Sci.* USA, **89**, 10979–10983.

Stebbings, H., Lane, J.D., and Talbor, N.J. (1995). mRNA translocation and microtubules : insect ovary models. *Trends Cell Biol.*, **5**, 361–365.

Stephenson, E., and Mahowald, A. (1987). Isolation of *Drosophila* clones encoding maternally restricted RNAs. *Dev. Biol.*, **124**, 1–8.

Suter, B., and Steward, R. (1991). Requirement for phosphorylation and localization of the *Bicaudal–D* protein in *Drosophila* oocyte differentiation. *Cell*, **67**, 917–926.

Suter, B., Romberg, L.M., and Steward, R. (1989). *Bicaudal-D*, a *Drosophila* gene involved in developmental asymmetry: localized transcript accumulation in ovaries and sequence similarity to myosin heavy chain tail domains. *Genes Dev.*, **3**, 1957–1968.

Theurkauf, W. (1994). Premature microtubule-dependent cytoplasmic streaming in *cappuccino* and *spire* mutant oocytes. *Science*, **265**, 2093–2096.

Theurkauf, W., Alberts, B., Jan, Y., and Jongens, T. (1993). A central role for microtubules in the differenciation of *Drosophila* oocytes. *Development*, **118**, 1169–1180.

Theurkauf, W., Smiley, S., Wong, M., and Alberts, B. (1992). Reorganization of the cytoskeleton during *Drosophila* oogenesis: implications for axis specification and intercellular transport. *Development*, **115**, 923–936.

Wang, C., and Lehmann, R. (1991). *nanos* is the localized posterior determinant in *Drosophila*. *Cell*, **66**, 637–648.

Waring, G.L., Allis, C.D., and Mahowald, A.P. (1978). Isolation of polar granules and the identification of polar granulae-specific protein. *Dev. Biol.*, **66**, 197–206.

Wharton, R., and Struhl, G. (1991). RNA regulatory elements mediate control of *Drosophila* body pattern by the posterior morphogen *nanos*. *Cell*, **67**, 955–967.

Wieschaus, E., and Szabad, J. (1979). The development and function of the female germline in *Drosophila melanogaster*: a cell lineage study. *Dev. Biol.*, **68**, 29–46.

Wyman, R. (1979). The temporal stability of the *Drosophila* oocyte. *J. Embryo. Exp. Morph.*, 50, 137–144.

Xue, F., and Cooley, L. (1993). *kelch* encodes a component of intercellular bridges in *Drosophila* egg chambers. *Cell*, **72**, 681–93.

Yue, L., and Spradling, A.C. (1992). *hu-litai shao*, a gene required for ring canal formation during *Drosophila* oogenesis, encodes a homolog of adducin. *Genes Dev.*, **6**, 2443–2454.

Zalokar, M. (1976). Autoradiographic study of protein and RNA formation during early development of *Drosophila* eggs. *Dev. Biol.*, 49, 425–437.

Genome Analysis in Eukaryotes: Developmental and evolutionary aspects
R.N. Chatterjee and L. Sánchez (Eds)
Copyright © 1998 Narosa Publishing House, New Delhi, India

3. Egg Maturation and Events Leading to Embryonic Development in the Sawfly, *Athalia rosae* (Hymenoptera)

Kugao Oishi*, Masatsugu Hatakeyama* and Masami Sawa**

*Department of Biology, Faculty of Science, Kobe University,
Nada, Kobe 657, Japan

**Department of Biology, Aichi University of Education,
Kariya, Aichi 448, Japan

I. Introduction

All members of the insect order Hymenoptera reproduce parthenogenetically: most species reproduce by arrhenotoky (haploid males develop from unfertilized eggs and diploid females from fertilized eggs) and some do so by thelytoky (diploid females develop from unfertilized eggs) or cyclical parthenogenesis (White, 1973; Crozier, 1975; Suomalainen *et al.*, 1987; Luck *et al.*, 1993). Because of these unique modes of reproduction and also because large numbers of species in the order have unique and/or economically important characteristics such as sociality and parasitoidism, the Hymenoptera, especially those of the higher suborder Apocrita (bees, wasps, ants, etc.), have been extensively studied in terms of genetics, ecology, developmental biology, and evolution (see DuPraw, 1967; Rothenbuhler, 1975; Cassidy, 1975; Winston, 1987; Gauld and Bolton, 1988; Ross and Matthews, 1991; Moritz and Southwick, 1992; Wrensch and Ebbert, 1993). Less is known about species of the lower suborder, Symphyta (sawflies, horntails, etc.).

Egg activation and fertilization are subjects which have been little studied in insects but are being extensively understood in other animal groups (see Metz and Monroy, 1985; Dale, 1990 for review). We felt that the Hymenoptera is useful material for such studies in insects because in this order egg activation and fertilization are, indeed, separate events. We choose the turnip sawfly, *Athalia rosae* (Tenthredinidae, Symphyta), as an experimental animal for two reasons. Nearly 100% of the eggs (mature oocytes) dissected from the ovaries of this species can be activated *in vitro* (Naito, 1982), and studies on the lower suborder should provide a balanced understanding of the evolutionary aspects of the order. Here we describe a study on the processes of egg maturation, activation and fertilization in the turnip sawfly.

II. General Biology

A. Life Cycle

The turnip sawfly, *Athalia rosae,* is widely distributed throughout the Palearctic region. The form found in the western half is the subspecies, *A. rosae rosae,* and that in the east, including Japan is the subspecies, *A. rosae ruficornis* (Abe, 1988). *A. rosae ruficornis* (for brevity, simply referred to as *A. rosae* below) is a multivoltine species without aestivation, which in Japan, has 3–7 generations per year depending on the local climate. During the winter they hibernate as prepupae.

They can be maintained continuously in the laboratory at 25°C under a 16 hr light-8 hr dark cycle, one generation taking about a month (Sawa *et al.,* 1989). The adult female and male *A. rosae* measure about 8 and 6 mm in length, respectively. Females lay eggs individually in young leaves of the cruciferous plants along the edges. Eggs are about 800 μm in length and 350 μm in diameter. Under the laboratory conditions, embryonic development takes about 5 days. Embryonic development can be followed easily as the egg shells are quite transparent (Fig. 1). Caterpillar-like larvae (much like those of Lepidoptera in morphology) actively feed on fresh leaves of cruciferous plants; hence the species is a serious agricultural pest (for *A. rosae rosae,* see Sáringer, 1976). Larval development takes 9 days (6 instars) for the female and 8 days (5 instars) for the male. The last instar larvae do not feed much but wander around, dig into the soil, and construct cocoons. Within the cocoon, the larva forms a prepupa (5 days both for females and males), then molt to become a pupa (5 and 6 days for females and males, respectively). Adults stay in the cocoons for one day, then

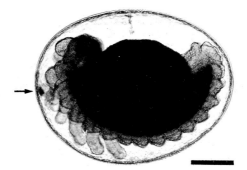

Fig. 1   A fresh embryo 2 days after sperm micro-injection through the anterior end (see Section IVB) and *in vitro* artificial activation (Section IVA). The egg shell is so transparent that embryonic development can be followed without any pretreatment. Arrow indicates the scar caused by inserting the micro-injection pipette. Bar, 200 μm.

emerge and come out of the soil. Both female and male adults live about 2 weeks if fed on diluted honey. Females lay up to 100 or more eggs each but only if given fresh leaves of cruciferous plants (Sawa *et al.,* 1989).

In nature, freshly emerged adults, both females and males, are attracted to the leaves of *Clerodendron trichotomum* (Verbenaceae), a non-host plant species, and feed on the glandular trichomes on the leaf surface (Shimokawa and Kitano, 1989). Feeding causes them to accumulate a repellent compound, clerodendrin D, which protects them from being attacked by predatory animals (Nishida and Fukami, 1990). Feeding also causes them to become sexually motivated (noted by various people, see for example Sawa *et al.,* 1989; Nishida and Fukami, 1990), although the chemical nature of this function remains to be elucidated.

B. Sex Determination

Sex in many arrhenotokous Hymenoptera is determined by the single-locus multiple-allele system (Crozier, 1975; Cook, 1993). Hemizygous and homozygous individuals develop as males, and those that are heterozygous, become females. Often, however, diploid males do not complete development.

In *A. rosae,* for which the single-locus multiple-allele mechanism has been established, homozygous diploid males are perfectly viable and produce diploid sperm (Naito and Suzuki, 1991). Fertilized eggs from brother-sister matings, between diploid females and diploid males produce either triploid heterozygous females or triploid homozygous males. These males are viable and apparently produce triploid sperm, but they are sterile when mated with diploid females (Naito and Suzuki, 1991). Diploid females produce haploid eggs through normal meiotic division, while triploid females produce aneuploid eggs as chromosome segregation becomes abnormal in meiosis and hence are quite sterile. Clearly, non-reductional maturation division in males proceeds in a sex-specific manner, independent of the ploidy level.

As is usual in insects, sex in *A. rosae* is determined cell autonomously. Sexually dimorphic characteristics in the external gross morphology are limited in *A. rosae.* Except for the genital structures, which are obviously sex-dimorphic, the body size is about the only differentiating character. Haploid males are smaller than diploid females. Diploid males are larger than haploid males but smaller than diploid females. Triploid females are larger than diploid females but the size distribution overlaps. Triploid males are about the same size as diploid males (Naito and Suzuki, 1991).

Cell autonomous marker mutations affecting such characters as bristle morphology and body color are not yet available in *A. rosae.* Hence, sex mosaics (gynandromorphs) can be recognized unequivocally only when female and male genital structures coexist (Hatakeyama *et al.,* 1990b). Gynandromorphism may be inferred when an adult sawfly has one part of the body (say, a left wing, or the left part of the abdomen) larger than the other part, but cannot be stated unequivocally.

III. Egg Maturation

A. Ovarian Development

The ovary of adult *A. rosae* is composed of about 14 meroistic polytrophic ovarioles. The ovaries first become distinguishable from the testes in the third instar larvae to the practiced eye. In the last (6th) instar larvae development of ovarioles begins, and in the prepupal stage differentiation to the oocyte and nurse cells becomes apparent (Hatakeyama *et al.*, 1990a).

Each ovariole contains up to 10 egg chambers. Older egg chambers tend to have about 60 nurse cells and younger chambers about 30 cells, both with considerable variation in numbers (Hatakeyama *et al.*, 1990a). This violation of the $2^n - 1$ rule known in the meroistic polytrophic ovarioles may possibly be rather common in the Hymenoptera as the honey bee (*Apis mellifera*), which belongs to the higher suborder Apocrita, also shows the variation (Büning, 1994).

B. Vitellogenesis

As in most other insect species, *A. rosae* has two vitellogenin polypeptides [L (for large)-Vg with an apparent molecular mass of 180 kDa as measured on SDS-polyacrylamide gel electrophoresis (PAGE), and S (for small)-Vg of 50 kDa] in the vitellogenic female haemolymph, and two corresponding vitellins (L-Vn and S-Vn) in the egg (there are no major yolk proteins other than the Vns derived from Vgs) (Hatakeyama and Oishi, 1990; Hatakeyama *et al.*, 1990a; Kageyama *et al.*, 1994). Vitellogenesis as determined by detecting Vgs in the haemolymph upon Western blotting using anti-*A. rosae* L-and S-Vn antisera, begins in the late pupal stage at the last day of pupal development and proceeds practically until the last day of adult life (Hatakeyama *et al.*, 1990a). Unlike *Apis mellifera* which has a single Vg polypeptide of 180 kDa on SDS-PAGE that is also present in the male haemolymph though in a small quantity (Trenczek and Engels, 1986; Trenczek *et al.*, 1989), Vgs are not detected in the male haemolymph of *A. rosae*.

We cloned cDNAs for the *A. rosae* Vgs and characterized the gene expression (Kageyama *et al.*, 1994). The longest cDNA clone was 6.0 kb in length and a 6.5 kb mRNA was detected in the female fat body extract by Northern blotting.

The gene is transcribed as a single 6.5 kb mRNA, which is translated into a single long polypeptide (> 200 kDa), cleaved into two pro-vitellogenins (S from the N-terminal and L from the C-terminal) in the female fat body cells, and they are then secreted into the haemolymph (Fig. 2). The Vg gene begins to express in the female fat body cells in the late pupal stage two days before adult emergence. The gene is neither expressed in the male fat body cells, nor in the ovaries.

Partial nucleotide sequencing and the deduced amino acid sequences revealed complete matches with the 28 N-terminal amino acid sequence

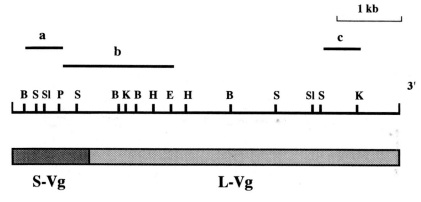

Fig. 2. A restriction map of the *A. rosae* Vg cDNA (center) with information on the Vg polypeptide (bottom) and on the probes (top) used for Northern blotting. B, *Bam*HI: S, *Sac*I; Sl, Sal I; P, *Pst*I; K, *Kpn*I; H, *Hind*III; E, *Eco*RI.

and 5 N-terminal amino acid sequence chemically determined for S-Vn and L-Vn respectively. The deduced amino acid sequence in the S-L boundary region contains unusually long stretches of serine residues, a characteristic [similar to the vertebrate phosvitins and presumed to be vertebrate-specific (Wahli, 1988)] also noted in the mosquito *Aedes aegypti* (Chen *et al.*, 1994) and the silk moth *Bombyx mori* (Yano *et al.*, 1994), but not in the boll weevil *Anthonomus grandis* (Trewitt *et al.*, 1992; Heilmann *et al.*, 1993). Another feature of the Vg cDNA of *A. rosae* is the absence of charged amino acids in the putative signal peptide as is also true of *Anthonomus grandis* (Trewitt *et al.*, 1992) and the locust *Locusta migratoria* (Locke *et al.*, 1987) but not *Bombyx mori* (Yano *et al.*, 1994).

We examined the antigenic similarities by using anti- *A. rosae* L-and S-Vn antisera, and mRNA sequence similarities by using fragments of cloned *A. rosae* Vg cDNA (Fig. 2) from among the Symphyta (Takadera, *et al.*, 1996). Unlike a report (Shirk, 1987) which showed that the antigenic similarity in the yolk proteins of pyralid moth species extends only to the subfamily level, our results demonstrated that similarities in the Vn antigenicity and in the Vg mRNA sequence extended well beyond the family level. In all the species examined [21 species belonging to the largest and most diverse group, the superfamily Tenthredinoidea: 4 species (1 genus) of the family Argidae and 17 species (4 subfamilies, 11 genera) of the family Tenthredinidae], there are two Vns, L and S, each reacting with the anti-*A. rosae* L-Vn and S-Vn antisera, respectively. Of these, 14 species (two families, 8 genera) were examined and they had a single 6.5 kb Vg mRNA under the high stringency conditions regardless of the probes used (Fig. 2). In a few species there was a third group of Vns, M (for medium) in addition to L and S. For example, *Tenthredo hokkaidonis* had two M-Vns, M1 (115 kDa) and M2 (100 kDa). The L- and M1 Vns reacted with the anti-*A. rosae* L-Vn antiserum, while the M2 and S reacted with the anti-S-Vn antiserum.

Among the three fragments of cloned *A. rosae* Vg cDNA used as probes (Fig. 2), probes a and c each representing the 5′(S-Vn) and the 3′(L-Vn) regions detected a single 6.5 kb mRNA, whereas probe b representing the S-L boundary region did so but only very weakly. It thus appears that a single, long pre-pro Vg polypeptide (S-L) is cleaved at either of the two sites, one at the ordinary S-L boundary and the other at somewhere down toward the C-terminal. These results warrant further detailed molecular studies on the Vg genes in the Hymenoptera.

C. Ovarian Transplantation

Previtellogenic immature ovaries can be removed from adults just after eclosion while still in cocoons, and transplanted into the abdomens of adult males (Hatakeyama and Oishi, 1990). The transplantation itself causes the appearance of Vgs in the male haemolymph to some extent. The topical application of juvenile hormone III increases the Vg concentration in the male haemolymph to a considerable degree. The previtellogenic ovaries transplanted into the abdomens of adult male hosts which were given topical application of juvenile hormone III then took up the Vgs and matured. Maturation in *A. rosae* can be shown by the artificial activation of the eggs, which develop into haploid adult males (see below, Section IVA). About 10% of the apparently mature eggs obtained by intraspecific ovarian transplantation, completed embryonic and post-embryonic development upon activation (Hatakeyama and Oishi, 1990).

Ovarian transplantation can be extended to the male hosts other than the original species. Interspecific transplantation was first applied to the closely related species, *Athalia infumata,* adult males of which respond to the topical application of juvenile hormone III as in *A. rosae* and show Vg production (Hatakeyama *et al.,* 1995). The Vgs and Vns of *A. infumata* are indistinguishable from those of *A. rosae* on SDS-PAGE and on Western blotting by using anti-*A. rosae* L-and S-Vn antisera. The *A. infumata* Vg mRNA (6.5kb) is however, not identical to the *A. rosae* Vg mRNA (6.5 kb). A fragment of the cloned *A. rosae* Vg cDNA (probe a, Fig. 2) hybridized equally well with the *A. rosae* and *A. infumata* mRNAs on Northern blots, while probe c did so but only very weakly with *A. infumata*. The Vg gene in the transplanted *A. rosae* ovaries is not expressed, hence there are no contributing Vgs. The immature *A. rosae* ovaries thus sequestered and accumulated only the heterospecific *A. infumata* Vgs and matured. A fraction of these mature eggs (4.5%), upon artificial activation, completed embryonic and post-embryonic development to become haploid adult males. As noted above, the Vg genes appear to be similar among the Symphyta. Whether the interspecific ovarian transplantation can be extended beyond the genus and even beyond the family level remains to be determined.

Intraspecific ovarian transplantation into adult male hosts has been achieved in *Bombyx mori.* The eggs that developed were devoid of Vn derived from

Vg as no Vg production was induced, but they accumulated other major yolk proteins [egg specific protein (contributed from the ovarian follicle cells) and 30 kDa proteins (sex non-specific haemolymph proteins) (Yamashita and Indrasith 1988)], and upon artificial parthenogenetic activation, these eggs completed embryonic and post-embryonic development (Yamashita and Irie, 1980). Intra- and interspecific ovarian transplantation and maturation of eggs (measured by the accumulation of yolk proteins; parthenogenetic activation of mature eggs is not possible) has also been achieved in *Drosophila* (Kambysellis, 1970; Lamnissou and Zouros, 1989). However, the same yolk protein genes are active both in the fat body and in the ovarian follicle cells in the higher Diptera (Wyatt, 1991). *A. rosae* is therefore a unique system in which the contribution of yolk proteins strictly of heterospecific origin can be examined.

IV. Egg Activation and Fertilization

A. *In Vitro* Activation

The artificial activation of eggs has been achieved only in a few insects (Sander, 1985, 1990). Although in the arrhenotokous Hymenoptera, males develop from unfertilized, parthenogenetically activated eggs, generally only laid eggs are activated and develop. The symphytan Hymenoptera, at least the Tenthedinidae sawflies, differ in that the eggs (mature oocytes) dissected from the ovaries can be activated to develop simply by immersing them in water (Naito, 1982; Sawa and Oishi, 1989a). These *in vitro* activated eggs complete embryonic and post-embryonic development into haploid adult males.

Mature eggs dissected from the ovaries of *A. rosae* and placed in saline (0.15 M NaCl), are not activated and can retain the ability, to be activated upon exposure to water for an hour in the saline. If the eggs in the saline are removed and placed in distilled water for 20 min, they can continue development after returning to the saline solution. This 20 min period corresponds to the time required for the arrested metaphase nucleus of the egg (mature oocyte) to proceed to telophase (Sawa and Oishi, 1989a) (see below). The spindle of meiotic division up to this time is oriented parallel to the dorsal surface of the egg (Sawa and Oishi, 1989a). The spindle orientation then gradually becomes perpendicular to the egg surface and the second meiotic division follows which is completed in 120–130 min (Sawa *et al.*, unpublished).

Dissected eggs can be activated also by various other means (Sawa and Oishi, 1989a). About 20% of the eggs in the saline become activated if pricked by a needle at any site, or if taken out and dried for 3 min and returned to the saline solution. Ambient pH has a dramatic effect. When placed in 0.15 M phosphate buffer at pH 7.0 or 8.0 no eggs are activated, while about 60% are activated in the same buffer at pH 5.8. In 0.15 M citrate buffer, 30% were activated at pH 6.0, 80% at pH 5.0 and 50% at pH 4.0.

Mature insect eggs (occytes) are arrested at metaphase of the first meiotic division and wait for fertilization or for activation (Counce, 1973). Upon fertilization or activation, meiosis resumes and produces four nuclei without cell division, all four nuclei being in close proximity in the same cytoplasm. One nucleus becomes a female pronucleus and participates in the development, three others being degenerated as polar body nuclei. The *in vitro* activated *A. rosae* eggs normally develop into haploid males. When explanted eggs are activated not at 25°C as is usual but at higher temperature (35–37°C), all four nuclei produced following the completion of meiosis can then participate in development with or without fusion among themselves (Hatakeyama *et al.*, 1990b). Cytological examinations of 2-day-old embryos revealed that a large number of individuals were diploid, triploid, and some were even tetraploid and haploid-diploid mosaics. Eggs from +/*yfb* (yellow fat body, a color mutation, see below Section IVB) heterozygous mothers should ordinarily develop into either + or *yfb* haploid males. Following activation at higher temperature, these eggs produced, in addition to haploid males, diploid homozygous (+/+ and *yfb/yfb*) and heterozygous (+/*yfb*) males and females, triploid females, gynandromorphs, haploid-haploid male mosaics (+ ↔ *yfb*). There must be a cytoplasmic factor(s), which is responsible for preventing the polar body nuclei from participating in development and which is easily incapacitated in *A. rosae*.

B. In vitro Fertilization

Artificial, *in vitro* fertilization in insects has been attempted from time to time but without success (Clarke, 1990; Sander, 1990). Even the artificial insemination has only been possible in honey bees and a few other species (Leopold, 1991). Thus, the *A. rosae* system in which artificial fertilization was achieved by sperm micro-injection (Sawa and Oishi, 1989b, c; Hatakeyama *et al.*, 1994a, b) represents a unique system in insects.

Mature eggs were dissected from ovaries and kept in saline. Up to 5 eggs at a time, were placed on double-sided adhesive tape on a slide and dried by touching with a piece of filter paper. Sperm bundles from adult male seminal vesicles were suspended in a solution (see below) and dispersed by flushing with a micro-injection pipette. Under the inverted phase-contrast microscope, the eggs were injected (within 3 min) through the anterior pole with sperm, each egg receiving one dozen or more sperm. In this system it is not possible to inject individual sperm as is practicable in other systems such as mammals. The injected eggs were then placed on a filter paper wet with distilled water in small Petri dishes and activated.

A mutation, *yellow fat body* (*yfb*), which alters the color of the fat body cells to bright yellow from the normal dark blue-green, and which can be seen through the still unpigmented integument in the last instar larval-to-pupal stages, was used to confirm fertilization: +/*yfb* heterozygotes show intermediate coloration which can be readily distinguished from +/+ and *yfb*/*yfb* homozygotes. Eggs from *yfb*/*yfb* females when injected with + sperm would develop as +/*yfb* diploid females if fertilized and as *yfb* haploid males if unfertilized.

Fertilization took place only when the sperm was suspended in solutions at a low salt concentration, the best results (about 10% of the injected eggs developed into diploid heterozygous adult females) being obtained when the sperm were suspended in distilled water (Sawa and Oishi, 1989b). Apparently the sperm membrane has to be damaged to form the male pronucleus. Eggs can be first activated, then injected with sperm. More than 20% of the eggs 20 min post-activation were fertilized, but none of the eggs 60 min post-activation were fertilized (Sawa and Oishi, 1989c). Clearly by then, which is well before the completion of meiotic divisions in the egg (mature oocyte) nucleus, cytoplasmic changes occurred and the eggs were determined to proceed towards parthenogenetic development.

Fertilization by means of sperm micro-injection may be extended to obtain interspecific hybrids between species with premating isolation. There may be species-pairs which show strong premating but little postmating isolation. Limited attempts have been made between *A. rosae* and *A. infumata,* and between *A. rosae* and *A. kashmirensis,* with a strong premating isolation and each with a different karyotype. Hybrid embryos were obtained but apparently they did not develop further (Sawa, 1991; Oishi *et al.*, 1993).

In these experiments, sperm were introduced into the eggs through the anterior pole. In an earlier study (Sawa and Oishi, 1989b) we obtained 2 putative haploid-haploid male chimeras (+ ↔ *yfb*, haploid-size males with intermediate coloration probably produced by the independent participation of a haploid egg nucleus and a haploid sperm-derived male nucleus) but never thereafter. We then examined the scar caused by inserting the injection pipette on the embryos 2 days later to confirm the correct pole of injection (Fig. 1). It became apparent that about 1 in every 200 to 300 eggs injected developed with the scar at the pole opposite to the pole where the injection was intended. Most probably the earlier result reflected such an error.

Attempts have been made to inject sperm through the posterior pole of the eggs (Hatakeyama *et al.*, 1994a). Fertilization never occurred, but instead a smaller fraction (about 1%) of the eggs injected developed as haploid-haploid (+ ↔ *yfb*) male chimeras. Progeny tests indicated that the cells derived from the sperm nucleus deposited at the posterior as well as those from the egg nucleus positioned anteriorly contributed to form haploid male gametes in these chimeras.

As the system uses sperm injected into the cytoplasm of mature oocytes, the sperm do not have to be alive and motile. In fact, sperm suspended in distilled water for micro-injection show abnormal morphology. Sperm suspended in 0.15 M NaCl can be frozen by placing them directly in liquid nitrogen without adding any cryoprotectants, and stored in a deep-freezer at -80°C for weeks (Fig. 3). They can still fertilize mature oocytes upon injection (Hatakeyama *et al.*, 1994b). Attempts at cryopreservation in insects have mostly been directed towards embryos, larval ovaries and testes, but rarely gametes, because of the difficulty of artificial insemination (Leopold, 1991). Our results, and the finding that the *A. rosae* previtellogenic ovaries can mature solely with the heterospecific *A infumata* vitellogenin as described above (Section IIIC) suggest that in extreme cases, a species can be recovered from cryopreserved immature ovaries and sperm in the absence of the original species.

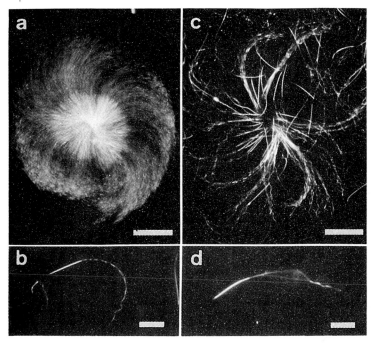

Fig. 3.   Fresh sperm bundles taken from the male seminal vesicle (a); single fresh sperm (b); head = 15 μm, tail = 40 μm; (c) frozen-thawed sperm in loose bundles; and (d) individualized frozen-thawed sperm. Mitochondrial derivative is separated from the axoneme (d). Bars in (a) and (c) indicate 20 μm, and in (b) and (d) 10 μm.

V. Future Direction

We selected *Athalia rosae* as an experimental animal with which to pursue egg maturation, activation and fertilization, which cannot be achieved or can be only with great difficulty in other experimental systems such as *Drosophila* (Sawa *et al.*, 1989). Although our results are still preliminary, we believe we are developing a useful experimental system with which to examine the molecular mechanisms of egg maturation, activation and fertilization, which have been largely ignored in insects mainly because of technical difficulties. Obviously we need more basic biology to increase understanding of the processes described above. Embryogenesis and spermatogenesis are currently being studied.

The ease of *in vitro* activation provides the *A. rosae* system with useful and unique means for studying various aspects of developmental biological questions. It should be possible to obtain exactly timed mature oocytes as well as embryos (of one sex, male) in a relatively large quantity, a feature particularly suited to biochemical and molecular studies.

Meiosis in general and in relation to parthenogenesis has been a subject of intensive research (Moens, 1987; Suomalainen *et al.*, 1987). The occurrence of female-specific reductional maturation divisions in *A. rosae*, which most probably a common phenomenon in the Hymenoptera, provides an important feature to be studied. Mechanisms of sex determination, which have been increasingly understood (Ryner and Swain, 1995), are extremely varied. Simple extrapolation from the systems such as *Drosophila* may not prove realistic. Thus, molecular analyses of the single-locus multiple-allele sex determination mechanism in the Hymenoptera have to be performed, and the relationship to the sex-specific reductional maturation division must be studied. To this end, some genetics are required. Development of formal genetics with *A. rosae*, may be difficult, if not impossible. We should perhaps first attempt to construct a linkage map using RAPD molecular markers, as has been achieved with the honey bee (Hunt and Page, 1994, 1995). Along with this line, attempts should be made to develop a system to obtain transgenic stocks: injection of cloned cDNAs (from, for example, *Drosophila*) with sperm may work (Lavitrano *et al.*, 1989; Francolini *et al.*, 1993), or we may be able to recover appropriate transposable elements as reported in the honey bee (Kimura *et al.*, 1993).

Elucidation of the mechanism which allows only a single nucleus among the four produced by meiosis and placed in close proximity in the same cytoplasm to participate in development is another subject of future study. This may be related to the formation of the pronucleus, the prevention of the polar body nuclei and supernumerary sperm nuclei from participating in development. The responsible cytoplasmic factor(s) may be distributed in a particular manner so that sperm nuclei introduced at the posterior end of the egg can only participate in development infrequently. An initial study that should be relatively simple, may be the manipulation of cytoplasm such as removal from particular areas and introduction to a new site by micro-injection.

Hopefully the basic biology (egg activation, fertilization, gamete preservation, etc.) and the molecular biology (introduction of specific genes into the genome, etc.) as being developed in *A. rosae* may be extended to other economically important Hymenoptera.

Acknowledgments

Work in our labs has been supported in part by grants from the Ministry of Education, Science and Culture, Japan, and by Research Fellowships of the Japan Society for the Promotion of Science for Young Scientists to M.H., support that is gratefully acknowledged.

REFERENCES

Abe, M. (1989). A biosystematic study of the genus *Athalia* Leach of Japan (Hymenoptera: Tenthredinidae). *Esakia (Fukuoka, Japan)* **26**, 91–131.

Büning, J. (1994). "The Insect Ovary: Ultrastructure, Previtellogenic Growth and Evolution." Chapman and Hall, London.

Cassidy, J.D. (1975). The parasitoid wasps, *Habrobracon* and *Mormoniella*. In "Handbook of Genetics" Vol. 3. "Invertebrates of Genetic Interest" (R.C. King, ed.), pp. 173–203. Plenum Press, New York.

Chen J-S., Cho, W-L., and Raikhel, A.S. (1994). Analysis of mosquito vitellogenin cDNA. Similarity with vertebrate phosvitins and arthropod serum proteins. *J. Mol. Biol.* **237**, 641–647.

Clarke, C.A. (1990). *In vitro* fertilization — some comparative aspects. *J. Roy. Soc. Med.* **83**, 214–218.

Cook, J.M. (1993). Sex determination in the Hymenoptera: a review of models and evidence. *Heredity* **71**, 421–435.

Counce, S.J. (1973). The causal analysis of insect embryogenesis. In "Developmental Systems: Insects" (S.J. Counce and C.H. Waddington, eds.), Vol. 2, pp. 1–156. Academic Press, New York.

Crozier, R.H. (1975). Hymenoptera. In "Animal Cytogenetics" (B. John, ed.) Vol. 3: Insecta 7, pp. 1–95. Gebrüder Borntraeger, Berlin.

Dale, B., ed. (1990). "Mechanism of Fertilization: Plants to Humans." NATO ASI Series H: Cell Biology, Vol. 45. Springer-Verlag, Berlin.

DuPraw, E.J. (1967). The honeybee embryo. In "Methods in Developmental Biology" (F.H. Wilt and N.K. Wessells, eds.), pp. 183–217. Thomas Y. Crowell, New York.

Francolini, M., Lavitrano, M., Lamia, C.L., French, D., Frati, L., Cotelli, F., and Spadafora, C. (1993). Evidence for nuclear internalization of exogenous DNA into mammalian sperm cells. *Mol. Reprod. Dev.* **34**, 133–139.

Gauld, I., and Bolton, B., eds. (1988). "The Hymenoptera." Oxford Univ. Press, Oxford.

Hatakeyama, M., and Oishi, K. (1990). Induction of vitellogenin systhesis and maturation of transplanted previtellogenic eggs by juvenile hormone III in males of the sawfly, *Athalia rosae. J. Insect Physiol.* **36**, 791–797.

Hatakeyama, M., Sawa, M., and Oishi, K. (1990a). Ovarian development and vitellogenesis in the sawfly, *Athalia rosae ruficornis* Jakovlev (Hymenoptera, Tenthredinidae). *Invert. Reprod. Dev.* **17**, 237–245.

Hatakeyama, M., Sawa, M., and Oishi, K. (1994a). Production of haploid-haploid chimeras by sperm injection in the sawfly, *Athalia rosae* (Hymenoptera). *Roux's Arch. Dev. Biol.* **203**, 450–453.

Hatakeyama, M., Sawa, M., and Oishi, K. (1994b). Fertilization by micro-injection of cryopreserved sperm in the sawfly, *Athalia rosae* (Hymenoptera.) *J. Insect Physiol.* **40**, 909–912.

Hatadeyama, M., Kageyama, Y., Kinoshita, T., and Oishi, K. (1995). Completion of development in *Athalia rosae* (Hymenoptera) eggs matured with heterospecific *Athalia infumata* yolk protein. *J. Insect Physiol.* **41**, 351–355.

Hatakeyama, M., Nakamura, T., Kim. K.B., Sawa, M., Naito, T., and Oishi, K. (1990b). Experiments inducing prospective polar body nuclei to participate in embryogenesis of the sawfly *Athalia rosae* (Hymenoptera). *Roux's Arch. Dev. Biol.* **198**, 389–394.

Heilmann, L.J., Trewitt, P.M., and Kumaran, A.K. (1993). Proteolytic processing of the vitellogenin precursor in the boll weevil, *Anthonomus grandis*. *Arch. Insect Biochem. Physiol.* **23**, 125–134.

Hunt, G.J., and Page, R.E., Jr. (1994). Linkage analysis of sex determination in the honey bee (*Apis mellifera*). *Mol. Gen. Genet.* **244**, 512–518.

Hunt, G.J., and Page, R.E., Jr. (1995). Linkage map of the honey bee, *Apis mellifera*, based on RAPD markers. *Genetics* **139**, 1371–1382.

Kageyama, Y., Kinoshita, T., Umesono, Y., Hatakeyama, M., and Oishi, K. (1994). Cloning of cDNA for vitellogenin of *Athalia rosae* (Hymenoptera) and characterization of the vitellogenin gene expression. *Insect Biochem. Mol. Biol.* **24**, 599–605.

Kambysellis, M.P. (1970). Compatibility in insect tissue transplantations. I. Ovarian transplantations and hybrid formation between *Drosophila* species endemic to Hawaii. *J. Exp. Zool.* **175**, 169–180.

Kimura, K., Okumura, T., Ninaki, O., Kidwell, M.G., and Suzuki, K. (1993). Transposable elements in commercially useful insects. I. Southern hybridization study of silkworms and honeybees using *Drosophila* probes. *Jpn. J. Genet.* **68**, 63–71.

Lamnissou, K., and Zouros, E. (1989). Interspecific ovarian transplantations in *Drosophila:* vitellogenin uptake as an index of evolutionary relatedness. *Heredity* **63**, 29–35.

Lavitrano, M., Camaioni, A., Fazio, V., Dolci, S., Farace, M.G., and Spadafora, C. (1989). Sperm cells as vectors for introducing foreign DNA into eggs: Genetic transformation of mice. *Cell* **57**, 717–723.

Leopold, R.A. (1991). Cryopreservation of insect germplasm: Cells, tissues and organisms. *In* "Insects at Low Temperature" (R.E. Lee, Jr. and D.L. Denlinger, eds.), pp. 379–407. Chapman and Hall, New York.

Locke, J., White, B.N., and Wyatt, G.R. (1987). Cloning and 5′ end nucleotide sequences of two juvenile hormone-inducible vitellogenin genes of the African migratory locust. *DNA* **6**, 331–342.

Luck, R.F., Stouthamer, R., and Nunney, L. P. (1993). Sex determination and sex ratio patterns in parasitic Hymenoptera. *In* "Evolution and Diversity of Sex Ratio in Insects and Mites" (D.L. Wrensch and M.A. Ebbert, eds.), pp. 442–476. Chapman and Hall, New York.

Metz, C. B., and Monroy, A., eds. (1985). "Biology of Fertilization." Vols. 1–3. Academic Press, New York.

Moens, P.B., ed. (1987). "Meiosis." Academic Press, Orlando.

Moritz, R.F.A., and Southwick, E.E. (1992). "Bees as Superorganisms — An Evolutionary Reality." Springer-Verlag, Berlin.

Naito, T. (1982). Chromosome number differentiation in sawflies and its systematic implication. (Hymenoptera, Tenthredinidae). *Kontyû (Tokyo)* **50**, 569–587.

Naito, T., and Suzuki, H. (1991). Sex determination in the sawfly, *Athalia rosae ruficornis* (Hymenoptera): Occurrence of triploid males. *J. Hered.* **82**, 101–104.

Nishida, R., and Fukami, H. (1990). Sequestration of distasteful compounds by some pharmacophagous insects. *J. Chem. Ecol.* **16**, 151–164.

Oishi, K., Sawa, M., Hatakeyama, M., and Kageyama, Y. (1993). Genetics and biology of the sawfly, *Athalia rosae* (Hymenoptera): Review. *Genetica* **88**, 119–127.

Ross, K.G., and Matthews, R.W., eds. (1991). "The Social Biology of Wasps." Comstock Publishing Associates, Ithaca.

Rothenbuhler, W.C. (1975). The honey bee, *Apis mellifera*. *In* "Handbook of Genetics" Vol. 3. "Invertebrates of Genetic Interest" (R.C. King, ed.), pp. 165–172. Plenum Press, New York.

Ryner, L.C., and Swain, A. (1995). Sex in the '90s. Meeting review. *Cell* **81**, 483–493.

Sander, K. (1985). Fertilization and egg cell activation in insects. *In* "Biology of Fertilizaiton" (C.B. Metz and A. Monroy, eds.), Vol. 2. pp. 409–430. Academic Press, Orlando.

Sander, K. (1990). The insect oocyte: Fertilization, activation and cytoplasmic dynamics. *In* "Mechanism of Fertilization: Plants to Humans" NATO ASI Series H: Cell Biology, Vol. 45 (B. Dale, ed.), pp. 605–624. Springer-Verlag, Berlin.

Sáringer, Gy. (1976). Problems of *Athalia rosae* L. (Hym.: Tenthredinidae) in Hungary. *Acta Agronomica Academiae Scientiarum Hungaricae* **25**, 153–156.

Sawa, M. (1991). Fertilization of hetero-specific insect eggs by sperm injection. *Jpn. J. Genet.* **66**, 297–303.

Sawa, M., and Oishi, K. (1989a). Studies on the sawfly, *Athalia rosae* (Insecta, Hymenoptera, Tenthredinidae). II. Experimental activation of mature unfertilized eggs. *Zool. Sci.* **6**, 549–556.

Sawa, M., and Oishi, K. (1989b). Studies on the sawfly, *Athalia rosae* (Insecta, Hymenoptera, Tenthredinidae). III. Fertilization by sperm injection. *Zool Sci.* **6**, 557–563.

Sawa, M., and Oishi, K. (1989c). Delayed sperm injection and fertilization in parthenogenetically activated insect egg (*Athalia rosae*, Hymenoptera). *Roux's Arch. Dev. Biol.* **198**, 242–244.

Sawa, M., Fukunaga, A., Naito, T., and Oishi, K. (1989). Studies on the sawfly, *Athalia rosae* (Insecta, Hymenoptera, Tenthredinidae). I. General biology. *Zool. Sci.* **6**, 541–547.

Shimokawa, K., and Kitano, H. (1989). Feeding habit of the turnip sawfly adult *Athalia rosae ruficornis*, *A. lugens infumata* and *A. japonica* (Hymenoptera, Tenthredinidae) on the glandular trichome of *Clerodendron trichotomum* leaves and its biological significance. *Jpn. J. Ent.* **57**, 881–888.

Shirk, P.D. (1987). Comparison of yolk production in seven pyralid moth species. *Int. J. Invert. Reprod. Dev.* **11**, 173–188.

Suomalainen, E., Saura, A., and Lokki, J. (1987). "Cytology and Evolution in Parthenogenesis." CRC Press, Boca Raton, Frorida.

Takadera, K., Yamashita, M., Hatakeyama, M., and Oishi, K. (1996). Similarities in vitellin antigenicity and vitellogenin mRNA nucleotide sequence among sawflies (Hymenoptera: Symphyta; Tenthredinoidea). *J. Insect Physiol.* **42**, 417–422.

Trewitt, P.M., Heilmann, L.J., Degrugillier, S.S., and Kumaran, A.K. (1992). The boll weevil vitellogenin gene: Nucleotide sequence, structure, and evolutionary relationship to nematode and vertebrate vitellogein genes. *J. Mol. Evol.* **34**, 478–492.

Trenczek, T., and Engels, W. (1986). Occurrence of vitellogenin in drone honeybees (*Apis mellifica*). *Int. J. Invert. Reprod. Dev.* **10**, 307–311.

Trenczek, T., Zillikensm, A., and Engels, W. (1989). Developmental patterns of vitellogenin haemolymph titre and rate of synthesis in adult drone honey bees (*Apis mellifera*) *J. Insect Physiol.* **35**, 475–481.

Wahli, W. (1988). Evolution and expression of vitellogenin genes. *Trends Genet.* **4**, 227–232.

White, M.J.D. (1973). "Animal Cytology and Evolution." 3rd. ed. Cambridge Univ. Press. London.

Winston, M.L. (1987). "The Biology of the Honey Bee." Harvard Univ. Press, Cambridge.

Wrensch, D.L., and Ebbert, M.A., eds. (1993). "Evolution and Diversity of Sex Ratio in Insects and Mites." Chapman and Hall, New York.

Wyatt, G.R. (1991). Gene regulation in insect reproduction. Review. *Invert. Reprod. Dev.* **20**, 1–35.

Yamashita, O., and Indrasith, L.S. (1988). Metabolic fates of yolk proteins during embryogenesis in arthropods. *Dev. Growth Differ.* **30**, 337–346.

Yamashita, O., and Irie, K. (1980). Larval hatching from vitellogenin-deficient eggs developed in male hosts of the silkworm. *Nature (Lond.)* **283**, 385–386.

Yano, K., Toriyama-Sakurai, M., Watabe, S., Izumi, S., and Tomino, S. (1994). Structure and expression of mRNA for vitellogenin in *Bombyx mori. Biochim. Biophys. Acta* **1218**, 1–10.

Genome Analysis in Eukaryotes: Developmental and evolutionary aspects
R.N. Chatterjee and L. Sánchez (Eds)
Copyright © 1998 Narosa Publishing House, New Delhi, India

4. Developmental Aspects of Mulberry and Nonmulberry Silkworm Species: A comparative study

K.P. Gopinathan, Omana Joy and Amit Singh
Microbiology and Cell Biology Department, Indian Institute of Science
Bangalore-560012, India

I. Silkworms: An Experimental Model System

The mulberry silkworm, *Bombyx mori* has been exploited by man for more than 4000 years, for the production of the exotic silk yarn used in making fabric. The silk fibre, proteinaceous in nature synthesised by *B. mori* larvae, is unmatched in beauty and elegance by any of the man made fibres. The silk industry is reported to have its origin in China dating back to 2600 B.C. and found its way to India over the Himalayas. According to the recorded history, a young Chinese Empress by name Si Ling-Chi has been accredited with the development and exploitation of the silk cocoon for the production of exotic silk fibre. The silk industry spread widely in Europe in the sixth and seventh centuries A.D.

There are many different races and strains of silkworm exploited for silk production. About 90% of the silk produced in the world is from the mulberry silkworms and the remaining 10% originates from the nonmulberry silkworms such as Eri (*Philosomia cynthia ricini*), Muga (*Antheraea assama*) and Tasar (*Antheraea mylitta*) which feed on host plants other than mulberry. India is the second largest producer of silk in the world and has the unique distinction of producing all four varieties of silk.

The mulberry silkworm has been a target of intensive scientific investigations right from the ancient times. Genetic breeding to get better yielding varieties of silkworms has been practiced for a long time. The genetic legacy of this organism, dates back even earlier to that of the fruitfly *Drosophila*, the Cinderella of modern genetics. The achievements in *B. mori* genetics, however, have been very different from that of *Drosophila* because the attempts have been mainly directed towards economic benefits to improve the quality and yield of silk. In the following sections we have provided a detailed review on the developmental biology of *B. mori* with occasional references to the nonmulberry species wherever possible.

Bombyx mori belongs to Phyllum: Invertebrata; Class: Arthropoda; Order: Lepidoptera; Suborder: Heterocoera; Family: Bombycidae.

In the past decade, basic studies on the developmental biology, genetics and molecular biology of the silkworm have been intensified (Willis *et al.*, 1995). The central aim of developmental biology is to elucidate the mechanism of generation of the highly complex metameric adult from a single celled embryo. *B. mori is* now coming up as an important model system because more than 200 Mendelian mutations affecting a wide range of developmental, morphological and biochemical traits have been already mapped (Doira, 1992). Besides, many practical breeding strains of *B. mori* that differ in complex polygenic traits affecting qualitative and quantitative characters such as silk yield, disease resistance and feeding behaviour are also available although the underlying genetic basis of these phenomena is not very clear.

Members of the order Lepidoptera are eumetabolous and their body is covered with overlapping flat scales (Mani, 1968). The integument is coloured cryptically and is densely clothed with setae and flat overlapping scales. The head is small and is associated with long slender antennae with numerous segments, often clevate, pectinate or hooked apically or plumose in males. Generally Lepidopteran larvae are plant feeders except a few which are predaceous and scavengers, or feed on stored products. Most of them feed externally on foliage while a fair number of the minute species mine through leaves or leaf petioles, stems, trunks and roots.

B. mori is a member of *Bombycidae*, a family of 15 species of small dull moths having bipectinate antennae. The proboscis is absent and the legs are hairy and without spurs. The larval body is elongated with dorsal hinges or a terminal horn and is hairless. *B. mori* is supposed to have originated in the southern districts of China and has been differentiated from *Bombyx (Theophila) mandarina*. *B. mori* strains have been classified into various categories on the basis of colour, size, shape of the egg, cocoon and other morphological characteristics. The most commonly used classification is on the basis of the number of broods or voltinisms (Tazima, 1979), namely:

(i) Univoltine or One Brooded Silkworms: This category includes most of the improved races which go for a single brood in a year. The worms are sensitive to irregular environment, especially high temperatures. Eggs of univoltine insects hatch after an elaborate and graduated storage in cold. The weight of the cocoon, weight of the shell, ratio of the shell to cocoon and weight of slime are high.

(ii) Bivoltine or Two Brooded Silkworms: They are characterized by two broods in a year. Rearing period is short and the consumption of leaves is less as compared to the univoltines. The worms are strong and the weight of the cocoon, weight of the shell, ratio of shell to cocoon and weight of slime are less than univoltine but more than multivoltines. The members of this class-NB4D2, NB18, NB1 are high yielding varieties and are extensively used for sericulture.

(iii) Multivoltine or Polyvoltine Silkworms: Multivoltines are tropical denizens and their eggs hatch if kept at a temperature higher than 21°C for 10–15 days, giving rise to many broods a year. Their life cycle is short and these sturdy worms are especially suitable for hot climates. The weight of the cocoon, weight of the shell, ratio of shell to cocoon and weight of slime are less than the uni- or bivoltines. Examples of this class are Pure Mysore, *C. nichi* and Nistari. In spite of being poor yielders of silk, they are used in sericulture because of their sturdy nature to generate cross breeds with the bivoltines possessing the high yielding quality.

Another commonly used mode of classification is on the basis of the country to which the race belongs (*e.g.* the Japanese, Chinese, European and Indian races). These races have strains suitably adapted to the climatic conditions of that geographical belt.

Fig. 1. Scanning electron micrograph of a first instar larva of *Bombyx mori* (magnification 50×).

II. Life cycle

Bombyx mori is a holometabolous insect with a life cycle of 45 days comprising all four stages—the embryo, larva, pupa and adult. The scanning electron micrographic view of *B. mori* larva at an early stage of development is presented in Fig. 1. Life cycle of *B. mori* follows a cyclic pattern (Anderson; 1972) (Fig. 2a). The development of silkworm is marked by an extended period of 10 days of embryonic development followed by about 20 days of

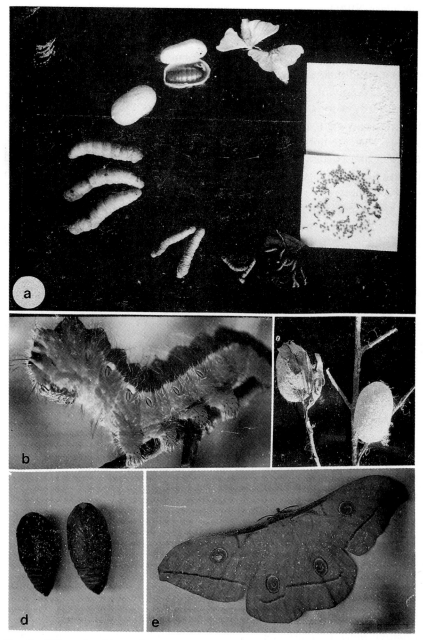

Fig. 2. Life cycle of the mulberry and nonmulberry silkworm. (a) Life cycle of *B. mori* showing different stages of development from the embryo to the adult. Clockwise from top, adult moths (male and female), the eggs soon after oviposition, the larval development through instars I to V and the pupa (within the cocoon) present all stages of *B. mori* life cycle. (b-e); Life cycle of nonmulberry silkworm, Tasar, (*A. mylitta*). Larva in the fifth instar (b); pupal stage (cocoon) (c); pupae dissected from the cocoon (d); and adult moth (e).

larval development, 10 days of pupal life and 2–3 days of adult life. The larval stage is the only period during which the insect is able to feed. Under natural conditions the larva feeds on fresh mulberry leaves, although they can be adapted to be reared on artificial diet blended with mulberry leaf powder (Yanagawa *et al.*, 1988; Tazima, 1989). Larval development is interrupted with moults during which a new cuticle is produced and the old one is shed. This phenomenon of shedding the old skin is called ecdysis and the time interval between consecutive moults is called an instar. There are five instars in the larval development of *B. mori*. The larval and adult stages are separated by a distinct pupal stage during which metamorphosis occurs.

The most important consequence of this mode of development is that two separate systems develop side by side (Tazima, 1979; Ransom, 1982). The embryo forms both the larva and the islands of cells called imaginal discs, the progenitors for the adult structures (Poodry, 1980). The first instar larva is about 3 mm in length and after 24 h the body length reaches 7 mm and the skin surface becomes glossy. In the fifth instar, the larva attains a maximum length of about 75 mm and feeds voraciously. The mature larva at the end of the fifth instar spins a cocoon of silk thread for nearly two days within which it transforms itself into the pupa. The adult moth emerges from the cocoon after 10 days. On emergence, the male and female moths mate and the female lays 300–500 eggs.

The life cycle of nonmulberry silkworms is similar to that of *B. mori* (Chowdhary, 1970). The life cycles of Tasar, *Antherea mylitta* and Eri, *Philosomia cynthia ricini* are presented in Fig. 2b and 3.

III. Gametogenesis

Embryos are generated by the fusion of gametes (mature eggs and sperms) which are generated by the process of gametogenesis. Both spermatogenesis and oogenesis in silkworm have been extensively studied. Primordial germ cells in the embryo give rise to the gonads or reproductive organs of the larvae and imago. The germ cells which are initially 20 in number, increase by division and give rise to testis in the male and ovary in the females (Tazima, 1964; Sakaguchi, 1979a).

A. Spermatogenesis

A newly hatched larva has a large number of germ cells present in the testis which gradually increases in number by division (Sakaguchi, 1979a). Sperms are produced in the testis which is divided into four follicles or chambers (Fig. 4a). Testicular follicles are lined with a layer of epithelium, whose cells rest upon a basement membrane surrounded by connective tissue. Each follicle consists of four zones of development, *viz.*,

(i) **Germarium**, the region containing primordial germ cells.

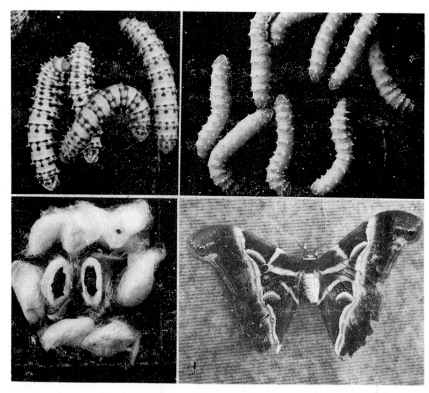

Fig. 3.   Life cycle of Eri silkworm. The variability of *P. cynthia ricini* strains, is apparent from the larval markings (a, b). The loose structure of Eri cocoon (c) as compared to the *Bombyx* or Tasar cocoons is clearly seen. Eri pupa within the cocoon (c) and the adult moth (d).

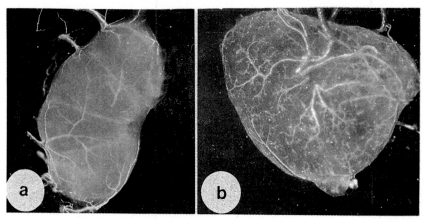

Fig. 4.   Gonads of *B. mori*. Larva in the fifth instar was dissected and the tissues were viewed under a microscope. (a) Kidney shaped testis showing all the four chambers. (b) Triangular ovary.

(ii) **Zone of growth**, where spermatogonia increase in size and give rise to spermatocytes of the first and second order.

(iii) **Zone of division and reduction**, where spermatozoa undergo meiosis and give rise to spermatids.

(iv) **Zone of transformation**, where these haploid spermatids form spermatozoa without undergoing any morphological transformation. Each haploid cell contains 0.52 picogram of DNA (Rasch, 1974).

The spermatozoa collect in vesicular seminalis. During spermatocyte formation, a group of cells develop a layer of cells called the gonocyst outside the spermatogonia. The spermatogonia continue to reside in the gonocyst and form spermatocyte, which begins to increase in size. The gonocyst is now known as cytocyst, each of which contains 2–64 spermatocytes. Maturation division of the spermatocytes results in four spermatids which by change of form generates spermatozoa. All nonnucleated spermatozoa are absorbed near the end of the follicle and only the nucleated ones reach the vas deferens. The number of nucleated spermatozoa in each testis is about 1.4–2 million. A nucleated sperm is a long filamentous body with an acrosome followed by a chromatic head and axial filament. The sperm bundle varies with the race of the silkworm (Sakaguchi, 1979a).

B. Oogenesis

A pair of ovaries is located on either side of the dorsal vessel under the epidermis of the eighth larval segment (Fig. 4b). Larval ovaries can be distinguished from testis only after the third instar when the ovary becomes triangular and the testis gets kidney shaped. The ovary is divided into four compartments in the young larva which elongates and assumes a tubular shape. It is then called the ovariole. At the end of each ovariole is present a spindle shaped apical cell which increases in number by direct division and forms eight cells named cytocysts (Tazima, 1979). One of the eight cytocysts forms the primary oocyte and the others form the nurse cells, which are irregular in shape. The oocyte is always at the lower end of the group and the nucleus which was at the centre assumes a position at the upper end of the cell (for review see, Tazima, 1979). Nurse cells and oocytes are covered by a common follicular epithelium. After absorption of the nurse cells, follicular cells envelop the oocyte completely. By the fourth day of the pupal life the eggs increase in number and fill the abdomen, and by the fifth day, oocytes become enveloped by shells formed by the follicular cells by the process of the choriogenesis (Kafatos *et al.*, 1995). Chorion formation is found only in the members of *Bombycidae* family of the order Lepidoptera and is thought to have evolved to protect the embryo during the extended period of embryogenesis and the prolonged period of diapause.

Towards the end of oogenesis, follicular epithelial cells synthesize and secrete onto the oocyte's surface, a complex set of proteins. These proteins assemble to form the tripartite structure of the egg shell, consisting of an inner vitelline membrane, a chorion and a very thin outer sieve layer (Kafatos *et al.*, 1995; Papanikolaou *et al.*, 1985). The vitelline membrane is typically < 2 μm in thickness, whereas the chorion can vary depending on the species, from < 1 μm to > 50 μm.

B. mori eggs are enveloped in a protective membrane, the chorion, which covers them in a net like pattern in which the microfibrils are arranged in a helicoid manner (Fig. 5a, b). Chorion formation begins with the formation of the layer of tubules on the outersurface of vitelline membrane. These tubules then coalesce to form the inner edge of a thin striated layer. In subsequent stages, compartmented layers are secreted above the striated

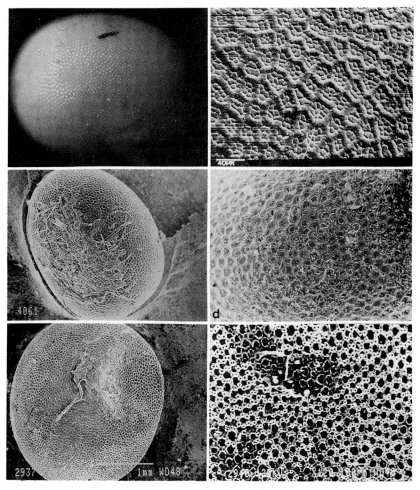

Fig. 5. Scanning Electron Micrographs of the silkworm egg surface (a, b) *B. mori;* (c, d) *A. mylitta;* (e, f) *P. cynthia ricini;* (b, d and f) show magnified view of the egg surface, depicting the characteristic netlike patterns of the chorion.

layer. These layers form the chorion. The egg consists of the protoplasm forming a reticulum on the outer side within the vitelline membrane and enveloping the inner deutoplasm or yolk. Yolk consists of vitelline spheres and globules of fat.

The egg has an anteriorly situated aperture called micropyle through which the sperms enter, marking the head end of the larva. Near the micropyle there is a chrysanthemum like marking (Nambiar *et al.*, 1991). The rest of the egg is covered with a net like pattern (Fig. 5b), a characteristic print of the shape of the follicle cells which secretes the chorion (Mazur *et al.*, 1980). The egg undergoes maturation division to form the ovum. Maturation starts when the eggs are laid and the first polar body is released. Second maturation division follows in about 60 min after the eggs are laid and is completed in about 20 min releasing the second polar body. Conjugation of the egg and sperm takes place in the next 40 min to form the embryo (Hinton, 1981).

In ooctye cells of *B. mori* most of the proteins are not synthesized in the ovarian cells but are taken up from the blood. The yolk proteins include female specific factors which are present in the haemolymph of *B. mori* (Kawaguchi *et al.*, 1988a; b; Koyabashi *et al.*, 1988; Raikhel and Dhadialla, 1992). During larval stages, these factors are secreted into the blood by tissues other than the ovary. They are subsequently transferred to the yolk through the basement lamina. Carbohydrates and lipids also occur in the yolk sphere but they are synthesized in the ovary. RNA is synthesized in the nurse cells and then migrates into the oocyte in the form of ribosomes (for review see, Goldsmith, 1995) as in the case of *Drosophila*.

C. Fertilization

B. mori eggs provide several advantages for experimental embryological studies because the number of eggs laid by one female moth is about 300–500; the egg laying is highly synchronous and 80% of the total eggs is deposited within four hours. Besides, an experimentally convenient size of the egg (1.20×0.80 mm^2), availability of techniques for artificial hatching, extended period of embryogenesis and many genetic characters specific to the egg (e.g., shape, shell colour, yolk colour, serosa colour, volitinism etc.) also make the system suitable for experimental embryological studies (Ueno *et al.*, 1995).

During fertilization the spermatophores enter the bursa copulatrix of the female during coitus. Sperms emerge from the spermatophores and enter the spermatheca by the combined action of the secretions of the spermatheca, peristalic movements of the vestibulum as well as their own motility. The sperm tail gets separated from the head, a centrosome appears near the 'top of the egg and an aster is formed. The sperm head gradually swells into a spherical body forming the male pronucleus (Sakaguchi, 1979a). Soon after the maturation division, the male and female pronuclei fuse together and this phenomenon is termed as impregnation or fertilization.

Copulated females begin to lay eggs within a few hours of emergence. Eggs are covered with a substance secreted from the accessory glands opening at the egg tube or vestibulum. The property of the egg to stick to the surface on which it is laid is due to a glue like secretion by the mother. The Mysore and the Japanese races possess the highest sticking character.

The fertilized eggs give rise to the embryo which undergoes development. Unfertilized eggs are pale yellow in color, and can be distinguished from the fertilized eggs which turn black by the penultimate day of hatching. The dead eggs are also pale yellow in colour but their shells collapse by the second subsequent day of laying.

Occasionally some unfertilized eggs develop by parthenogenesis. This is common in *B. mori* (for review see, Tazima, 1979). The nucleus and the yolk of the unfertilized eggs do not undergo any change. There is no regular nuclear division in the daughter cells of the parthenogenetic eggs. In some, the size of the daughter cells may be uneven and in some multinuclear cells are formed.

Polyspermy is another variation, quite usual in *B. mori*. One to eleven sperms may enter an ovum but usually 2–3 sperms are found. Of the many sperms that enter only one fertilizes the female nuclei, thereby resulting in the production of polyploid or dispermic androgenic individuals (Sakaguchi, 1979a).

D. Gametogenesis in Nonmulberry Silkworms

The nonmulberry silkworms also follow the same pattern of developmental changes in spermatogenesis and oogenesis as *B. mori* except for small variations in terms of duration of larval and pupal development. Tasar (*A. mylitta*) silk moths deposit about 200 eggs in batches of 5–10, over a period of 6–7 days (Chowdhary, 1960). The eggs are ovoid, dorsoventrally flattened and bilaterally symmetrical along the antero-posterior axis with a characteristic net like pattern (Fig. 5c, d). They are about 3 mm in length and 2.5 mm in diameter and weigh 10 mg each. Two brown parallel lines along the equatorial plane divide the surface of the egg into three zones, *viz.*, edge zone, disc zone and equatorial zone. The Muga (*A. assama*) moths lay altogether 150–200 eggs in clusters of 8–10. The eggs are ovoid, slightly elongated, 2.8 × 2.3 mm in size and weigh 9 mg each. Eri (*P. cynthina ricini*) moths lay 400–500 eggs for 3–4 days as a single layer or in 2-3 layers. (Chowdhary, 1970). A gummy secretion from the female moth dries up quickly, fastening the closely laid eggs together. The eggs are ovoid, measure 1.5 × 1.0 mm and weigh about 6 mg each (Fig. 5e).

In *P. cynthia ricini*, there is a uniform distribution of follicular imprints on the surface of the egg (Fig. 5f). Micropyle is on the broader end where the head of the embryo is seen, and has a petal like pattern (Nambiar *et al.*, 1991). In Antheraea species, the follicular imprints change according to whether the eggs are streaked, partially streaked or streakless (Jolly and

Sen, 1969; 1974). The follicular imprints can be round, polygonal or irregular in shape (Fig. 5d).

IV.   Embryogenesis

B. mori embryos develop and hatch in about 10 days at 25°C after fertilization. Various stages of the embryonic development of *B. mori* (Ueno *et al.*, 1995) are schematically presented in Fig. 6 and Table 1.

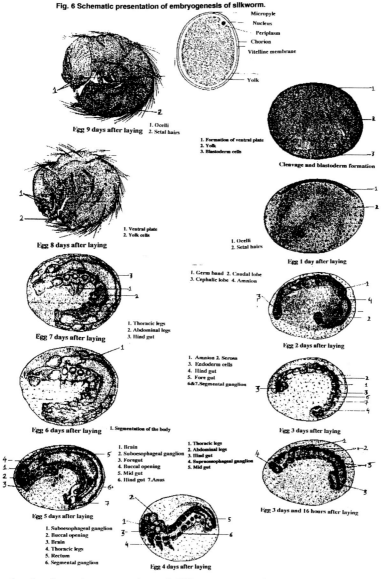

Fig. 6 Schematic presentation of embryogenesis of silkworm.

Fig. 6.   A schematic presentation of different stages of embryonic development of *B. mori* (see Text and Table 1 for details).

Table 1 Stages of Embryogenesis

	Time	Morphological Landmarks
1.	0–2 h	Second maturation division and syngamy.
2.	2–10 h	Progress of synchronous cleavage. Mitosis is followed by uniform invasion of periplasm. The remainder persist within the yolk mass as primary vitellophages.
3.	12 h	Establishment of syncitial blastoderm.
4.	20 h	Formation of germ band.
5.	25 h	Gastrulation takes place accompanied by elongation of the germ band.
6.	35 h	Segmentation of mesoderm. 17 segments can be clearly observed.
7.	40 h	Appendage at head and thoracic region develop. Eggs enter into diapause in univoltine and bivoltine strains at this stage.
8.	48 h	Differentiation of appendages.
9.	60 h	Invagination of proctodeum reaching borderline between 18th and 19th segment.
10.	3 days	Invagination of the trachea initiates.
11.	3.5 days	Invagination of the proctodeum proceeds and reaches the 11th segment.
12.	4 days	Blastokinesis (Embryonic movement) begins.
13.	5 days	Blastokinesis is completed.
14.	6 days	External processes are formed.
15.	7 days	Taenidium in spiral band is formed in the tracheal tube.
16.	8 days	Pigmentation begins in head.
17.	9 days	Pigments formed at seta and epithelium.
18.	9.5 days	Hatching of embryo.

A. Early Development

After fertilization, the centrally located nucleus generates many nuclei by mitotic divisions which migrate towards the surface of the egg. The metaphase of the first cleavage in *B. mori* takes place at about 160–190 min (Table 1). Migration of the cleavage nuclei to the periplasm results in the formation of syncytical blastoderm whereas a small number of the remaining nuclei in the yolk mass act as vitellophages (Fig. 6). After 12 h of development these nuclei and the associated cytoplasm undergo cellularization resulting in the formation of the blastoderm (Toyama, 1902; Miya, 1978; Tazima, 1979; Nagy *et al.*, 1996).

B. Late Development

About 15 h after egg laying a dense embryonic primordium follows, whose cells increase in thickness and get concentrated. These cells also increase in number through cell division (Tazima, 1979). In the blastoderm this area extends broadly on the ventral to ventrolateral side of the egg and is called the germ band. It is formed along the periphery on the side of the shorter radius (Fig. 6). The cells outside the germ band become flat and get connected with each other. After gradual thickening the germ band undergoes gastrulation and segmentation.

The germ band, on cell multiplication and differentiation generates three germinal layers, ectoderm, mesoderm and endoderm. The amniotic folds appear from the edges of the germ band parallel to the long axis of the egg partly covering it. These folds grow inwards towards each other and coalesce. The outer wall of the amniotic fold is known as serous membrane or serosa, and the inner wall, amion. The cavity formed by the two amniotic folds in which the germ band lies is known as the aminotic cavity.

After 24 h of development, an elliptic disc called the ventral plate is formed which becomes guitar shaped at the end of 40 h (Fig. 6). Due to cell movement the germ band curves inwards at the edges and sinks into the yolk. The yolk cells undergo division during this period and form the primary yolk cells. Secondary yolk cells are recruited from the blastoderm (Miya, 1978; Keino and Takesue, 1982).

At later stages of development (around 50 h) the anterior (procephali or protocephali) and posterior (caudal lobe) ends of the germ band become enlarged, accompanied by caving in of the middle region (ventral side) to form the primitive groove (Fig. 6). The embryo in hibernating eggs attain the germ band stage soon after the eggs are laid and remains quiescent till the completion of the hibernation period. This stage of development is known as "Bandelette stage" and the process of hibernation is called "Diapause" (Yamashita and Hasegawa, 1985).

In the bandellete stage embryo, a new group of mesodermal cells appears below the primitive groove which deepens and gets closed from the head thereby forming an open tube at the head end. This pore is known as blastopore. After hibernation, the embryo elongates and occupies 80% of the circumference of the egg. This stage of the embryo is known as "maximum length period". It is marked by well defined segmentation (Keino and Takesue, 1982). During the "maximum length period" the mesoderm cells below the primitive groove begin to grow laterally and envelope the embryo as ectoderm. These mesodermal cells form the appendage of the head and the thorax. About three days after development the embryo shortens and widens. As the mesoderm grows, the colon sac appears and the appendages of the abdominal segments begin to develop.

There are 20 embryonic segments in normal silkworm embryo, 6 in the head region, 3 in the thorax and 11 in the abdomen. Head comprises of (i) Occular Segment of Protocephalon (ii) Antennary Segment (iii) Intercalary or Tritocerebral Segment (iv) Mandibular Segment (v) Maxillary segment and (vi) Labial segment.

Blastokinesis or periods of body position reversal, a phenomenon commonly observed in the Lepidopterans (Ueno *et al.*, 1995), involves movement of the germ band cells on the ventral side facing the outer side of the egg moving through an arc at the head region till the dorsal side come to face towards the outside of the egg (Fig. 6). This is again reversed after some time through the same arc to its original ventral position. Blastokinesis of the embryo is followed by development and generation of the rest of the organs from the ectodermal and the mesodermal cells. Thoracic appendages develop and become three segmented. Although completely formed, the embryo hatches out only after 10 days of development (Miya, 1985, Nagy *et al.*, 1995).

C. Diapause

Diapause is a period of quiescence to tide out the unfavourable conditions during which visible activity and many physiological processes are suspended. It is an adaptation of the insects to do away with the unfavourable conditions (Lees, 1955). In silkworm, diapause occurs in embryos soon after the germ band stage during embryogenesis and further development of the embryos is arrested till the period of hibernation is over. Diapause, like eclosion, is a hormonally triggered phenomenon (Denlinger, 1985; for review see, Goldsmith, 1995).

Unlike other insects where the activity ceases only at the onset of the unfavourable period, in some strains of silkworm irrespective of growth conditions the life cycle is programmed to undergo diapause or hibernation. In other words the quiescence is a heriditary characteristic triggered by an internal "timing" mechanism which brings about cessation of activity in advance of the unfavourable condition. The diapause factor from. *B. mori* has been purified recently (Imai *et al.*, 1991). On the basis of diapause, *B. mori* strains have been classified into uni-, bi- and multivoltine (see above).

Experiments on a large variety of insects have revealed that the diapause is linked with seasonal changes in day length or photoperiodism. In *B. mori* a lengthening photoperiod triggers diapause whereas in most diapausing insects the shortening of photoperiod is responsible for it. There are artificial means such as acid treatment (*e.g.* exposure to HCl of specific gravity 1.075 at 46°C for 15 min) or cold storage of eggs, for breaking diapause.

The Eri silkworm, *P. cynthia ricini* belongs to the nonhibernating multivoltine insects whereas the *Antheraea* species undergo diapause which can be broken by changes in the photoperiods.

D. Embryonic development of Eri silkworm

In *P. cynthia ricini*, within 2 h of oviposition, the cleavage division of the nuclei and cellular movement towards the periphery begins and by 24 h germ band formation gets completed. At 48 h the amnion and serosa are formed and segmentation begins. The embryo reaches the "maximum length stage" which continues to form a visible segmented body on the 3rd day. By the 4th day some of segments fuse and blastokinesis begins. On the 5th day of development the embryo assumes "S" shape and appendages start appearing. Slowly, the embryo moves to the periphery and formation of the internal organs is completed. On the 7th day, the silk gland appears and from the 8th day onwards till the hatching on the 10th day, the embryo increases in size (Krishanappa, 1989).

V. Larval Structures

Larval body consists of metamerically arranged segments mainly divided into head, thorax and abdomen (French, 1990). The head is formed by the fusion of six embryonic segments. The second, fourth and sixth segments carry the antennae, mandibles, maxilla and labium (Fig. 7a). Clypeus and labium are prominent. Six pairs of ocelli are situated just behind and a little above the base of the antennae. Antennae are formed of five segments. The mouth parts are located downwards in front of the face and are composed of a pair of mandibles and maxillae with labium and labrum. Mandibles are used for mastication and consist of two hard pieces. The maxilla consists of a single maxillary lobe and the palpi which is made up of three segments. The labrum hangs down from the frontal portion to form a flap of the mouth. The labium is situated on the ventral region of the head with a pair of labial palps which form a sensory organ. A spinneret through which the silk fibre is extruded, is situated at the proximal position between the two labial palpi (Fig. 7b).

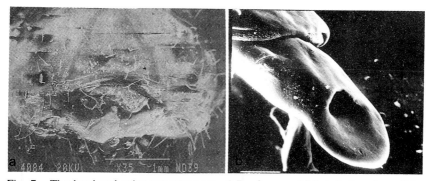

Fig. 7. The head and spinneret structures of *Bombyx mori* (a) Anterior view of the head showing the mouthparts, antenna and the spinneret.; (b) Scanning electron microscopic view of the spinneret.

The morphological characters of the spinneret differ between the mulberry and nonmulberry species and even amongst the members of the nonmulberry class (Fig. 8). The head and spinneret structures of the Tasar (*A. mylitta*) and Eri (*P. cynthia ricini*) are presented in Fig. 8a, b and c, d panels respectively.

The thorax of *B. mori* comprises of three segments, pro-, meso-, and meta thorax carrying a pair of legs ventrally. Each leg is formed of three segments (Fig. 9 a, b, c). Generally all the silkworms have an eye spot on the dorsal aspect of the mesothorax.

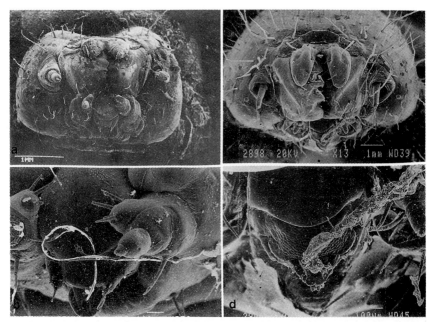

Fig. 8. The head and the spinneret of the nonmulberry silkworms. Eri, (*P. cynthia ricini*) (a, b) and Tasar (*A. mylitta*) (c, d). (a) and (c), anterior view of the head showing the mouth parts and the sensory organs; (b) and (d) magnified view of the spinneret with the silk fibre coming out.

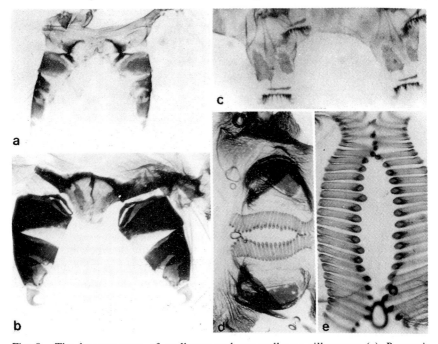

Fig. 9. The leg structure of mulberry and nonmulberry silkworms (a) *B. mori;* (b) *P. cynthia ricini;* (c, d and e) Differential interference contrast images of the pseudopods of *B. mori* at different magnifications.

In the larval abdomen the last three of the eleven abdominal segments fuse to form a single segment. The hatched larvae therefore possess only nine segments. Ventrally, the third, fourth, fifth, sixth and ninth segments bear the abdominal legs or pseudopods (Fig. 9d, e) which help in locomotion by providing the anchor of the worm body. There is a caudal horn on the dorsal surface of the eighth abdominal segment. There are nine pairs of spiracles of which one pair is located ventrolaterally on the first thoracic segment and a pair in each of the first to eight abdominal segments. The ninth or last abdominal segment bears the anal plate and caudal legs (Sakaguschi, 1979b).

The larval exoskeleton consists of an outer cuticle and an inner hypodermis. The epidermis is derived from the primary and secondary cuticle of the larva and has trichogenous cells which secrete 'setae.' The integument consists of exoskeleton or the outer chitinous framework and endoskeleton or internal framework. Individual parts are known as apodemes which arise as invaginations of the bodywall between adjacent sclerites. Endoskeleton of the head is called the tentorium and that of thorax, endothorax.

VI. Internal Anatomy of Larva

The digestive organ of *B. mori* larva comprises of an alimentary canal which is a straight tube from the mouth to the anus and may be divided into three main divisions, the foregut, the midgut and the hindgut. Foregut consists of the mouth, a funnel shaped mouth cavity, pharynx and oesophagus. Midgut is a long cylindrical tube. Hindgut consists of small intestine, colon and rectum. There is a pylorus near the anterior end of the small intestine (Sakaguchi, 1979b).

Excretory system consists of a colon, called ileum and a pair of malphigian tubules situated at the junction of the small intestine. There are six regions in each tubule, namely, region inside rectum, region outside rectum, winding region, ascending limb, descending limb and urinary bladder.

The respiratory system comprises of a network of tracheal bush which radiates from the spiracle and serves the function of both inspiration and expiration. Spiracle consists of peritrene, sieve plate and atrium. Except the forepart of the silk gland, trachea permeates all sense organs by frequent branching.

Circulatory system consists of an open dorsal vessel which is open at the anterior head end and is blind at the posterior end. Its finely drawn out anterior end is known as aorta and the posterior end, heart. Each segment from the second thoracic to the ninth abdominal segment has a pair of ostia. The aorta does not pulsate but the heart does, and the rate of pulsation varies with age, movements, temperature and race of the larvae. Systole and diastole of the heart are caused by eight pairs of alary muscles and the musculature of the heart. Silkworm has a unique adaptation of ocassional reversal of blood flow indicating the absence of a valvular system or the presence of a dual valve (Sakaguchi, 1979b). Normally the blood flows in the antero-posterior direction. Reversal occurs during embryonic stage, period of hatching or larval eclosion, prepupal and pupal stages. Blood, a clear liquid of pH 6.3–6.5 and 90–95% water consists of plasma and corpuscles. Corpuscles are leucocytes of four kinds, namely, proleucocytes, phagocytes, globulated cells and giant cells. The colour of the blood is pale yellow in the white races and dark yellow in the yellow races. Blood turns dark on exposure to air due to melanization.

Silkworm has a well developed muscular system which helps in locomotion and other physiological activities such as digestion and circulation. There are two types of muscles, the skeletal and the visceral muscles. The skeletal muscles help in locomotion. The visceral muscles control movements of the internal organs such as the alimentary canal. Muscles of the head consist of the visceral muscles in the mouth and antenna. Wing muscles consist of two sets, the direct and the indirect muscles, the former being pleural muscles and the latter being attached to the thorax. Meso- and metathorax have a series of longitudinal muscles in the abdomen as well as dorsoventral and pleural muscles in the leg.

The fat body is arranged as a loose mesh of cells, inserted by delicate membranous connective tissue. It has a storage function for reserve materials such as glycogen and proteins, and carries out the intermediary metabolism and biosynthesis.

The central nervous system consists of the brain or cerebral ganglion, the suboesophageal ganglion and the ventral nerve cord. The brain is situated just above the oesophagus and between the apodemes of the tentorium. The suboesophageal ganglion is situated ventrally and is the ganglionic centre of the head. It gives off paired nerves to the mandibles, maxillae and labium. Nerve cords from the brain supply the ocelli, antennae and labium. The ventral nerve cord consists of a series of ganglia situated on the floor of the thorax and abdomen.

Sympathetic nervous system comprises of the oesophageal sympathetic and the visceral sympathetic system. The former consists of a frontal ganglion situated in front of the brain which gives off anteriorly, the frontal nerves going to the clypeus and two roots which connect it to the brain. Posteriorly, the frontal ganglion gives off the recurrent nerve which extends along the mid-dorsal line of the oesophagus for some distance and enlarges into hypocerebral ganglion. A delicate plexus of the nerve fibres and multipolar nerve cells are situated in the hypodermis. These innervate the sense organs of the body.

Sense organs comprising of tactile sensillae are distributed over the integument and are particularly abundant on the antennae, palpi and legs. Olfactory and gustatory sensillae are present in the mouth parts or pharynx and act as chemoreceptors for taste. Six pairs of ocelli are present in the larva. The ocelli can distinguish between light and dark and convey only coarse image of a nearby object.

Reproductive organs consist of a pair of ovaries situated on the dorsal aspect of the fifth abdominal segment from each of which a blind tube develops and ends at the seventh abdominal segment in the female, or a pair of testis and a blind tube situated dorsally in the fifth abdominal segment in the male. Sexual dimorphism is not very apparent during early larval stages. The male has Herold imaginal bud in the central line on the ventral side of the twelfth segment. The female has four spots of Ishiwata Imaginal bud at the ventral side of the 11th and 12th segment. These spots are visible only during early fifth instar.

VII. Glands

Experiments on ligating insects and exchanging body fluids and tissues of individuals in different stages of development have shown that most of insect development is governed by the hormonal milieu, beginning from the first instar larva (Norak, 1975). During development of *B. mori,* the retention of the immature and juvenile characteristics, development of adult structures and moulting are controlled by hormones secreted by the brain and several glands associated with it (Koyabashi, 1979, 1988 Gupta, 1979; Ohnishi and Ishizaki, 1990). The principal glands are:

(i) Exuvial Glands: These are epidermal glands which produce moulting hormone. They are also known as moulting or Versonian glands. There are 15 pairs of unicellular glands of this kind, two pairs in each thoracic segment, a pair on each abdominal segment from the first to the seventh and two pairs on the eighth segment. These glands are made up of three cells and perform the excretory function in eclosion.

(ii) Tracheal Glands: An oval or spherical narrow gland which goes into spiracle is formed of three cells as in the exuvial glands.

(iii) Salivary Glands: A pair of filamentous tubular glands which secrete saliva and open near the articulations of mandibles.

(iv) Prothoracic Glands: These are present on the inner aspect of the fifth thoracic spiracle. These secrete ecdysone responsible for moulting and development of adult structures. The pupal prothoracic gland secretes another substance responsible for the initiation of imaginal or adult characters (Koyabashi, 1979; Sakurai, 1983).

(v) Suboesopharyngeal Gland: It is a band like gland situated on the mid ventral aspect of the oesophagus. It secretes factors responsible for the hibernation of silkworm eggs. It also secretes juvenile hormone necessary for the maintenance of larval characteristics during preadult life.

(vi) Peritracheal Glands: They are associated with the pericardium and help in digestion of proteinaceous substances in the body cavity and purification of the blood.

(vii) Corpus Allatum: It is a pair of small ovoid whitish bodies lying behind the supraoesophageal ganglion and closely associated with the oesophageal ganglia of the sympathetic nervous system. The removal of the corpora allata in the third or fourth larval instar brings about precocious pupation with elimination of the intermediate moulting. Silkworms deprived of the corpora allata give rise to small cocoons before pupation (Tobe and Stay, 1985).

(viii) Brain Hormone: Brain regulates activation of both prothoracic glands and corpora allata, the former by hormone secretion and the latter by direct nerve connection. It appears that in response to a hereditary pattern of tissue and organ behaviour, the brain secretes growth hormones. This causes rhythmic occurance of moulting and development of the proper adult structures. The brain hormone and ecdysone (prothoracic gland hormone) are steroids in nature (Sehnal and Rembold, 1985).

(ix) Silk Glands: Silk is secreted by a pair of long, tubular organs called the silk glands which are the second largest organs in the silkworm (Shimura, 1983a). They are considered as modified salivary glands because of their origin from the labial segments. Even though, salivary glands are present in the silkworm, they are derived from mandibular segment. The silk gland originates as an ectodermal invagination of the basal part of the labium during germ band stage and its morphological development is completed after blastokinesis stage (Nunome, 1937). They are cylindrical and tubular (Fig. 10a) with walls that are only one cell thick and with characteristically branched nuclei (Tazima, 1979). On the 6th or 7th day of embryogenesis silk glands can be seen as a 1 mm long elongated mass. The organization of the silk gland begins by the 8th day and is completed by the 10th day before the larva hatches out. In the fully grown larva, it occupies most of the ventro-lateral side of the body from the fourth to the eighth segment. A peculiar feature of the silk gland development is that all the cycles of cell division in the tissue are confined to the embryonic stage and further growth is related to the increase in tissue mass and not in cell number (Gage, 1974b). In the early stages, relative rate of growth and development are parallel to the insect body which increase by 20 fold at the onset of fifth instar resulting in silk glands contributing nearly half of the body weight (Shimura, 1983a). The continued growth with the absence of cell division results in 16000 times increase in size (Suzuki, 1977). Growth in the silk glands is caused by the polyploidization of the nuclei in the absence of the nuclear division which results in 200,000 times more DNA as compared to

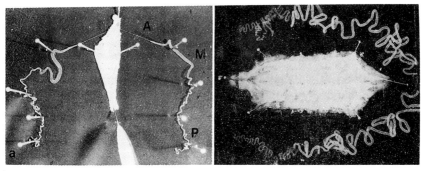

Fig. 10. Silk glands of the mulberry and nonmulberry species (a) *B. mori*, (b) *A. mylitta*. The three distinct regions anterior, middle and posterior silk glands (denoted as A, M and P) are clearly discernible.

a normal diploid cell with no evidence of selective amplification of any specific sequences. Each silk gland is made up of 1000 cells (Tazima, 1979; Akai, 1984) and comprises of three physiologically distinct regions:

Anterior Silk Gland (ASG), made up of about 200 cells, is a straight tube opening at the anterior end into a duct and posteriorly into the middle silk gland (Fig. 10a).

Middle Silk Gland (MSG), is made up of about 250 cells and is further divided into fore, mid and hind parts. MSG is a straight tube opening anteriorly at the fore end and posteriorly into the middle region which is the largest of the three definite flexions. The beginning of the middle region of MSG is narrow but widens suddenly and the hind part is narrow and uniform in diameter. The last part is curved and folded up between the dorso-visceral muscles and the trachea. The MSG synthesizes and secretes the glue protein, sericin(s) (Shimura, 1983b; Hui and Suzuki, 1995).

Posterior Silk Gland (PSG), is a long convulated region, 15–20 cm long and is made up of 550 cells. The PSG is specialized for the synthesis of silk fibre proteins, fibroin(s).

The silk glands from both the sides join anteriorly to form a common duct which opens outside through a spinneret, a retractile tubular membranous structure, conical in shape and bluntly pointed at the apex (Fig. 10a and 7b). In the nonmulberry silkworm, the basic plan of silk gland is similar to *B. mori* and consists of three different parts, viz., anterior, middle and posterior silk gland (Fig. 10b). The clear distinction between the middle and posterior silk glands though discernible is not as clear as in *B. mori*.

The spinneret is surrounded by a chitinous ring and localized scleretoid bars and a lambda shaped silk press are visible within the anterior common duct (Joy, 1986). A pair of glands known as Fillipi's or Layonneti's glands are situated at the junction of the silk gland. A viscous fluid is secreted by them. The wall of the silk gland is composed of three layers, the tunica propria, gland cells and the tunica intima enclosing the lumen of the gland. Tunica intima is made up of thickly laid chitin throughout but the anterior portions are removed at each ecdysis.

The number of cells in the silk gland varies according to the strain and volitinism, but does not differ between the two glands in a pair from single individual or between sexes, or during stages of postembryonic development. The silk glands degenerate at the pupal stage.

VIII. Silk Secretion

B. mori ingests about 50 g of mulberry leaves per larva and 60% of the nitrogen contained in the leaves is used for the silk protein synthesis (Ito, 1979). On an average, a larva produces about 0.6 g of silk protein, as estimated from the cocoon shell weight. From a cocoon, a single uninterrupted fibre of 1 to 1.2 kilometer length is obtained. The silk cocoon is basically made up of two classes of proteins, the fibroin(s) and sericin(s) (Shimura, 1979; 1983b; Gamo, 1987; Hui and Suzuki, 1995). In order to reel the silk fibre, the cocoons are boiled and therefore, most of the sericin(s) being water soluble, are removed. The water insoluble fibre protein is composed of a heavy chain fibroin (fibroin H; molecular weight 350,000) linked through a disulfide bond to a light chain fibroin (fibroin L; molecular weight 25,000). A third protein P25 synthesized in the posterior silk glands (Couble *et al.*, 1983) and coordinately regulated with fibroin H and L chains is also known but the nature of its association with the fibroin chains is not yet clear. The fibroin(s) secreted into the lumen of PSG are transported to and stored in the middle silk gland until the secretion of silk fibre (Prudhomme *et al.*, 1985). The synthesis of silk fibroin by *B. mori* is taken as a classical example for tissue specific and developmental stage specific gene expression (Suzuki, 1977; Hui and Suzuki, 1995).

Sericin(s) are glue proteins which make the fibroin chains stick. Different variants of sericin are synthesized within specific territories of the middle silk gland and are secreted into the lumen. Here they cover the accumulated fibroin. The mixture of fibroin(s) coated with sericin(s) is transported through the ASG. The highly polymerized mixture of fibroin(s) and sericin(s) still in the liquid state is passed through the silk press and extruded through the spinneret. Solidification of the fibre results in consequence to the release of mechanical pressure and the fibre gets hardened in contact with air.

Significant qualitative and quantitative differences have been observed in the amino acid composition of the silk fibroins of mulberry and nonmulberry silkworms. *B. mori* fibroin H is glycine rich (44–46% glycine and 29–30% alanine) whereas the fibroins from Eri or Tasar silkworms are generally alanine rich (34–39% alanine and 24–26% glycine). Silk formation during the spinning period is largely controlled by the level of the reserves supplied from the degenerating tissues.

IX. Metamorphosis into Pupa

During metamorphosis of the larva to pupa, the larval skin becomes pale and transluscent in the white cocoon races and golden yellow in the yellow cocoon races. At this stage, the worm gets seated on a support platform by the prolegs and starts spinning. The head is waved in the form of figure "8" extruding liquid silk from the jet of spinneret and spins a cocoon around itself. The spinning period is only about 24 h. Ecdysis of the larval skin takes place in the cocoon (Sakurai, 1983; Ohnishi and Ishizaki, 1990). The body of the larva becomes contracted and distended, the hypodermis secretes a fresh layer of chitin beneath the old cuticle and ecdysis takes place. The secretion of the exuvial gland gradually loosens the old and the new cuticle which results in dehiscence of the larval skin along the mid-dorsal aspect of the thorax. The exuvia is gradually slipped off from behind, thus liberating the pupa.

Larva transforms into pupa in 51–53 h from the time spining commences in *B. mori* (Pure Mysore strain). This period is marked by the reduction in length, weight, loosening of cuticle, change in colour, reversal of the blood flow in dorsal vessel and loss of active habits. The larval head is flexed and the prepupa formed has a tendency to lie on its side.

Pupal life starts with the last ecdysis and extends till it becomes an imago which is generally 6–8 days depending upon the climatic conditions. It is shorter in summer than in winter. The changes during transformation from prepupa to pupa are marked by reduction in length, increase in weight, formation of pale yellow cuticle, change in color of spiracles from black to brown and reduction in the number of spiracles from nine to seven pairs. Metamorphosis of the six pairs of simple larval eyes to a pair of large compound eyes, prefiguring of the imaginal appendages, *viz*: antennae, legs and wings, and lowering of the peristaltic contraction of the dorsal vessel also take place. The body is navicular and capable of wriggling movements of 2–3 post segments only. The pupal body is divided into head, thorax and abdomen. Larval mouth parts drop off during ecdysis leaving behind labrum, the labium and the maxilla with two palpi. Setiform antennae appear. The thorax is divided into pro-, meso-and metathorax. Abdomen is segmented. The visible number of segments on the dorsal aspect of the pupa is nine and on the ventral aspect is eight (Tazima, 1979).

The female pupa is usually larger than the male. In the pupal body histolysis of the larval organs and histogenesis of the imaginal organs occur actively. The larval genital organs differentiate into the imaginal organs during the pupal stage.

X.   The Adult Stage (Imago or Moth)

Imago emerges from the pupal case by rupturing the cuticle. Due to a secretion discharged from the mouth of imago at one of the poles, the cocoon softens and the moth emerges, head foremost through the apperture formed by pushing aside the filaments at the softened pole of the cocoon. On emergence, the moth has its wings folded and the body and wings are still moist. It rests for 10–30 min till the wings get unfolded on drying. During the resting period, there is a brownish discharge from the anus (meconium) which is the product of pupal metabolism (Sakaguchi, 1979b).

The body of the moth can be clearly divided into three parts: head, thorax and abdomen. The appendages of the head are the antenna, mandibles, maxillae, labium and labrum. On the head capsule there is a pair of compound eyes and no ocelli. Antennae are inserted on the epicranium. They have a pectinate type of structure, composed of 35–40 small segments, which differ between strains and with volitinism (Fig. 11a). Each compound eye is an aggregation of similar units known as ommatidia, which are hexagonal in shape (Fig. 11b). Each ommatidium consists of the corneal lens, crystalline cone, corneagen cell, pigment cell, retinula cell, rhabdom and nerve fibre.

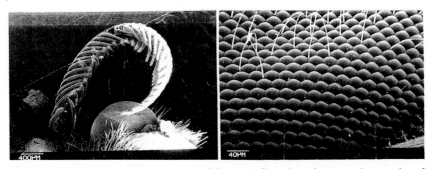

Fig. 11. Antenna and Compound eye of *B. mori*. Scanning electron micrographs of Antenna (a); compound eye (b) of *B. mori*.

The thorax is composed of three segments, the prothorax, mesothorax and metathorax (as in larva). These are covered with chitinous sclerites of the head. The dorsal sclerite is termed tergite, and the ventral sternite and the sclerite on both the sides, pleurites. There are three pairs of thoracic legs. Each leg is formed of coxa, trochanter, femur, tibia and tarsus. In female, the tibia has only one spur while in male, it has two spurs. Tarsus, the ultimate segment of the leg, consists of five small segments and the last segment has two claws. Between these claws, the pretarsus or the penultimate segment is supported by a median flex or plate on the ventral side. In front of and above this plate, the pretarsus expands into a median lobe or aerolium. On the underside of the tarsal joints, there are pulvilus like organs or aerolium. The aerolium and the pulvillae are pad like organs which enable the moths to climb steep or smooth surfaces.

Meso- and metathorax bear two pairs of wings which are covered by scales on both sides. The shape and size of the scales are varied. Thoracic scales differ from the scales on the head and abdomen.

The abdomen consists of eight segments in males and seven segments in females. The terminal segments of the abdomen of males and females are greatly modified at the genitalia.

XI. Internal Anatomy of the Adult Moth

The integument of the imago consists of a cuticular layer, ectodermal cells, trichogens and basement membrane. Scales are produced by trichogens. The alimentary canal undergoes a remarkable change through pupal- adult development as the imago does not take food. The crop, at the posterior portion of the oesophagus, develops into a spherical chamber containing the cocoon-digesting enzyme which aids the adult to escape. The rectum differentiates into an enlarged sac which accumulates the brownish waste solution named the moth urine. Waste solution contains uric acid, salts and other substances.

Respiratory organs of the adult are more simplified than that of the larva. The trachea form a branched series of tubes spreading throughout the body in the form of a fine network and supply oxygen to the various organs. Trachea communicate with the exterior by six pairs of openings called stigmata situated laterally on the abdominal segments only.

Regarding musculature, some larval muscles pass unchanged onto the adult while some are destroyed without replacement. Thoracic muscles are well developed.

Sense organs of the moth comprise of a pair of bipectinate antennae (Fig. 11a), large compound eyes in which there are 3000 ommatidia (Fig. 11b) and U-shaped labial palpi bearing sense organs. Reproductive system changes markedly through metamorphosis from larva to adult. In males, there is a pair of testis situated in the fifth abdominal segment, dorsolaterally on either side of the ventral nerve cord. Each testis leads into vas deferens opening into the seminal vesicle, an enlarged sac which is broader at the anterior end than at the posterior. The seminal vesicle leads into the ductus ejaculatorous, a long thin tube which ends in the aedeagus, a chitinous tube with a slightly upward curve at the distal end. At the proximal end it is enlarged into a sac which functions during eversion of the aedagus. A pair of acessory glands is associated with male reproductive system.

Copulatory apparatus consists of "oo" shaped chitinous plate on the eighth abdominal sternite situated ventral to the abdomen. There is a pair of bent processes which is large and embraces the vinculum on both sides, and which is also a part of the ninth abdominal sternite. Curved claspers are found on both sides of the aedagus.

In females there are four egg follicles in each ovary. Ovaries are paired and occupy a very large part of the abdomen. A pair of oviducts proceeds from the ovaries and unite to form a single, thick large oviduct which terminates at the orifice, at the posterior extremity of the body. A sac like structure known as the bursa copulatrix is situated ventral to the common oviduct. The bursa opens posteriorly by a short tube, the vagina. Paired accessory glands occur antero-dorsal to the oviduct. These glands open into the ovipositor, a specialized region containing the egg laying apparatus which is a continuation of the oviduct. The spermatheca, a sac for the reception and storage of sperms, opens into the ovipositor anterior to the accessory gland opening. The ductus seminalis and the spermathecal duct are closely located on a raised ridge on the dorsal side of the ovipositor.

The anal apparatus is situated dorsally. The orifice of the ovipositor is in the middle of the bursa copulatrix ventral. Eggs are extruded by muscular contractions of the walls of the ovipositor.

XII. Genetics of Silkworm

The genetics of the silkworm dates back to the pre Mendelian era. *B. mori* females are heterogametic in sex chromosome constitution and the males are homogametic (Robinson, 1971; for review see Goldsmith, 1995). Numerous natural mutations and mosaics have been observed in the silkworm cultures (Doira, 1992). Although the genetic analysis of silkworm had been initiated quite early, the efforts were mainly confined to the silk producing countries of the east and the majority of the reports were in the regional languages, thereby imposing restrictions on readership.

A major handicap of the *B. mori* system has been the large number of chromosomes ($n = 28$) with a haploid genome size of 530 million base pairs (Gage, 1974a; Rasch, 1974) in contrast to *Drosophila* ($n = 4$) which has a haploid genome size of 140 million base pairs (Rasch *et al.*, 1971). This made *Drosophila* a favoured choice for genetic studies. Besides, the small chromosome size, dispersed centromeres and the lack of other distinctive cytological landmarks in *B. mori* limited its usefulness in cytogenetics. Only a few chromosomes show consistent morphological features (Traut, 1976; Kawazoe, 1987). Neverthless, the study of silkworm genetics is heading towards rapid development today. The large repository of the mutant stocks available will be a valuable resource that can be exploited to their full potential.

Genetics of the silkworm has evolved along two lines, with stocks carrying distinctive morphological, developmental and behavioral mutations, or large collections of inbred geographic and improved races used for practical breeding. Many of the geographic races found in most countries practising sericulture, have been characterized in terms of quantitative and complex traits useful for breeding but the stock collections remain largely as untapped resources for modern genetic and molecular studies. Most markers assigned to the genetic linkage maps are spontaneous, involving visible characters found during mass rearing for silk production. Mutations have also been picked up by mutagenesis primarily by X-ray and γ-irradiation as well as chemical mutagenesis. There is substantial divergence in the genes controlling the characters associated with silk production, growth rate, survival, fecundity, fertility and disease resistance. Many of these characters are of immense importance in sericulture.

Silkworm mutations affect a broad range of developmental, physiological, biochemical and behavioral characters. They include stage specific lethals, abnormalities in larval body shape, segment identity, organ and tissue formation, pigmentation patterns, biosynthetic pathways, variant isozymes and haemolymph proteins, life history traits such as stage duration, number of molts, entry into diapause, resistance to fungal and viral diseases, altered food discrimination and aberrant cocoon spinning. Some of these traits and variety of silkworm mutants have been elegantly reviewed by Goldsmith (1995).

Of particular relevance to the synthesis of silk are the Naked pupa mutations, *Nd* and *Nd-S*, presumably harboring mutations in the genes coding for fibroin H and L chain genes, which are unlinked. The *Nd* and *Nd-S* spin the cocoons containing primarily sericins, and their posterior silk glands are abnormally small or missing. The molecular nature of the defect will become clear only after the alleles of *Nd* and *Nd-S* are cloned. Likewise, there are flimsy cocoon mutants (*fle*) which spin light cocoon, containing both sericins and fibroins, but disproportionately reduced in fibroin content.

Homeotic mutations have been known for a long time in the case of silkworm. An example is the *Extra leg* (*E*) mutant which is characterized by supernumerary or deficient legs on specific segments with abnormalities in the position and number of the crescent markings or stars on abdominal segments. The *E* locus has been recently identified as the silkworm homologue of *bithorax* homeotic gene complex of *D. melanogaster.* Several mutations affecting segment identity have also been mapped. A description of the roles of the homeotic genes in the body plan as well as in the silk gland development of *Bombyx mori* has been provided by Ueno *et al.* (1995) and Hui and Suzuki (1995).

The molecular genetics and cytogenetic approaches making use of *in situ* hybridization with fluorescent probes on the chromosome spread are still at infant stages when it comes to silkworms. Most of the conventional molecular strategies have been used for cloning silkworm genes. With the addition of the molecular markers to the genetic linkage maps, it has now become possible to use positional cloning to isolate a gene whose product is unknown.

Attempts are already under way to construct linkage maps of *B. mori* using RAPDs (random amplified polymorphic DNA) (Promboon *et al.*, 1995; Nagaraja and Nagaraju, 1995) as well as micro- and mini satellite mappings (Nagaraju *et al.*, 1995). Probes generated from such studies may form essential tools for chromosomal *in situ* hybridization and will aid in developing a molecular map of the silkworm (Ashburner, 1992; Goldsmith and Shi, 1994). Recently, a technique for *in situ* hybridization using fluorescent probes on silkworm chromosome spreads has been developed. These approaches would significantly enhance the opportunities for correlating cloned genetic markers with specific chromosomes and perhaps localizing the genes in relation to one another on an extended chromosome (Trask, 1991).

The regulation of expression of silk protein genes, the chorion genes as well as *tRNA* genes from *Bombyx mori* have been extensively investigated and these aspects have been extensively reviewed by Kafatos *et al.*, (1995), Hui and Suzuki (1995) and Sprague (1995).

XIII.   Future Prospects

The excitement and promise of developmental biology has never been greater as researchers close in today on the secrets of how a single fertilized egg cell goes through the complex and beautifully orchestrated series of changes that create an organism. Methodologies are becoming available for answering many of the pertinent questions. The most important unanswered question in developmental biology, perhaps is how the body's specialized organs and tissues are formed (Barinaga, 1994). This topic, known as *morphogenesis* encompasses the formation of all tissues and organs, from the first embryonic tissue layer to the adult heart, brain or kidney. Equally interesting are the questions on how patterns form in the embryo that tell different parts what to become; how cells respond to signals during development and how individual cells become committed to particular developmental fates. Before any tissue or organ can form, earlier steps must occur that tell the cells, who they are and what tissues they should form. Mutated fruit flies, manipulating frog embryos in culture, gene knockouts and transgenesis approaches in mouse embryos have become the favourite tools for such studies today. Many developmentally important genes from lower animals such as fruit flies *Drosophila* and worms *Caenorhalditis* have counterparts in mice or human. Therefore, a most fundamental question relates to how organisms as disparate in evolutionary terms as worms and mammals use similar strategies, often the very same genes during development.

The antibody based staining techniques, extensively applied in the *Drosophila* system to decipher the pattern of expression of developmentally regulated genes or the nature of their function have not been applied to the *B. mori* system. This has been largely due to the non-availability of techniques for dechorionation or devitellinization of *B. mori* eggs. Likewise, the powerful technique of "transgenesis" is also yet to be developed in the case of silkworms. However, the authors have made some advances in this direction in recent years (Joy and Gopinathan, 1994; Singh and Gopinathan, 1997). It is anticipated that such approaches will contribute significantly in elucidating various mechanisms in developmental biology of silkworm. A large variety of transposable elements including some retrotransposons (*e.g.* R1Bm, R2Bm, Dong, Pao, Mag, Hermes and Minos) have been reported from *Bombyx* and other Lepidopterans (see review Eickbush, 1995), which may prove to be helpful in developing gene transfer system in silkworm.

B. mori is an ideal specimen for developmental biology research because of its easily distinguishable characteristics in the various stages of the life cycle. The eggs hatch out several times a year with or without artificial stimulus depending upon the race; silkworm rearing is easy and non hazardous and the reproductive ability of the organism is very high (a single female on an average produces 300–400 egg and the male is polygamous, and if needed, can be mated to eight females). The silkworm being an insect of great economic value, the basic biological studies on the organism will have significant practical implications.

Acknowledgments

We thank the Department of Biotechnology, and the Department of Science and Technology, Govt. of India for financial support to our laboratory. We also thank the Indo-French Centre for Promotion of Advanced Research and Indo-EEC cooperation for financial support through collaborative research programmes. We are thankful to Dr. Madhusoodan Nambiar for helping in preparation of electron micrographs.

REFERENCES

Akai, H. (1984). The ultrastructure and functions of the silk gland cells of *Bombyx mori*. *In* "Insect Ultrastructure," 2 (R.C. King and H. Akai, eds.), pp. 323–364; New York Plenum.

Anderson, D.T. (1972). The development of holometabolous insects. In "Developmental Systems: Insects, "(S.J. Counce and C.H. Waddington, eds.), 165–242; London Academic Press.

Ashburner, M. (1992). Mapping insect genomes. *In* "Insect Molecular Science", (J.M. Crampton and P. Eggleston eds.), pp. 51–75; London: Academic Press.

Barinaga, M. (1994). Looking to development's future *Science* **266**; 561–564.

Chowdhary, S.N. (1960). Genetics of Tasar silkworm. *2nd Int. Tech. Seri. Conf.* Murcia, Spain, pp. 1–6.

Chowdhary, S.N. (1970). Ericulture. Assam, Directorate of Sericulture and Weaving, pp. 11–40.

Couble, P., Moine, A., Garel, A. and Prodhomme, J.C. (1983). Developmental variation of a non-fibroin mRNA of *B. mori* silk gland encoding for a low molecular weight protein. *Dev. Biol* **97**; 398–407.

Denlinger, D.L. (1985). Hormonal control of diapause. *In* "Comprehensive Insect Physiology, Biochemistry and Pharmacology", 8 (G.A. Kerkut and L.I. Gilbert, eds.), pp. 353–412; Oxford Pergamon.

Doira, H. (1992). Genetical stocks and mutations of *Bombyx mori*. Important Genetic Resources. Fukuoka, Japan: Institute of Genetic Resources, Kyushu University.

Eickbush, T.H. (1995). Mobile elements of Lepidopteran genome. *In* "Molecular Model Systems in the Lepidoptera" (M.R. Goldsmith and A.S. Wilkins, eds.), pp. 20–76; Cambridge University Press, Cambridge CB2 1RP, UK.

French, V. (1990). The development of segmentation in the invertebrates. *Seminars in Dev. Biol.* **1**; 89–100.

Gage, L.P. (1974a). The *Bombyx mori* genome analysis by DNA reassociation kinetics. *Chromosoma*, **45**; 27–42.

Gage, L.P. (1974b) Polyploidization of silk gland of *Bombyx mori. J. Mol. Biol.* **86**; 97–108.

Gamo, T. (1987) Components of silk proteins and their genetic loci in silkworm. *Sericologia,* **21**; 53–58.

Goldsmith, M.R. and Shi, J. (1994). A molecular map for the silkworm: constructing new links between basic and applied research. *In* "Silk polymers: Material Science and Biotechnology". **544** (D. Kaplan; W.W. Adams; B. Farmer and C. Viney, eds.), American Chemical Society Symposium Series, pp. 45–58. Washington, DC, ACS.

Goldsmith, M.R. (1995). Genetics of the Silkworm: revisiting an ancient model system *In* "Molecular Model Systems in the Lepidoptera" (M.R. Goldsmith and A.S. Wilkins, eds.), pp. 21–76; Cambridge University Press, Cambridge, CB2 1RP, UK.

Gupta, A.P. (1979). Morphogenetic Hormones of Arthropods., **1**. New Brunswick, NJ; Rutgers University Press.

Hinton, H.E. (1981). *In* "Biology of Insect Eggs". **1**. Oxford Pergamon.

Hui, C.C. and Suzuki, Y. (1995). Regulation of the silk protein genes and the homeotic genes in silk gland development. *In* "Molecular Model Systems in the Lepidoptera" (M.R. Goldsmith and A.S. Wilkins, eds.), pp. 249–271, Cambridge University Press, Cambridge, CB2 1RP, UK.

Imai, K., Konno, T., Nakazawa, Y., Komiya, T., Isobe, M., Koga, K., Goto, T., Yaginuma, T., Sakakibara, K., Hasegawa, K. and Yamashita, Y. (1991). Isolation and structure of diapause hormone of the silkworm, *Bombyx mori. Proc. Jap. Acad.* **67 B**, 98–101.

Ito T. (1997). Silkworm nutrition. In "The silkworm an important laboratory tool", (Tazima, Y. ed.), pp. 121–157; Publ. Kodansha Ltd. Tokyo, Japan.

Jolly, M.S. and Sen, S.K. (1969). Patterns of follicular imprints in egg shell. a species specific character in Antheraea, **1**. *Bull. Ent.* **10**; 32–38.

Jolly, M.S. and Sen, S.K. (1974). Patterns of follicular imprints in egg shell, a species specific character in Antheraea, II. *Indian J. Seric.* **13**; 36–43.

Joy, O. (1986). Spinning apparatus of silkworm larvae, *Bombyx mori L.* The spinneret. *Curr. Sci.* **55**; 872–873.

Joy, O. and Gopinathan, K.P. (1994). Expression of microinjected foreign DNA in silkworm, *Bombyx mori. Curr Sci* **66**; 145–150.

Kafatos, F.C., Tzertinis, G., Spoerel, N.A. and Ngriyen, H.T. (1995). Chorion genes: an overview of their structure, function and transcriptional regulation. *In* "Molecular Model Systems in the Lepidoptera" (M.R. Goldsmith and A.S. Wilkins, eds.), pp. 181–215, Cambridge University Press, Cambridge, CB2 1RP, UK.

Kawaguchi, Y., Miyaji, Y., Doira, H. and Fuji, H. (1988a). Changes of proteins during development of ovary in the *small egg 2* mutant of *Bombyx mori. J. Seric. Sci. Jap.,* **53**; 448–455.

Kawaguchi, Y., Nho, S.K., Miyaji, Y. and Fuji, H. (1988b). Characteristics of the *sm-2* egg in *Bombyx mori. J. Seric. Sci. Jap,* **57**; 157–164.

Kawazoe, A. (1987). Comparative karyotype analysis of the silkworm, *Bombyx mori linnaeus* and *B. Mandarina Moore* (Lepidoptera: Bombycidae). *La Kromosoma II* **46**; 1521–1532.

Keino, H. and Takesue, S. (1982). Scanning electron microscopic study on the early development of silkworm eggs (*Bombyx mori. L*) *Development Growth and Differentiation,* **24**; 287–294.

Koyabashi, M. (1979). Insect endocrinology. In "The silkworm an important laboratory tool", (Y. Tazima, ed), pp. 159–187; Publ. Kodansha Ltd. Tokyo, Japan.

Koyabashi, Y., Miyaji, Y., Doira, H. and Fuji, H. (1988). Changes of proteins during development of ovary in the *small egg 2* mutant of *Bombyx mori*. *J. Seric. Sci. Jap.* **53**; 448–455.

Krishnappa, B.L. (1989). Embryolgy of *Samia cynthia ricini* Boisduval (Lepidoptera: Saturniidae), PhD. Thesis.

Lees, A.D. (1955). "The physiology of diapause in Arthropods". Cambridge University Press, Cambridge.

Mani, M.S. (1968). *In* "General Entomology" pp. 415–421 Oxford and IBH Publishing Co.

Mazur, G.D., Reiger, J.C. and Kafatos, F.C. (1980). The silkmoth chorion; morphogenesis of surface structures and its relation to synthesis of specific proteins. *Develop. Biol.* **76**; 305–321.

Miya, K. (1978). Electron microscopic studies on the early embryonic development of the silkworm, *Bombyx mori* I. Architecture of the newly laid egg and the changes of the sperm entry. *J. Faculty of Agriculture,* Iwate University **3**; 436–467.

Miya, K. (1985). Determination and formation of the basic body pattern in embryo of the silkmoth *Bombyx mori* (Lepidoptera, Bombycidae). *In* "Recent Advances in Insect Embryology in Japan". (H. Ando and K. Miya, eds), 107–123, Tsukuba, Japan, ISEBU

Nagaraja, G.M. and Nagaraju, J.E. (1995). Genome fingerprinting of the silkworm *Bombyx mori* using random arbitrary primers. *Electrophoresis* **16**; 1633–1638.

Nagaraju, J., Sharma, A., Sethuraman, B.N., Rao, G.V. and Singh, L. (1995). DNA fingerprinting in silkworm *Bombyx mori* using banded krait satellite DNA-derived probe. *Electrophoresis* **16**, 1639–1642.

Nagy, L., Riddiford, L.M. and Kiguchu, K. (1994). Morphogenesis in the early embryo of the Lepidoptera *Bombyx mori. Dev. Biol.* (in press)

Nambiar, P.M., Prakash, N.A. and Gopinathan, K.P. (1991). Ultrastructure of the head, spinneret, silk gland and egg of Eri silkworms, *Philosomia ricini. Sericologia*, **31**; (3) 493–507.

Norak, V.J.A. (1975). "Insect hormone". Chapman Hall London.

Nunome, J. (1937). The silk gland development of *Bombyx mori. Bulletin of Applied Zoology.* **9**; 69–91.

Ohnishi, E. and Ishizaki, H. (1990). *In* "Molting and metamorphosis." Japan Scientific Press, Tokyo.

Papanikolaou, A.M., Marjaritis, L.H. and Homodrakas, S.J. (1985). Ultrastructural analysis of chorion formation in silkmoth *Bombyx mori. Canad. J. Zool.* **64**; 1158–1173.

Poodry, C.A. (1980). Imaginal discs: morphology development. *In* "The Genetics and Biology of *Drosophila*" **2d** (Ashburner, M. and Wright, T.R.F., eds.) pp. 407–432. Academic Press, London, New York, San Francisco, Toronto, Syndey.

Promboon, A., Shimada, T., Fujiwara, H. and Koyabashi, M. (1995). Linkage map of random amplified polymorphic DNAs (RAPDs) in the silkworm, *Bombyx mori. Genet. Res. (Camb.)* **66**; 1–7.

Prudhomme, J.C., Couble, P., Garel, J.P. and Daillie, J. (1985). Silk synthesis. *In* "Comprehensive Insect Physiology, Biochemistry and Pharmacology", **10** (G.A. Kerkut and L.I. Gibert, eds.) 571–594 New York, Pergamon.

Raikhel, A.S. and Dhadialla, T.S. (1992). Accumulation of the yolk protein in insect oocytes. *Annu Rev Entom* **37**; 215–218.

Ransom, R. (1982). Techniques. *In* "A handbook of *Drosophila* development". (R. Ransom ed.), 1–27. Elseiver Biomedical Press Amsterdum, New York, Oxford.

Rasch, E.M., Barr, H.J. and Rasch, R.W. (1971). The DNA content of sperm of *Drosophila. Chromosoma* 33; 1–18.

Rasch, E.M. (1974). The DNA content of sperm and haemocyte nuclei of the silkworm, *Bombyx mori, L. Chromosoma,* 35; 1–18.

Robinson, R. (1971). *In* "Lepidoptera Genetics". Oxford. Pergamon.

Sakaguchi, B. (1979a). Gametogenesis, fertilization and embryogenesis of the silkworm. In "The silkworm an important laboratory tool", (Tazima, Y. ed.) pp. 5–30. Publ. Kodansha Ltd. Tokyo, Japan.

Sakaguchi, B. (1979b). Postembryonic development of the silkworm. In "The silkworm an important laboratory tool", (Tazima, Y. ed.) pp. 31–51. Publ. Kodansha Ltd. Takyo, Japan.

Sakurai, S. (1983). Temporal organization of endocrine events underlying larval ecdysis in the silkworm, *B. mori. J. Insect Physiol.,* 29; 919–932.

Sehnal, F. and Rembold, H. (1985). Brain stimulation of juvenile hormone production in insect larvae. *Experientia,* 41; 684–685.

Shimura, K. (1979). Synthesis of silk proteins. In "The silkworm an important laboratory tool", (Tazima, Y. ed.), pp. 184–211, Publ. Kodansha Ltd. Tokyo, Japan.

Shimura, K. (1983a). The physiology and biology of spinning in *Bombyx mori. Experientia,* 39; 441–450.

Shimura, K. (1983b). Chemical composition and biosynthesis of silk proteins. *Experientia,* 39; 455–465.

Singh, A. and Gopinathan, K.P. (1991). A novel whole mount antibody staining technique for silkworm embryos. Curr. Sci., 72; 214-219.

Sprague, K.U. (1995). Control of transcription of *Bombyx mori* RNA polymerase III. *In* "Molecular Model Systems in the Lepidoptera" (M.R. Goldsmith and A.S. Wilkins, eds.), pp. 273–291; Cambridge University Press, Cambridge, CB2 1RP, UK.

Suzuki, Y. (1977). Differentiation of the silk gland a model system for the study of differential gene action. *In* "Results and Problems in Cell differentiation", 8 (W. Beermann, ed.), 1–44; Berlin, Heidelberg and New York. Springer Verlag.

Tazima, Y. (1964). "The Genetics of Silkworm" Logos Press Ltd, Great Britain.

Tazima, Y. (1979). Introduction. *In* "The silkworm an important laboratory tool", (Tazima, Y. ed.) pp. 1–4; Publ. Kodansha Ltd. Tokyo, Japan.

Tazima, Y. (1989). Alteration of food habits of domesticated silkworm, *Bombyx mori. Sericologia* 29; 437–453.

Tobe, S.S. and Stay, B. (1985). Structure and function of corpus allatum. *In* "Advances in Insect Physiology" 181 (M.J. Berridge, J.E. Treherrance and V.B. Wigglesworth, eds.) pp. 305–432; Academic Press, London, New York.

Toyama, K. (1902). Constitution of the study of the silkworm. I. On the embryology of the Silkorm. *Bulletin of the College of Agriculture, Tokyo Imperial University,* 5; 73–118.

Traut, W. (1976). Pachytene mapping of the female silkworm *Bombyx mori* L. (Lepidoptera). *Genetica.* 47; 135–142.

Trask, B.J. (1991). Fluorescence *in situ* hybridization: Applications in cytogenetics and gene mapping. *Trends Genet,* 7; 149–154.

Ueno, K., Nagata, T. and Suzuki, Y. (1995). Role of the homeotic genes in the Bombyx body plan. *In* "Molecular Model Systems in the Lepidoptera" (M.R. Goldsmith and A.S. Wilkins, eds.) pp. 165–180; Cambridge University Press.

Willis J.H.; Wilkins, A.S. and Goldsmith, M.R. (1995). A brief history of Lepidoptera as model system . *In* "Molecular Model Systems in the Lepidoptera" (M.R. Goldsmith and A.S. Wilkins, eds.), pp. 1–20; Cambridge University Press, Cambridge, UK.

Yamashita, O. and Hosegawa, K. (1985). Embryonic diapause. *In* "Comparative Insect Physiology, Biochemistry and Pharmacology" **1** (G.A. Kerdut, and C.I. Gilbert eds.), pp. 407-434; Pergamon Press, Oxford.

Yanagawa, H., Watanabe, K. and Nakamura, M. (1988). Composition of artifical diets from the original strains of the silkworm, *Bombyx mori,* by applying a linear programme method. *Bulletin of the Sericultural Experiment Station of Japan,* **30**; 569–588.

Genome Analysis in Eukaryotes: Developmental and evolutionary aspects
R.N. Chatterjee and L. Sánchez (Eds)
Copyright © 1998 Narosa Publishing House, New Delhi, India

5. Early Events Associated with Sex Determination in *Drosophila melanogaster*

Lucas Sánchez, Pedro P. López and Begoña Granadino

Centro de Investigaciones Biológicas, Velázquez 144, 28006 Madrid, Spain

I. Introduction

Sex determination is the commitment of an embryo to either the female or the male developmental pathway. In *Drosophila melanogaster,* 2X;2A individuals (X,X chromosome; A, autosomal set) are females and XY;2A individuals (Y, Y chromosome) are males. A series of results led to the discovery that in *Drosophila melanogaster* sex is determined by the ratio of the X chromosomes to sets of autosomes (reviewed in Baker and Belote, 1983; Nöthiger and Steinmann-Zwicky, 1985). Firstly, XXY and XO flies are female and male, respectively. This indicates that the Y chromosome plays no role in sex determination. Secondly, gynandromorphs are sexually mosaic individuals with some portions of the body typically male and others typically female. Such individuals arise from the loss of an X chromosome during the early development of XX flies. The sharp borderline between female and male areas indicates that sex hormones do not control sexual development as a whole, but that each individual cell chooses its sex autonomously, according to its genotype. Thirdly, 2X;3A flies are mosaic individuals with male and female structures. Moreover, clones of cells with one X chromosome and one set of autosomes develop into female structures (Santamaria and Gans, 1980). This indicates that sex is not determined by the absolute number of X chromosomes but by the ratio of X chromosmes to sets of autosomes. In the 2X;3A sexual mosaics, the X:A ratio is at a threshold level between a normal female and a normal male signal. Some cells interpret this ambiguous signal as female while others interpret it as male.

Since females and males differ in the number of sex chromosomes, a process has evolved to eliminate the difference in the doses of the sex chromosome-linked genes in the two sexes. This process is called dosage compensation. In *Drosophila melanogaster,* the two X chromosomes in females are active and dosage compensation is achieved in males by hypertranscription of the single X chromosome (Lucchesi and Manning, 1987; Kuroda et al., 1993).

In *Drosophila melanogaster,* the ratio of the X chromosomes to sets of autosomes (X;A) is the primary genetic signal that triggers sex determination and dosage compensation by acting on the gene *Sex-lethal (Sxl)* to set its state irreversibly into either the female mode (which represents its functional ON state), or the male mode (which represents its non-functional OFF state). Activation of *Sxl* also requires the maternal product of the gene *daughterless (da) (Cline, 1978).* Once the state of activity of *Sxl* is defined, the X;A signal is no longer needed, and both sex determination and dosage compensation come under the control of *Sxl*, whose functional state is stably maintained by an autoregulatory process.

The gene *Sxl* controls the expression of two independent sets of regulatory genes (Lucchesi and Skripsky, 1981). One set is formed by the sex determination genes; mutations in these genes affect sex determination while having no effect on dosage compensation. The control of these genes throughout development occurs by sex-specific splicing of their products. A hierarchical interaction exists between these genes: the product of a gene controls the sex-specific splicing of the pre-mRNA from the downstream gene in the genetic cascade (reviewed in Baker, 1989). The other set of genes is formed by the *male-specific lethal* genes (*msl's*); mutations in these genes affect dosage compensation whilst having no effect on sex determination (Uenoyama et al., 1982; Belote, 1983; Bachiller and Sánchez, 1989). Since both sex determination and dosage compensation sets of genes are subordinated to *Sxl,* misexpression of *Sxl* can produce sex-specific lethality and/or sexual transformation to either males or females. Two sets of *Sxl* mutations have been isolated. One set is formed by loss-of-function mutations, generically named as Sxl^f, which are characterised by their recessive female-specific lethal phenotype (Cline, 1978; Marshall and Whitte, 1978; Sánchez and Nöthiger, 1982; Granadino et al, 1991a). Loss-of function mutations at *Sxl* cause female-specific lethality due to hypertranscription of the two X chromosomes. Moreover, clones of cells with the 2X;2A chromosome constitution that are homozygous for Sxl^f mutations develop male, instead of female, structures. The other set of *Sxl* mutations, generically named as Sxl^M, is formed by gain-of-function mutations. These are characterised by their dominant male-specific lethal phenotype (Cline, 1978; Maine et al., 1985a; b). Gain-of-function mutations at *Sxl* cause male-specific lethality due to hypotranscription of the single X chromosome. Moreover, X;2A clones of cells mutant for Sxl^M mutations develop female, instead of male, structures.

II.   Genetic basis of the X:A signal

A.   The "balance concept" of sex determination

On the basis of genotypes with variable X: A ratios, Bridges (1921; 1925) formulated his "balance concept" of sex determination. This hypothesis considers sex as a quantitative character, with continuous variation, under the control of two opposing polygenic systems: the female and the male determining systems, located on the X chromosome and autosomes respectively. These would be made up by many elements, each one having a small effect. The opposing action of both systems would determine the sex of the individual, according to the stechiometric assumption that two doses of the female factors (2X chromosomes) outweigh the effect of two doses of male factors (2 sets of autosomes), leading to female development; whereas two doses of male factors (2 autosomal sets) outweigh the effect of one dose of female factors (1X chromosome), leading to male development. In the case of 2X;3A individuals, the stechiometry of the interaction between female and male determining factors is such that, within the same individual, in some cells the male development is imposed and in others the female factors prevail leading to female development.

B.   A hypothesis based on noncoding DNA

Gadagkar et al (1982) and Chandra (1985) proposed a model based on noncoding DNA to explain how cells would assess the X:A ratio signal. They proposed the existence of noncoding DNA sequences in the X chromosome (locus π), that would bind a repressor encoded by an autosomal gene ϖ. Moreover, they incorporated in their model the proposition of Cline (1978) that *Sxl* is the key gene which responding to the X:A signal, controls the processes of sex determination and dosage compensation, in such a way that *Sxl* would be activated in females (2X;2A) but not in males (X;2A). The model also states that the maternal product of the gene *daughterless (da)* (DA product) is required for *Sxl* activation. The model consists of the following components: (a) the DA product produced in the mother and stored in the egg, (b) the Ω repressor from the autosomal gene ω, which would be activated in the zygote by the DA product, and (c) the gene *Sxl* and the locus π, both of which have affinity for the Ω repressor and RNA polymerase. A central assumption in this model is that the Ω repressor and RNA polymerase compete with each other for binding sites on both *Sxl* and π loci in such way that RNA polymerase binds to both loci with a lower affinity than that of the Ω repressor. It is further assumed that *Sxl* has a lower affinity than π for repressor as well as polymerase. Following this model, the amount of repressor in either male or female zygotes would be the same, since both have two copies of the gene ω and both inherit the same amount of maternal DA product. The male embryo has only one X chromosome and therefore only one dose of the low affinity *Sxl* gene and one dose of the high affinity locus π. The repressor would bind significantly to both sites thus preventing the binding of RNA polymerase to *Sxl*; consequently, little or no SXL product would be produced. In the female embryo, there are two doses of both *Sxl* and π loci. Most of the repressor will be sequestered by the π sites, so that the *Sxl* loci will be free to bind RNA polymerase. The synthesis of the SXL product will then take place. However, results obtained so far have shown that the X:A signal is made up of conventional genes. This signal results from the interaction between X-linked (numerator elements) and autosomal (denominator elements) gene products.

C. Numerator elements of the X:A signal

The isolation of X-linked genes involved in the formation of the X:A signal, and then involved in determining the state of activity of *Sxl*, could, in principle, be approached by selection of sex-specific lethal mutations. This selection must be based on the expected properties that any X-linked gene must fulfil to be a numerator element of the X:A signal.

A numerator element should display several properties:

1. Reduction of its zygotic doses should specifically kill females as a consequence of a failure to activate *Sxl*. This female lethality should be suppressed by *Sxl^M* mutations. These are constitutive mutations that express female *Sxl* function independently of the X:A signal (Cline, 1978).
2. Increase of its zygotic doses should specifically kill males since *Sxl* is inappropriately activated. This male lethality should be suppressed by loss-of-function mutations at *Sxl*.
3. The activation of *Sxl* requires the maternal product of the gene *da* (Cline, 1978). Therefore, mutations at *da*, and at any numerator element, are expected to display female-specific lethal synergistic interaction. Such interactions are also expected between mutations in different numerator elements. In both cases, the female lethality should be suppressed by *Sxl^M* mutations.
4. A variation of the zygotic doses of numerator elements should alter the sexual phenotype of triploid intersexes (2X;3A); an increase should feminize, whereas a reduction should masculinize these individuals.

To search for X-linked genes that function as "numerator elements"of the X:A signal, a systematic screening has been performed (Sánchez et al., 1994). This screening was based on an expected property that a "numerator element" of the X: A signal should exhibit, namely, the female lethal synergistic interaction between *Sxl* and the putative "numerator element". This property is due to the dose-sensitive character of the X:A signal and on the role of this signal in the activation of its target gene *Sxl*. Therefore, a search was made for X chromosome regions whose deficiency in transheterozygosis with a single Sxl^+ allele caused lethality and/or sex transformation in females. For this purpose, crosses were made between $Df(1)X, Sxl^+/$ *Balancer, Sxl^+ females* and Sxl^{fl}/Y males. The $Df(1)X$ refers to any deficiency of the X chromosome, and Sxl^{fl} is a loss-of-function mutation of Sxl. Deficiencies were selected that, in this cross, presented a reduced female progeny of genotype $Df(1)X, Sxl^+/Sxl^{fl}$ with respect to their *Balancer, Sxl^+/Sxl^{fl}* daughters. By this procedure, a set of regions in the X chromosome that interact synergistically with Sxl were identified as the cause of a female-specific lethal phenotype. The selected deficiencies can be grouped into two classes. Class I deficiencies are characterised because the *Balancer, Sxl^+/Sxl^{fl}* females show a normal viability, whereas in Class II deficiencies these females present a reduced viability, relative to the viability of their *Balancer, Sxl^+/Y* brothers. Class I deficiencies comprise $Df(1)$ svr, $Df(1)N71$, $Df(1)HF$ 396 and $Df(1)N19$. Class II deficiencies are $Df(1)HC244$ and $Df(1)$ RA2. For those X chromosome regions for which no deficiencies were available, a different approach was followed through also based on an expected property for a putative "numerator element", namely, the male-specific lethality and/or the feminization of males carrying duplications for the "numerator elements". Using this procedure, an analysis was made of the viability of males carrying a duplication of the X chromosome region under study in combination with a duplication for any of the already known "numerator elements" of the X: A signal as *sis-a* and *sc*. None of the regions tested by this procedure contained "numerator elements" of the X:A signal. Taken together, the analyses of deficiencies and duplications, involved the investigation of around 80% of the X chromosome cytogenetic bands.

Among the selected X chromosome regions that synergistically interact with Sxl, $Df(1)$ $N71$ corresponds to the gene *sisterless-a* (Cline, 1986), $Df(1)svr$ corresponds to the gene *scute* (Torres and Sánchez, 1989; Parkhurst et al., 1990; Erickson and Cline, 1991) and $Df(1)HF396$ corresponds to the gene *runt* (Duffy and Gergen, 1991; Torres and Sánchez, 1992). The deficiencies $Df(1)N19$ and $Df(1)RA2$ define, respectively, the X chromosome regions 17A9-10; 17A12-B1 and 7D10; 1F12 for which no gene has been identified so far (Sánchez et al., 1994). The $Df(1)HC244$ corresponds to the gene *sans-fille* (Oliver et al., 1988; Steinmann-Zwicky, 1988), which is not a "numerator element" of the X:A signal, but whose function is required for correct splicing of the Sxl RNA (Salz, 1992; Brown and Salz, 1993).

(a)　The gene sisterless-a

Cline (1986) identified the gene *sisterless-a (sis-a)* during the examination of a group of putative female-lethal mutations that arose during an EMS mutagenesis designed to identify X-linked maternal effect mutations. The identification of this gene was straightforward, since its basic mutant phenotype was female-specific lethality. This lethal effect is recessive and is suppressed both by gain-of-function Sxl^M mutations, and, to a lesser extent, by a duplication of Sxl^+. There exists a female-lethal synergistic interaction between loss-of-function mutations at *sis-a, da* and *Sxl*. This lethality is also suppressed by Sxl^M. Homozygous *sis-a⁺* daughters of heterozygous *da* mothers exhibited no masculinizing differentiation when homozygous for *mle*, one of the male-specific lethal mutations that specifically affect hypertranscription of the male X chromosome (dosage compensation). In contrast, with daughters heterozygous for *sis-a*, more than half exhibited mosaic intersexual forelegs; i.e. cells exhibiting either pure female or pure male development, but never true intersexual phenotypes. Moreover, heterozygosis for *sis-a* causes masculinization of the triploid intersexes (2X;3A). Importantly, somatic clones homozygous for *sis-a* induced during female larval development do not show sexual transformation. This indicates that the role of *sis-a* in female sexual development is restricted to the early stages of development and does not affect the maintenance of the female sexual pathway throughout development. The *sis-a* mutant phenotype is zygotic and not maternal.

(b) The gene scute

The involvement of the gene *scute (sc)* in sex determination was not so straightforward and was largely ignored due to the pleiotropic effect displayed by *sc* mutations. In fact, the gene *sc*, which is a component of the *achaete-scute* complex (AS-C) (Garcia-Bellido, 1979; Campuzano et al., 1985), was well known for a long time for its role in the development of the central and peripheral nervous systems (Garcia-Bellido, 1979; Jiménez and Campos-Ortega, 1979; Dambly-Chaudiére and Ghysen, 1987). The role of *sc* in sex determination was revealed through the synergistic lethal interaction in females transheterozygous for loss-of-function mutations at both *sc* and *Sxl* genes.

Different regions included in *Df(1)svr* were tested to determine the chromosomal region or gene(s) responsible for the female lethal synergistic interaction with *Sxl*. All deficiencies with deleted sequences delimited by the distal breakpoint of $In(1)sc^{L8}$ and the breakpoint of the terminal *Df(1)RT650* show lethal interaction with Sxl^{f1} (Torres and Sánchez, 1989). Independently, Cline (1988) also identified this region of the *achaete-scute* complex (AS-C) that he named *sisterless-b (sis-b)*. A detailded genetic analysis of that region through the use of point mutations, sc^{10-1} and Hw^{49cR5} led to the discovery that the proposed *sis-b* function corresponds to the gene *sc* of the AS-C complex (Torres and Sánchez, 1989; Parkhurst et al., 1990; Erickson and Cline, 1991, 1993). There are female-specific lethal synergistic interactions between loss-of function mutations at *sc* and either *Sxl* or *sis-a*. In both cases, the lethality is suppressed by the gain-of-function Sxl^{M1} mutation. Simultaneous duplications of *sis-a* and *sc* are lethal to males. This lethality is suppressed by the loss-of-function Sxl^{f1} mutation. There are female-specific lethal synergistic interactions between loss-of-function mutations at sc and the maternal product of the gene da. This lethality is also suppressed by Sxl^{M1}. The loss-of-function sc^{10-1} mutation masculinizes triploid intersexes (2X;3A).

The constitutive Sxl^{M1} mutation, which suppresses the lethality of females caused by loss-of-function sc mutations, does not suppress, however, their scute phenotype (Torres and Sánchez, 1989). Thus, the gene sc has a dual function: the scute function involved in neurogenesis and the sis-b function involved in sex determination through its participation in the formation of the X:A signal. The female-lethal synergistic interaction between Sxl and sc is thermosensititve; 18°C being the permissive temperature 29°C the restrictive. The specific male lethality caused by the simultaneous duplication of both sc and $sis-a$ is also themosensitive, but in this case 18°C is the restrictive temperature and 29°C the permissive. Temperature shifts throughout development revealed that the thermosensitive phase of both female and male lethal interactions lies around the blastoderm stage (Torres and Sánchez., 1991; Erickson and Cline, 1993). Furthermore, by means of a hsp $70-sc$ chimeric gene (HSSC-3) (Rodriguez et al., 1990), sc was expressed at different developmental times and checked for when this expression suppressed the sis-b mutant phenotype in females and caused lethality in males due to expression of Sxl (Torres and Sánchez, 1991). The expression of sc activates Sxl only at a very specific stage in development, coinciding with the thermosensitive phase delimited by temperature-shift experiments and corresponding to the syncytial blastoderm stage when the gene sc undergoes a homogeneous expression (Romani et al., 1987; Cabrera et al., 1987; Parkhurst et al., 1993; Erickson and Cline, 1993). Thus, the sc expression at the syncytial blastoderm stage is responsible for the sis-b function of this gene and coincides with the time in development when the X:A signal acts on Sxl (see below in the section "Developmental meaning of the X:A signal").

The analysis of the mutations sc^{10-1} and Hw^{49cR5} suggested that $achaete$ (ac) can partially substitute of the sis-b function of $scute$: sc^{10-1}, a double mutant affecting both sc and ac genes, is more defective for sis-b function than Hw^{49cR5} (Torres and Sánchez, 1989), a mutation that overexpresses a normal ac gene (Campuzano et al., 1985; Balcells et al., 1988). Parkhurst et al. (1993) examined the effects of $lethal-of-scute$ ($l'sc$) and ac genes on sex determination using two fusion genes, $hb-l'sc$ and $hb-ac$, that misexpress both genes under the control of the $hunchback$ (hb) promotor. They found that both $l'sc$ and ac genes of the AS-C complex display weak feminizing activities, enhancing male lethality, and rescuing the female lethality of sc mutations

(c)   The gene runt

Females doubly heterozygous for Sxl^{f1} and $Df(1)HF396$ are less than 10%

of the total of their *Sxl^{f1}/Balancer, Sxl^+* daughters. The analysis of different deficiencies included in *Df(1)HF396* led to the finding that all the tested deficiencies that delete the gene *runt (run)* show the female lethal interaction (Torres and Sánchez, 1992). Reduced function of *run* results in female-specific lethality and sexual transformation of XX flies that are heterozygous for *Sxl, sc* or *sis-a*; this lethality being suppressed by *Sxl^{M1}* (Duffy and Gergen, 1991; Torres and Sánchez, 1992). The *run^{YP17}* is a temperature-sensitive mutation. The thermosensitive-phase of the female-specific lethal interaction between *sis-a* and *run^{YP17}* mutations lies at the time when the X:A signal determines the activity of *Sxl* (Torres and Sánchez, 1992). Moreover, loss-of-function mutations masculinize triploid intersexes (2X;3A) (Duffy and Gergen, 1991; Torres and Sánchez, 1992), and there is a weak female-specific dominant synergism between *run* and the maternal product of *da* (Torres and Sánchez, 1992). In contrast, simultaneous duplications of *sc* and *run* show a limited ability to induce *Sxl*-dependence male-specific lethality (Torres and Sánchez, 1992). All these results suggest that the gene *run* is needed for the initial step of *Sxl* activation by the X:A signal but that it has no major role as an X-counting "numerator element" of this signal as has *sc* or *sis-a*. The gene *run* is also distinguished from *sc* and *sis-a* by its non uniform expression in the embryo; i.e., *run* is only needed in the central region of the embryo for the activation of *Sxl* (Duffy and Gergen, 1991); however, *sc* (Romani et al., 1987; Cabrera et al., 1987; Parkhurst et al., 1993; Erickson and Cline, 1993) and *sis-a* (Erickson and Cline, 1993) are expressed throughout the embryo.

(d)   Chromosome region 17A9-10;17A12-B1

Females doubly heterozygous for a *Sxl* deficiency and *Df(1)N19* show reduced viability. This lethality is suppressed by *Sxl^{M1}*. There exists a female-lethal synergistic interaction between *Df(1)N19* and *sc* or *sis-a*, which is also suppressed by *Sxl^{M1}*. Moreover, the Df(1) *N19/+* females show reduced viability when their mothers are heterozygous for the *da* mutation. This synergistic interaction exclusively involves the maternal, not the zygotic, DA product. The analysis of other deficiencies and duplications revealed that the chromosome region 17A9-10;17A12-B1 is responsible for this lethal interaction. On the other hand, males doubly heterozygous for a duplication of this chromosome region and either *sc* or *sis-a* are viable. All these results suggest that the chromosome region 17A9-10;17A12-B1 is involved in the initial step of *Sxl* activation but has no major role as a "numerator element" of the X:A signal as has *sc* or *sis-a* (Sánchez et al., 1994).

The cross between *Df(1)N19/FM6* and *Df(1)run ^{1112}, y f^{36a}/y^+Ymal^{106}, run^+* males yielded 51 *Df(1)N19/Df(1)run^{1112}, y f^{36a}* experimental females (15% of those viable) and 340 *FM6/Df(1)run^{1112}, y f^{36a}* control females. This synergistic interaction has been confirmed when using another *run* allele (*run^{YE96}*). Therefore, there exists a female-specific lethal synergistic interaction between $Df(1)N19$ and run, contrary to previously reported results (Sánchez et al., 1994).

D.   Denominator elements of the X:A signal

The expected properties for a "denominator element" of the X:A signal are the opposite of those expected for a "numerator element" of this signal (see Section IIC). So far, only one "denominator element" of the X:A signal has been identified, the gene *deadpan (dpn)*. This gene, which is involved in neurogenesis, was identified using the enhancer-trap method (Bier et al., 1989). Males with decreased copies of *dpn* and increased copies of *sc* are lethally affected. The lethality is higher as the number of *sc* copies increases. However, this lethality is supressed if the males are deficient for *Sxl* (Younger-Shepherd et al., 1992). On the other hand, females with increased copies of *dpn* and decreased copies of *sc* are lethally affected (Younger-Shepherd et al., 1992). By using antibodies against the female-specific SXL protein, it was found that *Sxl* is misexpressed in male embryos mutant for *dpn* and also in male embryos with increased copies of *sc* and decreased copies of *dpn*. *Sxl* expression is suppressed in females embryos with reduced copies of *sc* and increased copies of *dpn* (Younger-Shepherd et al., 1992). Moreover, females with three doses of dpn^+ and either one dose of Sxl^+, or sc^+, or $sis-a^+$, have reduced viability. This interaction is temperature-sensitive (29°C is the restrictive temperature and 18°C the permissive). In all three cases the temperature-sensitive period lies early in development (Sánchez et al., 1994). All these results support the idea of *dpn* being a "denominator element" of the X:A signal. In this context, it is worth mentioning that the DPN transcript is detectable early in development (Younger-Shepherd et al., 1992), when the X:A signal acts on *Sxl* (see Section "Developmental meaning of the X:A signal").

The cross between yw; $dpn^1/CyO,P(y^+)$ females and $Df(1)run^{1112},yf^{36a}$ run/y^+ $Ymal^{106}$, run^+ males yielded 138 y w/y^+Ymal^{106}, run^+; CyO, $P(y^+)/+$ males (97.2% of those viable), 140 yw/y^+Ymal^{106}, run^+, $dpn^1/+$ males (98.6% of those viable) and 142 y $w/Df(1)run^{1112}$, yf^{36a} run; $dpn^1/+$ females (used as viability reference). This indicates that the duplication of run^+ does not affect the viability of males with one dose of dpn^+, contrary to that which occurs in the case of duplication for sc^+ (Younger-Shepherd et al., 1992). This might indicate that the genes *run* and *dpn* do not interact with each other. In this respect, it is worth mentioning that although *run* is needed to activate *Sxl*, it does not behave as a proper "numerator" element of the X:A signal as males with a duplication of *run* and either *sc* or *sis-a* are fully viable (Torres and Sánchez, 1992).

III.   Maternal effect genes acting on *Sxl* activation

The X:A ratio is necessary but insufficient to activate *Sxl*. A set of maternal products act as either activators or repressors of this gene.

A. The gene *daughterless (da)*

The gene *da* shows a dual expression corresponding to its dual function: a maternal function required for *Sxl* activation by the X:A signal, and a zygotic function involved in neurogenesis. The ambiguous X:A signal of the triploid intersexes (2X;3A) causes them to develop as mosaic flies with male and female tissues. However, triploid intersexes from mothers with reduced maternal *da*$^+$ function develop as male flies. Mutations at *da* and at *Sxl* (Cline, 1978), or *sis-a* (Cline, 1986), or *sc* (Cline, 1988; Torres and Sánchez, 1989), display a female-specific dominant synergism. This is in good agreement with the role of the maternal DA product in the initial step of *Sxl* activation. Females heterozygous for *run* deficiencies show reduced viability when coming from *da*/+ mothers, independently of their zygotic phenotype for *da* (Torres and Sánchez, 1992). However, this synergistic interaction is less strong than the observed between *da* and *sis-a*, or *da* and *sc*. Similarly, there is a female-specific dominant lethal synergistic interaction between the *Df(1)N19* and the maternal DA product (Sánchez et al., 1994).

B. The *Df(1) RA2*

The cross between *Df(1)RA2/FM7* females with *Sxl*$^-$/Y males yields either *Df(1)RA2/Sxl*$^-$ or *FM7/Sxl*$^-$ females and *FM7* males. In this cross, both types of females have drastically reduced viability compared to their brothers. However, both types of females recover their viability when they carry the *Dp(Sxl*$^+$*)*, or if the mother carries two doses of the chromosome region deleted by the *Df(1)RA2*. These results indicate that there is a synergistic interaction between *Df(1)RA2*, and *Sxl*, and suggest that the females *FM7/Sxl*$^-$ from *Df(1)RA2/FM7* mothers would die since their *Sxl* gene is not properly activated. This indicates a maternal effect of that chromosome region in the activation of *Sxl*. It is possible that there is a zygotic lethal interaction between *Df(1)RA2* and *Sxl*, as transheterozygous females are slightly more lethal than their *Sxl*/+ sisters, but this zygotic effect may be masked by the very strong maternal effect. The *Df(1)RA2* interacts with *sc* and wih *sis-a;* and in both cases the maternal effect is also observed as the *sc*/+ and *sis-a*/+ females have reduced viability comparable to their *Df(1) RA2/sc*$^-$ and *Df(1)RA2/sis-a*$^-$ sisters respectively. Thus, the interaction between *Df(1)RA2* and either *sc* or *sis-a* has also a maternal effect. These results suggest the existence of a maternal component in the *Df(1)RA2* needed for *Sxl* activation.

The cross between *Df(1)RA2/FM7* females and *Df(1)run*^{1112},*y f*^{36a}/*y*$^+$ *Ymal*^{106,} *run*$^+$ *males* yielded 10 *Df(1)RA2/Df(1)run*^{1112}, *yf*^{36a} females (5% of those viable), 20 *FM7/Df(1)run*^{1112,} *yf*^{36a} females (10% of those viable) and 180 *FM7/y*$^+$ *Ymal*^{106}, *run*$^+$ males (used as viability reference). This result indicates the existence of a female-specific lethal synergistic interaction between *Df(1)RA2* and *run*. This interaction is mainly due to the maternal effect of *Df(1)RA2* because females that do not carry this deficiency are also lethally affected. The possibility of a zygotic interaction between *Df(1)RA2* and *run* cannot be discarded since females that carry this deficiency are slightly more affected than their sisters without the deficiency. As mentioned above, the strong maternal effect of *Df(1)RA2* might mask the existence of any zygotic interaction.

C. The gene *extramacrochaetae*

The gene *extramacrochaetae (emc)* functions as a negative regulator of adult sensory organ development, possibly through its interaction with proteins encoded by the AS-C complex and/or *da* (Botas et al, 1982; Moscoso del Prado and Garcia-Bellido, 1984; Ellis et al., 1990; Garrell and Modolell, 1990; Cubas et al., 1991; Skeath and Carroll, 1991; Van Doren et al., 1991). On the other hand, the poor viability of the *dpn* heterozygous males carrying a duplication of sc^+ is further decreased when the maternal *emc* function is reduced; most of the scaper males lacked terminalia (Younger-Shephard et al., 1992). The effect of *emc* is maternal, since both $Dp(sc^+)$ *dpn/+; emc^-/+* and $Dp(sc^+)$ *dpn/+; +/+* males are lethal (Younger-Shephard et al., 1992). Thus, the maternal EMC product functions as a negative regulator of *Sxl*.

IV. The X:A signal controls *Sxl* expression at the level of transcription

The initial regulation of *Sxl* by the X:A signal seems to occur at the transcriptional level (Salz et al., 1989). To test this hypothesis, early *Sxl* transcription was examined in a sample of embryos in which all of the females were simultaneously heterozygous for loss-of-function *sc* and *sis-a* mutations and a deficiency for *Sxl* (these females die during the embryonic stage due to their inability to activate *Sxl*). Oregon-R wild-type embryos. were used as controls (Torres and Sánchez, 1991). The *Sxl* transcripts were detected by *in situ* hybridization of embryos with a probe that detects all the known *Sxl* transcripts. In Oregon-R embryos, two types of embryos were found around the blastoderm stage which differed in their capacity to hybridise with the probe. These embryos fall into a bimodal distribution with a 1:1 ratio of the two types. This bimodal distribution most likely reflects differences between females and males, with respect to the early expression of *Sxl*. Where all female embryos are heterozygous for *sc, sis-a* and *Sxl* (Cross: Oregon-R females with *sc^-, Sxl^-, sis-a^-/sc^+Y* males), the embryos fall into a unimodal distribution, which corresponds to the class that does not hybridise with the *Sxl* probe. This variation in the pattern of *Sxl* expression cannot be attributed to the presence of a single dose of *Sxl* in the experimental females, since Sxl^-/Sxl^+ females containing two doses of both *sc* and *sis-a* are fully viable. These results support the transcriptional control of *Sxl* by the X:A signal.

The early *Sxl* transcripts have a unique exon (E1) that is located within the first intron of the late *Sxl* transcripts (Keyes et al., 1992). Blastoderm embryos fell into two distinct classes after hybridisation with the E1 probe: half hybridised with E1, while the other half did not (Keyes et al., 1992). Moreover, none of the embryos from the cross of wild-type females to *sc⁻ Sxl⁻ sis-a⁻sc⁺/Y* males hybridised with the E1 probe (Keyes et al., 1992). Further, Keyes et al. (1992) made a reporter construct with a 3 kb *Sxl* DNA fragment containing the embryonic promotor fused to LacZ. They generated transgenic flies for this construct and LacZ expression was assayed. They found that the embryonic promotor drives LacZ expression in early embryos, and that only 50% of them expressed LacZ. These results indicate that the embryonic promoter is active in females, but not in males. Thus, all these results indicate that the X:A signal controls *Sxl* expression at the level of transcription.

V. Molecular nature of the X:A signal

The identification of a set of genes involved in the initial step of *Sxl* activation indicates that a conventional genetic system is the basis of the X:A signal. By definition, this signal is strictly zygotic and is the key element in the *Sxl* activtion process since the maternal products needed to activate *Sxl* are also present in the male zygote in which early activation of *Sxl* does not occur.

The genes *sc, dpn* and *da* encode basic-helix-loop-helix proteins (bHLH) (Villares and Cabrera, 1987; Caudy et al., 1988), whereas the gene *emc* encodes an HLH protein (Ellis et al., 1990). The basic domain confers the DNA-binding capacity, and the HLH domain confers the capacity to interact with other HLH proteins. Thus, bHLH proteins have the capacity to form homo-or hetero-dimer complexes with capacity to bind to DNA and act as transcriptional regulators (Murre et al., 1989a,b). Particular associations of different bHLH proteins form different complexes that vary in their affinity to distinct DNA-binding sites (Murre et al., 1989b; Benezra et al., 1990; Sun and Baltimore, 1991). Based on these characteristics, a molecular model for the X: A signal has been formulated (Parkhurst and Ish-Horowicz, 1992). This model proposes that heterodimer complexes between DA an SIS ("numerator elements") proteins are formed that activate transcription of the gene *Sxl*. The interaction between SIS ("numerator") and AUTOSOMAL ("denominator") proteins will sequester SIS proteins, so that an effective concentration of DA-SIS complexes will only be attained in XX zygotes, since these have twice the amount of SIS proteins than XY zygotes. Consequently, only in XX zygotes will activation of *Sxl* will take place. It is possible that the "denominator elements", as *dpn,* exert their negative effect on *Sxl* by direct interaction with DNA, this interaction being prevented in females in which the amount of SIS ("numerator") products is twice the amount found in males. To this respect, evidence has been provided for the formation of heteromeric complexes between DA and SC (Deshpande et al., 1995; Liu and Belote, 1995), and between DA and SIS-A, DA and DPN and DPN and SIS-A (Liu and Belote, 1995). Moreover, Hoshijima et al (1995) have reported the existence of E boxes (binding sequences for b-HLH proteins) and D boxes (binding sequences for DPN) in the upstream region of the early Sxl-promotor that is activated by the X:A signal. This is a simplifed view of the *Sxl* activation process. Certainly the interactions are more complex since the "numerator" products do not play strictly equivalent roles in the formation of the X:A signal. Thus, for example, the interaction between *sc* and *run* is stronger than betwen *run* and *sis-a* (Torres and Sánchez, 1992), and the interaction between *sis-a* and *dpn* is also stronger than between *run* and *dpn* (Sánchez et al., 1994). Moreover, asymmetric relationships exist between *sc* and *sis-a* in the ability of a duplication of either element to suppress the lethal effect due to a decreased dose of the other element (Cline, 1988; Torres and Sánchez., 1989).

Erickson and Cline (1993) have cloned and characterised the gene *sis-a*, finding that it encodes a protein with motifs of the basic leucine zipper class of transcription factors. Based on the fact that *sc* and *sis-a* encode two different classes of transcription factors, they further proposed the possible existence of different binding sites at the gene *Sxl* for these two types of "numerator" factors. Those binding sites might be present in multiple copies that would contribute to amplification of the twofold difference in SC and SIS-A products in females and males. This would result in an all-or-nothing *Sxl* activation response. Furthermore, they suggested a possible difference in the developmental time of action of the negative factors on *Sxl*: the maternal factors would primarily act to define the initial threshold of SIS products needed to activate *Sxl*, whereas the zygotic factors, acting immediately after the maternal factors, would raise that threshold above a level that prevents *Sxl* activation in males since in these, the single *sc* and *sis-*a gene doses do not supply the amount of SIS products required to overcome the effect of the negative factors on *Sxl*.

VI. Developmental meaning of the X:A signal

The role of the X:A signal could be visualised in two different ways. One possibility is that the X:A signal may be continuously needed by the cells during development to stay in the chosen sexual pathway and to maintain properly adjusted the dosage compensation process. Under this hypothesis, XO clones induced at any time during development of XX flies would survive and differentiate male structures (Figure 1A). Alternatively, the X:A signal could be used by the cells at a certain time in their development ("register time", t_R) to set up their sex and dosage compensation processes. Under this second hypothesis, XO clones induced before that time t_R would survive and differentiate male structures, whereas XO clones induced later would die because their dosage compensation process would be upset (Figure

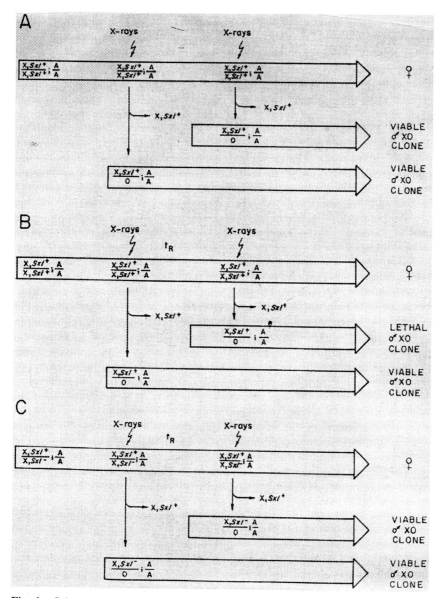

Fig. 1. Scheme showing the alternative hypothesis for the time specificity of the X:A signal. For details, see text.

1B). To answer this question a clonal analysis strategy was used. Genotypes were constructed that allowed the removal, by mitotic recombination induced by X-irradiation, of one of the X chromosomes from a cell at different times in development, and thus to produce XO clones in XX female flies (Sánchez and Nöthiger, 1983; Bachiller and Sánchez, 1991).

The results demonstrated that XO clones induced at around the blastoderm stage survive and differentiate males structures, while clones induced later in development are lethally affected. However, the XO clones survive and differentiate male structures if they carry a *Sxl⁻* mutant allele, independent of the time in development when the clones are induced (Figure 1C). These results indicate that the X:A signal irreversibly sets, in a cell-autonomous manner, the state of activity of *Sxl* sometime around the blastoderm stage. Once this is achieved the X:A signal is no longer needed and both sex determination and dosage compensation come under the control of *Sxl.* Accordingly, XO clones generated after the blastoderm stage would die because the SXL product continues to be made which results in a fatal deficit of the X chromosomal gene products in XO clones. If, on the other hand, the XO clones contain a *Sxl⁻* mutation, the rate of transcription of that single X chromosome increases, allowing those XO clones to survive and form male structures. Furthermore, these results are compatible with the idea of *Sxl* being the only gene that responds to the X:A signal.

The gene *Sxl* produces two temporally distinct sets of transcripts which correspond to the function of the female-specific early and the non-sex specific late promoters respectively (Salz et al., 1989). The early set is produced as a response to the X:A signal which controls *Sxl* expression at the transcriptional level (see Section IV). The control of *Sxl* after the blastoderm stage and in adult life occurs by sex-specific splicing of its primary transcripts. Male and female transcripts differ by the inclusion of a male-specific exon that places a stop codon in the open reading frame, giving rise to truncated, presumably non-functional, proteins. In females, this exon is spliced out and functional proteins are produced (Bell et al., 1988; Bopp et al., 1991). The capacity of *Sxl* to function as a stable switch is due to a positive autoregulatory function of its own product which is involved in the female-specific splicing of its RNA (Cline, 1984; Bell et al., 1991; Sakamoto et al., 1992; Horabin and Schedl, 1993a, b). In this splicing control, the SXL protein participates together with the function of the genes *fl(2)d* (Granadino et al., 1990; 1991b: 1992), *vir* (Hilfiker and Nöthiger, 1991) and *snf* (Salz, 1992; Brown and Salz, 1993).

The temporal restriction of the X:A signal function is related to the existence of the two *Sxl* promotors. In females, the gene *Sxl* is expressed from its early promotor that responds to the X:A signal; consequently, early SXL proteins are produced. In males, the early expression of *Sxl* does not occur because the X:A signal is not formed, resulting in the absence of early SXL proteins. Later, when the constitutive *Sxl* promotor begins to function, the presence of the early SXL proteins in females directs the SXL RNAs to the female mode of splicing giving rise to the late SXL proteins, and consequently the female-splicing state of *Sxl* is set up. In males, on the contrary, the first late SXL RNAs follow the male mode of splicing since no early SXL proteins are available, resulting in the establishment of the male-splicing state of *Sxl.* Therefore, the developmental meaning of the X:A signal is to "open", at a specific time at the beginning of development, the early *Sxl* promotor which provides females with the early SXL proteins needed to establish female-specific control of *Sxl,* once the late constitutive promoter of this gene starts functioning.

The time specificity that characterises the function of the X:A signal on the early *Sxl* promotor could be due to the formation of the X:A signal only at syncytial blastoderm stage, or to the exclusive presence, at this stage, of any of the maternal products involved in *Sxl* activation. Among the zygotic (X:A signal) and maternal genes involved in *Sxl* activation, some, such as *sc, run, dpn, da* and *emc,* are not only expressed at the syncytial blastoderm stage, but also at later stages. This later expression only occurs in certain groups of cells, contrary to the more generalised early expression that is involved in *Sxl* activation. Genes *sc, dpn, da* and *emc* are expressed in the neurogenic regions of both sexes (reviewed in Ghysen and Dambly-Chaudiére, 1988; Campuzano and Modolell, 1992), whereas *run* is expressed in addition in the same stripes of every other segment in females and males (Kania et al., 1990; Duffy et al., 1991; Gergen and Butler, 1988). In males, activation of *Sxl* is not induced in the cells where these genes are expressed. Moreover, by means of a *hsp 70-sc* chimeric gene (HSSC-3) (Rodriguez et al., 1990), *sc* was expressed at different developmental times and checked for when this expression suppressed the sis-b mutant phenotype in females, and caused lethality in males due to expression of *Sxl* (Torres and Sánchez, 1991). The expression of *sc* activates *Sxl* only at a very specific stage in development, coinciding with the thermosensitive phase delimited by temperature-shift experiments and corresponding to the syncytial balstoderm stage when the gene *sc* undergoes an homogeneous expression (Romani et al., 1987; Cabera et al., 1987; Parkhurst et al., 1993; Erickson and Cline, 1993). Thus, the time specificity that characterises the function of the X:A signal is not explained by the late restricted expression of those genes. In contrast, Erickson and Cline (1993) reported the developmental profile of *sis-a* expression: this gene is only expressed during early developmental stages, and its only essential function seems to be sex determination. Therefore, the time specificity of the X:A signal can be explained by the temporally restricted expression of the gene *sis-a.* Other genes, defined by *Df(1)N19* and *Df(1)RA2,* which still need to be characterised, may also show a restricted temporal expression.

VII. The X:A signal in the germ line

The germline exhibits sexual dimorphism as does somatic tissue. Cells with the 2X;2A chromosomal constitution will follow the oogenic pathway and X;2A cells will develop into sperm. However the sex of *Drosophila* germ cells is determined by a mechanism different to that acting in somatic cells. XX cells enter the male pathway when developing in a male host animal. XY and XO cells, in contrast, form spermatocytes even when developing in a host ovary. Thus, the sex of germ cells is determined by cell- autonomous

(X:A) and inductive signals. (Steinmann-Zwicky et al 1989; Nöthiger et al 1989). Both signals regulate *Sxl* whose activity is required for normal female germ cell development (Cline 1983; Schüpbach 1985; Steinmann-Zwicky et al 1989). In fact two transcripts have been discovered, one specific to germline cells, and a second found in both somatic and germline cells though in greater quantity in the latter (Salz et al 1989).

The X:A signal in somatic and germline tissues occurs differently. Cline (1986) reported that clones of germ cells homozygous for *sis-a* develop into functional oocytes. XX germ cells lacking *sis-b* function produce functional eggs when allowed to develop in a host female. (Steinmann-Zwicky, 1993). Simultaneous heterozygosity for *sc, sis-a, run,* and a deficiency for *Sxl* is a severe constitution (even more severe than, for example, homozygosity for *sc* alone), which causes somatic cell lethality as a consequence of a failure to activate *Sxl* (Torres and Sánchez, 1991). Germ cells heterozygous for *sc, sis-a, run* and a deficiency for *Sxl* transplanted into female hosts develop into functional oocytes (Granadino et al 1993). These transplants were performed in host embryos lacking their own pole cells because they came from mothers homozygous for the *oskar* mutation (Lehmann and Nüsslein-Volhard, 1986). Thus the genes *sc, sis-a,* and *run* needed to activate *Sxl* in the soma seem not be needed for the activation of this gene in the germline. As mentioned above, the genes *sc* and *da* encode bHLH proteins (Villares and Cabrera 1987; Caudy et al, 1988) that interact for the activation of *Sxl* in the soma. The exemption of the gene *sc* for the activation of *Sxl* in the germline would agree with the fact that *da* is not required for oogenesis (Cronmiller and Cline 1987). Therefore, although the X:A signal ratio provides a sex-determining signal in the germ cells, the elements forming this signal are different from those forming the X:A ratio in somatic cells. Consequently, in germ cells, *Sxl* is activated by a mechanism that is different from that acting in the soma. On the other hand, the X:A signal determines the state of activity of *Sxl* in the somatic line around the blastoderm stage (see section VI). Nevertheless, in the germline, the time period during development when this signal activates *Sxl* is different since no Sxl transcripts are detected in pole cells at the blastoderm stage. (Keyes et al, 1992; Sánchez and Torres unpublished data).

Acknowledgments

This work was supported by grant PB92-0006 from D.G.I.C.Y.T. from the Ministerio de Educación y Ciencia, España.

REFERENCES

Bachiller, D. and Sánchez, L. (1989). Further analysis on the *male-specific lethal* mutations that affect dosage compensation in *Drosophila melanogaster. Roux's Arch. Dev. Biol.* **198**, 34–38.

Bachiller, D. and Sánchez, L. (1991). Production of XO clones in XX females of *Drosophila. Genet Res* **57**, 23–28.

Baker, B.S. (1989). Sex in flies: The splice of life. *Nature* **340**, 521–524.

Baker, B.S. and Belote, J.M. (1983). Sex determination and dosage compensation in *Drosophila melanogaster. Ann. Rev. Genet.* **17**, 345–393.

Balcells, Ll., Modolell, J. and Ruiz-Gómez, M (1988). A unitary basis for different *Hairy wing* mutations of *Drosophila melanogaster. EMBO J.* **7**, 3899–3906.

Bell, L.R., Maine, E.M., Schedl, P. and Cline, T.W. (1988). *Sex-lethal*, a *Drosophila* sex determination switch gene, exhibits sex-specific RNA splicing and sequence similar to RNA binding proteins. *Cell* **55**, 1037–1046.

Bell, L.R., Horabin, J. I., Schedl, P. and Cline, T.W. (1991). Positive autoregulation of *Sex-lethal* by alternative splicing maintains the female determined state in *Drosophila. Cell* **65**, 229–239.

Belote, J.M. (1983). Male-specific lethal mutations of *Drosophila melanogaster.*II. Parameters of gene action during male development. *Genetics* **105**, 881–896.

Benezra, R., Davis, R.L., Lockstone, D., Turner, D.L. and Weintraub, H. (1990). The protein Id: a negative regulator of the helix-loop-helix DNA binding proteins. *Cell* **61**, 49–59.

Bier, E., Vaessin, H., Shepherd, S., Lee, K., McCall, K., Barbel, S., Ackermenn, L., Carretto, R., Uemura, T., Grell, E., Jan., L. Y. and Jan, Y. N. (1989). Searching for pattern and mutation in the *Drosophila* genome with a P-lacZ vector. *Genes Dev.***3**, 1273–1287.

Bopp, D., Bell, L.R.. Cline, T.W. and Schedl, P. (1991). Developmental distribution of female-specific *Sex-lethal* proteins in *Drosophila melanogaster. Genes Dev* **5**, 403–415.

Botas, J. Moscoso del Prado, J. and Garcia-Bellido, A. (1982). Gene-dose titration analysis in the search of transregulatory genes. *EMBO J.* **1**, 307–310.

Bridges, C.B. (1921). Triploid intersexes in *Drosophila melanogaster. Science* **54**, 252–254.

Bridges, C.B. (1925). Sex in relation to chromosomes and genes. *Amer. Nat.* **59**, 127–137.

Brown, E.A. and Salz, H.K. (1993). The *Drosophila* sex determination gene *snf* is utilized for the establishment of the female-specific splicing pattern of *Sex-lethal. Genetics* **134**, 801–807.

Cabrera, C.V., Martinez-Arias, A. and Bate, M. (1987). The expression of three members of the *achaete-scute* complex correlates with neuroblasts segregation in *Drosophila. Cell* **50**, 425–433.

Campuzano, S., Carramolino, L., Cabrera, C.V., Ruiz-Gómez, M., Villares, R., Boronat, A. and Modolell, J. (1985). Molecular genetics of the *achaete-scute* gene complex of *D. melanogaster. Cell* **40**, 327–338.

Campuzano, S. and Modolell, J. (1992). Patterning of the *Drosophila* nervous system: the *achaete-scute* gene complex. *Trends Genet.* **8**, 202–208.

Caudy, M., Grell, E.H., Dambly-Chaudiére, C., Ghysen, A., Jan, L.Y. and Jan, Y. N. (1988). The maternal sex determination gene *daugtherless* has zygotic activity necessary for the formation of peripheral neurons in *Drosophila. Genes Dev* **2**, 843–852.

Chandra, H.S. (1985). Sex determination: A hypothesis based on noncoding DNA. *Proc. Natl. Acad. Sci. USA* **82**, 1165–1169.

Cline, T.W. (1978). Two closely-linked mutations in *Drosophila melanogaster* that are lethal to opposite sexes and interact with *daughterless. Genetics* **90**, 683–698.

Cline, T.W. (1983). Functioning of the genes *daughterless (da)* and *Sex-lethal (Sxl)* in *Drosophila* germ cells. *Genetics* **104** (Suppl), s16–17.

Cline, T.W. (1984). Autoregulatory functioning of a *Drosophila* gene product that establishes and maintains the sexually determined state. *Genetics* **107**, 231–277.

Cline, T.W. (1986). A female specific lethal lesion in an X-linked positive regulator of the *Drosophila* sex determination gene *Sex-lethal*. *Genetics* **113**, 641–663.

Cline, T.W. (1988). Evidence that *"sisterless-a"* and *"sisterless-b"* are two of several discrete "numerator elements" of the X:A sex determination signal in *Drosophila* that switch *Sex-lethal* between two alternative stable expression states. *Genetics* **119**, 829–862.

Cronmiller, C. and Cline, T.W. (1987). The *Drosophila* sex determination gene *daugtherless* has different functions in the germline versus the soma. *Cell* **48**, 479–487.

Cubas, P., de Celis, J.F., Campuzano, S. and Modolell, J. (1991). Proneural clusters of *achaete-scute* espression and the generation of sensory organs in the *Drosophila* imaginal wing disc. *Genes Dev.* **5**, 996–1008.

Dambly-Chaudiére, C. and Ghysen, A. (1987). Independent subpatterns of sense organs required independent genes of the *achaete-scute* complex in *Drosophila* larvae. *Genes Dev.* **1**, 297–306.

Deshpande, G., Stukey, J. and Schedl, P. (1995). *scute* (sis-b) function in *Drosophila* sex determination. *Mol Cell Biol* **15**: 4430–4440.

Duffy, J.B. and Gergen, J.P. (1991). The *Drosophila* segmentation gene *runt* acts as a position-specific numerator element necessary for the uniform expression of the sex-determinating gene *Sex-lethal*. *Genes Dev* **5**, 2176–2187.

Duffy, J.B., Kania, M.A. and Gergen, J.P. (1991). Expression and function of the *Drosophila* gene *runt* in early stages of neuronal development. *Development* **113**, 1223–1230.

Ellis, H.M., Apann, D.R. and Posakony, J.W. (1990). *extramacrochaetae*, a negative regulator of sensory organs development in *Drosophila*, defines a new class of helix-loop-helix protein. *Cell* **61**, 27–38.

Erickson, J.W. and Cline, T.W. (1991). Molecular nature of the *Drosophila* sex determination signal and its link to neurogenesis. *Science* **251**, 1071–1074.

Erickson, J.W. and Cline, T.W. (1993). A bZIP protein, Sisterless-a, collaborates with bHLH transcription factors early in *Drosophila* development to determine sex. *Genes Dev.* **7**, 1688–1702.

Gadagkar, R., Nanjundiah, V., Joshi, N. V. and Chandra, H.S. (1982). Dosage compensation and sex determination in *Drosophila*: mechanism of measurement of the X/A ratio. *J. Biosci.* **4**, 377–390.

García-Bellido, A. (1979). Genetic analysis of the *achaete-scute* system of *Drosophila melanogaster*. *Genetics* **91**, 491–520.

Garrel, J. and Modolell, J. (1990). The *Drosophila extramacrochaetae* locus, an antagonist of proneural genes that, like these genes, encodes a helix-loop-helix protein. *Cell* **61**, 39–48.

Gergen, J.P. and Butler, B.A. (1988). Isolation of the *Drosophila* segmentation gene *runt* and analysis of its expression during embryogenesis. *Genes Dev.* **2**, 1179–1193.

Ghysen, A. and Dambly-Chaudiére, C. (1988). From DNA to form: the *achaete-scute* complex. *Genes Dev.* **2**, 495–501.

Granadino, B., Campuzano, S. and Sánchez, L. (1990). The *Drosophila melanogaster* *fl(2)d* gene is needed for the female-specific splicing of *Sex-lethal* RNA. *EMBO J.* **9**, 2597–2602.

Granadino, B., Torres, M., Bachiller, D. Torroja, E., Barbero, J.L. and Sánchez L. (1991a). Genetic and molecular analysis of new female specific lethal mutations at the gene *Sxl* of *Drosophila melanogaster*. *Genetics* **129**, 371–383.

Granadino, B., San Juán, A.B. and Sánchez, L. (1991b). The gene *fl(2)d* is required for various *Sxl*-controlled processes in *Drosophila* females. *Roux's Arch. Dev. Biol.* **200**, 172–176.

Granadino, B., San Juan, A.B., Santamaría, P. and Sánchez, L. (1992). Evidence of a dual function in *fl(2)d*, a gene needed for *Sex-lethal* expression in *Drosophila melanogaster. Genetics* **130**, 597–612.

Granadino, B., Santamaria, P. and Sánchez, L. (1993). Sex determination in the germ line of *Drosophila melanogaster*: activation of the gene *Sex-lethal. Development* **118**, 813–816.

Hilfiker, A. and Nöthiger, R. (1991). The temperature-sensitive mutation *vir^{ts}* (*virilizer*) identifies a new gene involved in sex determination of *Drosophila. Roux's Arch. Dev. Biol.* **200**, 240–248.

Horabin, J.I. and Schedl, P. (1993a). Regulated splicing of the *Drosophila Sex-lethal* male exon involves a blockage mechanism. *Mol. Cell. Biol.* **13**, 1408–1414.

Horabin, J.I. and Schedl, P. (1993b). *Sex-lethal* autoregulation requires multiple *cis*-acting elements upstream and downstream of the male exon and appears to depend largely on controlling the use of the male exon 5' splice site. *Mol. Cell Biol.* **13**, 7734–7746.

Hoshijima, K., Kohyama, A., Watakabe, I., Inoue, K., Sakamoto, H. and Shimura, Y. (1995). Transcriptional regulation of the *Sex-lethal* gene by helix-loop-helix proteins. *Nucl. Acid Res.* **23**; 3441–3448.

Jiménez, F. and Campos-Ortega, J.A. (1979). On a region of the *Drosophila* genome necessary for central nervous system development. *Nature* **282**, 310–312.

Kania, M.A., Bonner, A.S., Duffy, J.B. and Gergen, J.P. (1990). The *Drosophila* segmentation gene *runt* encodes a novel regulatory protein that is also expressed in the developing nervous system. *Genes Dev.* **4**, *1701–1713*.

Keyes, L.N., Cline. T.W. and Schedl, P. (1992). The primary sex determination signal of *Drosophila* acts at the level of transcription. *Cell* **68**, 933–943.

Kuroda, M.I., Palmer, M.J. and Lucchesi, J.C. (1993). X chromosome dosage compensation in *Drosophila. Seminars Dev. Biol.* **4**, 107–116.

Lehman, R. and Nüsslein-Volhard, C. (1986). Abdominal segmentation, pole cell formation, and embryonic polarity requires the localized activity of *oskar*, a maternal gene in *Drosophila. Cell* **47**: 141–152.

Liu, Y. and Belote, J.M. (1995). Protein-protein interactions among components of the *Drosophila* primary sex determination signal. *Mol Gen Genet* **248**: 182–189.

Lucchesi, J.C. and Skripsky, T. (1981). The link between dosage compensation and sex differentiation in *Drosophila melanogaster. Chromosoma* **82**, 217–227.

Lucchesi, J.C. and Manning, J.E. (1987). Gene dosage compensation in *Drosophila melanogaster. Adv. Genet.* **24**, 371–429.

Maine, E.M., Salz, H.K., Cline, T.W. and Schedl, P. (1985a). The *Sex-lethal* of *Drosophila*: DNA alterations associated with sex-specific lethal mutations. *Cell* **43**: 521–529.

Maine, E.M., Salz, H.K., Schedl, P., and Cline, T.W. (1985b). *Sex-lethal*, a link between sex determination and sexual differentiation in *Drosophila melanogaster. Cold Spring Harbor Symp. Quant. Biol.* **50**; 595–604.

Marshall, T. and Whitte, J.R. (1978). Genetic analysis of the mutation *female-lethal* in *Drosophila melanogaster. Genet. Res.* **32**: 103–111.

Moscoso del Prado, J. and García-Bellido, A. (1984). Genetic regulation of the *achaete-scute* complex of *Drosophila melanogaster. Roux's Arch. Devl. Biol* **193**, 242–245.

Murre, C., McCaw, P.S. and Baltimore, D. (1989a). A new DNA binding and dimerization motif in inmunoglobulin enhancer binding, *daugtherless, MyoD* and myc proteins. *Cell* **56**, 777–783.

Murre, C., McCaw, P.S., Vässin, H.,Caudy M., Jan, Y.N., Cabrera, C.V., Buskin, J.N., Hauschka, S.D., Lassart, A.B., Weintraub, H. and Baltimore, D. (1989b). Interactions between heterologous helix-loop-helix proteins generate complex that bind specifically to a common DNA sequence. *Cell* **58**, 537–544.

Nöthiger, R. and Steinmann-Zwicky, M. (1985). Sex determination in *Drosophila. Trends Genet.* **1**, 209–215.

Nöthiger, R., Jonglez, M., Leuthold, M., Meier-Gerschwiller, P. and Weber, T. (1989). Sex determination in the germline of *Drosophila* depends on genetic signals and inductive somatic factors. *Development* **107**, 505–518.

Oliver, B., Perrimon, N. and Mahowald, A.P. (1988). Genetic evidence that the *sans-fille* locus is involved in *Drosophila* sex determination. *Genetics* **120**, 159–171.

Parkhurst, S.M., Bopp, D. and Ish-Horowicz, D. (1990). X:A ratio, the primary sex determination signal in *Drosophila,* is transduced by helix-loop-helix proteins. *Cell* **63**, 1179–1191.

Parkhurst, S.M. and Ish-Horowicz, D. (1992). Common denominators for sex. *Current Biol* **2**, 629–631.

Parkhurst, S.M., Lipshitz, H.D. and Ish-Horowicz, D. (1993). *achaete-scute* feminizing activities and *Drosophila* sex determination. *Development* **117**, 737–749.

Rodríguez, I., Hernández, R., Modolell, J. and Ruiz-Gómez, M. (1990). Competence to develop sensory organs is temporally and spatially regulated in *Drosophila* epidermal primordia. *EMBO J.* **9**, 3583–3592.

Romani, S., Campuzano, S. and Modolell, J. (1987). The *achaete-scute* complex is expressed in neurogenic regions of *Drosophila* embryos. *EMBO J.* **6**, 2085–2092.

Sakamoto, H., Inoue, K., Higuchi, I., Ono, Y. and Shimura, Y. (1992). Control of *Drosophila Sex-lethal* pre-mRNA splicing by its own female-specific product. *Nucleic Acids Res* **20**, 5533–5540.

Salz, H.K., Maine, E.M., Keyes, L.N., Samuels, M.E., Cline, T.W. and Schedl, P. (1989). The *Drosophila* female-specific sex-determination gene, *Sex-lethal,* has stage, tissue-, and sex-specific RNAs suggesting multiple modes of regulation. *Genes Dev* **3**, 708–719.

Salz, H.K. (1992). The genetic analysis of *snf:* a *Drosophila* sex determination gene required for activation of *Sex-lethal* in both the germline and the soma. *Genetics* **130**, 547–554.

Sánchez, L. and Nöthiger, R. (1982). Clonal analysis of *Sex-lethal,* a gene needed for female sexual development in *Drosophila melanogaster. Wilhem Roux's Arch. Dev. Biol.* **191**: 211–214.

Sánchez, L. and Nöthiger, R. (1983). Sex determination and dosage compensation in *Drosophila melanogaster*: production of male clones in XX females. *EMBO J.* **2**, 485–491.

Sánchez, L., Granadino, B. and Torres. M. (1994). Sex determination in *Drosophila melanogaster:* X-linked genes involved in the initial step of *Sex-lethal* activation. *Dev. Genet.* **15**, 251–264.

Santamaria, P. and Gans, M. (1980). Chimeras of *Drosophila melanogaster* obtained by injection of haploid nuclei. *Nature* **287**, 143–144.

Schüpbach, T. (1985). Normal female germ cell differentiation requires the female X chromosome to autosome ratio and expression of *Sex-lethal* in *Drosophila melanogaster. Genetics* **109**, 529–548.

Skeath, J.B. and Carroll, S.B. (1991). Regulation of *achaete-scute* gene expression and sensory organ pattern formation in the *Drosophila* wing. *Genes Dev.* **5**, 984–995.

Steinmann-Zwicky, M. (1988). Sex determination in *Drosophila*: the X chromosomal gene *liz* is required for *Sxl* activity. *EMBO J.* **7**, 3889–3898.

Steinmann-Zwicky, M. (1993). Sex determination in *Drosophila*: *sis-b,* a major numerator element of the X:A ratio in the soma, does not contribute to the X:A ratio in the germ line. *Development* **117**,763–767.

Steinmann-Zwicky, M., Schmid, H. and Nöthiger, R. (1989). Cell-autonomous and inductive signals can determine the sex of the germ line of *Drosophila* by regulating the gene *Sxl. Cell* **57**, 157–166.

Sun, X.H. and Baltimore, D. (1991). A inhibitory domain of E12 transcription factor prevents DNA binding in E12 homodimers but not in E12 heterodimers. *Cell* **64**, 459–470.

Torres, M. and Sánchez, L. (1989). The *scute (T4)* gene acts as a numerator element of the X:A signal that determines the state of activity of *Sex-lethal* in *Drosophila melanogaster. EMBO J.* **10**, 3079–3086.

Torres, M and Sánchez, L. (1991). The sisterless-b function of the *Drosophila* gene *scute* is restricted to the state when the X:A ratio signal determines the activity of *Sex lethal. Development* **113**, 715–722.

Torres, M. and Sánchez, L. (1992). The segmentation gene *runt* is needed to activate *Sex-lethal*, a gene that controls sex determination and dosage compensation in *Drosophila. Genet Res* **59**, 189–198.

Uenoyama, T., Uchida, S., Fukunaga, A. and Oishi, K. (1982). Studies on the Sex specific lethals of *Drosophila melanogaster.* IV. Gynandromorph analysis of three male-specific lethals, *mle, msl-2* and *mle(3) 132. Genetics* **102**: 223–231.

Van Doren, M., Ellis, H. M. and Posakony, J.W. (1991). The *Drosophila* extramacrochaetae protein antagonize sequence-specific DNA binding by daughterless/achaete-scute protein complexes. *Development* **113**, 245–255.

Villares, R. and Cabrera, C.V. (1987). The *achae-scute* gene complex of *D. melanogaster:* conserved domains in a subset of genes required for neurogenesis and their homology to *myc. Cell* **50**, 415–424.

Younger-Shepherd, S., Vaessin, H., Bier, E., Yeh Jan, L. and Nung Jan, Y. (1992). *deadpan,* an essential pan-neural gene encoding an HLH protein, acts as a denominator in *Drosophila* sex determination. *Cell* **70**, 911–922.

Genome Analysis in Eukaryotes: Developmental and evolutionary aspects
R.N. Chatterjee and L. Sánchez (Eds)
Copyright © 1998 Narosa Publishing House, New Delhi, India

6. The Development of Male- and Female-Specific Sexual Behavior in *Drosophila melanogaster*

Laurie Tompkins

Department of Biology, Temple University, Philadelphia,
Pennsylvania 19122, U.S.A.

The wren goes to 't' and the small gilded fly
Does lecher in my sight.
Let copulation thrive.

Shakespeare, *King Lear*

I. Introduction

Alfred Sturtevant, one of the fathers of modern genetics, was the first to describe the complex courtship ritual that males from the common pomace fly species *Drosophila melanogaster* (then called *Drosophila ampelophila*) perform in response to conspecific females. In his 1915 paper, Sturtevant showed that the courtship behaviors are male-specific, in the sense that females do not perform them. In addition, he noted that the ability to elicit vigorous, prolonged courtship from males ("sex appeal") is, for the most part, a female-specific characteristic.

Why do males and females perform and elicit different behaviors? In this review, I will summarize our current understanding of the answers to this question. Since other recent reviews focus on the effects of environmental factors (Hirsch and Tompkins, 1994) and sexual experience (Hall, 1994; Kubli, 1996) on flies' sexual behavior, these topics will not be addressed here.

II. Behaviors performed and elicited by sexually mature flies

A. Courtship

The salient courtship behaviors that sexually mature males perform in response to virgin females are illustrated in Figure 1. The male initiates courtship by briefly vibrating his abdomen (Tompkins, 1990), then approaches the female, orienting his body so that he faces her abdomen (*orientation*). If the female

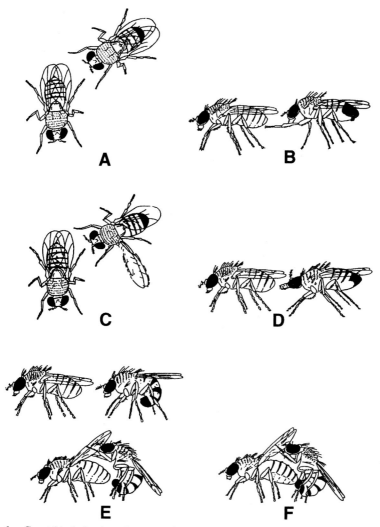

Fig. 1. Courtship behaviors that normal, sexually mature males perform in response to virgin females. (A), orientation; (B), tapping (attempted) and following; (C), wing vibration; (D) licking (attempted); and (E) attempted copulation. Copulation is shown in (F).

responds to the male by running away, as she usually does, the male will either pivot his body to maintain the orientation posture (the pivoting behavior is also called orientation) or pursue her from behind (*following*). While the male is orienting to the female or following her, he will attempt to tap the female's abdomen with one of his forelegs (*tapping*) and extend the wing that is closest to the female's body (*wing extension*), vibrating the wing to produce a courtship song (*singing* or *wing vibration*). Later in the courtship bout or in a subsequent bout, the male will extend his proboscis in an attempt to touch the female's genitalia (*licking*) and curl his abdomen in an attempt to touch the female's genitalia with his own (*attempted copulation*). Performance of one behavior does not preclude simultaneous performance of another behavior; for example, males usually sing while following or orienting to females. Nor is the sequence of behaviors rigidly fixed, although some sequences are more likely than others (e.g., Markow and Hanson, 1981; Welbergen et al., 1992). For example, males that initiate courtship in response to a female always orient, follow, tap, and sing before performing the other behaviors, and they often attempt to copulate immediately after licking a female's genitalia. However, males can and usually do perform "early" behaviors again after they have performed "advanced" behaviors.

It is often useful to quantitate courtship by calculating a male's *courtship index*, which is the percentage of time that the male spends performing one or more of the courtship behaviors until mating ensues or the observation period ends, whichever occurs first. For three- to five-day-old wild-type males that have been collected within a few hours of eclosion, subsequently maintained in isolation for 3–5 days, transferred without anesthesia to small (.2 cm^3) chambers with one virgin female each, then observed for ten minutes, courtship indices typically range from 70–95 (see Tompkins et al., 1980). Most males will initiate several bouts of courtship and attempt to copulate at least three times before the female will indicate her receptivity to copulation by opening her vaginal plates, thus allowing intromission to occur (Tompkins and Hall, 1983). The time elapsing between the beginning of the observation period and the time at which the flies begin to copulate, if the flies mate during the observation period, is their *copulation latency*. The percentage of pairs that begin to mate during the observation period is their *copulation frequency*.

B. Sex Appeal

Virgin females elicit much higher courtship indices than normal, sexually mature males (the courtship index elicited by a mature male is typically 0-3, Tompkins et al., 1980; see Tompkins, 1989 for a discussion of factors that affect the extent to which homosexual courtship occurs). Moreover, the courtship that a male does perform in response to another mature male is qualitatively different from the courtship that he performs in response to a virgin female. Specifically, a bout of homosexual courtship typically consists of a few seconds of orientation and tapping, after which time the courting male abruptly stops courting the male "sex object."

Why do mature males elicit much less courtship than virgin females? Mutant males with visual defects (i.e., *glass^3* and *no-receptor-potential A* males, which are blind; and *optomotor-blind* males, which do not respond normally to horizontally moving stimuli) perform half as much courtship in response to virgin females as wild-type (normal) control males (Tompkins et al., 1982). These observations imply that the sight of a moving female stimulates normal males to perform courtship. However, the visual stimulus associated with movement of a male fly is as attractive as that of a female. Evidence for this is provided by observations of mutant *Sex-lethal* females that are transformed into phenotypic males but synthesize the sex pheromones that are characteristic of females, which elicit as much courtship from wild-type males as female controls do (Tompkins, 1984). Thus, the inability of mature males to elicit vigorous courtship is not mediated by an inhibitory stimulus that is associated with the males' movement. Rather, mature males elicit very little courtship because they synthesize pheromones that inhibit the courtship that would otherwise be stimulated by their movement and, conversely, do not synthesize courtship-stimulating pheromones.

Evidence for males' inhibitory pheromones is provided by the observation that hydrocarbons and other compounds that are soluble in organic solvents, when extracted from sexually mature males, inhibit the courtship that males perform in response to virgin females (Tompkins and Hall, 1981). By fractionating the extracted hydrocarbons, then testing the subfractions to see whether they inhibit males' courtship of females, two compounds that have inhibitory activity have been identified (Jallon et al., 1981; Scott, 1986). One of the inhibitory pheromones is octadecenyl acetate (also known as *cis*-vaccenyl acetate), a lipid that is synthesized in the ejaculatory bulb of the male reproductive system (Butterworth, 1969). The other inhibitory pheromone is (Z)-7-tricosene, a cuticular hydrocarbon (Scott, 1986). Virgin females do not synthesize detectable quantities of octadecenyl acetate, but they do synthesize (Z)-7-tricosene, although in quantities that are ca. 10-fold less than those made by males (see Tompkins and McRobert, 1995). Thus, octadecenyl acetate and (Z)-7-tricosene are classified as male-specific and male-predominant pheromones, respectively.

Virgin females, unlike males, do synthesize a courtship-stimulating pheromone, as shown by the observation that hydrocarbons extracted from virgin females stimulate a five- to ten-fold increase in the courtship that two sexually mature males perform in response to each other (Tompkins et al., 1980; Tompkins and Hall, 1984). By fractionating the extracted hydrocarbons, then testing the subfractions to see whether they stimulate normal males to court each other, (Z, Z)-7, 11-heptacosadiene, a cuticular hydrocarbon, has been identified as the courtship-stimulating pheromone that is synthesized by virgin females from most wild-type strains (Antony et al., 1985). Since males do not, in general, synthesize detectable quantities of this compound (Antony et al., 1985; cf. Tompkins and McRobert, 1995), (Z, Z)-7, 11-heptacosadiene is classified as a female-specific pheromone.

C. Receptivity to copulation

As noted in section II-A, a sexually mature female that has been sufficiently stimulated by a male's courtship will indicate her receptivity to copulation by opening her vaginal plates in response to one of the male's attempts to copulate. What are the stimuli associated with males' courtship that cause females to be receptive to copulation? Wild-type males whose wings have been surgically amputated, males from *raised* mutant stocks, and mutant *nonA*[diss] males, all of which produce abnormal courtship songs or no songs at all, have significantly longer copulation latencies than control males (Schilcher, 1976; Kulkarni et al., 1988; Wheeler et al., 1989; McRobert et al., 1995). Conversely, mutant *para*[sbl1] and *para*[sbl2] females, which respond abnormally to chemical stimuli (Lilly and Carlson, 1990), have significantly longer copulation latencies than control females (Tompkins et al., 1982; Gailey et al., 1986). Thus, acoustic stimuli that normal males produce when they vibrate their wings, in conjunction with as yet unidentified pheromones that are associated with courting males, stimulate normal, sexually mature females to be receptive to copulation.

III.   Behaviors performed and elicited by immature flies

A.   Courtship and male sex appeal

Males that only a few hours old are like females in that they are incapable of performing courtship in response to sexually attractive flies. The ages at which males become competent to perform the various courtship behaviors have been determined for males from a Canton-S wild-type strain. Most of the Canton-S males oriented, tapped, followed, and extended their wings in response to virgin females when they were 24 hours old, but only half of the one-day-old males vibrated their wings or licked the females, and very few of the males attempted copulation. When they were 36 hours old, all of the males vibrated their wings; when they were 48 hours old, all of the males licked the females' genitalia. However, not until they were five days old did almost all of the males attempt to copulate during a 10-minute observation period (Ford et al., 1989).

With regard to their sex appeal, males that have recently eclosed from their pupal cases elicit as much courtship from older, sexually mature males as virgin females do (Jallon and Hotta, 1979; McRobert and Tompkins, 1983). Males from the aforementioned Canton-S wild-type strain begin to lose their sex appeal when they are ca. 3–4 hours old. After that time, the males' sex appeal gradually diminishes until they are 2–3 days old, at which time they are maximally unattractive (Curcillo and Tompkins, 1987).

Why do immature males elicit much more courtship than mature males? Hydrocarbons extracted from immature males' cuticles stimulate a ten-fold increase in the courtship that two mature males perform in response to each other (Tompkins et al., 1980). By fractionating hydrocarbon extracts, then testing the fractions to see whether they stimulate two mature males to court each other, (Z)-11-tritriacontene and (Z)-13-tritriacontene, which are cuticular hydrocarbons, have been identified as courtship-stimulating pheromones (Schaner et al., 1989). However, unlike the sex pheromones that mature males synthesize, (Z)-11-tritriacontene and (Z)-13-tritriacontene are neither male-specific nor male-predominant (Antony and Jallon, 1981; see below).

As males become sexually mature, they stop synthesizing (Z)-11-tritriacontene and (Z)-13-tritriacontene and begin to synthesize (Z)-7-tricosene and octadecenyl acetate (Jallon, 1984). With regard to the times at which these changes in pheromone synthesis occur, mutant *olfactory-C* males, which respond abnormally to chemical stimuli, perform as much courtship in response to one-to nine-hour-old wild-type males as control males do. However, the mutant males perform significantly more courtship, compared to controls, in response to males that are at least ten hours old (Tompkins and Hall, 1981; Curcillo and Tompkins, 1987). Since *olfactory-C* males respond abnormally to acetates (Rodrigues and Siddiqi, 1978) and to (Z, Z)-7, 11-heptacosadiene (Tompkins et al., 1980), which is structurally similar to (Z)-7-tricosene, this observation suggests that immature males begin to synthesize quantities of octadecenyl acetate, (Z)-7-tricosene, or both inhibitory pheromones that normal, sexually mature males can detect when the immature males are ten hours old.

B. Female sex appeal and receptivity to copulation

Newly eclosed females, like immature males, synthesize (Z)-11-tritriacontene and (Z)-13-tritriacontene and thus elicit vigorous courtship. During the first few days of adult life, females stop making (Z)-11-tritriacontene and (Z)-13-tritriacontene and begin to make (Z, Z)-7, 11-heptacosadiene (Antony and Jallon, 1981). Thus, as they become sexually mature, virgin females retain their sex appeal, but the females begin to synthesize a sex-specific courtship-stimulating pheromone.

Young females also differ from sexually mature females in that immature females do not mate, despite the fact that sexually mature males perform vigorous courtship and repeatedly attempt to copulate with recently eclosed females. Females do not become receptive to copulation until they 18–24 hours old (Manning, 1967), after which time virgin females can be stimulated to open their vaginal plates and thus permit intromission to occur in response to courting males' attempts to copulate.

IV. Cells that function differently in mature males and females

As described in section II, sexually mature males and virgin females differ with regard to at least five attributes that are related to their sexual behavior. Males perform courtship in response to sexually attractive flies, and they synthesize a male-specific pheromone and a male-predominant pheromone that inhibit the courtship that their movement would otherwise stimulate. Females, conversely, synthesize a courtship-stimulating pheromone, and they are receptive to copulation with courting males. In addition, normal males can distinguish virgin females from sexually mature males. However, since females do not perform courtship, it is impossible to determine, by observing wild-type flies, whether the ability of courting males to distinguish mature males from females is a male-specific characteristic.

Why do sexually mature males and females perform and elicit different behaviors? The fact that they do implies that there are cells that function differently in the two sexes. Identification of the cells that must be of male or female genotype for a fly to perform a sex-specific behavior (the *foci* for that behavior) has been facilitated by the fact that XO flies, which have a single X chromosome and no Y chromosome, behave like normal (XY) males (Hall, 1977). Thus, it is possible to identify the focus for a sex-specific behavior by correlating the behavior of gynandromorphs (sex mosaics), each of which has a unique distribution of XX (female) and XO (male) tissues, with the presence of female or male cells in a specific part of the fly (Hotta and Benzer, 1972). Whether it is the *presence* of cells of one sex or the *absence* of cells of the opposite sex that is required in a focus for performance of the behavior cannot be determined by analyzing normal gynandromorphs. However, analysis of other types of mosaics and observations of mutant flies have provided information in this regard for some of the behavioral foci.

It is theoretically possible to analyze gynandromorphs to identify foci for male- and female-specific aspects of pheromone synthesis, since octadecenyl acetate, (Z)-7-tricosene, and (Z, Z)-7, 11-heptacosadiene can be accurately measured in single flies (see Tompkins and McRobert, 1995). Although this has not yet been done, one can deduce where the foci for sex-specific aspects of pheromone synthesis must be by analyzing sex mosaics' ability to elicit vigorous courtship from mature males since, as discussed in section II-B, a fly's sex appeal is determined by the sex pheromones that it synthesizes.

A. Foci for courtship

1. The "early behaviors"

Analysis of gynandromorphs in which the genotype of neuronal cell bodies can be ascertained has revealed that a bilaterally symmetrical cluster of neurons in the dorsal posterior brain must be male on at least one side of the brain for a mosaic to perform what will henceforth be called the early courtship behaviors—following, tapping, and wing extension—in response to a virgin female (Hall, 1977). In this analysis, orientation was not scored as a separate behavior; rather, "following" includes the two behaviors that are defined in this review as following and orientation. With regard to whether male neurons or the absence of female neurons is required in one of the foci for a fly to perform courtship, all-male mosaics in which the dorsal posterior brain is structurally abnormal because neurons in that part of the brain do not synthesize acetylcholinesterase do perform the early courtship behaviors (Greenspan et al., 1980). This observation suggests that it is not the presence of normal male neurons in one of the bilaterally symmetrical foci that determines whether a fly can perform courtship. Rather, the presence of normal female neurons in both of the bilateral foci prevents a fly from developing or functioning in such a way that it can perform the early courtship behaviors.

What do the neurons that constitute the early courtship behavior foci do? The foci may include part of the mushroom bodies (Hall, 1979), the neuropil of which consists of two bilaterally symmetrical bundles of fibers at which input from olfactory, taste, visual, and acoustic receptors converges (Strausfeld, 1976; Heisenberg et al., 1985). Since males court females in response to visual and chemical stimuli, it is logical to assume that the mushroom bodies would constitute part of the neural circuitry that mediates courtship; however, they could be a non-sex-specific component of the circuitry. The mushroom bodies are sexually dimorphic, in that there are more fibers in the mushroom bodies of females (Technau, 1984), which is consistent with the idea that the mushroom bodies are the foci for the early courtship behaviors. If this is the case, the fact that mutant males that lack mushroom bodies court females (e.g., de Belle and Hesenberg, 1994) provides additional support for the hypothesis that normal males initiate courtship in response to virgin females because they lack female neurons in the bilateral foci for the early courtship behaviors.

2. Wing vibration, licking, and attempted copulation

Since some sex mosaics with male neurons in the early courtship foci are incapable of singing, licking, and attempting copulation, it is obvious that there is at least one additional sex-specific focus, distinct from the early courtship behavior foci, for these behaviors. With regard to the question of where the foci for the "advanced" behaviors are? Analysis of gynandromorphs has revealed that a fly cannot vibrate its wings to produce a courtship song unless it has male neurons in one or both of the foci for the early courtship behaviors and, in addition, male neurons in at least one side of the mesothoracic ganglia. For licking, a fly must have male neurons on both sides of the dorsal posterior brain (it is not clear whether the licking foci are congruent with the early courtship foci) and male neurons in the focus for wing vibration. To attempt copulation, a fly must have male neurons in the foci for the early behaviors, wing vibration, and licking, as well as male tissue in the thoracic ganglia. Unlike the foci for the other courtship behaviors, which are in discrete parts of the central nervous system, the focus for attempted copulation is *diffuse*, i.e., the probability that a gynandromorph will attempt to copulate with a female is determined by the amount of male tissue in the sex mosaic's thoracic ganglia, rather than the presence of male cells in a specific part of the ganglia (Hall, 1977, 1979; Schilcher and Hall, 1979).

3. An overview of the courtship foci

What has identification of the various courtship foci, shown schematically in Figure 2, contributed to our understanding of the neural basis for normal males' ability to perform the courtship behaviors and normal females' inability to perform them? First, it is noteworthy that all of the foci are in the central nervous system. It is possible that the photoreceptors that are stimulated by the sight of a moving female and/or the chemoreceptors that respond to her courtship-stimulating pheromone are required to be male for a fly to perform courtship optimally (see section V-B). However, it is obvious from the gynandromorph analysis that none of the sensory receptors that are stimulated

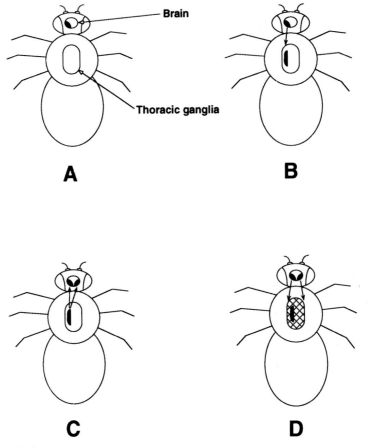

Fig. 2. A diagrammatic representation of the neural foci for the courtship behaviors. Discrete foci are shaded; the diffuse focus is cross-hatched. Arrows represent the order in which the foci function when a male initiates courtship (see text). (A), the early courtship behaviors (i.e., orientation, following, tapping, and wing extension); (B), wing vibration; (C), licking; (D), attempted copulation. Only one of the foci for the early courtship behaviors and one of the two foci for wing vibration are shown, since only one of the two bilateral symmetrical foci for these behaviors has to be male for a fly to perform the behaviors.

when a male performs courtship are required to be of male genotype for the performance of courtship *per se.*

The courtship foci are also of interest because they function sequentially, in the sense that a sex mosaic cannot vibrate its wings, lick, or attempt copulation unless it has male neurons in the foci for the previous behaviors, plus male neurons in another part of the central nervous system. Thus, in gynandromorphs, performance of the early behaviors is a prerequisite for wing vibration; performance of the early behaviors and wing vibration are prerequisites for licking; and performance of the early behaviors, wing vibration and licking are prerequisites for attempted copulation. This behavioral sequence is also displayed by normal males, which when initiating courtship in response to a virgin female always perform at least one of the early behaviors before they vibrate their wings, always sing before licking, and always lick before attempting copulation.

A model for the manner in which the sex-specific circuitry underlying normal males' courtship of virgin females functions is suggested by the foregoing observations. For two foci that control sequential behaviors, the essential features of this model are as follows: (1) if neural activity in the focus for the first behavior reaches a threshold, the fly performs the first behavior; and (2) if neural activity in the focus for the first behavior reaches a higher threshold, neural activity occurs in the focus for the second behavior, and the fly performs the second behavior. The fact that males with low courtship indices rarely perform the advanced behaviors (see Tompkins and McRobert, 1995) can be explained by evoking the dual threshold feature of the model to postulate that the foci for the early behaviors are stimulated to the extent that the fly is able to perform those behaviors, but not to the extent that the foci for subsequent behaviors release the motor outputs for the behaviors that they control. In addition, the model provides a framework for formulating hypotheses about the ontogeny of courtship. Specifically, males may become competent to perform the various courtship behaviors at different times after eclosion because the foci for the early courtship behaviors, wing vibration, licking, and attempted copulation become functional at progressively later times. Alternatively, all of the foci may become functional simultaneously, but males may spend more time performing the early courtship behaviors as they become sexually mature, which would increase the probability that the thresholds for activation of the foci for the advanced behaviors would be attained.

B. Foci for sex pheromone synthesis

In other insects, waxy cuticular hydrocarbons like (Z)-7-tricosene and (Z, Z)-7, 11-heptacosadiene are synthesized in the underlying epidermis (see Blomquist et al., 1993). If this is true for *D. melanogaster,* and the genotype of the epidermal cells themselves determines their pattern of pheromone synthesis, the foci for synthesis of large quantities of (Z)-7-tricosene in males and synthesis of (Z, Z)-7, 11-heptacosadiene in females are predicted to be in the epidermis of the abdomen, since mature males direct their courtship behaviors toward females' abdomens and, conversely, usually stop courting other mature males after they touch the potential sex-objects' abdomens. Octadecenyl acetate, in contrast, is synthesized in the ejaculatory bulb, which is a component of the male reproductive tract. Thus, the focus for synthesis of octadecenyl acetate in males is predicted to be the ejaculatory bulb itself. As noted above, foci for specific aspects of sex pheromone synthesis have not yet been identified. However, analysis of sexually mature gynandromorphs to identify tissues that must be female for a fly to elicit vigorous courtship from a mature male has revealed that sex appeal is correlated with female tissue in the abdomen (Jallon and Hotta, 1979; Tompkins and Hall, 1983). Thus, the focus for sex appeal is consistent with the predicted foci for sex-specific synthesis of the three pheromones that determine whether a sexually mature fly will elicit vigorous courtship from a male.

C.   Foci for receptivity to copulation

As discussed in section II-C, virgin females are stimulated by auditory and chemical cues that are associated with males' courtship to be receptive to copulation. Is the neural circuitry that mediates females' receptivity sex-specific? Analysis of gynandromorphs with normal female genitalia to which males responded by performing vigorous courtship and repeatedly attempting copulation revealed that the receptors that are stimulated when a sexually attractive fly elicits vigorous courtship need not be of female genotype for a sex mosaic to copulate with a male. However, a bilaterally symmetrical cluster of neurons in the dorsal anterior brain is required to be female on both sides of the brain for a sex mosaic to be receptive to copulation. Like the early courtship behavior foci in males, the female receptivity foci may include part of the mushroom bodies (Tompkins and Hall, 1983), although the female receptivity foci are not congruent with the foci for the early courtship behaviors in males (cf. Hall, 1977, 1979; Tompkins and Hall, 1983). As is the case for the foci for the early courtship behaviors, it is reasonable to assume that the mushroom bodies constitute part of the neural circuitry that functions when a female is stimulated by a male's courtship, since they integrate input from receptors that respond to chemical and acoustic stimuli. However, the mushroom bodies need not constitute a sex-specific component of the circuitry.

D.   Foci for avoidance of homosexual courtship

With regard to the neural circuitry underlying a courting male's ability to distinguish between sexually mature males and virgin females, it is logical to assume that a mature male that was unable to perceive or respond to the inhibitory pheromones that mature males synthesize would perform vigorous courtship in response to another mature male. Thus, if the sensory receptors that are stimulated by mature males' inhibitory pheromones or the interneurons at which input from those receptors converges functioned differently in males and females, males in which those receptors or interneurons were genetically transformed so that they functioned as they did in females might be expected to perform vigorous homosexual courtship. Evidence suggesting that this is the case was recently obtained by Ferveur et al. (1995), who identified three enhancer trap lines, males from which courted females and also performed vigorous courtship in response to normal, sexually mature males. In the males that performed homosexual courtship, some of the neurons in the antennal lobes, mushroom bodies, or both parts of the brain expressed the female-specific *transformer* gene product and were thus genetically female (see section V-B). Since axons from olfactory receptors on the antenna and maxillary palps project to the antennal lobes, whose input ultimately converges in the mushroom bodies, the observations of Ferveur and his coworkers suggest that there are foci in the olfactory centers of the brain that are required to be male to prevent a fly from developing or functioning in such a way that it will perform vigorous courtship in response to sexually mature males.

The question of whether it is the presence of male neurons or the absence of female neurons in the foci that, in normal males, suppress homosexual courtship is of particular interest, since the foci may be congruent with the foci for females' receptivity to copulation (Ferveur et al., 1995). Thus, neurons in the olfactory centers of the brain may mediate positive responses to mature males' pheromones if they are of female genotype and negative responses to mature males' pheromones if they are of male genotype. Alternatively, the foci for males' avoidance of homosexual courtship may not be congruent with the foci for females' receptivity to copulation. If this is the case, the neurons in the foci for avoidance of homosexual courtship may mediate negative responses to mature males' pheromones if they are of male genotype but be nonfunctional if they are of female genotype. It should be possible to determine whether it is the presence of male neurons or the absence of female neurons that suppresses normal males' homosexual courtship of mature males by observing mutant males in which the neurons that constitute the focus for avoidance of homosexual courtship are missing or abnormal (see Heisenberg et al., 1985). If males from Ferveur et al.'s enhancer trap lines perform homsosexual courtship because the neurons in the foci for avoidance of homosexual courtship are not male and are thus nonfunctional, mutant males in which these neurons are missing or abnormal should court mature males, assuming that the mutant males are capable of performing courtship. If, on the other hand, the males from the enhancer trap lines perform homosexual courtship because neurons in the foci for avoidance of homosexual courtship are genetically female and thus mediate an acceptance response to the pheromones associated with mature males, mutant males in which these neurons are missing or abnormal will not perform vigorous homosexual courtship in response to mature males.

V. Genes that confer sex-specificity on the behavioral and pheromonal foci

A. Sex-lethal

On a genetic level, why do the sex-specific foci function differently in males and females? Since sex is determined by the ratio of X chromosomes to autosomes in *D. melanogaster*, attention has been focused on the *Sex-lethal* gene, since its activity is directly regulated by the X:autosome ratio. Specifically, *Sex-lethal* transcripts are differentially spliced in response to signals from genes that are on the X chromosome(s) and the autosomes, resulting in the synthesis of full-length *Sex-lethal* proteins in female cells and short *Sex-lethal* peptides in male cells. Body size and differentiation of the sex-specific cuticular derivatives (the external genitalia, the analia, the sex combs on males' forelegs, and the pigmentation of the abdomen) are regulated by the female-specific *Sex-lethal* proteins, in that a fly will be female-sized and have the cuticular derivatives that are characteristic of females if it synthesizes the female-specific *Sex-lethal* proteins (Figure 3). The male-specific *Sex-lethal* peptides, in contrast, are not required for a fly to be male-sized or have the cuticular derivatives that are characteristic of males (see Salz et al., 1989).

Observations of chromosomal females expressing partial loss-of-function *Sex-lethal* alleles, chromosomal males expressing partial gain-of-function alleles, and chromosomal males with deletions of the locus revealed that the female-specific *Sex-lethal* proteins are required for synthesis of (Z, Z)-7, 11-heptacosadiene. In addition, expression of the female-specific *Sex-lethal* proteins renders flies incapable of courting virgin females and also inhibits synthesis of octadecenyl acetate and male-specific quantities of (Z)-7-tricosene. The *Sex-lethal* peptides that normal males make, in contrast, are not required for males to perform any of the courtship behaviors in response to virgin females, for synthesis of octadecenyl acetate or large quantities of (Z)-7-tricosene, or to suppress synthesis of (Z, Z)-7, 11-heptacosadiene (Tompkins and McRobert, 1989; 1995). Clonal analysis of the *Sex-lethal* gene has revealed that it must be expressed endogenously in the cells that develop differently in the two sexes to regulate the sex-specific differentiation of those cells (see Sanchez and Nothiger, 1982). Accordingly, the foci for courtship behavior and the foci for male- and female-specific aspects of sex pheromone synthesis can be defined as cells which, if they express the female-specific *Sex-lethal* proteins, develop and function in such a way that the fly is incapable of performing courtship and synthesizes the sex pheromones that are characteristic of females (Figure 3).

CUTICULAR DERIVATIVES

Female-specific *Sex-lethal* proteins

↓

Female-specific *transformer* protein   +   Non-sex-specific *transformer-2* protein

↓

Female-specific *doublesex* protein: NO male-specific *doublesex* protein

↓   Female-specific *intersex* gene product

↓

Tissue-specific target genes expressed in female-specific mode

COURTSHIP

Female-specific *Sex-lethal* proteins

↓

Female-specific *transformer* protein   +   Non-sex-specific *transformer-2* protein

↓

Tissue-specific target genes expressed in female-specific mode

SEX APPEAL

Female-specific *Sex-lethal* proteins

↓

Female-specific *transformer* protein   +   Non-sex-specific *transformer-2* protein

↓

Female-specific *doublesex* protein: NO male-specific *doublesex* protein

↓

Tissue-specific target genes expressed in female-specific mode

Fig. 3. Gene products that confer female specificity on the sex-specific cuticular derivatives; on the foci for the courtship behaviors, thus rendering the fly incapable of performing courtship; and on the foci for female sex appeal, thus rendering the fly capable of eliciting vigorous courtship. The *Sex-lethal* proteins regulate synthesis of all three of the pheromones that are relevant to mature flies' sex appeal. The presence of the female-specific *doublesex* proteins and/or the absence of the male-specific *doublesex* proteins regulate synthesis of (Z, Z)-7, 11-heptacosadiene in females but are not required to repress synthesis of octadecenyl acetate; it is not yet known whether they repress synthesis of male-specific quantities of (Z)-7-tricosene.

With regard to the focus for avoidance of homosexual courtship, chromosomal males in which the entire *Sex-lethal* gene has been deleted do not court normal, sexually mature males (Tompkins and McRobert, 1989). Thus, the male-specific *Sex-lethal* peptides are not required for the cells in this focus to develop in such a way that performance of vigorous homosexual courtship in response to mature males is suppressed.

B. The transformer, transformer-2, and doublesex

The question of whether the "downstream" genes whose expression is controlled by the female-specific *Sex-lethal* proteins regulate the manner in which the behavioral and pheromonal foci develop or function is of interest, since analysis of other traits has revealed that the downstream genes do not regulate all aspects of sex-specific development in the same way. Specifically, null mutations in the *transformer* or *transformer-2* genes transform chromosomal females into flies that look like males, while null mutations in the *intersex* or *doublesex* genes transform chromosomal females into intersexes whose cuticular derivatives express the characteristics of both sexes. Conversely, null mutations in the *transformer, transformer-2* or *intersex* genes have no effect on the differentiation of cuticular derivatives in chromosomal males, although null mutations in the *doublesex* gene do transform chromosomal male flies into intersexes. These observations, in conjunction with molecular analyses of the *transformer, transformer-2,* and *doublesex* genes, have led to the development of models for the differentiation of sex-specific cuticular derivatives in which the absence of the male-specific *doublesex* protein, the presence of the *transformer-2* and *intersex* gene products, and the presence of the female-specific *transformer* and *doublesex* proteins are required for normal female development (Figure 3). For normal male development, in contrast, only the presence of the male-specific *doublesex* protein and the absence of the female-specific *doublesex* protein are required (see Baker and Ridge, 1980; Baker and Wolfner, 1988; Baker, 1989). However, chromosomal females that express loss-of-function mutations in the *transformer, transformer-2, intersex,* or *doublesex* genes are female-sized flies (Lindsley and Zimm, 1992), and differentiation of the muscle of Lawrence, a male-specific abdominal muscle, is not affected by null mutations in the *doublesex* or *intersex* genes (Taylor, 1992). Thus, the *transformer, transformer-2, doublesex,* and *intersex* genes do not regulate body size or the development of the muscle of Lawrence in the same way as they regulate the differentiation of the sex-specific cuticular structures.

1. The transformer and transformer-2

Observations of chromosomal males expressing null mutations in the *transformer* gene to see whether the mutant flies courted females and/or elicited courtship from males revealed that the mutations have no behavioral effects on chromosomal males. In contrast, chromosomal females that express the *transformer* mutation are like normal males, in that they court virgin females, elicit very little homosexual courtship, synthesize octadecenyl acetate, and do not synthesize (Z, Z)-7, 11-heptacosadiene (Hall, 1978; Baker and Ridge, 1980; McRobert and Tompkins, 1985; Jallon et al., 1986). Similarly, chromosomal females that are homozygous for null mutations in the *transformer-2* gene court females vigorously and elicit relatively little courtship from males (Baker and Ridge, 1980). Chromosomal males that express null mutations in the *transformer-2* gene have not been observed to see whether they behave like normal males. However, the behavior of chromosomal females that express *transformer* and *transformer-2* mutations and chromosomal males that express *transformer* mutations is consistent with the hypothesis that these genes regulate the development of the foci for courtship and sex pheromone synthesis in the same way that they regulate the development of the sex-specific cuticular derivatives and the muscle of Lawrence (Figure 3).

2. The intersex

In contrast to *transformer* and *transformer-2*, mutations in the *intersex* gene do not affect the sex-specific behavioral and pheromonal foci in either sex (Figure 3). Chromosomal females that express null mutations in the *intersex* gene are like wild-type females, in that they do not perform any courtship in response to sexually attractive flies. Conversely, chromosomal males expressing null mutations in the *intersex* gene perform vigorous courtship (McRobert and Tompkins, 1985; Taylor et al., 1994). Thus, the *intersex* gene does not regulate the manner in which the courtship foci develop or function in males or females.

With regard to the effects of *intersex* mutations on sex pheromone synthesis, chromosomal females flies that express a null mutation in the *intersex* gene elicit vigorous courtship from males (McRobert and Tompkins, 1985). Moreover, although McRobert and Tompkins (1985) reported that chromosomal males that were homozygous for the ix^1 mutation elicited more courtship than male controls, other investigators have observed that chromosomal males that are homozygous for the ix^1 mutation or other null alleles and those that are heterozygous in *trans* for combinations of null alleles elicit very little courtship (Taylor et al., 1994). These observations suggest that genetic factors other than the *intersex* mutations themselves caused the ix^1/ix^1 males that McRobert and Tompkins observed to elicit courtship from normal males. Accordingly, it is unlikely that the *intersex* gene regulates any aspect of sex pheromone synthesis in males or females.

3. The doublesex

Like *intersex*, mutations in the *doublesex* gene do not affect flies' behavior as one would predict, based on their effects on the differentiation of sex-specific cuticular derivatives. With regard to their sex appeal, chromosomal males and females that are homozygous for null mutations in the *doublesex* gene elicit less courtship than normal females, but they are more attractive than normal males (McRobert and Tompkins, 1985; Jallon et al., 1988). The mutant males and females do not synthesize significant quantities of (Z, Z)-7, 11-heptacosadiene. Moreover, many chromosomal males that express null mutations in the *doublesex* gene synthesize relatively little octadecenyl acetate (Jallon et al., 1988). Thus, at least one aspect of sex pheromone synthesis in males and one aspect of sex pheromone synthesis in females are regulated by the *doublesex* gene. However, chromosomal females that are homozygous or heterozygous in *trans* for null mutations in the *doublesex* gene and mutant females that express the male-specific *doublesex* protein, rather than the female-specific *doublesex* protein, do not perform any courtship in response to virgin females (McRobert and Tompkins, 1985; Taylor et al., 1994). Thus, differentiation of the courtship behavior foci in females is not effected by the presence of the female-specific *doublesex* protein or the absence of the male-specific *doublesex* protein (Figure 3).

The role of the male-specific *doublesex* proteins, if any, in mediating males' ability to perform courtship is controversial. Some chromosomal males that are homozygous for null mutations in the *doublesex* gene perform vigorous courtship, while others have lower courtship indices or do not court virgin females at all (McRobert and Tompkins, 1985; Taylor et al., 1994). Chromosomal males that are heterozygous in *trans* for different *doublesex* null mutations have higher courtship indices than chromosomal males that are homozygous for a single mutation, which implies that genetic factors other than the *doublesex* mutations play a role in depressing the courtship of males from the *doublesex* stocks. However, the *trans* heterozygotes' courtship is not as vigorous nor as consistent as that of control males. Taylor et al. (1994) have argued that the fact that some chromosomal males that express null *doublesex* mutations perform normal courtship, in conjunction with the observation that chromosomal males that express null *doublesex* mutations sing normally, implies that the male-specific *doublesex* protein is not required for a male to perform courtship. However, since many chromosomal males with *doublesex* mutations, even those that are heterozygous in *trans* for different mutant alleles, do not perform normal courtship, the male-specific *doublesex* protein may be required to "fine-tune" the nervous system so that it functions optimally when a male is stimulated to perform courtship.

In what cells might the male-specific *doublesex* protein be required for a male to perform consistently vigorous courtship? Cells in the early courtship behavior foci may not function optimally in males if they do not express the male-specific *doublesex* protein. Alternatively, receptors or interneurons in courting males' visual and taste systems that are stimulated by visual and chemical cues associated with virgin females may be more responsive or more consistently responsive to those stimuli if the neurons express the male-specific *doublesex* protein. The latter possibility is suggested by observations of mutant males that cannot see or are not stimulated to court by (Z, Z)-7, 11-heptacosadiene (i.e., *glass, no-receptor-potential A,* and *para^{sbl}* males), which are like males with *doublesex* mutations in that their courtship indices are lower and more variable than those of controls (Tompkins, 1984). The neurons in which the male-specific *doublesex* protein is required for a male to perform vigorous, consistent courtship could be identified by generating enhancer trap lines, males from which express the male-specific *doublesex* protein in some neurons but are otherwise *doublesex*-null. If males from lines that express the *doublesex* protein in their visual or taste systems, in the early courtship behavior foci, or elsewhere have consistently high courtship indices, this would suggest that the male-specific *doublesex* protein is required in the cells in which it is expressed for normal males to perform optimal courtship.

VI.   Genes required for normal sexual behavior

Many mutations, in addition to those that have been described in previous sections, affect flies' sexual behavior (reviewed in Hall, 1994). Mutations that do not have overt pleiotropic effects are of particular interest, since they may define genes whose expression is regulated by the sex determination genes—the "target" genes, whose expression is tissue-specific, that mediate behavioral and pheromonal aspects of sexual differentiation (see Burtis and Wolfner, 1992). Alternatively, mutations that specifically affect sexual behavior may define genes that are expressed in the same way in both sexes, but are of interest nonetheless because they affect the manner in which the sex-specific foci function.

A. Fruitless

The most extensively analyzed mutations in the "relatively specific" category are those that are associated with *In(3)fru,* a chromosome inversion that, when homozygous, renders mature males behaviorally sterile because they do not attempt to copulate with virgin females (Hall, 1978). Moreover, *In(3)fru* males court normal, sexually mature males vigorously, elicit vigorous homosexual courtship when they are sexually mature, and synthesize hydrocarbons that stimulate mature males to court each other (Hall, 1978; Tompkins et al., 1980). Genetic analysis has revealed that the proximal breakpoint of the inversion is associated with its effects on the mutant flies' sex appeal and, presumably, synthesis of the courtship-stimulating hydrocarbons, while the distal breakpoint is associated with the aberrant courtship that *In(3)fru* males perform in response to virgin females and mature males. The loci that are affected by the proximal and distal breakpoints of the inversion have been designated as *fruitless-stimulation* and *fruitless,* respectively (Gailey and Hall, 1989).

With regard to the effects of *fruitless-stimulation* on sex appeal, it was possible that the proximal breakpoint of the *In(3)fru* inversion transformed some or all of the foci for pheromone synthesis such that they functioned as they do in females. Alternatively, the *fruitless-stimulation* breakpoint could prevent sexual maturation, causing older males to synthesize the courtship-stimulating pheromones that are characteristic of immature males. However, analysis of hydrocarbons extracted from the cuticles of mature *In(3)fru* males revealed that the mutant males do not synthesize (Z, Z)-7, 11-heptacosadiene or the courtship-stimulating pheromones that immature males make. Rather, the mutant males synthesize relatively small quantities of (Z)-7-tricosene and relatively large quantities of another cuticular hydrocarbon, (Z)-7-pentacosene, in comparison to Canton-S wild-type males (Jallon, 1984; Gailey and Hall, 1989). The cuticular hydrocarbon profile of *Df(3)fru* males is characteristic of mature males from *D. melanogaster* strains that are descended from West African populations (see Scott and Richmond, 1988). Although the significance of this hydrocarbon profile in relation to the flies' sex appeal is unclear (see Tompkins, 1989), the fact that mutant *Df(3)fru* males' pattern of cuticular hydrocarbon synthesis is like that of mature males from some wild-type strains suggests that the *fruitless-stimulation* breakpoint does not transform males' pheromone-producing cells so that they function as they do in females, nor does it arrest males' development.

Strictly speaking, the *fruitless* gene does not have a specific effect on flies' sexual behavior, since its product is required in males for differentiation of the muscle of Lawrence (Gailey et al., 1991). However, *fruitless* is nonetheless interesting with regard to its effects on the sex-specific foci, since mutations in this gene have adverse effects on several aspects of male sexual behavior. Specifically, the courtship songs of males that are homozygous for the *fru¹* allele, which is associated with the distal breakpoint of *Df(3)fru*, are abnormal (Wheeler et al., 1989). In addition, *fru¹*, as well as some of the other mutant *fruitless* alleles that have subsequently been isolated, are like *Df(3)fru* in that they render males incapable of attempting copulation (Taylor et al., 1994). These observations suggest that the mutant *fru* alleles may partially transform the neurons in one or more of the male courtship foci so that they function more like they do in females. Moreover, since mature males expressing all of the mutant *fruitless* alleles perform vigorous courtship in response to normal, sexually mature males (Taylor et al., 1994), the *fruitless* alleles may transform the foci for avoidance of homosexual courtship such that the foci do not function, as they do in wild-type males, to suppress courtship of mature males.

With regard to the question of whether *fruitless* mutations affect both sexes, mutant females that are homozygous for the *Df(3)fru* inversion elicit vigorous courtship from wild-type males and have normal copulation latencies (Hall, 1978). Thus, the *fruitless* gene does not appear to regulate the manner in which the sex-specific behavioral and pheromonal foci differentiate or function in female flies.

B. Other genes that affect male sexual behavior

The *he's not interested* mutation is of interest because males that express it perform little or no courtship in response to normal, sexually mature virgin females or males. Mutant *he's not interested* females elicit vigorous courtship from wild-type males and have normal copulation latencies, and the *he's·not interested* mutation has no overt effects on males' or females' non-sexual behaviors (R. Friedman, M. Harvey, and L. Tompkins, manuscript in preparation). Thus, the *he's not interested* mutation appears to have a specific effect on males' courtship, although it is not yet known whether the gene is expressed in the courtship foci or in other parts of the fly.

Mutations in two other genes also have relatively specific effects on males' courtship, although these mutations differ from *he's not interested* in that males that express them initiate courtship of females normally. Males expressing the *croaker* mutation sing and fly abnormally; in addition, *croaker* males' copulation freqencies are very low (Yamamoto et al., 1990). The aforementioned *raised* and *nonA*^{diss} mutations, which affect males' courtship songs, do not have such drastic effects on mutant males' copulation frequencies, which suggests that the *croaker* mutation affects other aspects of males' courtship in addition to their courtship songs or, alternatively, that *croaker* males' songs are uniquely repellent to virgin females. Males expressing the *no-bridge*^{KS49} mutation, in which the protocerebral bridge that separates the two mushroom bodies is disrupted, also sing abnormal songs. Moreover, although the mutant males do lick females, they rarely attempt to copulate (Bouhouche et al., 1993). In the *no-bridge* mutant, the protocerebral bridge defect may impede the flow of information from the licking foci in the brain to the attempted copulation foci in the thoracic ganglia, although it is not clear how such an anatomical defect could be responsible for the mutants' abnormal courtship song. Identification of the primary defects in the *croaker* and *no-bridge* mutants is thus of interest.

With regard to the foci for sex-specific aspects of pheromone synthesis, males that express the *nerd* mutation synthesize less (Z)-7-tricosene than controls. Moreover, males from *nerd* stocks have lower copulation frequencies than controls, and they attempt copulation less often. Females from *nerd* stocks also have abnormally low copulation frequencies; moreover, the effects of the mutations in the *nerd* stock on males and females' copulation latencies are additive, in that males from the *nerd* stock rarely mate with females from the *nerd* stock (Ferveur and Jallon, 1993). Although the effects of the *nerd* mutation on pheromone synthesis have been mapped to the third chromosome, it is not known whether the abnormal behaviors that males and females from the *nerd* stock exhibit are caused by the same mutation that affects pheromone synthesis. If they are, *nerd* gene function may be required in the female receptivity foci and, in males, in the foci for attempted copulation and synthesis of male-specific levels of (Z)-7-tricosene. Alternatively, the *nerd* mutation may exert its primary effect on the foci for synthesis of (Z)-7-tricosene, normal levels of which may be required for a male to attempt copulation quickly and for a female to be readily stimulated by a male's courtship to be receptive to copulation.

Finally, no mutant allele that has a specific effect on the foci for avoidance of homosexual courtship has yet been identified. However, males in which the wild-type allele of the *white* gene is expressed in all cells rather than in the compound eyes, ocelli, and testis sheaths, as it is in normal adult flies (Lindsley and Zimm, 1992), perform vigorous homosexual courtship in response to normal, sexually mature males. Males that express the *white* gene ectopically do elicit courtship from normal males; however, it is probably the stimulatory effects of the abnormal males' wing vibrations, rather than synthesis of a courtship-stimulating sex pheromone, that is responsible for their sex appeal (Zhang and Odenwald, 1995). With regard to the question of how ectopic expression of a gene that affects pigmentation could effect homosexual behavior, the *white* gene product has been identified as a tryptophan-guanine transporter. Accordingly, since tryptophan is a precursor for monoamine neurotransmitters, Zhang and Odenwald have suggested that expression of the *white* gene in the central nervous system may affect synthesis of serotonin and dopamine, perturbations of which are associated with homosexual behavior in vertebrates. Regardless of whether this hypothesis is correct, it is obviously of interest to determine whether the cells in which ectopic expression of the wild-type *white* allele is required for homosexual courtship are neurons in the sex-specific foci that mediate normal males' avoidance of homosexual courtship (see section IIID).

VII.   The development of the sex-specific foci

A. Developmental Stages when Sex-specificity is Conferred

When do the mutations that confer sex-specificity on the behavioral and pheromonal foci and those that are required for normal sexual behavior exert their effects on the foci? Is the wild-type gene product required when a sex-specific focus is differentiating; to maintain its sex specificity between the time that it develops and the time, shortly after eclosion, that it begins to function; or during the entire period in the life of the adult fly during which it functions? In this regard, nothing is known about *transformer, doublesex,* or any of the putative target genes that are required to be non-mutant for normal sexual behavior. However, temperature shift analysis of conditional alleles of the *Sex-lethal* and *transformer-2* genes has provided information about the times when these components of the sex determination cascade must be expressed as they are in normal females for a fly to be incapable of performing courtship when it becomes sexually mature.

With regard to *Sex-lethal,* chromosomal females expressing a heteroallelic combination of partial loss-of-function mutations in the *Sex-lethal* gene are transformed into phenotypic males at all temperatures that are permissive for survival (22° through 29°C). However, the mutations have a temperature-sensitive effect on the flies' ability to court females. Specifically, mutant flies that are raised and maintained as adults at 22° or 25°C rarely court virgin females; in contrast, approximately half of the mutant flies that are raised and maintained at 29°C do perform courtship (Tompkins and McRobert, 1995). Shifting the flies from permissive temperatures to restrictive temperatures and *vice versa* between the major developmental stages (embryo, larva, pupa, adult) and after the first three days of adult life revealed that expression of the female-specific *Sex-lethal* proteins in the embryo, pupa, or the first three days of adult life is, in most flies, necessary and sufficient to prevent a fly from being capable of performing courtship when it becomes sexually mature (Tompkins, 1986).

With regard to *transformer-2,* the times at which gene expression is required in females for normal behavior have been identified by analyzing chromosomal females that are heterozygous in *trans* for a deletion of the *transformer* gene and a temperature-sensitive allele, *tra-2^{ts1}*. If they are raised at 16°C, the mutant flies are transformed into intersexes that, in most cases, are unable to perform courtship. If the mutant flies are raised and maintained at 29°C, in contrast, they are transformed into phenotypic males that do court females. Temperature-shift experiments like those conducted with mutant *Sex-lethal* flies have revealed that expression of the *transformer-2* gene during the pupal period and in adult life is required to prevent a fly from being capable of performing courtship (Belote and Baker, 1987).

The results of the temperature shift experiments with *Sex-lethal* and *transformer-2* differ in two respects: there is an embryonic temperature-sensitive period for *Sex-lethal* but not for *transformer-2;* and development of the capability to perform courtship is blocked by expression of the female-specific *Sex lethal* proteins during any of the temperature-sensitive periods, whereas the *transformer-2* gene must be expressed during both of the temperature-sensitive periods to prevent a fly from being able to perform courtship. It is possible that the sex determination cascade is bifurcated, in that *Sex-lethal* regulates an as yet unidentified "gene X," whose product does not interact with the *transformer-2* or *transformer* gene products. Thus, if gene X were expressed in the embryo or pupa as it normally is in females, the fly would differentiate in such a way that it would not be capable of performing courtship after eclosion, regardless of whether the *transformer-2* gene was expressed in a later stage of development. Alternatively, expression during embryogenesis or during the pupal period of the one copy of the mutant *tra-2^{ts1}* allele that is present in chromosomal females that are heterozygous for the temperature-sensitive allele and a deletion of the locus, even at permissive temperatures, may not be sufficient to repress development of the courtship foci.

Regardless of the discrepancies in the data obtained for *Sex-lethal* and *transformer-2*, it is noteworthy that both genes exert their effects on the courtship foci during the pupal period, before males are capable of performing courtship, and after eclosion. In the pupa, the larval central nervous system is reorganized to form its adult counterpart, which is augmented by the addition of new neurons that differentiate during the pupal period (see Truman, 1990; Hartenstein, 1993). Moreover, the mushroom bodies develop throughout the larval and pupal periods (de Belle and Heisenberg, 1994), but the number of fibers in the mushroom bodies continues to increase throughout the first week of adult life (Technau, 1984). Thus, it is not surprising that genes that regulate the development of parts of the central nervous system that function differently in sexually mature males and females exert their effects in pupa and adults, when the adult central nervous system is developing. Nor is it unexpected that there is a temperature-sensitive period during embryogenesis for the female-specific *Sex-lethal* proteins' effects on the courtship foci, since the larval central nervous system, from which the adult brain and thoracic ganglia develops, differentiates during embryogenesis (Hartenstein, 1993).

Finally, the times during development at which the sex determination genes exert their effects on sex-specific aspects of pheromone synthesis are not yet known. However, the epidermis of the adult fly and the male reproductive system, in which the two cuticular pheromones and octadecenyl acetate are synthesized, respectively, differentiate during the pupal period (Bodenstein, 1950). Thus, the pupal period is the earliest stage at which the sex determination genes could regulate the manner in which the foci for sex-specific aspects of pheromone synthesis differentiate.

B. Hormonal control of sexual maturation

As discussed in sections V and VI, development of the sex-specific behavioral and pheromonal foci is regulated by the sex determination genes and their as yet unidentified targets. In addition, hormonal factors mediate the development of the sex-specific foci, effecting their maturation so that the cells that comprise the foci function as they do in sexually mature males and females. Evidence for this is provided by analysis of flies that express mutations in the *apterous* gene, the phenotypic effects of which can be ameliorated by topical application of juvenile hormone analogs. Four-day-old males that are homozygous for the *apterous*[4] mutation synthesize (Z)-11-tritriacontene and (Z)-13-tritriacontene, the courtship-stimulating cuticular pheromones that are characteristic of immature males, and elicit vigorous courtship from normal, sexually mature males (Jallon and Hotta, 1979, Jallon et al., 1986). However, three-day-old males that express another *apterous* mutation do synthesize octadecenyl acetate (Wicker and Jallon, 1995), and two-day-old *apterous*[4] males perform all of the courtship behaviors (Tompkins, 1990). Thus, juvenile hormone is required in males to terminate synthesis of the non-sex-specific courtship-stimulating pheromones that are characteristic of immature flies, but it is probably not required to initiate synthesis of the sex-specific inhibitory pheromone that is characteristic of mature males nor for the maturation of the courtship behavior foci.

With regard to the *apterous* gene's effects on females, sexually mature virgin females that express the *apterous*[4] mutation have lower copulation frequencies than control females (Ringo et al., 1992). However, it is not clear whether this phenotype is a primary effect of the *apterous*[4] mutation on the female receptivity foci or a secondary effect of the debilitating effects of the mutation on female flies. However, three day old *apterous*[4] females synthesize only trace amounts of (Z, Z)-7, 11-heptacosadiene (Jallon et al., 1986), and mutant *ecdysoneless-1*[ts] females, shifted to non-permissive temperatures immediately after eclosion, synthesize 40% less (Z, Z)-7, 11-heptacosadiene than control females (Jallon et al., 1986). Thus, juvenile hormone and ecdysone are required in females to initiate synthesis of the sex-specific courtship-stimlating pheromone.

VIII Conclusion

What have the observations described in previous sections contributed to our understanding of why normal, sexually mature males and females behave differently? We know that courting males respond differently to males and females because sexually mature males and females synthesize different sex pheromones. Conversely, sex pheromones that mature males synthesize, in conjunction with auditory stimuli that are associated with males' courtship songs, stimulate virgin females to be receptive to copulation. Analysis of sex mosaics has revealed that males and females synthesize different sex pheromones and respond differently to the sex pheromones that other flies synthesize because a relatively small number of cells in the central nervous system, the epidermis, and the reproductive system differentiate and function differently in males and females. Some of the genes that confer sex specificity on these cells have been identified, as have genes that may be the targets of the genes that determine behavioral and pheromonal aspects of sex. At least two of the genes that determine the sex specificity of the courtship behavior foci in the central nervous system confer or maintain the sex-specificity of the foci during the pupal period, before males are capable of performing courtship, and after eclosion.

Why do the behaviors that males and females perform and elicit change as the flies become sexually mature? We know that immature males and females are attractive because they synthesize the same courtship-stimulating pheromones. As the flies become sexually mature, males begin to synthesize courtship-inhibiting pheromones and become competent to perform courtship, while females begin to synthesize a sex-specific courtship-stimulating pheromone and become competent to be stimulated by males' courtship to be receptive to copulation. In females, two hormones effect the maturation of the foci for sex-specific patterns of cuticular pheromone synthesis. In males, one of these hormones is required to terminate synthesis of the non-sex-specific courtship-stimulating cuticular pheromones that are characteristic of immature flies.

Acknowledgments

I am grateful to Scott McRobert for Figure 1 and to Larry Yager for Figures 2 and 3. In addition, I thank Ward Odenwald and Eric Kubli for permission to cite manuscripts in press; Mariana Wolfner and John Carlson for assistance with references; and my husband and the people in my laboratory for their patience and understanding while I was writing this review. Support for my research on sexual behavior in *D. melanogaster* was provided by the Human Frontier Sciences Program.

REFERENCES

Antony, C., and Jallon, J.-M. (1981). Evolution des hydrocarbures comportalement actifs de *Drosophila melanogaster* au cours de maturation sexuelle. *CR Acad. Sci.* **292,** 239–242.

Antony, C., Davis, T.L., Carlson, D.A., Pechine, J.-M., and Jallon, J.-M. (1985). Compared behavioral responses of male *Drosophila melanogaster* (Canton-S) to natural and synthetic aphrodisiacs. *J. Chem. Ecol.* **11,** 1617–1629.

Baker, B.S. (1989). Sex in flies: the splice of life. *Nature* **340,** 521–524.

Baker, B.S., and Ridge, K.A. (1980). Sex and the single cell. I. On the action of major loci affecting sex determination in *Drosophila melanogaster. Genetics* **94,** 383–423.

Baker, B.S., and Wolfner, M.F. (1988). A molecular analysis of *doublesex*, a bifunctional gene that controls both male and female sexual differentiation in *Drosophila melanogaster. Genes Devel.* **2,** 447–489.

Belote, J.M., and Baker, B.S. (1987). Sexual behavior: its genetic control control during development and adulthood in *Drosophila melanogaster. Proc. Natl. Acad. Sci. U.S.A.* **84,** 8026–8030.

Blomquist, G.J., Tillman-Wall, J.A., Guo, L., Quilici, D.R., Gu, P., and Schal, C. (1993). Hydrocarbon and hydrocarbon derived sex pheromones in insects: biochemistry and endocrine regulation. *In* "Insect Lipids: Chemistry, Biochemistry and Biology" (D.W. Stanley-Samuelson and D.R. Nelson, eds.), pp. 317–351. University of Nebraska, Lincon.

Bodenstein, D. (1950). The postembryonic development of *Drosophila. In* "Biology of *Drosophila*" (M. Demerec, ed.), pp. 275–367. Hafner, New York.

Bouhouche, A., Benziane, T., and Vaysse, G. (1993). Effects de deux mutations neurologiques, *minibrain³* et *no-bridge^{KS49}*, sur la parade nuptiale chez *Drosophila melanogaster. Can. J. Zool.* **71,** 985–990.

Burtis, K.C., and Wolfner, M.F. (1962). The view from the bottom: sex-specific traits and their control in *Drosophila. Sem. Devel. Biol.* **3,** 331–340.

Butterworth, F.M. (1969). Lipids of *Drosophila*: a newly discovered lipid in the male. *Science* **163,** 1356–1357.

Curcillo, P.C., and Tompkins, L. (1987). The ontogeny of sex appeal in *Drosophila melanogaster* males. *Behav. Genet.* **17,** 81–86.

de Belle, J.S., and Heisenberg, M. (1994). Associative learning in *Drosophila* abolished by chemical ablation of mushroom bodies. *Science* **263**, 692–695.

Ferveur, J.-F., and Jallon, J.-M. (1993). *nerd*, a locus on chromosome III, affects male reproductive behavior in *Drosophila melanogaster. J. Naturwiss.* **80**, 474–475.

Ferveur, J.-F., Stortkuhl, K.F., Stocker, R.F., and Greenspan, R.J. (1995) Genetic feminization of brain structures and changed sexual orientation in male *Drosophila. Science* **267**, 902–905.

Ford, S.C., Napolitano, L.M., McRobert, S.P., and Tompkins, L. (1989). Development of behavioral competence in young *Drosophila melanogaster* adults. *J. Insect Behav.* **2**, 575–588.

Gailey, D.A., and Hall, J.C. (1989). Behavior and cytogenetics of *fruitless* in *Drosophila melanogaster:* different courtship defects caused by separate, closely linked lesions. *Genetics* **121**, 773–785.

Gailey, D.A., Lacaillade, R.C., and Hall, J.C. (1986). Chemosensory elements of courtship in normal and mutant, olfaction-deficient *Drosophila melanogaster. Behav. Genet.* **16**, 375–405.

Gailey, D.A., Taylor, B.J., and Hall, J.C. (1991). Elements of the *fruitless* locus regulate development of the muscle of Lawrence, a male-specific structure in the abdomen of *Drosophila melanogaster* adults. *Devel.* **113**, 879–890.

Greenspan, R.J., Finn, J.A., and Hall, J.C. (1980. Acetylcholinesterase mutants in *Drosophila* and their effects on the structure and function of the central nervous system. *J. Comp. Neurol.* **189**, 741–774.

Hall, J.C. (1977). Portions of the central nervous system controlling reproductive behavior in male *Drosophila melanogaster. Behav. Genet.* **7**, 291–312.

Hall, J.C. (1978). Courtship among males due to a male-sterile mutation in *Drosophila melanogaster. Behav. Genet.* **8**, 125–141.

Hall, J.C. (1979). Control of male reproductive behavior by the central nervous system of *Drosophila*: Dissection of a courtship pathway by genetic mosaics. *Genetics* **92**, 437–457.

Hall, J.C. (1994). The mating of a fly. *Science* **264**, 1702–1714.

Hartenstein, V. (1993). "Atlas of *Drosophila* Development." Cold Spring Harbor, Plainview, N.Y.

Heisenberg, M., Borst, A., Wagner, S., and Byers, D. (1985). *Drosophila* mushroom body mutants are deficient in olfactory learning. *J. Neurogen.* **2**, 1–30.

Hirsch, H.V.B., and Tompkins, L. (1994). The flexible fly: experience-dependent development of complex behaviors in *Drosophila melanogaster. J. Exp. Biol.* **195**, 1–18.

Hotta, Y., and Benzer, S. (1972). Mapping of behavior in *Drosophila* mosaics. *Nature* **240**, 527–535.

Jallon, J.-M. (1984). A few chemical words exchanged by *Drosophila* during courtship and mating. *Behav. Genet.* **14**, 441–478.

Jallon, J.-M., and Hotta, Y. (1979). Genetic and behavioral studies of *Drosophila* female sex appeal. *Behav. Genet.* **9**, 257–275.

Jallon, J.-M., Antony, C., and Benamar, O. (1981). Un anti-aphrodisiac produit par males de *Drosophila melanogaster* et transfere aux femelles lors de la copulation. *CR Acad. Sci.* **292**, 1147-1149.

Jallon, J.-M. Antony, T.P., Yong, C., and Maniar, S. (1986). Genetic factors controlling the production of aphrodisiac substances in *Drosophila melanogaster. In* "Advances in Vertebrate Reproduction" (M. Porchet, J.-C. Andries, and A. Dhainaut, eds.), Vol. 4, pp. 445–452. Elsevier, New York.

Jallon, J.-M., Lauge, G., Orssaud, L., and Antony, C. (1988). Female pheromones in *Drosophila melanogaster* are controlled by the *doublesex* locus. *Genet. Res.* **51,** 17–22.

Kubli, E. (1996). The *Drosophila* sex-peptide: a peptide pheromone involved in reproduction. *In* "Advances in Developmental Biochemistry" (P. Wasserma, ed.), Vol. 4, pp. 99–128. Academic Press, New York.

Kulkarni, S.J., Steinlauf, A.F., and Hall, J.C. (1988). The *dissonance* mutant of courtship song in *Drosophila melanogaster*: isolation, behavior, and cytogenetics. *Genetics* **118,** 267–285.

Lilly, M., and Carlson, J. (1990). *smellblind*: a gene reqired for *Drosophila* olfaction. *Genetics* **124,** 293–302.

Lindsley, D.L., and Zimm, G.G. (1992). "The Genome of *Drosophila melanogaster.*" Academic Press, New York.

Manning, A. (1967). The control of sexual receptivity in female *Drosophila*. *Anim. Behav.* **15,** 239–250.

Markow, T.A., and Hanson, S.J. (1981). Multivariate analysis of *Drosophila* courtship. *Proc. Natl. Acad. Sci. U.S.A.* **78,** 430–434.

McRobert, S.P., and Tompkins, L. (1983). Courtship of young males is ubiquitous in *Drosophila melanogaster*. *Behav. Genet.* **13,** 517–523.

McRobert, S.P., and Tompkins, L. (1985). The effect of *transformer, doublesex,* and *intersex* mutations on the sexual behavior of *Drosophila melanogaster*. *Genetics* **111,** 89–96.

McRobert, S.P., Schnee, F.B., and Tompkins, L. (1995). Selection for increased female sexual receptivity in *raised* stocks of *Drosophila melanogaster*. *Behav. Genet.,* **25,** 303–309.

Ringo, J., Werczberger, R., and Segal, D. (1992). Male sexual signalling is defective in mutants of the *apterous* gene of *Drosophila melanogaster*. *Behav. Genet.* **22,** 469–487.

Rodrigues, V., and Siddiqi, O. (1978). Genetic analysis of chemosensory pathway. *Proc. Indian Acad. Sci.* **87,** 147–160.

Salz, H.K., Maine, E.M., Keyes, L.N., Samuels, M.E., Cline, T.W., and Schedl, P. (1989). The *Drosophila* female-specific sex-determination gene. *Sex-lethal,* has stage-, tissue-and sex-specific RNAs suggesting multiple modes of regulation. *Genes Devel.* **3,** *708–719.*

Sanchez, L., and Nothiger, R. (1982). Clonal analysis of *Sex-lethal,* a gene needed for female sexual development in *Drosophila melanogaster*. *Wilhelm Roux's Arch.* **191,** 211–214.

Schaner, A.M., Dixon, P.D., Graham, K.J., and Jackson, L.L. (1989). Components of the courtship-stimulating pheromone blend of young male *Drosophila melanogaster:* (Z)-13-tritriacontene and (Z)-11-tritriacontene. *J. Insect Physiol.* **35,** 341–345.

Schilcher, F. (1976). The role of auditory stimuli in the courtship of *Drosophila melanogaster*. *Anim. Behav.* **24,** 18–26.

Schilcher, F., and Hall, J.C. (1979). Neural topography of courtship song in sex mosaics of *Drosophila melanogaster*. *J. Comp. Physiol.* **129,** 85–95.

Scott, D. (1986). Sexual mimicry regulates the attractiveness of mated *Drosophila melanogaster* females. *Proc. Natl. Acad. Sci. U.S.A.* **83,** 8429–8433.

Scott, D., and Richmond, R.C. (1988). A genetic analysis of male-predominant pheromones in *Drosophila melanogaster*. *Genetics* **119,** 639–646.

Strausfeld, N.J. (1976). "Atlas of an Insect Brain." Springer-Verlag, New York.

Sturtevant, A.H. (1915). Experiments on sex recognition and the problem of sexual selection in *Drosophila*. *J. Anim. Behav.* **5,** 351–366.

Taylor, B.J. (1992). Differentiation of the male-specific muscle in *Drosophila melanogaster* does not require the sex-determining genes *doublesex* or *intersex*. *Genetics* **132**, 179–191.

Taylor, B.J., Villella, A., Ryner, L.C., Baker, B.S., and Hall, J.C. (1994). Behavioral and neurobiological implications of sex-determining factors in *Drosophila*. *Devel. Genet.* **15**, 275–296.

Technau, G.M. (1984). Fiber number in the mushroom bodies of adult *Drosophila melanogaster* depends on age, sex and experience. *J. Neurogen.* **1**, 113–126.

Tompkins, L. (1984). Genetic analysis of sex appeal in *Drosophila*. *Behav. Genet.* **14**, 411–440.

Tompkins, L. (1986). Genetic control of sexual behavior in *Drosophila melanogaster*. *Trends Genet.* **2**, 14–17.

Tompkins, L. (1989). Homosexual courtship in *Drosophila*. In "Perspectives in Neural Systems and Behavior" (T. J. Carew and D.B. Kelley, eds.), pp. 229–248. Liss, New York.

Tompkins, L. (1990). Effects of the *apterous*[4] mutation on *Drosophila melanogaster* males' courtship. *J. Neurogen.* **6**, 221–227.

Tompkins, L., and Hall, J.C. (1981). *Drosophila* males produce a pheromone which inhibits courtship. *Z. Naturforsch.* **36c**, 694–696.

Tompkins, L., and Hall, J.C. (1983). Identification of brain sites controlling female receptivity in mosaics of *Drosophila melanogaster*. *Genetics* **103**, 179–195.

Tompkins, L., and Hall, J.C. (1984). Sex pheromones enable *Drosophila* male to discriminate between conspecific males from different laboratory stocks. *Anim. Behav.* **32**, 349–352.

Tompkins, L., and McRobert, S.P. (1989). Regulation of behavioral and pheromonal aspects of sex determination in *Drosophila melanogaster* by the *Sex-lethal* gene. *Genetics* **123**, 535–541.

Tompkins, L., and McRobert, S.P. (1995). Behavioral and pheromonal phenotypes associated with expression of loss-of-function mutations in the *Sex-lethal* gene of *Drosophila melanogaster*. *J. Neurogen.* **9**, 219–226.

Tompkins, L., Hall, J.C., and Hall, L.M. (1980). Courtship-stimulating volatile compounds from normal and mutant *Drosophila*. *J. Insect Physiol.* **26**, 689–697.

Tompkins, L., Gross, A.C., Hall, J.C., Gailey, D.A., and Siegel, R.W. (1982). The role of female movement in the sexual behavior of *Drosophila melanogaster*. *Behav. Genet.* **12**, 295–307.

Truman, J.W. (1990). Metamorphosis of the central nervous system in *Drosophila*. *J. Neurobiol.* **21**, 1072–1084.

Welbergen, P., Spruijt, B.M., and Dijken. F.R. (1992). Mating speed and the interplay between female and male courtship responses in *Drosophila melanogaster* (Diptera: Drosophilidae). *J. Insect Behav.* **5**, 229–244.

Wheeler, D.A., Kulkarni, S.J., Gailey, D.A., and Hall, J.C. (1989). Spectral analysis of courtship songs in behavioral mutants of *Drosophila melanogaster*. *Behav. Genet.* **19**, 503–528.

Wicker, C., and Jallon, J.-M. (1995). Hormonal control of sex pheromone biosynthesis in *Drosophila melanogaster*. *J. Insect Physiol.* **41**, 65–70.

Yamamoto, D., Sano, Y., Ueda, R., Togashi, S., Tsurumura, S., and Sato, K. (1990). Newly isolated mutants of *Drosophila melanogaster* defective in mating behavior. *J. Neurogen.* **7**, 152.

Zhang, S.-D., and Odenwald, W.F. (1995). Misexpression of the *white* gene triggers male-male courtship in *Drosophila*. *Proc. Natl. Acad. Sci. U.S.A.*, **92**, 5525–5529.

Note added in proof: Since this review was written, several papers that are germane to the topic have been published. Notably, Ferveur et al. (*Science* **276**, 1555-1558, 1997), utilizing enhancer trap lines in which the *transformer* gene is expressed in specific tissues of XY flies, demonstrated that the foci for synthesis of the female-specific and male-predominant cuticular sex pheromones are the oenocytes, which are rows of cells under the abdominal cuticle. This observation is in agreement with the prediction, based on analysis of gynandromorphs' ability to elicit vigorous courtship, that the foci are subcuticular, but the cells in which the pheromones are made are not, as posited in this review, epidermal. With regard to genes that regulate sex-specific behaviors, Villella and Hall (*Genetics* **143**, 331–334, 1996) resolved a disagreement described in this review by verifying that *doublesex* mutations do affect males' courtship. In this review, *fruitless* is described as a gene that has male-specific behavioral effects. However, Ryner et al. (*Cell* **87**, 1079–1089, 1996) identified female-specific *fruitless* transcripts in the central nervous system, suggesting the possibility that *fruitless* mutations have an as yet unidentified effect on female-specific behaviors. Finally, *dissatisfaction*, which when mutant affects male courtship, female receptivity, and males' ability to distinguish females from mature males, can be added to the list of genes whose products are required for normal sexual behavior (Finley et al., *Proc. Natl. Acad. Sci.* USA **94**, 913–918, 1997).

Genome Analysis in Eukaryotes: Developmental and evolutionary aspects
R.N. Chatterjee and L. Sánchez (Eds)

7. Elimination of X Chromosomes and the Problem of Sex Determination in *Sciara ocellaris*

A.L.P. Perondini

Departamento de Biologia, Instituto de Biociências, Universidade de São Paulo,
C. Postal 11461, CEP 05422–970, São Paulo, Brazil

I. Introduction

The genetic and molecular basis of ontogenetic processes is one of the most crucial problems in modern biology. Sex determination represents one of such process. Recently, a coherent picture about the complex gene interactions controlling sexual choice in some animals is emerging, at least for the worm *Caenorhabditis elegans* (see reviews Hodgkin, 1990; Ryner and Swain, 1995), the fly *Drosophila melanogaster* (Nöthiger and Steinmann-Zwicky, 1987; Baker, 1989; Steinmann-Zwicky et al., 1990; Cline, 1993; Sánchez *et al.*, 1994) and for the mouse and humans (Ryner and Swain, 1995).

Among insects, a multitude of sex determination mechanisms seems to exist. Taken as paradigm the mechanism of sex determination so far described for *Drosophila*, Nöthiger and Steinman-Zwicky (1987) proposed that an underlying unifying concept may exist for the insect species. Sex determination could be based on the interaction of three genetic elements which would form the initial step of a hierarchical control system: a primary signal that together with the product of maternal genes, regulate a key gene (*Sxl*), the state of activity of which would control a short cascade of genes that govern sexual development. In *Drosophila* details of this array of genes are already known (ref. cit.). Recently, genes that play a role in sex determination and which are homologues to the *Drosophila Sxl* have been found in other insects. However, differences in the mode of action of these genes seems to be present (Hilfiker-Kleiner *et al.* 1993; Puchalla, 1994; Traut, 1994; Müller-Holtkamp, 1995).

In sciarid species, the males are XO:2A, and the females are XX:2A (Metz, 1938). Thus, it seems that the first genetic signal that triggers sex determination is the ratio of the number of X-chromosomes to the number of set of autosomes (X:A ratio), like in *Drosophila*. Assuming that the X:A ratio in *Sciara* has the same role as in *Drosophila,* one has to consider that in sciarid species the process of sex determination involves an additional step comparable to that of *Drosophila*. In *Drosophila*, the chromosomal constitution of the embryo is fixed at fertilization, depending solely on the chromosome sets of the gametes (X:A for the oocytes, X:A or Y:A for the sperms, although the Y-chromosome playing no role in sex determination). In sciarid flies, the establishment of the male or female specific X:A ratio is a consequence of a differential elimination of X-chromosomes which occurs during early embryogenesis at the syncitial blastoderm stage (DuBois, 1932, 1933). Whether or not the sciarid flies possesses a cascade of genes governing sexual development downstream of the X:A ratio, is still not known. However, the fact that in *Sciara ocellaris* dosage compensation seems to operate according to the model of *Drosophila* (Cunha *et al.*, 1994), suggests that in *Sciara* there is a key gene homologous to *Sxl*, that in *Drosophila* regulates both, sex determination and dosage compensation (Sanchez *et al.*, 1994; Kelley and Kuroda, 1995; Ryner and Swain, 1995).

The complex genetic system of the sciarid flies, however, involves not only elimination of sex chromosomes but also differential elimination of autosomes in males (Metz, 1938). Although the males are diploid for the autosomes, they behave as haploid in terms of genetic transmission since they pass on only the set of chromosomes received from their mothers. Thus, the sciarid males are "functional haploids" (Hartl and Brown, 1970). The chromosomal inheritance in sciarid flies has being studied under two perspectives: (a) their role in determination of sex and (b) the possible adaptive advantage by incorporating the functional haploidy in males.

For a discussion of models, implications and meanings of the sciarid mode of reproduction the readers are referred to the papers of Hamilton (1967), Bull (1983) and Haig (1993). In relation to sex determination in the flies, one can find a rich literature. Most of the work has been directed to the understanding of the aspects of chromosome behavior and its relation to sex determination. Details of the processes has already been addressed in the review by Gerbi (1986).

Earlier, it was shown by DuBois (1932, 1933) and others (Metz, 1938; Crouse, 1943, 1960a), that the differential X-chromosome elimination and, hence, sex determination in *Sciara* depends on the mother's genotype and has to be controlled via the egg's cytoplasm. In the present article the role of cytoplasmic factor(s) on chromosome elimination in different species of *Sciara* will be reviewed extensively including some recent observations obtained with *Sciara ocellaris*.

Since the sciarid genetics is quite complex a brief overview of the chromosome segregation of *Sciara* will also be given, pointing some aspects necessary for the following discussion.

II.   Chromosome behavior in sciarids

The embryo of sciarid flies starts its development with two sets of autosomes and three X chromosomes, two of which are sister chromatids of paternal origin. Some species possesses, besides this set of chromosomes, usually three heterochromatic chromosomes (L-chromosomes). Figure 1 shows diagramatically the normal chromosome behavior in sciarids (Metz, 1938).

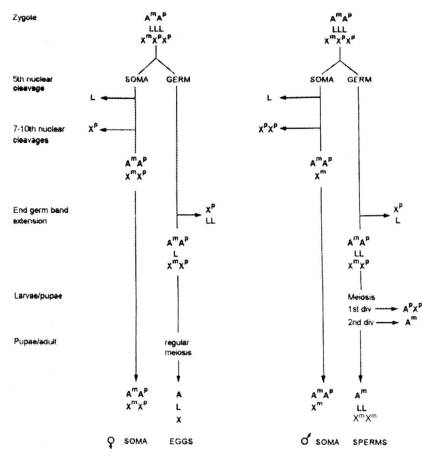

Fig. 1. Diagram showing chromosomal constitution of sciarid flies at different stages of development. The ontogenetic stages at which chromosome elimination takes place are shown by arrows. X^m and X^p denote the maternal or paternal origin of the X chromosomes (see text).

At the 5th nuclear cleavage (rarely at the 6th), usually two nuclei are set apart into the posterior tip of the egg being included in the germinal plasma where they form the pole cells. At this stage no elimination takes place in these nuclei. On the contrary, the nuclei that remained in the presumptive somatic region of the embryos eliminate the L-chromosomes (DuBois, 1932, 1933; Saint Phalle and Sullivan, 1996). Species like *Sciara ocellaris*, which do not possess L-chromosomes, show no elimination at this stage (Perondini et al., 1986).

The common point to all species so far studied is the differential elimination of X-chromosomes, which takes place during the 7th to 10th nuclear cleavages. One or two X-chromosomes of paternal origin are eliminated in the somatic nuclei of embryos that would develop as female or male, respectively. The nuclei in the soma of male individuals, therefore, would contain a maternal derived X and a set of autosomes from each parent, X^m/O; A^m/A^p. On the other hand, the female soma have two X chromosomes and a set of autosomes from both parents, X^m/X^p; A^m/A^p (DuBois, 1932, 1933; Gerbi, 1986; Perondini et al., 1986).

The next round of elimination occurs in germ cells. The primordial pole cells derivatives eliminate only one paternal X-chromosome, in both male and female embryos, but this elimination is postponed to a later stage, at the end of germ band extension when germ cells are already located in place where the gonads would be formed. Hence, regardless the sex of the embryo, the germ cells after elimination would have two X-chromosomes (X^m/X^p) and a regular set of pairs of autosomes. The development of these cells, bearing similar karyotypes, into ovaries or testes is brought about by interaction with or by induction of the surrounding somatic tissues (Crouse, 1960b; Strambi, 1987).

The final process of elimination occurs in male gametogenesis at the larval-pupal transition. In the meiotic mechanism of males synapsis of the homologues does not occur, preventing the occurrence of recombination. During the first division, the paternal derived haploid genome is eliminated in a bud of cytoplasm. In the second division, both the chromatids of the X-chromosome move precociously to one pole, followed by one chromatid of each autosome set. This set of chromosomes will be included in the functional spermatid. The other autosomal chromatids are eliminated in the same bud of cytoplasm into which the paternal derived chromosomes were eliminated during the first division. This unusual meiotic mechanism in male individuals has been found in all sciarid species so far studied (Metz et al., 1926; Basile, 1970; Amabis et al., 1979; Abott and Gerbi, 1981; Abott et al., 1981; Oliveira, 1982; Fuge, 1994; Esteban et al., 1997). Genetic analysis have shown that the eliminated chromosomes are, indeed, X-chromosome and autosomes derived from the male parents (Metz, 1938; Perondini, 1979; Gerbi, 1986). Oogenesis follows the regular, conventional pattern observed in most animal species with synapsis, recombination and segregation (Metz et al., 1926; Metz and Schmuck, 1929b; Berry, 1941).

III. Mechanism of X-chromosome identification

For the proper functioning of the sciarid chromosomic system, a mechanism by which the maternal and the paternal chromosomes can be distinguished one from another and the paternal ones selectively eliminated should exist.

Using X to autosomes translocations in S. coprophila, Crouse (1960b, 1977, 1979) showed that the controlling factor for unusual chromosome behavior is located in a heterochromatic block near the centromere of the X-chromosome. When the heterochromatic block is translocated to an autosome this translocated chromosome is always eliminated instead of the X. She called this factor as "Controlling Element" (CE). Since the chromosomes are eliminated by a mechanism of anaphase delay (DuBois, 1932, 1933), thus implying a modified or inactivated centromere, the CE was thought to exert its effect on the centromere of the chromosome where it is located (cis-acting), regardless whether this centromere belongs to the X or not (Crouse, 1960b, 1979). Recently, Saint Phalle and Sullivan (1996) provided evidences that the CE acts preventing sister chromatids separation, thus retaining the chromosome in the equatorial region of the cell and preventing the chromosomes to be included in either the daughter nuclei.

Crouse (1960ab) introduced the notion of "imprinting" to indicate a process by which the behavior of a pair of homologous chromosomes can be predetermined several cell generations before the stage of development in which the differential behavior manifests. One possibility (Crouse, 1960a), is that during male meiosis the chromosomes would acquire an "imprint" by which they are recognized as paternal chromosomes in the next generation. Brown and Chandra (1977) based on their observations in coccids, argued that the imprinting could occur in the fertilized egg before fusion of the pronuclei. It is known that the imprinting must be a reversible mechanism because a paternally-derived (imprinted?) chromosome, if transmitted to a female, would behave as a maternal chromosome in the next generation. Thus, the imprinting must be removed or modified in females to assure the proper functioning of the system (Crouse, 1960b).

Nothing is really known about the molecular nature of the imprinting in sciarid files nor about the developmental stage in which it is induced. So far no evidences for any process of methylation was found in the *Sciara's* genome (Gerbi, 1986). The hypotheses so far postulated to explain the chromosome imprinting in these flies (Rieffel and Crouse, 1966; Sager and Kitchin, 1975), leave ground for uncertainties (see Gerbi, 1986). It can explain the elimination of both paternally derived X-chromosomes in the male embryos, but does not explain how only one out of these two similar chromosomes (originated from sister chromatids during male meiosis) is eliminated in embryos destined to be females. How is this chromosome selected? It must be also recalled that in the germ cells of individuals of either sex only one X-chromosome of paternal origin is eliminated (Berry, 1941; Rieffel and Crouse, 1966). Furthermore, if both paternally-derived X chromosomes are imprinted, how can just one of them be selected and eliminated?

Errors in the process of chromosome elimination occur spontaneously albeit at low frequency in sciarid flies, either for the X-chromosome or the autosomes (DuBois, 1932, 1933; Metz and Lawrence, 1938; Crouse, 1943; Metz and Schmuck-Armstrong, 1961; Mori *et al.,* 1979; Mori and Perondini,

1980). Elimination of different number of paternally derived X chromosomes in different nuclei of a given embryo, produced individuals with a mixture of XO and XX tissues (gynandromorphs). Elimination of the maternally-derived X chromosome in some nuclei resulted in mosaic individuals, that bear tissues which are patroclinous for the X-chromosomes. Errors were detected also in flies exhibiting a homogeneous phenotype which is not expected in a given progeny. Although these flies, are morphologically regular male and female individuals, they are patroclinous for their X-chromosomes, a situation not regularly produced by the sciarid chromosomic system. This situation was observed in *S. coprophila* (Reynolds, 1938; Crouse, 1960a, b) and in *S. ocellaris* (Davidheiser, 1943; Mori *et al.*, 1979; Mori and Perondini, 1980, 1984; Strambi, 1987). The great majority of exceptional individuals bearing paternally derived X-chromosomes in the soma, in both species, are sterile. It may be noted here that in these individuals, usually the somatic parts of their gonads are formed, sterility being due to the absence of gametes (Davidheiser, 1943; Mori and Perondini, 1980).

IV. Sex ratio in sciarid flies

Another aspect of the sciarid sex determination mechanism that should be mentioned here is the progeny sex ratio. By crossing single males to different females, Moses and Metz (1928) in *S. coprophila* and Liu (1968) in *S. ocellaris*, have shown that the female is responsible for the determination of the sex of her offspring. In *S. coprophila* each female produces only unisexual progeny, being either female-producers (gynogenic) or male-producers (androgenic), in contrast to other species or strains which are "digenic", i.e., species or strains whose females produce bisexual progenies (Metz, 1938, Metz and Lawrence, 1938). Digenic strains of *S. ocellaris* present a widely variable sex-ratio hardly conforming to a 1:1 ratio (Metz, 1931, 1938; Davidheiser, 1943; Mori *et al.*, 1979).

The system in monogenic species, such as *S. coprophila,* functions as if regulated by a pair of alleles. Metz and Schmuck (1929a) proposed that gynogenic mothers are X'X while male-producing females are XX. Gynogenic mothers in S. *coprophila* are heterozygous for a long paracentric inversion on the X-chromosome which precludes recombination between the largest part of the X' and X (Metz, 1938; Crouse, 1960a; Gerbi, 1986). The X' chromosome is not transmitted regularly to males because all eggs produced by gynogenic X'X females are predetermined to give rise to female offspring. X'X and XX females in monogenic species occur at a 1:1 ratio. However, rare exceptional males (X'O) were observed in the offspring of X'X females as well as exceptional females (XX) which appear among the progenies of androgenic mothers (Metz and Moses, 1928; Metz and Schmuck, 1929b).

In order to explain the origin of the exceptional individuals and also the monogeny of sciarid flies, Metz (1938) hypothetically considered the problem in a quantitative basis. He postulated the existence of a threshold value that would separate the two types of females (male and female-producers) which could be set by a combination of different genes (or alleles) with different "strength" on the X as well as on the autosomes. In a gynogenic mother different gene combinations would set the threshold value in some eggs so close to the limit that elimination would follow the opposite pattern from that expected. This, for example, would explain the origin of X'O males from a gynogenic mother. Conditions that would prevent recombination between the X' and X would lock the sexual predetermination on the two modes, gynogeny or androgeny (Metz and Smith, 1931; Metz and Schmuck, 1931). The long paracentric inversion on the X', is the cause of the low level of recombination between these two chromosomes (Crouse, 1960a). *S. impatiens,* another monogenic species, has also a long paracentric inversion on the X' chromosome (Carson, 1946). Metz (1938) also invoke another possibility that X-chromosome elimination may be influenced, to some extent, by the chromosome constitution of the egg after fertilization, which would act in synergism with the predetermination of the eggs done by the mother's genotype. However, no further work has been carried out to substantiate the hypothesis.

A satisfactory model for regulation of digenic mode of progeny sex ratio of sciarid flies is still lacking. Metz (1938) suggested that gene(s) or alleles with different "strength" could exist in these species, the different combination of which would determine the specific sex-ratio of each female. The fact that in digenic *S. ocellaris,* recombination on the X-chromosome occurs more frequently than in *coprophila,* provide some support for the hypothesis of Metz. Yet, he recorded the difficulties in his model, due to the fact that in digenic females, males and females are generated at variable proportions. This may imply that X-chromosome elimination could be influenced by the chromosome constitution of the eggs after fertilization, similarly to the assumptions made for the monogenic mode of reproduction.

Nigro (1995) recently noted that when *S. ocellaris* were maintained at 20°C, the sex-ratio was extremely variable but with a median *circa* of 1:1. When the flies were cultured at 24° to 28°C, the distribution of sex ratios was significantly skewed toward more females, while the lower temperature (18°C) cause an opposite effect, that is, a sex-ratio deviation toward more males. Data reveal that the change in sex-ratio were not caused by differential mortality of a particular sex. Nigro also noted that temperature shift influenced the progeny sex ratio only when applied during the period of "sexual maturation" of the females. Thus, physiological alterations of the females, as is expected to be induced by temperatures within the range supported by the organism, cause modifications which are manifested by alterations on the maleness or femaleness predisposition of the eggs. These results indicate that during oogenesis, cytoplasmic factor(s) are produced and laid on the eggs to function as determinant of the sex of the females' offspring.

V. X-chromosome elimination in the soma

A. Factor(s) controlling elimination

DuBois (1932, 1933) showed that in *S. coprophila,* elimination of the X chromosomes occurs only after the cleavage nuclei migrated to the egg's periplasm. She concluded that the peripheral layer of egg cytoplasm must contain a "signal", produced by the female parent, that controls the process of chromosome elimination. Similar situation was noted by Perondini *et al* (1986) in *S. ocellaris.* Saint Phalle and Sullivan (1996) extended the observations of DuBois for *S. coprophila* and noted that although elimination do occur in the egg periplasm, occasionally elimination may occur in nuclei within the endoplasm. The initiation of chromosome elimination starts at different stages of nuclear cleavage in different embryos, from the 7th to 9th mitotic division in *S. coprophila* and from the 8th to the 10th in *S. ocellaris.* This may imply that elimination of chromosomes is not regulated by a "biological clock" (Perondini *et al.,* 1986) but elimination of X-chromosome in male embryos of *S. coprophila* seems to occur through multiple mitotic cycles while elimination in the female embryos is observed mainly at cycle 9 (Saint Phallel and Sullivan, 1996).

Reynolds (1938) analyzed androgenic females of *S. coprophila* which produced double-X eggs by non-disjunction besides the normal single-X ones. When the offsprings of these females were mated to normal males resultant zygotes carried three or four X-chromosomes which gave rise to males and females, respectively. In other words, irrespective of the number of X-chromosomes present in the zygotes, the eggs of androgenic mothers eliminated two paternally derived X chromosomes. A similar case of non-disjunction during oogenesis was detected in a digenic strain of *S. ocellaris.* Analysis of the sexual and the phenotypic ratios could be explained by a model which considers the number of X chromosome elimination as being one or two in different embryos, regardless of the fact that the zygotes had three or four X chromosomes (Guatimosim *et al.,* 1994).

Crouse (1960a) observed that when 3:1 disjunction occurs in eggs of female of *S. coprophila* bearing an X translocation, the eggs receive either two X or no sex chromosomes at all. After fertilization by wild type sperm (with two X chromosomes), zygotes originated from nullo-X eggs would have two paternal X-chromosome, while those originated from double-X eggs would have four X-chromosomes. When the nullo-X eggs were produced by a gynogenic female, male individuals (patroclinous for the X-chromosome) appeared, confirming that in eggs from a gynogenic mother only one X^p chromosome is eliminated. On the other hand, when the double-X eggs were produced by androgenic mothers, female individuals (matroclinous for their X-chromosomes) were observed, indicating that eggs from androgenic females are predetermined to eliminate two X^p chromosome. These studies demonstrated that the embryonic soma is unable to ascertain the total number of X in the zygote, rather eliminating just one or two X^P chromosomes according to the origin of the egg, from a gynogenic or androgenic mother, respectively.

Evidences so far indicate that the number of X chromosomes to be eliminated in a given embryo is predetermined by the mother, presumably by cytoplasmic factor or factors which regulate the differential elimination between the soma of males and females.

Direct proofs of the existence of these cytoplasmic factor are still lacking. Evidences additional to those described above have been provided by the analysis of an X-linked recessive mutation, *sepia*, in *S. ocellaris* (Mori *et al.*, 1979). This mutation, besides inducing alteration in the color of the cuticle, causes modifications in the sex ratio of the progenies of single pair of matings. It was noted earlier that *sepia* enhances the frequency of males (Mori *et al.*, 1979) but recently, it was observed that this effect depends on the temperature (Campos, 1996). The range of temperatures where the transition of maleness to femaleness in progeny sex ratio do occur in wild type flies, as observed by Nigro (1995), is modified in females bearing the *sepia* allele, skewness toward males occuring in a short interval and at higher temperatures (Campos, 1996). The gene *sepia* exerts its effects more drastically when the females are heterozygous. At appropriate temperature, the deviation in the sex ratio toward males is not accompanied by differential mortality of females. Rather, the deviation seems to be consequence of transformation to the male sex of a number of eggs that would be otherwise females. These results suggested that *sepia* mutation alters the process of chromosome elimination, favouring the elimination of two instead of one X^P chromosome. Its effect is not direct on the chromosomes because either of the X-chromosome bearing the *s* or the + allele are eliminated, provided that they are of paternal origin (Mori *et al.*, 1979; Campos and Perondini, 1993). The alterations in sex ratio produced by the *sepia* mutation are accompanied by errors in the process of X chromosome elimination, causing the appearance of gynandromorphs, mosaics, and other exceptional individuals (Mori *et al.*, 1979; Mori and Perondini, 1980; Strambi, 1987; Mori *et al.*, 1991; Campos and Perondini 1993; Nigro, 1995; Campos, 1996).

Heterozygosity for *sepia* seems to be sufficient to cause errors in the elimination of X-chromosomes, as shown by the results of crossing +/+ females to *s* males. However, when the egg cytoplasm is under influence of *sepia* mutation (in crosses involving heterozygous or homozygous sepia females) the frequency of the errors is higher. Heterozygous females (s/+) which received the *s* allele from s/s mothers produced 1.38% and 1.66% of gynandromorphs, respectively, when crossed to + and *s* males. However, the frequency of the aberrant flies was lower in the progenies of +/s females originated from crosses of +/+ females to *s* males. When crossed to + and *s* males these +/s females produced 0.37% and 0.66% of gynandromorphs respectively in their progenies (Mori *et al.*, 1979; Mori and Perondini,

1980). These results suggest that *sepia* acts via the egg's cytoplasm. It is therefore reasonable to believe that cytoplasmic factor(s) may have some role in the mechanism of X chromosome elimination.

To test this putative maternal influence, the effect of the *sepia* mutation was analyzed for five generations (Mori, 1978; Mori *et al.*, 1991). When *sepia* allele was transmitted from s/s mothers to heterozygous females and from these to their daughters it was called maternal transmission (MT). When the allele was transmitted from male to female or *vice-versa* alternatively at successive generations, it was called alternate transmission (AT). Sex ratio was significantly altered in MT, skewed toward the production of males in every generation, while in AT, the sex ratio deviation (favoring males) was larger when the *s* allele was transmitted by female parents. Nonetheless, when the *s* allele derived from the male parent a modification in the sex-ratio distribution was noted. The frequency of gynandromorphs increased from 0.59% (G1) to 5.85% (G5) in MT but remained low in AT, at about 0.70%. Similarly, the frequency of females giving rise to gynandromorphs was 32% at the initial generation, increased to 78% in MT but remained at the level of 20 to 40% in AT.

In another series of experiments the effect of *sepia* was again observed when it was present in the female parent and at a lower degree when it was transmitted by the male parent (Campos and Perondini, 1993; Campos, 1996).

The above results suggest that *sepia* seems to interfere on the processes of X-chromosome elimination not only via the egg cytoplasm but also by its presence on the zygote's genome. This data further support the hypothesis of Metz, who indicated that chromosome elimination may be influenced by the chromosomes present after fertilization.

B. Analysis of individuals with aberrant phenotypes

Gynandromorphs, mosaics and phenotypic exceptional individuals in *S. ocellaris* seems to resulted from two types of errors, which occur during the process of X-chromosome elimination (Mori and Perondini, 1980; 1984). The first type involves the number of paternally-derived X-chromosome elimination in different nuclei of a given embryo, producing a mixture of X^m/O and X^m/X^p nuclei and thus originating gynandromorphs. The other types occur in the mechanism of selection of the chromosomes that are to be eliminated, maternally derived X-chromosome being eliminated instead of the paternally-derived ones. Males (X^p/O; X^m/O) or females (X^p/X^p; X^m/X^p) mosaic individuals are thus generated bearing tissues which are patroclinous for the X-chromosome. When both types of errors in chromosome elimination occur in the same nuclei of a given individual, gynandromorphs are again formed (X^p/O; X^m/X^p). However, they are different from the former because the male tissues retained patroclinous X-chromosome (Mori, 1978; Mori and Perondini, 1980; Strambi, 1987). These observations seem to indicate that elimination of X-chromosomes is controlled by mechanims involving at least two steps, one of which regulates the number of chromosomes to be eliminated, while the other selects the chromosomes that are to be eliminated.

The body of flies bearing the *sepia*-enhanced errors in chromosome elimination, usually is composed of patches of wild type and sepia cell clones. The range of mosaicism is also variable, extending from individuals with small patches up to those with large clones of discordant tissues. Occasionally, the mosaics and the gynandromorphs are of the bilateral type (Mori and Perondini, 1984). Since in *Sciara,* the sex determination is specified by the differential X-chromosome elimination during syncitial blastoderm and since the mosaics or gynandromorphs are produced by errors in these mechanisms, it is impossible to ascertain in bisexual species, whether the errors had occurred in the embryos predetermined to be males or females. Strambi (1987) noted that in some gynandromorphs one side of the body was entirely of a given sex, while the other side showed patches of male and female tissues. Assuming that in the phenotypically homogeneous sides no error has occurred, their sexual phenotype would indicated the sexual predetermination of the eggs which gave rise to these gynandromorphs. Hence, the analysis of their complementary sides, presenting the male and female patches, would permit to score the frequency and the pattern of errors in embryos predetermined to develop as males or females. Strambi (1987) noted that out of 141 gynandromorphs, 24 individuals were completely female in one side while other 14 presented one side completely male. Errors in X-chromosome elimination would lead to the presence of XO tissues in the first group of flies (predetermined to be females), and of XX tissues in the second one (male predetermined). It is interesting to note here that in both situations the errors occurred more frequently in the anterior parts of the body. As it appears from Strambi's data, the mean frequency of male tissue in gynandromorphs derived from eggs with this putative female-predetermination was 40.6%, the proportion of female tissue in the male-determined gynandromorphs being 38.5%. Since, as pointed out by Mori and Perondini (1984), X-chromosome elimination occur in embryos having from 256 to 1,024 nuclei, it is expected that the average frequency of discordant phenotypes would be less than 1%, if each nucleus would autonomously exert the control of X-chromosome elimination, as is the case of chromosome loss generating *Drosophila* gynandromorphs (Janning, 1978; Portin, 1978). However, as it appears from the data, interference with the mechanism of chromosome elimination in *Sciara,* affects several nuclei simulatneously. Moreover, there were variation not only in the number of affected nuclei but also in the location within the body where the errors have occurred. The control mechanism for the differential elimination of X-chromosomes in *Sciara* seems, then, to be extrinsic to the nuclei. These observations further support the view that cytoplasmic factor(s) control the number of X-chromosomes to be eliminated in the soma of the Sciara embryos.

In order to search for such factors further, heterozygous embryos of *S. ocellaris,* originated from crosses of s/s females to + males, were irradiated with far UV (254 nm) at two preblastodermic stages, at early intravitelline cleavage (EC) and at initial syncitial blastoderm (SB) (Guatimosim and Perondini, 1994a; Guatimosim, 1996). These stages preceed the period of X-chromosome elimination by, approximately, 2 and 1 hours, respectively (Perondini *et al.,* 1986). At EC, irradiation affected mainly cytoplasmic structures and molecules since the nuclei are located deep within the endoplasm, while at SB, the nuclei are already at the periplasm and are also near to the UV energy (Guatimosim and Perondini, 1994b). Far UV fluences, low enough to yield high levels of survival, were nonetheless effective in promoting significant increase in the frequency of individuals showing errors of X-chromosome elimination. It was also noted that the frequency of these aberrant flies increased with increment of the UV fluence. However, irradiation at EC provoked an increase in the frequency of gynandromorphs (errors in the number of eliminated X-chromosomes) while at SB, the frequency of gynandromorphs also increases but mosaic individuals appeared in significant number. When embryos were exposed to near UV light (photoreactivation treatment) after being irradiated with far UV at both stages (EC and SB), survival was restored to control levels and the frequency of gynandromorphs as well as of mosaics decreased significantly. Since photoreactivation allows repair of UV induced damage in either DNA (Harm, 1980) or RNA (Kalthoff and Jäckle, 1982), it is reasonable to consider that the cytoplasm as well as the nuclei contains "targets" related to the mechanism of X-chromosome elimination. Moreover, the results also suggested that cytoplasmic mechanism would determine the number of X-chromosomes that are eliminated while the nuclear process would be involved in the identification of which chromosome are to be eliminated.

VI. X-chromosome elimination in embryonary germ cells

Elimination of X-chromosome in the embryonary germ cells of sciarid species takes place at a stage of development when the pole cells derivatives have migrated to their final location where the gonads are to be formed (Berry, 1941; Rieffel and Crouse, 1966; Oliveira, 1982; Perondini and Ribeiro, 1997). The karyotype of germ cells at stages preceding elimination is similar to that of the zygote: two sets of autosomes and three-X-chromosomes for the species without L-chromosomes. Regardless of the sex of the embryo, only one X-chromosome of paternal origin is always eliminated in germ cells (Berry, 1941; Rieffel and Crouse, 1966). This is a strictly controlled event corroborated by observations in individuals of *S. coprophila* (Reynolds, 1938) and *S. ocellaris* (Guatimosim *et al.,* 1994) that carried an extra maternal X-chromosomes in the oocytes. These females produced normal single-X as well as double-X eggs by a regular disjunction during meiosis. Fertilization of the double-X egg by normal sperms gave origin to zygotes and primordial germ cells that were quadrisomic for the X-chromosome ($X^m/X^m/X^p/X^p$). In these germ cells a single X^p chromosome was eliminated.

The question remains: since the germ cells contain two paternally-derived X-chromosomes, what is the mechanism by which only one X-chromosome is chosen and eliminated? Analysis of females with trisomic oocytes showed, similar to what happens in the somatic cells, germ cells unable to ascertain the total number of X-chromosome present: they always eliminate a single paternally derived X-chromosome.

The mechanism of X-chromosome elimination in the germ cells is very peculiar. Light microscopic studies have shown that during X-chromosome elimination in *S. ocellaris* (Berry, 1941; Rieffel and Crouse, 1966; Guatimosim and Perondini, 1994a; Perondini and Ribeiro, 1997) and in *S. coprophila* (Rieffel and Crouse, 1966), the germ cell nucleus is at an atypical interphase showing the chromatin condensed into the individual chromosomes, although the compaction seems to be less than the usual observed during cell division (Rieffel and Crouse, 1996; Perondini and Ribeiro, 1997). The X-chromosome which is eliminated, passes through the nuclear membrane and remains in the cytoplasm for several days before disappearing. In species like *S. coprophila*, in addition to X-chromosome, elimination of one or two L-chromosomes also occur by a process similar to that observed for the X-chromosome (Rieffel and Crouse, 1966).

Recently, an electron microscopic analysis of this process in *S. ocellaris*, showed that this unusual elimination is brought about by an active involvement of the nuclear envelope (Perondini and Ribeiro, 1997). Before being eliminated, the chromosome is first juxtaposed to the nuclear envelope. In the next step, the inner nuclear membrane forms invaginations which permeate the attached chromosome. Following evagination of this structure, a double-membranous vesicle containing the eliminated chromosome is found in the cytoplasm. However, the control mechanism of this peculiar behavior is unknown.

VII. Conclusion and future directions

The data reviewed in this chapter suggest that X-chromosome elimination in the somatic nuclei of the *Sciara* embryos is governed by processes which involve at least two mechanisms. One of them is responsible for the recognition of the chromosome as maternally or paternally-derived ones while the other would control the number of paternally derived chromosomes to be eliminated.

The mechanism of imprinting which makes possible the distinction between paternal and maternal chromosomes is unknown. However, data reveal that the site which control the chromosome behavior is located near the centromere of the X-chromosome. A similar situation is supposed to occur for the autosomes (Gerbi, 1986). Whatever this mechanism is, it imposes a different behavior on the paternal X, not only in relation to its elimination but also on the genetic functioning of the these "imprinted" chromosomes. As shown by the occurrence of exceptional individuals with patroclinous X chromosomes, somatic sex determination is not affected by the maintenance of the paternally-derived X-chromosomes in the soma. However, in the germ line the X^p-chromosomes are no longer able to maintain the regular maturation processes of the spermatocytes. Crouse (1966), using aneuploids produced by X to autosomes translocations in *S. coprophila*, noted that a single maternal X can promote a normal maturation of the sperms, while a single paternal X cannot. Crouse believed that abnormalities in the germ line resulted from the absence of a maternal X and not from the hemizygous condition of the chromosome *"per se"*.

These data together indicate that specific elimination of paternally-derived X in the embryonic soma can be uncoupled from the processes that triggers somatic (primary) sex determination. Differential elimination of X-chromosomes is necessary to establish the specific X:A ratio necessary for male or female development, but correct somatic sexual development can be accomplished regardless whether paternally or maternally-derived X-chromosomes are maintained in the soma. The meaning of the selective and specific elimination of paternally-derived X-chromosomes remains to be studied.

The females of *Sciara* produce eggs predetermined as to the sexual development of the future embryo. In terms of X-chromosome elimination, their eggs are predetermined to eliminate one or two paternal X-chromosomes. This is true either for monogenic or for digenic species. The data here further indicate that the predetermination mechanism is based in factor(s) housed in the egg cytoplasm, which are made during oogenesis. These factors seem to contain or involve nucleic acids in their moiety. It is not known whether they are organelles containing DNA or they are "sex determinants" (mRNA of maternal origin) present in the egg cytoplasm.

Additionally, preliminary observations suggest that the chromosome constitution of the embryo after fertilization may also play a role in the differential elimination of X-chromosomes in the somatic nuclei of the embryos.

The special mode of reproduction of sciarid flies, either monogeny or digeny, is also related or controlled by the predetermining mechanisms. In general, genic systems controlling sexual predetermination might be expected to be more flexible than those controlling sex itself. Regardless of alterations in the predetermination mechanism, functional adults of both sexes will be produced, provided these alterations involve only the predetermining mechanism. The consequence would be a variation in the progeny sex ratio with preponderance of one sex or the other. For the monogenic species of *Sciara,* the machanism of predetermination would have a more rigid genetic control causing any given female to produce only one type of eggs, either predetermined to develop as males or as females. Since a given female in digenic *Sciara* species, produces two kinds of eggs, some of them predetermined to develop as males and others as females, it seems likely that physiological conditions within the ovary are responsible for the predetermination of the offspring's sex. A variety of mechanisms could presumably exists to meet these requirements. Such examples are quite common among coccids where eggs for male or female development are predetermined by several mechanisms in different species, such as aging of the females, temperature, humidity, presence of symbionts, etc (Brown and Chandra, 1977). Indeed, the temperature effects on the sex-ratio in *S. ocellaris* (Nigro, 1995; Campos, 1996) are in line with these observation. As mentioned by Brown and DeLotto (1957) "... systems of predetermination would seem to offer mechanism permitting rapid adaptive changes in sex ratios without deleterious consequences to the individual". The adaptive values of the sciarid chromosomic system remains to be determined.

Acknowledgments

The studies reported in this chapter has been supported by the Brazilian Agencies (FAPESP, CNPq, CAPES and FINEP) Humboldt Foundation and by the European Union (Contract ALAMED Cl1*. CT94–0071). The author, who is a research fellow of CNPq, would like to thank Drs. A.B. da Cunha, Paulo A. Otto and Ann Stocker for critically reading the manuscript.

REFERENCES

Abott, A.G. and Gerbi, S.A., (1981). Spermatogenesis in *Sciara coprophila*. II. Precocious chromosome orientation in meiosis II. *Chromosoma* **83**; 19–27.

Abott, A.G., Hess, J.E. and Gerbi, S.A., (1981) Spermatogenesis in *Sciara coprophila*. I. Chromosome orientation on the monopolar spindle of meiosis I. *Chromosoma* **83**; 1–18.

Amabis, J.M., Reinach, F.C. and Andrews, N., (1979) Spermatogenesis in *Trichosia pubescens* (Diptera, Sciaridae). *J. Cell Sci.* **36**; 199–213.

Baker, B.S., (1989) Sex in flies: the splicing of life. *Nature* **340**; 521–524.

Basile, R., (1970) Spermatogenesis in *Rhynchosciara angelae* Nonato and Pavan. *Rev. Brasil. Biol.* **30**; 29–38.

Berry, R.O., (1941) Chromosome behavior in the germ cells and development of the gonads in *Sciara ocellaris*. *J. Morph* **68**; 547–583.

Brown, S.W. and Chandra, H.S., (1977) Chromosome imprinting and the differential regulation of homologous chromosomes. In: Cell Biology: a comprehensive treatise. vol I, (L. Goldstein and DM Prescott, eds) pp 109–189, Academic Press, NY.

Brown, S.W. and DeLotto, G., (1957) Cytology and sex ratios of an African species of armored scale insect insect (Coccoidea-Diaspididae). *Amer. Nat.* **93**; 369–379.

Bull, J.J., (1983) Evolution of sex determining mechanisms. WA Benjamin/Cummings, Menlo Park, Cal.

Compos, M.C.C., (1996) A mutação *sépia* e a determinação do sexo em *Sciara ocellaris* (Diptera, Sciaridae). MSc Dissertation, Departmento de Biologia, Instituto de Biociências, Universidade de São Paulo, Brazil, pp 89.

Campos, M.C.C. and Perondini, A.L.P., (1993) Influência do alelo sepia no processo de determinação sexual em *Sciara ocellaris*. *Rev. Brasil. Genet.* **16** (suppl). 280 (abstract).

Carson, H.L., (1946) The selective elimination of inversion dicentric chromatids during meiosis in the eggs of *Sciara impatiens*. *Genetics* **31**; 95–113.

Cline, T.W., (1993) The *Drosophila* sex determination signal: how do flies count to two? *Trends Genet* **9**; 385–390.

Crouse, H.V., (1943) Translocations in *Sciara*: their bearing on chromosome behavior and sex determination. *Univ. Mo. Res. Bull.* **379**; 1–75.

Crouse, H.V., (1960a) The nature of the influence of X-translocation on sex of progeny in *Sciara coprophila*. *Chromosoma* **11**; 146–166.

Crouse, H.V., (1960b) The controlling element in sex chromosome behaviour in *Sciara*. *Genetics* **45**; 1429–1443.

Crouse, H.V., (1966) An inducible change in state on the chromosomes of *Sciara*: its effects on the genetic components of the X. *Chromosoma* **18**; 230–235.

Crouse, HV, (1977) X heterochromatin subdivision and cytogenetic analysis in *Sciara coprophila* (Diptera, Sciaridae) I. Centromere localization. *Chromosoma* **63**; 39–55.

Crouse, H.V., (1979) X heterochromatin subdivision and cytogenetic analysis in *Sciara coprophila* (Diptera, Sciaridae). *Chromosoma* **74**; 219–239.

Cunha, P.R., Granadino, B, Perondini, ALP and Sánchez, L, (1994) Dosage compensation in sciarids is achieved by hypertranscription of the single X chromosome in males. *Genetics* **138**; 787–790.

Davidheiser, B., (1943) Inheritance of the X chromosome in exceptional males of *Sciara ocellaris* (Diptera). *Genetics* **28**; 193–199.

DuBois, A.M., (1932) A contribution to the morphology of *Sciara*. *J. Morph.* **54**; 161-195.

DuBois, A.M., (1933) Chromosome behavior during cleavage in the eggs of *Sciara coprophila* (Diptera) in relation to the problem of sex determination. *Z. Zellforsch. Mikrosk. Anat.* **19**; 595–614.

Esteban, M.R., Campos, M.C.C., Perondini, A.L.P. and Goday, C., (1997) Role of microtubules and microtubules organizing centers on meiotic chromosome elimination in *Sciara ocellaris*. *J. Cell Sci.* **110**; 721–730.

Fuge, H., (1994) Unorthodox male meiosis in *Trichosia pubescens* (Sciaridae). *J. Cell Sci.* **107**; 299–312.

Gerbi, S., (1986) Unusual chromosome movements in sciarid flies. In: Results and problems in Cell Differentiation, vol 13: Germ line-soma differentiation, (W. Hennig, ed), pp 71–104, Springer-Verlag, Berlin.

Guatimosim, V.M.B., (1996) Efeitos da radiação ultravioleta no processo de eliminação de cromossomos X nos núcleos somáticos do díptero *Sciara ocellaris* (Diptera; Sciaridae). Doctoral Thesis, Departmento de Biologia, Instituto de Biociências, Universidade de São Paulo, Brazil, pp 155.

Guatimosim, V.M.B., Boccato, M. and Perondini, A.L.P., (1994) Trissomia para o cromossomo X em *Sciara ocellaris* e a eliminação de cromossomos. *Rev. Brasil. Genet.* **17** (suppl); 271 (abstract).

Guatimosim, V.M.B. and Perondini, A.L.P., (1994a) Indução pela UV de erros de eliminação do cromossomo X em *Sciara ocellaris*. I. Tipos de erros. *Rev. Brasil. Genet.* **17** (suppl.); 270 (abstract).

Guatimosim, V.M.B. and Perondini, A.L.P., (1994b) Lethal effects of far UV on preblastodermic embryos of *Sciara ocellaris* (Diptera, Sciaridae). *Rev. Brasil. Genet.* **17**; 25–34.

Haig, D., (1993) The evolution of unusual chromosomal system in sciarid flies: intragenomic conflict and the sex ratio. *J. Evol. Biol.* **6**; 249–261.

Hamilton, W.D., (1967) Extraordinary sex ratios. *Science* **156**; 477–488.

Harm, W., (1980) Biological effects of ultraviolet radiation. Cambridge Univ. Press, NY.

Hartl, D.L. and Brown, SW, (1970) The origin of male haploid genetic systems and their expected sex ratio. *Theor. Popul. Biol.* **1**; 165–190.

Hilfiker-Kleiner, D., Dübendorfer, A., Hilfiker, A. and Nöthiger, R., (1993) Developmental analysis of two sex-determining genes, M and F, in the housefly, *Musca domestica. Genetics* **134**; 1187–1194.

Hodgkin, J., (1990) Sex determination compared in *Drosophila* and *Caenorhabditis. Nature* **344**; 721–728.

Janning, W., (1978) Gynandromorphs fate maps in *Drosophila*. In: Genetic mosaics and cell differentiation. (W.J., Gehring ed), pp. 1–28, Springer-Verlag, Berlin.

Kalthoff, K. and Jäckle, H., (1982) Photoreactivation of pyrimidine dimers generated by a photosensitized reaction in RNA of insect embryos (*Smittia* spec.). In: Trends Photobiology, (C. Helene and M. Charlier, eds), pp. 173–188, Plenum Press, NY.

Kelley, R.L. and Kuroda, M.I., (1995) Equality for X chromosomes. *Science* **270**; 1607–1610.

Liu, P.Y., (1968) Estudo biologico de cultura de *Bradysia tritici* (Diptera, Sciaridae) parasitada por gregarina. MSc Dissertation, Departmento de Biologia, Instituto de Biociências, Universidade de São Paulo, São Paulo Brazil, pp 42.

Metz, C.W., (1931) Unisexual progenies and sex determination in *Sciara. Quart. Rev. Biol.* **6**; 306–312.

Metz, C.W., (1938) Chromosome behavior, inheritance and sex determination in *Sciara. Am. Nat.* **72**; 485–520.

Metz, C.W. and Lawrence, E.G., (1938) Preliminary observations on *Sciara* hybrids. *J. Hered.* **29**; 179–186.

Metz, C.W. and Moses, M.S. (1928) Observations on sex-ratio determination in *Sciara* (Diptera). *Proc. Natl. Acad. Sci. USA* **14**; 931–932.

Metz, C.W. and Schmuck, M.L., (1929a) Unisexual progenies and the sex chromosome mechanism in Sciara. *Proc. Natl. Acad. Sci. USA* **15**; 863–866.

Metz, C.W. and Schrnuck, M.L., (1929b) Further studies on the chromosomal mechanism responsible for unusual progenies in *Sciara*. Tests of "exceptional" males. *Proc. Natl. Acad. Sci. USA* **15**; 867–870.

Metz, C.W. and Schmuck, M.L. (1931) Studies on sex determination and the sex chromosome mechanims in *Sciara. Genetics* **16**; 225–253.

Metz, C.W. and Schrnuck-Armstrong, L., (1961) Observations on deficient chromosome groups in developing *Sciara* larvae. *Growth* **25**; 89–106.

Metz, C.W. and Smith, H.B., (1931) Further observation on the nature of the X-prime (X') chromosome in *Sciara. Proc. Natl. Acad. Sci. USA* **17**; 195–198.

Metz, C.W., Moses, M.S. and Hoppe, E.N., (1926) Chromosome behavior and genetic behavior in *Sciara* (Diptera). I. Chromosome behavior in the spermatocyte division. *Z. Induk. Abstamm. Veresbungsl.* **42**; 237–270.

Mori, L., (1978) Ginandromorfos e determinação do sexo em *Sciara ocellaris*. Doctoral Dissertation, Departamento de Biologia, Instituto de Biociências, Universidade de São Paulo, Brazil, pp 103.

Mori, L. and Perondini, A.L.P., (1980) Errors in the elimination of X chromosomes in *Sciara ocellaris. Genetics* **94**; 663–673.

Mori, L. and Perondini, A.L.P., (1984) An analysis of *Sciara ocellaris* gynandromorphs and the morphogenetic fate map of presumptive adult cuticular structures. *J. Exp. Zool.* **230**; 29–35.

Mori, L., Dessen, E.M. and Perondini, A.L.P., (1979) A gene that modifies the sex ratio in a bisexual strain of *Sciara ocellaris. Heredity,* **42**; 353–357.

Mori, L., Guatimosim, V.M.B., Campos, M.C.C. and Perondini, A.L.P., (1991) Transmissão materna do gene *sepia* aumenta erros de eliminação do cromossomo X em *Sciara ocellaris*. *Rev. Brasil. Genet.* **14** (suppl); p 277 (abstract).

Moses, M.S. and Metz, C.W., (1928) Evidence that the female is responsible for the sex ratio in *Sciara* (Diptera). *Proc. Natl. Acad. Sci. USA* **14**; 928–930.

Müller-Holtkamp, F., (1995) The *Sex-lethal* gene homologue in *Chrysomia rufifacies* is highly conserved in sequence and exon-intron organization. *J. Mol. Evol.* **41**; 467–477.

Nigro, R.G., (1995) Efeito da temperatura na determinação sexual de *Sciara ocellaris* (Diptera, Sciaridae). MSc Dissertation, Departamento de Biologia, Instituto de Biociências, Universidade de São Paulo, São Paulo, Brazil, pp 79.

Nöthiger, R and Steinmann-Zwicky, M, (1987) Genetics of sex determination in eukarytes. In: Results and Problem in Cell Differentiation, vol 14, Structure and function of eukaryotic chromosomes. (W. Hennig, ed), pp. 271–300, Springer-Verlag, Berlin.

Oliveira, C.S., (1982) Estudo da eliminação de cromosomos na espermatogenese de *Sciara ocellaris*. MSc Dissertation, Departmento de Biologia, Instituto de Biociências, Universidade de São Paulo, Brazil, pp 61.

Perondini, A.L.P., (1979) On the transmission of the asymmetric polytene chromosome bands in *Sciara ocellaris*. *Caryologia* **32**; 365–372.

Perondini, A.L.P. and Ribeiro, A.F., (1997) Chromosome elimination in germ cells of *Sciara* embryos: involvement of the nuclear envelope. *Invert. Reprod. Develop.* **32**; 131–141.

Perondini, A.L.P., Gutzeit, H.O. and Mori, L., (1986) Nuclear division and migration during early embryogenesis of *Bradysia tritici* Coquillet (syn. *Sciara ocellaris*) (Diptera, Sciaridae). *Int. J. Insect Morphol. Embryol.* **15**; 155–163.

Portin, P, (1978) Studies on gynandromorphs induced with the *claret*-non-disjunctional mutation of *Drosophila melanogaster*. An approach to the timing of chromosome loss in cleavage mitosis. *Heredity* **41**; 193–203.

Puchalla, S., (1994) Polytene chromosomes of monogenic and amphogenic *Chrysomia* species (Calliphoridae, Diptera): analysis of banding patterns and *in situ* hybridization with *Drosophila* sex determining gene sequences. *Chromosoma* **103**; 16–30.

Reynolds, J.T., (1938) Sex determination in a "bisexual" strain of *Sciara coprophila*. *Genetics* **23**; 203–220.

Rieffel, S.M. and Crouse, H.V., (1966) The elimination and differentiation of chromosomes in the germ line of *Sciara*. *Chromosoma* **19**; 231–276.

Ryner, L.C. and Swain, A., (1995) Sex in the 90's. *Cell* **81**; 483–493.

Sager, R. and Kitchin, R., (1975) Selective silencing of eukaryote DNA. *Science* **189**; 426–433.

Saint Phalle, B. and Sullivan, W., (1996) Incomplete sister chromatid separation is the mechanism of programmed chromosome elimination during early *Sciara coprophila* embryogenesis. *Development* **22**; 3775–3784.

Sánchez, L., Granadino, B. and Torres, M., (1994) Sex determination in *Drosophila melanogaster*. X-linked genes involved in the initial step of *Sex-lethal* activation. *Develop. Genet.* **15**; 251–264.

Steinmann-Zwicky, M., Amrein, H. and Nöthiger, R., (1990) Genetic control of sex determination in *Drosophila*. *Adv. Genet.* **27**; 189–237.

Strambi, M.P.P., (1987) Utilização de ginandromorphos de *Sciara ocellaris* (Diptera, Sciaridae) na análise de alguns aspectos do desenvolvimento: primórdios do aparelho reprodutor, mapa morfogenético e padrão do mosaicismo. Doctoral Thesis, Departmento de Biologia, Instituto de Biociências, Universidade de São Paulo, São Paulo, Brazil, pp 188.

Traut, W, (1994) Sex determination in the fly *Megaselia scalaris*, a model system for primary steps of sex chromosome evolution. *Genetics* **136**; 1097–1104.

Genome Analysis in Eukaryotes: Developmental and evolutionary aspects
R.N. Chatterjee and L. Sánchez (Eds)
Copyright © 1998 Narosa Publishing House, New Delhi, India

8. Mechanisms and Evolutionary Origins of Gene Dosage Compensation

R.N. Chatterjee

Department of Zoology, University of Calcutta, 35 Ballygunge Circular Road, Calcutta-700 019, India.

I. Introduction

One of the consequence of sex chromosome heteromorphism is that there are differences in the amount of quality of the genetic material in the two sexes. The heterogametic sex (XX and ZW in female and XY and ZZ in males) has a single dose of some genes which are present in double dose in homogametic sex. In different animal groups where precise differentiation of the sex chromosomes in the two sexes have been established, the need for dosage compensation has been followed as an obligatory consequence depending on the functional significance of the genes in the inactivated or lost segment of the Y chromosome. Thus, dosage compensation of X linked genes can be considered as an evolutionary strategy required to equalize gene expression between individuals possessing different numbers of sex chromosomes for sex determination. The phenomenon of equalization of the X linked gene products therefore acts as a factor against the selection preference for a particular sex and restores the balance for the haplo-X in the sex against the diplo-X of the other. Therefore, it is reasonable to believe that strong selection forces favour it. Exceptions are however, evident in systems where females are heterogametic, for example, ophidians, avians and Lepidopterans. The phenomenon of dosage compensation was first identified by Muller *et al.* (1931) in *Drosophila*.

To understand the molecular solution of such compensatory mechanisms most of the investigators have so far been restricted mainly to the three animal groups—the nematodes, *Drosophila* and mammals. These animal groups, have the XY/XO type male heterogamety though the mode of sex differentiation is somewhat different in the three systems (Bridges, 1921; White, 1973; Lucchesi, 1978; Bull, 1983; Hodgkin, 1990; Villeneuve and Meyer, 1990; Steinmann-Zwicky *et al.*, 1990; Baker *et al.*, 1994). So far, four different ways of achieving dosage compensation have been recorded, such as (a) enhancing the transcriptional output of the single X chromosome in males, (b) reduction in the level of expression from the two X chromosomes in XX animals, (c) eliminating unwanted chromosomes in somatic cells and (d) silencing of one of the two X chromosomes in female. As dosage compensation mechanisms found so far in insects, nematodes and mammals, do not share a common ancestry it is generally believed that dosage compensation may have evolved apparently independently at least three evolutionary lineages. Yet, biochemical and genetical data support the hypothesis that fundamental programming for dosage compensation restores the genetic balance. The concept of genic balance means that the product level of sex linked genes bear the same relation to the average level of autosomal gene products in both sexes. Clearly, genes responsible for sex determination or sexual dimorphism are excluded from this requirement. However, a large number (nearly 500 to 1000 genes) of other X linked genes that code for 'house keeping' and specialized functions, respond to the genetic programming to compensate for two fold differences in the number between two sexes. Different organisms have evolved different mechanisms to compensate for the dosage differences of X chromosomes in the two sexes.

From an evolutionary stand point, it is an obvious question how genetic programming for dosage compensation is related in different organisms. Therefore, research on dosage compensation provides many important clues to fundamental biological problems. Converesely, research on dosage compensation raises many fundamental problems, such as tissue specific gene interaction, regulation of gene activity, autoregulation, sex specific co-ordinate regulation, evolution of sex and co-ordination of gene activities of different tissues etc. The genetics and molecular biology of dosage compensation have been addressed by various authors in recent reviews of *D. melanogaster* (Jaffe and Laird, 1986; Lucchesi and Manning, 1987; Chatterjee, 1992; Gorman and Baker, 1994; Baker *et al.*, 1994; Williams, 1995) for *C. elegans* (Hodgkin, 1990; Villeneuve and Meyer, 1990; Hsu and Meyer, 1993; Parkhurst and Meneely, 1994) and for mammals (Gartler and Riggs, 1983; Lyon, 1988; McBurney, 1988; Grant and Chapman, 1988; Migeon, 1994; Rastan, 1994; Graves, 1987, 1995). Evolutionry aspects of dosage compensation have also been discussed by Ohno (1967, 1983), Lyon (1974), Lucchesi (1978, 1989) and Charlesworth (1991, 1996). In the present article, attempt has been made to review first, dosage compensation in species where dosage compensation has been intensively investigated and then, consider the evolutionary aspects and other implications.

II. Chromosomal system and dosage compensation that exist between the sexes

As noted above, different animal groups do not share a common ancestry. Therefore, dosage compensation systems have evolved independently in different organisms. This issue is discussed in more detail below.

A. In Nematodes

An XX female (treating hermaphrodite and female as equivalent) and XO male system is found in most nematodes, including close relatives of *Caenorhabditis elegans*. The X:A ratio (0.67 in the male and 1.0 in the hermaphrodite) is the primary signal responsible for both sex determination and dosage compensation (Hodgkin, 1990; Hsu and Meyer, 1993). In this animal, multiple factors on the X chromosome contribute to the X:A ratio. To put it more precisely, there exist multiple numerator elements which contribute additively to the X part of the X:A ratio. However, dosage compensation of *C. elegans* seems to work by down regulating the level of transcription of genes on both chromosomes in the XX hermaphrodite (Hodgkin, 1990). In *C. elegans,* the assessment of the X:A ratio is believed to set the functional state of some master genes that co-ordinately control both dosage compensation and sexual phenotypes. The genetic data supports the hypothesis that control is unified by the *xol-1* (named after suspected function XO lethal) which is the master sex switch and controls both sex determination and dosage compensation (Hsu and Meyer, 1993). The *xol-1* gene which acts as an upstream negative regulator of the *sdc* genes (named after their suspected functions sex and dosage compensation) responds to the X:A ratio and controls the down regulator gene in XO animals. In reality, wild type *xol-1* activity is needed for male but not for female development. Expression of *xol-1* at higher levels cause inhibition of the activities of *sdc-1*, *sdc-2* and *sdc-3* genes. On the other hand, in hermaphrodite, where the *xol-1* activity is low, activity of *sdc* genes is high. Thus, *xol-1* is required for correct setting of the *sdc* genes. If *xol-1* does not act as a switch gene then X:A ratio acts negatively on *xol-1* rather than positively as *sdc-1* and *sdc-2*. It is therefore, believed that either *xol-1* gene product itself or a factor regulates *xol-1* activity. A related concern stems from the fact that three of the four master regulator genes, *xol-1, sdc-1* and *sdc-2* are X linked (Hodgkin, 1990, Villeneuve and Meyer, 1990) and *sdc-3* is autosomal. *sdc-1* is likely to function as an embryonic transcription factor and *sdc-3* gene encodes a protein with two zinc fingers near the COOH terminus (Nonet and Meyer, 1991). Mutations in the *sdc-3* result in dosage compensation defects. It is believed that *sdc-3* regulates dosage compensation by altering the DNA binding specificity of the zinc fingers.

There is evidence that a number of trans-acting genes is involved in the process of dosage compensation, possibly acting by modifying the chromatin structure. Various lines of data (see Villeneuve and Meyer, 1990) indicated that the reduction level of X linked gene expression in 2X animals requires at least four autosomal genes, *dpy*-21, -26, -27, -28. They are collectively known as DCD set. A fifth gene, *dpy*-30 has been identified but its properties have not been published (see Parkhurst and Meneely, 1994). The peculiarity of the DCD set genes is that four of the five genes, the exception being *dpy*-21, have very strong maternal effects on both gene expression and viability (see Table-1). This may imply that these gene products are present but inactive during the time in which the X:A ratio is being read and are then activated in XX animals for dosage compensation. A recent study has further demonstrated that in *C. elegans*, there are cis-acting compensation sequences for regulating the gene dosage in the X chromosome. Recently Meyer and his co-workers (Chuang *et al.*, 1994) cloned the *dpy*-27 gene. The protein product of the gene was also isolated and antibodies to it were raised. Using tracers of *dpy*-27, it has been noted that the protein is localised on both X chromosomes in XX hermaphrodite but not on the X chromosome in XO males. Meyer and his co-workers suggested that *dpy*-26 as well as the protein product of the *sdc*-3 gene binds to the X chromosome along the *dpy*-27. Chuang *et al.* (1994) suggested that the *dpy*-27 complex modulates chromosome structure. It is therefore reasonable to believe that like *Drosophila* (see below), dosage compensation in *C. elegans* may also involve a multi-protein complex that binds all along the X chromosome and changes its chromatin for lowering the gene expression in hermaphrodites.

B. In *Drosophila*

a. Mechanism of dosage compensation

Many years of painstaking analysis of dosage compensation in *Drosophila*, clearly indicate that molecular mechanism of dosage compensation in these species group is a product of complex evolutionary processes. Here I summarize first, the recent progress in dosage compensation in *Drosophila melanogaster* and then, analyse the dosage compensation mechanisms found in other species of *Drosophila* to understand tentalizing insights into the evolution of this mechanism.

In *D. melanogaster*, males are heterogametic sex, they carry one X and one Y chromosome—whereas females have two X chromosomes. In *D. melanogaster* about 20% of known genes lie on the X chromosome. Dosage equivalence for these genes between males and females of this species is achieved by hyper-transcriptive activity of the male X chromosome (Mukherjee and Beermann, 1965). However, not all genes on the X chromosome are dosage compensated. These exceptions itself have provided some insights into the selective advantage of dosage compensation. Given data indicate that the *yolk* protein genes are not expressed in males. Furthermore, those genes that are present in the both the X and Y chromosomes e.g., *bobbed* are not dosage compensated. The gene LSP 1α which code for α subunit of the larval serum protein, is also not dosage compensated (Roberts and Evans-Roberts, 1979; Ghosh *et al.*, 1989; Chatterjee *et al.*, 1992). In addition, alleles of salivary gland secretion polypeptide-4 (*sgs*-4) gene found in exceptional wild type strains are also not dosage compensated (Korge, 1981; Hofmann and Korge, 1987). This gene acts together with seven other *sgs* genes, all of them being autosomal. They

Table 1. Mutations affecting dosage compensation in *C. elegans*

Gene		Phenotype (loss of function allele)	Gene product	Functional activity	References
xol-1	XO	: Feminized, lethal	NI		Villeneuve and Meyer, 1990
	XX	: Hermaphrodites			
sdc-1	XO	: Male	1203 aa protein with 7 zinc fingers	Maternal, not sex specific	Villeneuve and Meyer, 1990; Nonet and Meyer, 1991
	XX	: Masculinized and high X linked gene expression			
sdc-2	XO	: Male	NI		Villeneuve and Meyer, 1990
	XX	: Pseudo-males, rare intersexes, some alleles lethal in XX			
sdc-3	XO	: Male	250 kD protein with two zinc fingers and cofactor binding motif.	Maternal, not sex-specific?	Delong *et al.*, 1993; Klein and Meyer, 1993
	XX	: Sex determination and dosage compensation defects genetically separable. Null alleles have no overt sex determination defects			
dpy-21	XO	: Male	NI		Villeneuve and Meyer, 1990
	XX	: *dpy*, hermaphrodite, dead			
dpy-26	XO	: Male	NI	Maternal	Villeneuve and Meyer, 1990
	XX	: Dead, few *dpy* hermaphrodites			

Table 1 Contd...

Gene		Phenotype (loss of function allele)	Gene product	Functional activity	References
dpy-27	XO	: Male	A protein has substantial homology to the *Xenopus* chromosome assembly proteins XCAP-C and XCAP-E that are putatively involved in chromosome condensation.	Maternal	Villeneuve and Meyer, 1990; Williams, 1995
	XX	: Dead, few *dpy*, hermaphrodites			
dpy-28	XO	: Male	NI	Maternal	Villeneuve and meyer, 1990
	XX	: Dead, few *dpy* hermaphrodities			
dpy-30	XO	: Male	NI	Maternal	Villeneuve and Meyer, 1990
	XX	: Dead, hermphrodite			

NI = Not Identified

together produce the larval saliva used for the subsequent attachment of the pupae to the substrate. It is worth pointing out that selective advantage for these genes is not absolutely necessary. Thus, dosage compensation seems to be necessary to avoid selection preference for a particular sex by eliminating aneuploidy effect of the X chromosome.

In· *Drosophila*, chromosomal signal is the ratio of the number of X chromosome to the number of sets of autosomes (X:A ratio). How X:A ratio is assessed in soma and how transduction of the X:A signal initiates the dosage compensation in *D. melanogaster* have been extensively reviewed (Jaffe and Laird, 1986; Lucchesi and Manning, 1987; Gorman and Baker, 1994; Baker *et al.*, 1994). Yet, for present review the key features of the process are discussed for comparison. In soma, the X:A ratio is assessed by means of dispersed chromosomal genes termed counting elements—X chromosomal genes are referred to as numerators and the autosomal genes are referred to as denominators. So far, three genes that fit the criteria for numerators, have been identified, *sisterless-a (sis-a), sisterless-b (sis-b)* and *runt*. In this review, the *sis-b/scute-α*/T4 gene has been referred as *sis-b* in recognition of its sex determination activity. A weaker fourth numerator element has recently been identified (for detail see the chapter of Sanchez *et al.* in this book). On the other hand, only one autosomal gene, *dead pan* (*dpn*), has been identified as denominator, so far. It may possible that other genes fulfilling the criteria for denominator elements exist in other autosomes. Table-2 summerized the phenotypes and properties of sex determination and dosage compensation regulatory genes in *D. melanogaster.* The X:A ratio (numerator/denominator) is counted in the early embryo by a mechanism involving the relative concentration of a set of maternal zygotic gene products primarily by basic helix-loop-helix (bHLH) proteins that are encoded by both X linked numerator elements and autosomal denominator elements. Clearly, the protein encoded by autosomal gene *dpn* in conjunction with a maternally provided gene product *emc* titrate the gene products of the X linked gene *sisterless* (*sis*) by formation of heterodimers (Erickson and Cline, 1993). The limiting factor in these interactions is the amount of SIS protein present, as both XX and XY organisms have equal amounts of DPN and EMC. Males which have only one X chromosome have too little free SIS protein to bind all the available DPN and EMC. SIS products also bind to the product of another maternally provided gene *doughterless (da)*—a bHLH protein product. Females have enough SIS protein (either a leucine zipper protein or bHLH protein) to allow them to accumulate sufficient DA-SIS heterodimers to go and initiate activation of the sex determining gene *Sex lethal* (*Sxl*) from its early promoter.

A combination of genetic and molecular methods (Cline 1984, 1985, 1986, 1988, 1993; Bell *et al.* 1991; Torres and Sánchez 1992; Bernstein and Cline 1994), has led to a fairly clear outline of the function and regulation of *Sxl*. *Sxl* encodes an RNA binding protein (SXL) that regulates sex specific

Table 2. Mutants affecting dosage compensation in *Drosophila melanogaster*

Gene	Phenotypes	Gene product(s)	Functional activity	References
a. Maternal effect genes				
daughterless (da)	XY : No effect	bHLH transcription factor	Activity required in female embryos as a positive activator of *Sxl*	Cormmiller *et al.*, 1988
	XX : Embryonic lethal			
	XY : Reduced male viability			
extramacrochaetae (emc)	XX : No effect	bHLH transcriptional repressor	Inhibits *Sxl* activation in males	Ellis *et al.*, 1990
	XY : No effect			
sansfille (snf/liz)	XX : No effect in presence of two *Sxl*+ genes. Trans-heterozygous with *Sxl* mutation leads to somatic sex transformation and reduced viability	NI	Positive regulator of *Sxl*, functions in females. Affects late. Female specific spliced *Sxl* RNA's and proteins. Possibly involved in *Sxl* autoregulatory feedback loop.	Steinmann-Zwicky, 1988; Oliver *et al.*, 1988
b. Zygotic effect genes				
sisterless-a (sis-a)	XY : No effect	b-ZIP transcriptional activator	Numerator. Functions in all somatic nuclei to activate *Sxl* transcription.	Cline, 1986; Cline, 1988; Erickson and Cline, 1993
	XX : Embryonic lethal			
sisterless-b (sis-b)	XY : No effect	b-HLH transcriptional activator	Acts as a Numerator. Functions in all somatic nuclei to activate *Sxl* transcription	Torres and Sánchez, 1989; Erickson and Cline, 1991
	XX : Embryonic lethal			
runt	XY : No effect	*Runt* domain (heterodimeric) transcription factor	Acts as numerator. Weaker than *sis* genes. Activates *Sxl* transcription in a spatially restricted domain in female embryos.	Duffy and Gergen, 1991; Torres and Sanchez, 1992
	XX : Embryonic lethal			

Table 2 Contd...

Gene	Phenotypes	Gene product(s)	Functional activity	References
dead pan (dpn)	XY : Reduced male viability XX : No effect	bHLH transcriptional repressor	Acts as denominator. Inhibit *Sxl* activation in males	Younger-Shepherd *et al.,* 1992
female lethal 2d [fl (2d)]	XY : Semi-fully lethal XX : Lethal	NI	Interacts with *Sxl.* Mutant females express male specific *Sxl* transcripts. Possibly involved in *Sxl* auto-regulatory feed back loop.	Granadino *et al.,* 1990; Granadino *et al.,* 1992
c. Larval lethal genes				
male less (mle)	XY : Lethal XX : No effect	MLE protein shows sequence similarity to four RNA helicase like proteins	May increase the rate of transcription by increasing the rate of elongation of transcripts or removing RNA from the transcription start site.	Belote and Lucchesi, 1980; Kuroda *et al.* 1991
male specific lethal-1 (msl-1)	XY : Lethal XX : No effect	MSL-1 protein contains PEST sequences	MSL-1 protein may regulate MSL-3 protein expression; may also associate with other MSLs to form a heteromeric complex.	Belote and Lucchesi, 1980; Palmer *et al.,* 1993, 1994
male specific lethal-2 (msl-2)	XY : Lethal XX : No effect	A protein with a RING finger and a metallothionine like cysteine cluster	Causes assembly of dosage compensation regulators by associating with one another in a heteromeric complex	Belote and Lucchesi, 1980; Zhou *et al.,* 1995; Kelley *et al.,* 1995; Baker *et al.,* 1994
male specific lethal-3 (msl-3)	XY : Lethal XX : No effect	A protein	MSL-3 proteins may associate with one another MSL proteins in a heteromeric complex	Uchida *et al.,* 1981; Baker *et al.,* 1994

NI = Not Identified

RNA splicing of its own RNA, as well as the splicing of at least one down stream sex regulatory RNA. The *Sxl* gene has separate promoters that are distinctly regulated and that establish or maintain *Sxl* activity (Bopp *et al.,* 1991). The initiation of *Sxl* activity is controlled at the level of transcription through the early promoter, *Sxl* PE. In females, the early promoter is stably activated and is only activated transiently before cellular blastoderm formation. Hence, active early SXL protein which is present only in females, initiate the production of active late SXL proteins from a different promoter within the *Sxl* gene. The presence or absence of the late SXL protein directs (by using autoregulatory loop) the initiation of a cascade of sex specific splicing interactions which results in a female mode of development. In males, in the absence of *Sxl* function, dosage compensation mechanism is operative. It is likely that although the intermediate dosage compensation target(s) of *Sxl* is not known, *Sxl* negatively regulates dosage compensation at the level of splicing.

A critical evaluation of the function of the *Sxl* gene further indicates that if dosage compensation is able to affect the numerator activity and if *Sxl* would be able to respond continuosly to the X:A signal then the system would not be stable. A male embryo with *Sxl* inactive would adopt hyper transcriptive activity of X chromosome thus leading to an apparent increase in the X:A ratio, which in turn would activate *Sxl,* thereby leading to oscillations in gene activity and no clear difference between the sexes. Therefore, autoregulation is necessary as autoregulation loop may stabilize the mechanism of maintainance of X chromosome activity. Furthermore, bifunctional activity of *Sxl* gene may increase the apparent accuracy of the reading of the X:A ratio. Embryonic cells that read the ratio wrongly and adopt an inappropriate *Sxl* state will die as a result of incorrect dosage compensation. However such mistakes can be edited out by autoragulatory loop of *Sxl* and the *Drosophila* embryo has enough cells to compensate for the loss of a few.

Most notably all but one of the genes involved in activating the *Sxl* gene which act as a sex specific function are also required for nerogenesis in both sexes (Erickson and Cline 1991). Whether the bifunctional genes are representative of an evolutionary or developmental intermediate in the sex determination and dosage compensation process is open to speculation (see Section V).

Although, the molecular mechanism of initiation and maintenance of hypertaranscriptive activity of the male X chromosome is still unclear, a number of transacting genes that regulate dosage compensation by modifying chromatin structure have been identified. These genes are *male specific lethal-1 (msl-1), male specific lethal-2 (msl-2)* and *male specific lethal-3 (msl-3)* and *male less (mle).* They are collectively referred to as *male specific lethals (msl's).* Loss of function mutation of these genes appear to lead to decreased level of X chromosome trascription in male and the individuals die at late larval early pupal stages. The *msl's* have no overt phenotypic effects in females. Subsequent studies have further indicated that as *msl's* are down stream to the *Sxl* gene, these genes are under the control of *Sxl*. However, exactly how this control is exercised is not understood. In this context, it may be noted here that out of the three cloned compensation genes (*msl-1, msl-2* and *mle*) only *msl-2* is known to be controlled at the splicing level. The *mle, msl-1* and *msl-3* genes are expressed in both sexes. Yet, their protein products associate with the X chromosome only in males (Kuroda *et al.* 1991, 1993; Palmer *et al.* 1993, 1994). Thus chromosome specific localization but not the expression of the genes themselves is dependent on the function of the other dosage compensation genes (possibly *msl-2*) and is prevented by functional *Sxl* product (see Baker *et al.* 1994; Kelley *et al.* 1995). Although the significance of the difference is not clear, MLE and MSL appear to differ in their localization in females. MLE is found dispersed at a number of consistant but low level sites on both the autosomes and the X chromosome whereas MSL-1 does not appear to bind to the polytene chromosome in females at all. Although *msl-1* encodes an uncharacterized protein, the *mle* product is likely to be an RNA helicase. This may indicate an effect on transcriptional elongation or translational initiation, but the exact physiological role of RNA helicase is unclear. Observations that *male specific lethal* genes cause the defect in dosage compensation and failure of *mle* to associate with the male X chromosome suggest that the four gene products work together in a chromosomal transcriptional complex.

It is interesting to note here that the *msl* based system of dosage compensation is likely to be common to the whole *Drosophila* genus (see Baker *et al.* 1994). The *msl-3* gene of *D. virilis*, a *Drosophila* sub-genus species separated from *D. melanogaster* about 62 million years ago (see below) has been cloned.

Interestingly, a particular acetylated isoform of histone H4, H_4Ac_{16} is associated with hundreds of sites along the length of the male X chromosome (Turner *et al.* 1992). The acetylated histone may itself result in a chromatin conformation more accessible to transcription or it may serve as a recognition site for other proteins that increase transcription. The co-localization at most X linked sites of the acetylated form of histone H4 and the MSL's raises the possibility of a link between MSL function in hypertranscription and chromatin structure (Bone *et al* 1994).

Various lines of data indicate that in *Drosophila*, male X chromosome has a different chromatin structure from the female X chromosomes and autosomes (Dobzhansky, 1957; Chatterjee *et al.* 1980). The male X chromosome has a more open chromatin structure as is evidenced by its more diffuse puffy appearence. It also has an increased binding affinity for exogenous RNA polymerase under *in situ* condition (Khesin, 1973; Chatterjee and Mukherjee, 1981; Chatterjee, 1985).

Replication-expression model of gene regulation suggests that the genes which replicated early in S phase have a higher probability of binding transcription factors such as TF II D,B,E,H, (Spreadling and Orrweaver, 1987; De Pamphilis, 1993; Zawel and Reinberg, 1995) and other non-histone proteins such as HMG 17 (Dorbic and Witting, 1987) and therefore have a higher probability of expression. Changes in the time of replication and transcriptional activity of genes may also correlate with changes in the polarity of replication (Smithies, 1982). Active H5 histone genes are replicated from the 5' site of the gene while inactive H5 genes are replicated from the 3' end (Trempe *et al.* 1988). This result may well indicate that different origins of replication may be used at different times according to the functional activity of the genes. Furthermore, as cis-acting control region affects both the transcription and replication processes, it is likely that the control of replication and transcription is linked. So, the observed faster rate of replication of the male X chromosome (Chatterjee and Mukherjee, 1977) might be sufficient to sustain a hyper-activation of the male X chromosome and is reinforce the maintenance of open chromatin structure of the male X chromosome.

Furthermore, to elucidate the mechanism of regulation of dosage compensation on a gene-by-gene basis, several authors have attempted to relocate the cloned genes by germ line transformation. The conclusion drawn from the experiments is that compensatory sequences are often tightly linked to individual X linked genes. The cis-acting sequences which act as two fold enhancers perhaps function under the control of dosage compensation. Conversely, the sequences may increase the affinity of an upstream regulatory site for an increased level of transcription of the male X chromosome by binding transacting positive regulators. However, a comparison of the dosage compensated X linked genes could not establish any clustering of particular sequences within the body of the gene itself, upto 4 Kb start site (see Baker *et al.* 1994).

Molecular data further indicate that X chromosomes carry more repetative sequences than that of autosomes (Waring and Pollack, 1987; DiBartolomeis *et al.* 1992). It was also reported that certain dinucleotides (i.e., CA/TG and C/G) are present at higher levels on the X chromosome than on the autosomes (Pardue *et al.* 1987; Huijser *et al.* 1987; Lowenhaupt *et al.* 1989). The three families of sequences that have been identified so far do not map at the same site, although the two families are localised at the X chromosome. However, there is no evidence that any of these sequences have a role in dosage compensation.

b. Evolution of dosage compensation in *Drosophila*
Works discussed so far have been mainly restricted to the regulation of dosage compensation in *D. melanogaster.* Examination of dosage compensation mechanisms in certain selected species of *Drosophila* can

also be informative regarding the evolution of dosage compensation in *Drosophila.* In genus *Drosophila,* there are a number of species where in addition to the original X chromosome (homologous to the X chromosome of *D. melanogaster*) one or more autosomal arms have emerged or are in process of establishing themselves as the neo-X chromosome. According to Patterson and Stone (1952) 54 different fusions have established themselves, 32 among autosomes and 12 between the X and autosomes. Throckmorton (1975) by analysing the *Drosophila* phylogenies suggested that the line leading to the *melanogaster* and *obscura* group separated in the middle Oligocene and that those leading to the *melanogaster* and *virilis* group separated in the Eocene, corresponding to a total evolutionary separation of approximately 60 to 80 million years respectively (Fig. 1). Yet, in the family Drosophilidae, the chromosomes maintained their fundamenatal integrity in course of evolution despite the fixation of multiple rearrangements. In Drosophilidae, the basic cytolotgical coniguration is five rod and one dot chromosomes i.e., six separated chromosomes elements (Muller, 1940; Sturtevant and Novitski, 1941). These elements have been modified by rearrangements such as inversions and centric fusions. A new species group usually evolves from the existing species group. A surprisingly concordant linkage and biochemical marker is also noted even among distantly related *Drosophila* (Foster *et al.* 1981; Malacrida *et al.* 1984). As chromosomes maintain their fundamental integrity throughout *Drosophila* evolution despite the fixation of multiple rearrangements, Muller (1940) suggested a system of nomenclature for identifying chromosome arms. Since the gains in chromosome number are relatively rare, and since the gene system of *D. melanogaster* has been studied more extensively than that of any other species, *D. melanogaster* chromosome element is taken as a standard. A brief survey of selected species from the family Drosophiladae provides some insight into the evolutionary significance of dosage compensation (Table-3). Two important points have emerged from the data in Table-3. Firstly, predominant location of the X chromosome has been maintained for more than 60 million years i.e., the X chromosomal element has been less mobile during evolution than autosomes. Secondly, in the process of sex chromosome evolution, at least three out of the four major autosomal elements (chromosomal arms) contributed a great deal, although the elements have been modified by rearrangements, such as inversions, translocations, so called Robertsonion fusion during the course of their emergence. This may imply that between species, an element can exist either as an autosome or as an X chromosome in the genus *Drosophila.* Taken together, these observations suggest that in *Drosophila,* each chromosome arm behaves inherently as an independent unit. Therefore, to facilitate the acquisition of the necessary regulatory machinery of a sex chromosome, the whole chromosome element has a selective advantage.

Investigations on the activity of genes located on the newly evolved sex

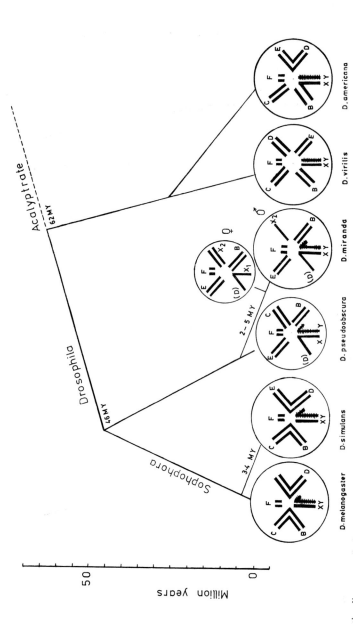

Fig. 1.   Schematic diagram of chromosome evolution of *Drosophila* species. Diagrams show the metaphase configuration of several *Drosophila* species. Y chromosome are marked by banded appearance. The chromosome configuration of *D. virilis* is thought to be that of ancestral species. The partially dosage compensated X_2 in *D. miranda* is indicated by arrows in both male and female karyotypes. Estimates of divergence time for different subgenera of the genus *Drosophila* and for the *melanogaster* and *obscura* species group within the subgenus *Sophophora* are shown in the figure.

Table 3. Summary of homologies of chromosomal elements of *Drosophila*

Species Group	Species	Element						References
		A	B	C	D	E	F	
Sub-genous Drosophila								
immigrans species group	*D. nasuta*	X	3(B+C)		2L	2R	4	Wakhoma and Kitagawa (1972)
	D.n. albomicana	X	(Y)+3(B+C)		2L	2R	4(+C)	Rao and Ranganath (1991)
repleta species group	*D.hydei*	X	4	3	5	2	6	Spencer, 1949
	D. neo-hydei	X	4	3	5	2	6	Wasserman, 1962
robusta species group	*D. robusta*	XL	3	2R	XR	2L	6	Metz, 1916
virilis species group	*D. americana*	X(X-4)	4(X-4)	5	3(2-3)	2(2-3)	6	Hughes, 1939
	D. littoralis	X	4(3-4)	5	3(3-4)	2	6	Hsu, 1952
	D. montana	X	4	5	3	2	6	Stone, Grif and Patt, 1942
	D. novomexicana	X	4	5	3	2	6	Patt, Stone and Grif, 1942
	D. texana	X	4	5	3(2-5)	2(2-3)	6	Stone and Patternson, 1947
	D. virilis	X	4	5	3	2	6	Chino, 1936
	D. lacicola	X	4	5	3	2	6	Patterson, 1944
Sub-genous Sophophora								
melanogaster species group	*D. ananassae*	X(—)	3R	3L	2R	2L	4(+XT)	Kikkawa, 1938
	D. melanogaster	X	2L	2R	3L	3R	4	Muller, 1940
	D. simulans	X	2L	2R	3L	3R	4	Sturtevant, 1921
	D. yakuba	X	2L	2R	3L	3R	4	Lemeunier and Ashburner, 1976
	D. montium	X	3L	3R	2R	2L	4(+XT)	Sturtevant and Nov. 1941
obscura species group	*D. affinis*	XL	4	3	XR	2	5	Sturtevant, 1940
	D. algonquin	XS	C	A	XL	B	D	Miller, 1939
	D. athabasca	XL	X_2	2	3	XR	4	Sturtevant and Dobzhansky, 1936

Table 3 Contd...

Species Group	Species	Element						References
		A	B	C	D	E	F	
	D. azteca	XS	C	A	XL	B	D	Dobzhansky and Socolov, 1939
	D. miranda	XL	4	X_2	XR	2	5	Dobzhanzky and Tan, 1936
	D. persimilis	X_2	4	3	XR	2	5	Tan (1935)
	D. pseudoobscura	XL	4	3	XR	2	5	Lancifield, 1922
	D. obscura	AL-AR	BLBR	JL	JR	EL-ER		Mainx et al., 1953
willistoni species group	*D. insularis*	XL (A-D)	2L(BC)	2R(BC)	XR	3 (F)		Dobzhansky et al., 1957
	D. tropicalis	XL (A-D)	2L(BC)	2R(BC)	XR	3 (F)		Burla et al. 1949
	D. willistoni	XL (A-D)	2(B-C)	2(B-C)	X(A-D)	3 (F)		Sturtevant and Nov. 1941
Endemic Hawaiian Drosophiloids grimshawi species	*D. grimshawi*	X	3	2	5	4	—	Carson and Stalker, 1968

R and L symbolize right and left arm

chromosome may further provide information on the stepwise molecular mechanism by which certain loci slowly or abruptly acquire the hypertranscriptive activity mechanism of the male X chromosome during the course of evolution. For example, *D. pseudoobscura* has a metacentric X chromosome with the left arm (XL) homologous to the *D. melanogaster* telocentric X chromosome and the right arm (XR) homologous to the *D. melanogaster* autosomal arm 3L (D element). The new X chromosome has evolved the ability to compensate the gene dosage difference between sexes (Abraham and Lucchesi 1974; Mukherjee and Chatterjee 1975). In *D. miranda,* another autosomal arm (C element) has been attached to the Y chromosome some time within the last 5 million years. The attached autosome is gradually acquiring the characteristics of the Y chromosome, displaying changes in the chromosome structure as well as in the process of degeneration of gene activity. The chromosome is now referred to as the neo-Y chromosome. In contrast, the homologue of the autosomal part of the neo-Y appears to be evolving into an X chromosome (referred as X_2). Some genes on the X_2 are now capable of dosage compensation by enhancing transcription in males, which compensates for the inactivity of the homologous genes on the neo-Y. Curiously, the X_2 of the male is comparatively thinner in comparison to autosomes.

A series of work of Steinemann and Steinemann (1991, 1992, 1993) clearly indicated that evolutionary changes during the process of sex chromosome differentiation in *D. miranda* are associated with massive DNA rearrangements. They have analysed the region around the larval cuticle protein (*lcp*) gene from the X_2 and compared this to the homologous region from the neo-Y chromosome. They noted that the striking difference between the regions of the two chromosomes is the large number of insertion sequences found on the neo-Y but not on the X_2. They therefore argued that the inactivation of the genes on neo-Y could be a consequence of insertion of repititive DNA sequences within the regulatory or coding sequences of the genes. The insertion sequences on the neo-Y chromosome are simply repeated in nature. Two of the insertions have sequences similar to retrotransposons—however, the sequences of most of the insertions yield no clues to the transposition mechanism (Steinemann and Steinemann, 1993). The Y chromosome of *D. melanogaster* also contains a distinctive sub-class of Het-A related repeats (Danilevskaya *et al.* 1993). It is therefore believed that the sequences have jumped into the transposed segment of the Y chromosome only to inactivate the chromatin environment and be trapped. Fragments of transposable elements recovered from heterochromatin are thought to result from accidental degradation of this excess baggage and trapped elements. Thus massive introduction of repitive DNA sequences and acquisition of these sequences by the neo-Y chromosome probably resulted in the gradual loss of genetic activity of this element and it became heterochromatinized. Although the hitch-hiking effect observed seems to be a typical characteristic of the evolution of the neo-Y in the *D. miranda*, it would not substantiate our hypothesis that dosage compensation mechanism of neo-X of the *D. miranda* has evolved by gradual gene-by-gene evolution. In summary, the above results indicate that degenration of the Y chromosome is characterised by two phenomena: (a) breakdown of the genetic activity of the Y chromosome and (b) change from a euchromatic chromosome state into a heterochromatic one (see Section V).

Second concern in the evolution of the neo-X chromosome of *D. miranda* is that concomitant with the reduction of the activity of the neo-Y, is the induction of the hyperactivity of the X_2 linked genes and progressive recruitment of mono- and di-nucleotide repeats on the X_2. Thus, the levels approach the higher levels seen in all other *Drosophila* X chromosomes. The ability of dosage compensation by the X_2 element appears to be correlated with acquisition of higher levels of CA/GT (Pardue *et al.* 1987). Furthermore, when one of the *D. miranda* neo-Y insertion sequences was used as a probe for *in situ* hybridization to *D. miranda* polytene chromosomes, it was observed that there was heavy hybridization over the neo-Y and low hybridization on the X_2. These results clearly suggest that in the relatively short evolutionary time, the sequences of neo-Y has evolved very differently. This data clearly indicate that the evolution of a genetically inert Y is an active process rather than simply the accumulation of non-functional loci, and that it is accompanied by the evolution of dosage compensation (see Section V).

c. Dosage compensation in Mammals

X chromosome inactivation is the means of regulating gene dosage by which mammals compensate for the difference in the number of X chromosomes between the two sexes. It may be noted here that the existing mammals are of three groups (a) Prototheria (egg laying monotremes), (b) Metatheria (Marsupials, which undergo part of their development in an external female pouch), and (c) Eutheria (placental mammals, which have their development in utero). The divergence of the prototheria and therian groups has been estimated to have occurred at 150–200 million years ago and the therian mammals split into marsupials (Infra-class Metatheria) and placental mammals 130–150 million years ago (Fig. 2). In all the groups, there is effectively a single active X chromosome in both sexes. The sex chromosome variants and functional status (cytological appearance) of X and Y chromosomes in males and females of different mammalian groups have been extensively reviewed elsewhere (Ohno, 1983; Jones, 1984; Graves and Foster, 1994; Graves, 1995). These studies have provided unexpected insights on the evolution of mammalian sex chromosomes and dosage compensation. Molecular cloning and gene mapping studies have further provided a valuable resource for understanding evolution of dosage compensation in mammals. These data are discussed below.

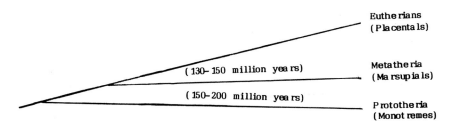

Fig. 2. Branching order of three groups of existing mammals showing the time of divergence.

(a) X chromosome inactivation in monotremes

The prototherians are represented today by three monotreme species. In these species groups, the X and Y chromosomes are almost entirely homologous (Murtagh, 1977; Wrigley and Graves, 1988) and differ only in length of the short arm. Recently, it has been shown that in some monotremes, the entire short arm of the X chromosome pairs with the long arm of the Y chromosome. In female, the inactive X chromosome has been identified by the functional status of X chromosome genes and cytologically by the appearance of allocyclic replication. However, the late replication pattern is tissue specific (Wrigley and Graves, 1988). The significance of the shift in replication patterns of the paternal and maternal X chromosome is not entirely certain. However, some authors argue that the regulatory factors in different tissues may have different affinities to regulate the paternally (X^p) or maternally (X^m) derived X linked loci (Wrigley and Graves, 1988; Graves and Foster, 1994). It is possible that X^p and X^m have different behaviour in different tissues as the regulation of X linked genes depend both on the paternally or maternally marked X chromosome and on the cellular factor(s) which may be different in different tissues. Different cell types differ on the basis of their heritage of protein composition and their ability to regulate X expression. It is therefore, believed that, in the first stage of evolution of X inactivation, not all loci are expected to be inactivated. However, it is still not clear whether the tissue specific maintenance of X inactivation in monotremes is the representative of an evolutionary or developmental intermediate in the X chromosome inactivation process in mammals or not.

(b) X chromosome inactivation in marsupials

Marsupials have heteromorphic X and Y chromosomes. However, their size, pairing relationships and gene content differ from those of eutherian mammals in revealing ways (Graves and Watson, 1991; Graves and Foster, 1994; Graves, 1995). Marsupials have a smaller basic X (about 3% of the haploid complement) and a tiny Y which do not appear to undergo homologous pairing and recombination. Gene mapping studies show that genes from the long arm and pericentric region of the human X map to marsupials' (monotremes also) X (see Fig. 3). This conserved segment of the X (XCS) is therefore likely to represent the original mammalian X which has been retained for at least 170 million years. Other regions of the mammalian genome may show minimal rearrangement. Fredga (1970, 1983) listed 27 species in which fusions between the autosomes and sex chromosome have been established (12 with the X and 15 with the Y). Thus, the conservation of X chromosome is a dominant trend in mammals.

Studies of X chromosome expression in marsupials (and in the extraembryonic tissues of eutherian mammals, see below) indicate that the pattern of X inactivation is preferential. There is clear evidence for absolute paternal X chromosome inactivation, both cytogenetically (Sharman, 1971) and from studies using electrophoretic polymorphisms of X linked genes (Cooper et al., 1971; Richardson et al., 1971). The preferential inactivation of the paternal X chromosome is one of the developmental processes that involves genomic imprinting. It is believed that non-random inactivation may be a consequence of differences that are present in oocyte and sperm X chromosomes at the time of fertilization. The sperm genome undergoes a process of DNA condensation which is associated with replacement of histones with protamines that facilitate interstrand cross linking of DNA (Kistler et al., 1973). These protamines are released from the sperm DNA during fertilization event (Ecklund and Levine, 1975). Histones are also replaced from egg histone pools. Consequently, it is observed that sperm and oocyte genomes are different cytogenetically. Sperm derived chromosomes are less condensed in the initial stage of mitosis of the first cleavage than the chromosomes of the oocyte (Donahue, 1972). It is possible that some of these differences could persist through the initial cleavage division as a consequence of conserving chromosome structure during chromatid replication. However, how such reactions could play any role in differential genomic expression it not clear.

Differential gene expression in paternally or maternally derived X chromosome must depend not only on different imprinting resulting from different ways in which chromatin is restructured in spermatogenesis and oogenesis, but also on the nature of the factors in the oocyte which bind to the chromosomal loci and regulate their expression (Chandra and Brown, 1975 and Solter, 1988). In case of non-random X chromosome inactivation, the differential marking of the sex chromosomes dictates the future expression of the chromosome (Zuccotti and Monk, 1995). Thus it would seem that in marsupials, X inactivation is imposed on X^p.

One of the consequences of paternal X inactivation is that most tissues of the female are functionally hemizygous for X linked genes. Since the male is hemizygous and since paternal X is inactive in female marsupials, one may argue that in both sexes there would be strong selection for the

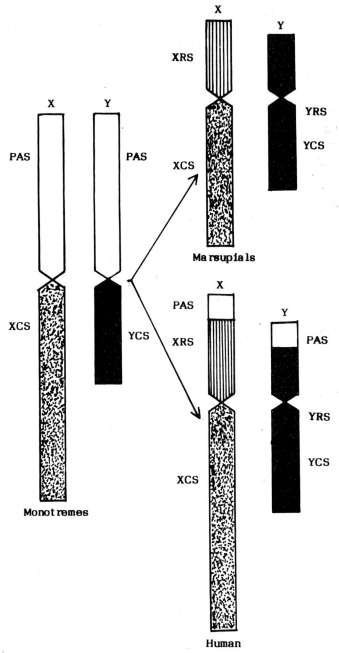

Fig. 3. Evolution of sex chromosome in mammals. Comparative gene mapping shows that X conserved segment (XCS) has been retained for at least 170 million years. The recently added segment of X (XRS) is only recorded on the X in eutherians. The Y chromosome also consists of conserved segment of Y (YCS) and recently added segment YRS. The X and Y pair only over the pseudoautosomal segment (PAS).

elimination of X linked detrimental mutations. In females, there is a selection of combinations of X linked genes which are favourable for females, similarly in males there must be a selection of combinations of genes favourable for males. Thus, non-random X inactivtion system in marspials reduces the fitness of both sexes equally.

(c) X chromosome inactivation in Eutherian Mammals

Eutherian mammals display a variety of evoltionary adaptations for dosage compensation. In most mammals, X chromosome inactivation is the means for regulating gene dosage (Lyon, 1961) as well as sex determination. From the available evidences it is clear that there must be multiple events involved in regulation of X chromosomes in mammals. However, the complexity of the X chromosome regulation is apparent when one considers the evolutionary aspects of this process. For example, in mammals, assessment of X:A ratio is the initial step for regulating dosage compensation, since polyploid cells can have more than one active X chromosome. Furthermore, in most eutherian mammals, X chromosome inactivation occurs randomly with respect to parental origin of the chromosome. However, one X chromosome subsequently differentiates so that it no longer responds to the signals that regulate the transcription of genes on its homologue. This differentiation step is initiated in the mouse at least at the time of implantation during blastocyte stage of development. It occurs in trophoectoderm and then in cell lineages that are continued to the embryo proper. It is assumed that in eutherian mammals both X chromosomes are equally accessible to the regulatory factors, either because of the time factors which allow to earse the imprinting or because of the concentration of the regulatory factors that are sufficiently high to override any difference in the binding affinity between two X chromosomes.

While researchers make progress on the mechanisms of dosage compensation in mammals, it appears that change in chromatin structure may be the key to dosage compensation. The active and inactive X chromatin differ in many properties including, methylation and acetylation pattern and timing of replication. DNA methylation certainly seems to play a role in silencing genes in the inactive X chromosome. The inactive X chromosome in female also lacks in acetylation of histone H4 (Jeppesen and Turner, 1993). Another possible signs of inactivation of one of the X chromosome in female is that inactive X chromosome is the last to initiate DNA synthesis. This late replicating chromosome is relatively condensed during interphase and is cytologically visible as Barr body or a sex chromatin mass at the nuclear periphery (Barr and Bertram, 1949). However, any of these properties could be the cause of dosage compensation or may simply be consequences of overall dosage compensation mechanism.

A great deal of evidence indicates further that inactivation of the X chromosome depends on a particular region of the X chromosome, termed inactivation centre or *Xic* (Russell, 1963; Therman *et al.*, 1974; Mattei *et al.*, 1981). The process of inactivation begins at the *Xic* which is required in *cis* and spreads from *xic/XIC* to the adjacent chromosomal regions. A gene termed *XIST/xist* (X-inactive specific transcript) has been identified and mapped to the *XIC/xic* region. The gene has the unique property of being expressed by the inactive X but not the active X chromosome. The xist gene has been proposed as the master regulatory switch locus that controls X-chromosome inactivation (Penny *et al.*, 1996). *Xist* is also expressed in male meiosis before inactivation of the X in spermatocytes (Graves and Foster 1994). It is believed that the original role of *Xist* may have been the control of X-inactivation in spermatocytes and that X-inactivation in females may have evolved from imprinting the Xist locus (Graves and Foster 1994). The *XIST* gene of the inactive X produces Xist RNA—a 15 kb RNA in mouse (Brockdorff *et al.*, 1992) and 17 kb RNA in human (Brown *et al.*, 1992). The xist RNA has no protein coding activity. However, it is stable and is associated with the total length of the X chromosome just before it becomes inactive. The role of *Xist* RNA in X chromosome inactivation is still not clear. It is believed that it could initiate X chromosome inactivation by interacting with a protein that controls inactivation or merely be a reflection of the unique chromosomal properties of the *Xic* region on the inactive X chromosome. Although the initial events of X chromosome inactivation involve the expression of *Xist* locus, the next step(s) must involve spreading of the inactivation throughout the chromosome (Rastan, 1994). It is still not known how the signal is spread. Although relevant studies have not yet been carried out, it seems likely that spreading of inactivation results from changes in chromatin structure induced by condensation or heterochromatinization of the X chromosome and DNA methylation. DNA methylation has long been considered as a means for controlling allele specific transcription, as it is stable and heritable through many cell division. Wolf *et al.* (1984) and Monk (1986) indicated that there is a clear connection between specific hypomethylation of the 5' CpG cluster and the gene activity of the X linked DNA. It is therefore considered that the body of the gene of the active X and the entire gene in the inactive X is variably and non-uniformly methylated. Therefore, differences in genetic activity of the X chromosomes can be modified by the state of activity of CpG clusters located at 5' end of the genes. CpG clusters on the active X are unmethylated as they are in most autosomal genes, whereas those on the inactive X are extensively methylated (Lyon, 1993; Migeon, 1994). On the basis of foot printing studies it was further postulated that methylation status of CpG clusters plays a role in controlling differential expression of paternal and maternal allele (Lyon, 1993; Norris et al 1994 Zuccotti and Monk, 1995). The absence of methylation at CpG clusters of X linked genes in oogonia and spermatogonia could account for the reversible nature of X inactivation in these cells. This data clearly suggests that DNA methylation of CpG cluster is not the primary inactivator of the X chromosome. On the contrary, it appears that DNA methylation of CpG islands helps to lock the silence at the locus once it has been inactivated. At any rate, methylation of CpG clusters is very important since it is responsible for other maintenance mechanisms (Grant and Chapman 1988; Rastan, 1994). For instance, DNA methylation is the major mechanism responsible for the faithful transmission of the inactive state through meiosis and for the mediation of cell memory. However, certain loci on the inactive X chromosome escape inactivation (Migeon, 1994). A more likely explanation is that there is either no CpG island near the gene or the islands are not methylated. Curiously, it has been recorded that some of these genes are recent additions to the mammalian X chromosome, since they are not present in the marsupials X chromosome (Megeon, 1994).

In a recent study, Zuccotti and Monk (1995) demonstrated that methylation of the *Xist* gene in sperm and eggs correlates with imprinted *Xist* expression of paternal alleles in early development. They also claimed that upon differentiation, one *Xist* allele becomes activated and hypomethylated concomitant with random X inactivation in eutherians.

(d) Evolutionary and other implications of X inactivation

Although molecular basis of X chromosome inactivation in mammals has been the subject of considerable speculation (Gartler and Riggs, 1983; Lyon, 1992, 1993; Migeon, 1994; Rastan, 1994), it is demonstrated by both qualitative and quantitative measurements of X linked gene expression of non-eutherian and eutherian mammals that genetic inactivation of one of the X chromosome in females are of two types: random and non-random. In certain tissues of adult female marsupials, there is clear evidence for absolute paternal X chromosome inactivation. On the other hamd, in placental mammals, inactivation occurs at random. As mentioned earlier, the non-random inactivation is imposed on X^P in marsupials. As a significantly overwhelming majority of nutrient transfer occurs after birth, during lactation for post-natal development, Moore and Haig (1991) argued that X chromosome inactivation mechanism is initiated in embryonic cells lineages of marsupials to avoid conflicting interest of maternal and paternal genes within the offsprings, resulting in preferential paternal X inactivation in all somatic tissues. On the contrary, although the regulatory mechanism in eutherian mammals may be derived from the marsupial mechanism, the choice of which X chromosome is to be active in the cell lineage would be random. Moore and Haig (1991) further argued that although there is a genetic conflict over the amount of resources an offspring obtain from its parent, it is necessary to cooperate between the two, to produce a viable offspring because both sets of genes have a common interest in the offspring surviving and to reproduce. There are however some evidences of non-random inactivation in extraembryonic lineages of rodents. Under such conditions, imprinting operates at the mirgin of the control with placental genes programmed to obtain as much nourishment as possible for the embryo and the maternal genes programmed to counter the effect. Therefore, the passive inactivation mechanism proposed for random inactivation does not readily allow for selective inactivation of X^P in extraembryonic cells. Subsequently, the transient X chromosome differences are either progressively lost during cleavage or become altered at some critical stage of development to a produce random inactivation in the foetus of eutherians. In eutherians, therefore, the random X inactivation may be self-imposed. It is, therefore, believed that the random inactivation mechanism has been progressively acquired by eutherians for their optimum fitness and has been acquired in stages.

In terms of the possibility to link the dosage compensation process and sex determination system in mammals, one should note an intriguing series of papers on identification and characterisation of dosage-sensitive-sex reversal (DSS) element on the X chromosome of human (Berstein *et al.,* 1980; Bardoni *et al.*, 1994). It is now anticipated that remnants of dosage dependent sex determination mechanism are still present in mammals. Thus, some authors claimed that necessity of dosage compensation in mammals has a stronger basis (Chandra, 1985; Bardoni *et al.*, 1994; King *et al,*. 1995).

d. In Lepidopterans and Birds

In some Lepidopterans, amphibians, fishes and birds females are heterogametic (Ohno, 1967). The sex chromosome in females of these animals are usually designated as Z and W instead of X and Y, respectively. To overcome the potentially deleterious effects of dosage imbalance between the sexes, these animal groups have adopted different strategies to cope with the problem of dosage differences. So far, Z chromosomes have been found to be euchromatic in somatic cells and also in the gametocytes of both males and females. In reality, the Z chromosome does not show cytological indications of dosage compensation (such as puffy appearance of sex chromosome as observed in the single X of male *Drosophila* or heterochromatinization of one of the X chromosome as in female mammals). Johnson and Turner (1979) noted that the Z linked gene coding for the enzyme 6GPD in *Heliconius* butterflies are not dosage compensated. Baverstock *et al.* (1982) have also noted that *aconitase-1* gene in bird is non-dosage compensated.

Grula and Taylor (1980a, b) have noted that, in some butterflies a large proportion of Z linked genes control female mate selection behavior and male courtship signals. On the other hand, evolutionary studies of other butterflies have failed to show any evidence for effects of sex linked genes on morphological characters (Clarke *et al.,* 1977). At any rate, the evidences suggest that in organisms where the number of genes concerned to basic functions in the sex chromosomes is low and/or sexual dimosphism is conspicuous, the selective pressure that drives the evolution of the dosage compensation mechanism would be attenuated. Curiously, it may be noted here that Z chromosome of the species group is relatively small in size (Bull, 1983 and Ohno, 1983). In addition, in both Lepidopterans and birds, the haploid number of chromosome are usually very high.

Although the occurrance of chromosomal basis of dosage compensation system is absent in the species group, a number of sex linked genes could be dosage compensated in some animals of the groups (see Baker *et al.*, 1994). One might therefore expect that at least the cis-acting dosage compensation sequences may be evolved in a gene-by-gene basis of these group of animals for dosage compensation. Alternatively, the dosage compensation mechanism is operated in these species group at post-transcriptional level. In this context, it may be noted here that the amount of yolk in the egg is usually large in oviparous taxa. One possibility is that due to large amount of nutrients present within the eggs, significant effects of dosage differences are not expected during early development of these animal groups. Thus, the dosage compensation systems comparable to those found in species with male heterogamety could not evolve in species with female heterogamety. However, at present, it is difficult to make any firm conclusion because sufficiently detailed information is not available from these species group.

e. Dosage compensation in species with impaternate males

In some animals, males are haploid and females are diploid. Therefore, dosage of all genes is twice as great in females compared to that in males. Thus, in haplo-diplo species, phenotypic differences are expected between males and females. More likely, haploidy results in a reduction of cell size. In reality, although haplo-diplo species are characterized by very marked sexual dimorphism, the cells, wings and eyes of haploid *Habrobracon* males are almost as large as those of females (Speicher, 1935; Whiting 1943; Cock, 1964). Thus, there is a problem of compensation or adaptation in haplo-diplo species analogous to the dosage compensation problem for sex linked genes in other species. The regulatory switch between haploidy and diploidy may not automatically give optimally adapted phonotypes. It is therefore reasonable to consider that haplo-diplo systems may have evolved only as a result of selection. For instance, in some haplo-diplo insects some genes are expressed only in diploid phase (females) and not in haploid phase (males). Data on the fraction of genes whose expression is limited to females were summarised by Kerr (1962) for various haplo-diplo species. These estimate range from 14% to 46% depending on the species and kind of phenotype considered. He also calculated that 14% of the deleterious alleles in bee (*Apis*) population is limited to females. Nevertheless, somatic polyploidy and polytene is related to the state of physiological function of the cell and to the cell growth without cell division. The expected evidence was actually observed in *Apis* where many male tissues became diploid or polyploid in the later stages of development. In *Habrobracon*, however, haploid males are indeed somatically haploid. Available sets of data further indicate that all species where the males occurring in natural populations are predominantly haploid irrespective of whether diploid males are fully viable as in *Mormoniella* (Whiting, 1960) have reduced viability as in *Habrobracon* or are inviable as in *Apis* (Mackenson, 1955; Rothenbuhler, 1957). Stern (1960) argued that somatic diploidy or polyploidy is a mechanism of dosage compensation in species with impaternate males.

In summary, it appears therefore that, natural selection must have acted at least as often to increase the phenotypic differences between the two sexes as to decrease or eliminate selectively disadvantageous consequences of the haploidy in males. Somatic diploidy or polyploidy in male is thus a mechanism of dosage compensation of the haplo-diplo species (Cock, 1964, 1993).

In mealy bugs and some other coccids, the set of chromosomes inherited from the father becomes heterochromatic during early embryogenesis in males and is eliminated during spermatogenesis (Nur, 1982; Nur *et al.*, 1988). This preferential inactivation of the paternal X chromosome is one of the classic examples of genomic imprinting i.e., the phenomenon whereby the expression or transmission of a gene, a chromosome, or a whole set of chromosomes depends on the sex of the parent from which it is inherited (Crouse, 1960; Monk, 1988). In Sciara also there is selective elimination of the paternal X chromosomes in male somatic cells and elimination of the entire parental chromosome set occur during spermatogenesis (Crouse, 1960).

III. Are the regulatory switches for dosage compensation in all taxa identical?

In species with heterogametic males, dosage compensation mechanism is operative. It is clear from the taxonomic distribution of sex chromosome heteromorphism that natural selection plays an important role to evolve the mechanism of dosage compensation independently in the various taxa, many times over.

As mentioned above, mechanisms that could compensate for two fold differences might include slight adjustments in the initiation of transcription, the rate of transcriptional elongation, the stability of the mRNA, the transport of mRNA to the ribosomes and so on. This compensation could occur in either sex or in both. Although the outcome of compensation is the same, the means by which dosage compensation is accomplished varies greatly from species to species. As noted above, in *Drosophila*, the male X chromosome is upregulated, whereas in nematode, there is a reduction in the level of expression of the two X chromosome in hermaphrodites. In mammals, on the other hand, X chromosome inactivation is the means of regulating gene dosage. Eliminating unwanted chromosomes in somatic cells is also the means of regulating gene dosage of coccids.

The primary events associated with gene dosage compensation in somatic cells is the assessment of the ratio of X chromosomes to the set of autosomes. This primary signal acts to set the functional state of a collection of major regulatory genes that co-ordinately regulate both dosage compensation and sexual phenotypes (except mammals see below).

As mentioned above, in *D. melanogaster* and *C. elegans* compensation control is unified by master switch genes (such as *Sxl* gene in *Drosophila* and *xol-1*, *sdc-1*, *sdc-2* and *sdc-3* in nematodes) that may regulate the function of a number of trans-acting dosage compensation genes. It is notable that in these species the master regulators that respond most directly to the X:A ratio are mostly located on the X chromosomes. Conceivably, their location on the X chromosome is of ancestral significance. Subsequent evolution might have co-opted other sex linked sequences as additional means, until eventually the X:A counting mechanism became separate from the regulator genes themselves. Thus, the balance between regulators determines whether genes will be activated or repressed (see Section V).

Although in mammals, morphological sex is determined independently of dosage compensation, the latter is achieved by X chromosome inactivation and depends on the X:A ratio. Recent evidences however, rise the possibility that X:A ratio may play a critical role in mammalian sex determination and therefore, dosage compensation and sex determination in mammals are linked (see Section II). Chromosomal duplications that include a region of a 160 kb stretches within Xp21 of the X chromosome cause XY humans to develop as females, despite having an apparently normal Y chromosome. Duplication of the X chromosomes does not inhibit male development as XXY individuals are male. The simplest explanation of these results is that when there are two doses of X linked genes -DSS, the effect of the Y chromosome is inhibited and female differentiations is initiated. Normally in XXY individuals, this gene would be subjected to X inactivation allowing male development. Therefore, in mammals, sex differentiation is likely to occur later and depends on a dominant Y linked gene. One obvious possibility is that DSS is required for ovarian development and must be repressed in order to allow formation of testes, which synthesized male promoting hormones (Berstein *et al.*, 1980, Bardoni *et al.*, 1994).

It is possible that *Sry* (named after suspected function of sex determining region of Y) has overriden the ancestral mechanism of dosage compensation. Thus, X inactivation is one of the developmental processes that has been acquired by the mammals. As mentioned above, the X inactivation is dependent on a major switch gene on the X, inactivation centre, *Xist* (see Section II).

In summary, the comparison of the three systems (see Table-4) clearly indicates that there are some similarities on an overall basis in the dosage compensation mechanisms in the three species, but no resemblance in terms of molecular biology.

Table 4. Comparison of systems of dosage compensation

Similarities	C. elegans	D. melanogaster	Homo sapiens
X/A ratio as basic signal	XX and XO sexes	XX and X(Y) sexes	XX and XY sexes
Master regulator of compensation and/or sex determination	xol, sdc	Sxl	DSS ?/ Xist ?
Compensation mechanism	Negative	Positive	Negative
Dosage compensation genes	DCD sets dpy-21, dpy-26, dpy-27, dpy-28, dpy-30	MSL sets mle, msl-1, msl-2 msl-3	Xist gene
Acetylation of H4 mechanism	NI	Acetylated histone H4 in the region of the genome for hyperactive genes	Acetylated histone H4 in the region of active X chromosome
Multiple repetitive sequences on X	NI	CA/TG, CT/AG and C/G repeats on the X chromosome	CpG islands present on the X chromosome
Multiple numerator elements on X	Octamers ?	sis-a, sis-b, sis-c etc.	NI

NI = Not Identified

IV.   Relationship between X linked and autosomal dosage compensation

Dosage compensation of X linked genes may be considered as an evolutionary strategy required to cope with the genetic imbalance that occurs during sex determination. Conversely, the evolution of compensatory mechanisms has enabled males to survive with the deterious effects of the monosomic condition of the X chromosome. Therefore, the existence of a compensatory regulatory mechanism is to be expected. X-linked dosage compensation may evolve by acquisition of the necessary regulatory elements on the new X chromosome. Natural selection has utilized different mechanisms in different species to evolve this process. Therefore, mechnisms for dosage compensation in *Drosophila, C. elegans* and mammals are very different. More generally, the control switch for dosage compensation operates at many levels: positive and negative control, control at the transcriptional level, processing of transcripts, translational control etc. X chromosomes also evolved in such a way that the genes within the X chromosome are co-ordinately controlled. Thus, X chromosomal dosage compensation appears to be a necessary regulatory adaptation of the X chromosome haploidy.

In contrast, a different type of compensation mechanism operates on autosomal genes. When whole autosomal arms are made trisomic, the activity of many genes is reduced to diploid levels, apparently as a consequence of general systems for maintaining the balance between gene products in the cell (Devlin *et al.*, 1982, 1988). In fact, a reduction in gene product levels by autosomal hyperploids reflects a homeostatic regulatory system for maintaining a balanced rate of gene expression in diploids (Devlin *et al.*, 1988). Several lines of evidence indicate that although the autosomal dosage compensation does not involved primary level of gene action (Bhadra and Chatterjee, 1986) there are some evidences that the mechanism that operates to bring out autosomal dosage compensation may result in altered expression of unlinked genes. An analogous situation has also been noted for X linked dosage compensation. Thus, the balanced relation has been maintained between the X chromosome and autosomes in the course of evolution of X chromosomal dosage compensation. The ability of the gene-dosage-compensation in autosomes is not a prerequisite for dosage compensation in sex chromosomes. Birchler *et al.* (1990) argued that some organisms have exploited and perfected the form of autosomal dosage compensation for regulation of compensation for the neo-X chromosome. This view is strengthened by the discovery that dosage differences can be tolerated at least for a million years or two in some species of *Drosophila*, such as *D. pseudoobscura, D. miranda* and others.

V.   Models for the origin and maintenance of dosage compensation

It is now clear that originally in all lineages the ancestral sex chromosomes were morphologically identical. The heteromorphic sex chromosomes were therefore derived from a pair of autosomes for sexual differentiation (Fig. 4). The evolution of hetermorphic sex chromosomes (XY and ZW) generally involved two process (a) the partial or complete supression of crossing over between the heterologous chromosomes in the heterogametic sex, and (b) an enhancement of the accumulation of repetitive sequences leading to degeneration of Y (or W) chromosome. Thus, the Y chromosome became functionally degenerated and heterochromatic. Charlesworth (1978; 1996) suggested that the absence of recombination in heterogametic sex could lead to the building up of the deterious mutations on the Y chromosome through the action of 'Muller's ratchet (Muller, 1964). The ratchet operates because in a finite population, mutant free Y chromosome can be lost by random drift, and if there is no recombination, they can not be regenerated. When this process can continue, it would result in a gradual increase in the average number of unfavourable mutations present on the Y chromosome, although a dozen or so male specific functional genes have presumably remained active on the Y chromosome. However, it is not known, whether the unique male specific functional genes have been added recently or they have not yet had their time to be degenerated or lost. Rice (1987) proposed another mechanism for the evolution of the Y chromosome. According to Rice, genetic hitch-hiking may operate either in conjunction with Muller's ratchet in the circumstances in which the ratchet mechanism would not apply on its own. He suggested that, in absence of crossing over, selection of a beneficial allele on the Y may lead to the accumulation of dysfunctional genes in the Y chromosome. Both Rice and Charlesworth have described how their models can explain not only the evolution of structurally and functionally degenerated Y chromosome, but also the evolution of dosage compensation.

John (1988) proposed some other features of the sequence of events of Y chromosome differentation. According to John, at least in some cases, the evolution of differential sex chromosome involved a change in chromatin conformation which initially was not caused by changes in DNA sequences. A brief survey of selected species of orthopterans (Arora and Rao, 1979; Ali and Rao, 1982; Rao and Ali, 1982) provides the logical basis of a scheme of the evolution of the X (or Y) chromosome via conformational route. Conformational heteromorphism may have been initiated in more than one way : (a) Firstly, the inactivation of the region containing the sex determining gene involved its late replication (b) Secondly, conformational heteromorphism between the sex chromosomes can be initiated by structural rearrangements of the chromosomes by inversion or transposition. Jones and his co-workers (Jones, 1984, 1989; Jones and Singh, 1985) proposed that if the sex determining locus jumped by a transposable element to a site where the locus responsible for normal mitotic chromosome condensation cycle was located, the sex determining genes then took over the control of

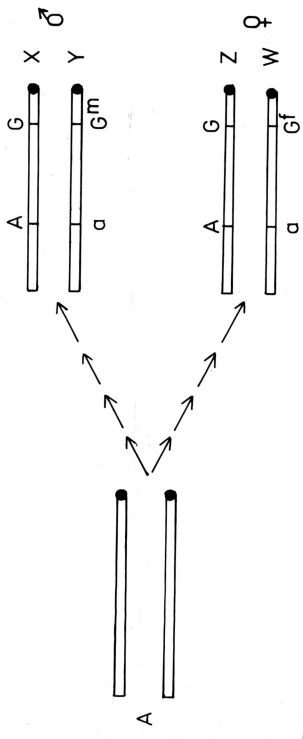

Fig. 4. A model for the formation of heteromorphic sex chromosome. The selection of suppression of recombination between the homologous chromosomes bearing the sex determining locus is the driving force in the evolution of Y chromosome. The sex limited allele at this locus is a. The hypothetical gene G has two alleles. G^m confers a benefit to males (Aa individuals) in male heterogametic (XY) individuals. Similarly, G^f allele confers a benefit to female heterogametic (ZW) individuals. In both situation, the G allele is favoured.

chromosome condensation. If the condensation pattern of the Y or W chromosome persisted into meiotic prophase and if there were one active and one inactive chromosomes, one would expect that there would be pairing failure. Conversely, if the condensation pattern of the chromosomes in the somatic cells of the heterogametic sex was modified by hijacking of a chromosome (due to transposition of sex determining gene to the control site of condensation cycle), it must result in the heterogametic sex suddenly becoming functionally hemizygous for the genes on the sex chromosomes. Since monosomy of a chromosome is lethal, it is obvious that hijacking should reduce fitness. Jones argued that as the widespread functional hemizygosity already existed in different chromosomes in the nature, the usual defects associated with monosomy were not detrimental during hijacking of a chromosome. However, the postulated inactivity of the W chromosome is contradicted by evidences from birds, Lepidopterans and snakes. Therefore, Jones hypothesis of "chromosomal hijacking" can not be account for the degeneration of Y or W chromosome and evolution of dosage compensation.

Whatever the initiating event (developmentally regulated late replication of a chromosomal region or association of a transposable element), it appears that conformational heteromorphism would result in pairing failure during meiosis in the heterogametic sex. Consequently, genetic isolation of Y or W chromosome led the evolution of their inertness largely through the accumulation of repetitive DNA sequences. Thus, the evolution of sex chromosome heteromorphism involved selective acquisition of the DNA sequences on the Y or the W chromosomes. Some of these sequences may be those which cause meiotic changes in condensation which evolve in response to selection to avoid the consequence of pairing failure. There is evidence for *D. miranda* that the neo-Y chromosome gradually acquiring the characteristics of the Y chromosome, displaying changes in DNA sequences as well as in the process of degeneration of gene activity (Steinemann, 1982; Steinemann and Steinemann, 1992, 1993; Steinemann *et al.*, 1996, see section II).

A great deal of evidence also indicates that differentiation between the sex determining chromosomes would occur as a consequence of genetic isolation of the homologue which remains restricted to one mating type of sex. Conversely, the consequence of the transformation of one of the isomorphic pair (of ancestral karyotype) of chromosome into a structural and functional degeneration of Y chromosome led to severe imbalance in the gene dosages presented on the X or Z chromosomes in males and females. As X chromosome contains genetic information concerned with basic metabolic or developmental functions equally important to males and females, a compensatory mechanism has evolved presumably for the purpose of preventing differential selection between the sexes. However, the precise routes by which dosage compensation has evolved are not clear. As discussed above, current ideas are based on the following observations: (a) the evolution of a genetically non-homologous pair of heteromorphic sex chromosomes (XX and ZW in female and XY and ZZ in males), (b) a progressive spreading of reduced recombination and (c) transformation of genetically inert Y or W chromosome. A possible route for evolution of compensation could then be illustrated as follows (Fig. 5).

It is clear from the taxonomic distribution of sex chromosome heteromorphism that the dosage compensation mechanism may have been evolved independently of its evolutionary status in different taxa, and may have originate 300 million years ago. Lucchesi (1978) argued that the evolution of sex chromosome heteromorphism is the direct consequence of evolution of dosage compensation. He suggested that gradual accumulation of functionally degenerated loci near sex determining locus on the Y chromosome would cause gene dosage differences between the sexes and that this would lead to the simultaneous evolution of the compensation mechanism. Thus, according to Lucchesi, the evolution of dosage compensation and the evolution of functionally degenarate Y chromosome are both gradual processes. On the other hand, Charlesworth (1978, 1996) claimed that the evolution of genetically inert Y is an active process rather than simply the accumulation of non-functional loci and that it is accompanied by evolution of dosage compensation. One common feature is to be noted from both the models that the increasing number of dysfunctional genes on the Y, created a selective pressure for favouring compensatory mutations on the X linked homologue. As there is a great variety of genetic and environmental modes of sex determination in different taxa, it is obvious that different animals make use of different control points to regulate dosage compensation. Interestingly, it is observed that the dosage compensation mechanism is found only male group of male heterogamety. It is possible that male heterogamety species may have exploited the meiotic inactivation mechanism which occurs in spermatogenesis (Lifschytz and Lindsley, 1972; Lifschytz, 1972). Once a mechanism of meiotic inactivation had been established it could be adjusted to operate in somatic cells modulating the activity of X chromosomes in various ways. This modulation has taken places in different species-subgroups of eukaryotes. Therefore, the adaptive modifications for dosage compensation reflects changes in several interacting gene products.

In *Drosophila*, the problem of differential dosage of X linked genes between males and females has been solved by two fold increase in the expression of X linked genes in males. Therefore, in *Drosophila*, positive regulator could cause an increase in gene activity in a sex specific manner. As a result, selection acting in males would lead to an increase in activity of their X linked genes so that eventually the cellular level of X linked gene products would be the same in both sexes. The secondary increase could be because too low a level of gene product like too high a level is

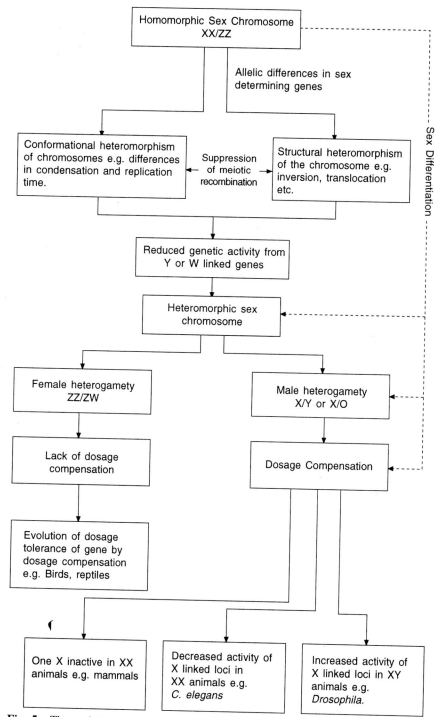

Fig. 5. The evolutionary sequence resulting for conformational and structural changes in sex chromosomes.

detrimental in itself. Analogous situation has also been noted in *C. elegans*. In these animals, the secondary decrease in X linked gene activity in females could lead to sex limited reduction in the activity of X linked genes in the homogametic sex relative to their activity in the heterogametic sex. Thus, when the gene product is higher than required level, it is energetically wasteful and consequent selection for reduction to the minimum required levels occurs.

In mammals, X chromosome inactivation results in dosage compensation. Lyon (1974) proposed that due to translocation, males had double doses of the X linked genes and a deleted Y. On the other hand, females had four copies of X linked genes. So, inactivation of one X chromosome in female somatic cells is imposed to maintain the normal double dose of each gene. Although the model was attractive at the time it was proposed, subsequent work produced results which were inconsistent with the hypothesis. Graves and her coworkers (Graves, 1987; McKay *et al.*, 1987; Wrigley and Graves, 1988; Graves and Foster, 1994) suggested that the Y chromosome of mammals has became progressively heterochromatinized and reduced during evolution and that inactivation gradually spread into the unpaired region of the X chromosome. The event of evolutionary changes for inactivation of the X chromosome was therefore initiated (McCarrey and Dilworth, 1992; Kay *et al.*, 1994) during evolution of the Y chromosome by progressive attrition and additions, so that the few genes it bears are relics of sex determination process. It is believed that the Y borne 'testis determining factor' which was identified as the *Sry* has been evolved recently. As the genes on the Y chromosome took over the male specific function of sex differentiation in mammals, it is possible that mammals may have exploited the X chromosome inactivation systems for the selective advantages of both sex determination and dosage compensation. Curiously, a series of paper has claimed that remmants of a dosage dependent sex determining (Dosage-sensitive-sex reversal gene) mechanism is still present in mammals (Bardoni *et al.*, 1994; King *et al.*, 1995).

However, the evolutionary routes for adaptive modifications for dosage compensation in groups with female heterogamety are more limited. The reason is that oocyte are usually large and long lived. Therefore, they need the product Z linked genes for maintaining their cellular physiology. Meiotic inactivation of these chromosomes is therefore selectively disadvantageous. In reality, in the oocytes of ZW females, meiotic pairing failure is avoided through euchromatinization of W rather than heterochromatinization of the Z chromosomes. Chandra (1991) also argued that as the egg size and egg content of the group of animals are usually large, the animals of these groups can survive without dosage compensation. Furthermore, as discussed above (see Section II) the number of active genes in the sex chromosomes of these species groups is low and the sexual dimorphism is conspicuous. Moreover, dosage compensation sequences have also been evolved in these species group (see Section II) in a gene-by-gene basis for dosage compensation (see Baker *et al.*, 1994). Thus, the selective pressure that drive for the evolution of dosage compensation mechanism has been attinuted in these species groups.

VI. Significance of dosage compensation

As it was noted earlier, sex determination often involves structurally distinct sex chromosomes. In heterogametic sexes, males and females exhibit severe imbalance in gene dosage imposed by heteromorphic sex chromosomes. As a result, dosage compensation mechanisms have been evolved to cope with the deleterious effects of the monosomy condition of the X chromosome in males. Obviously, the evolution of dosage compensation system allows the selective advantage of sexual dimorphism in phenotypes between males and females. Thus, there are at least some X linked genes (those concerned with sex determination) that would not be subject to dosage compensation because such genes are presumably required for gene dosage effect rather than for compensation (see Section II). Evolution of dosage compensation is therefore, allow dosage differences between the sexes and these differences could be useful to the organism by emphasizing and reinforcing mating type distinctions.

Although the evolution of heteromorphic sex chromosomes were the consequence of the evolution of dosage compensation and not *vice versa*, current knowledge on the molecular basis of sex determination and compensation events in mammals indicate that X chromosome inactivation might play a role in both the process (Chandra, 1985; King *et al.*, 1995). It may be significant in this respect that the random inactivation of one of the X chromosomes of the females in eutherian mammals could be advantageous from the point of view of selective advantages. Normally, it would be deterimental to a cell to have both the mutant and wild type alleles functioning within same cells. However, the females that have mosaic heterozygous expression do not show full effects of the deleterious recessive genes. It is possible that the normal allele of the cell populations often provides enough product of an essential gene to correct the defect of the cells that coused by the mutant allele. Alternatively, the cells that express normal allele grow in higher rate causing the elimination of cells that express the mutant allele. For example, if a female is heterozygous for mutant alleles of photopigments, her eye would be made up of a mosaic cone cells which when taken together could cover a far wider range of spectrum than any single cell would. She would effectively have three genes for colour vision. In contrast, males, being hemizygous for the X chromosome, and possessing mutant alleles of photopigments, would be colour blind. However, X inactivation can create problems when cells with mutation have a growth advantage (reminiscent of cancer cells). As a consequence, female manifest the disease usually found in males. Collectively, it appears that sex determination and dosage compensation result gain of fitness of both males and females in different aspects.

VII. Concluding remarks

It is generally believed that sexual reproduction speeds up adaptation by promoting the spread of favourable alleles and elimination of deleterious or damaged alleles (Muller, 1932). As discussed earlier, the mechanisms of sex determination in animals are greatly diverse (White, 1973; Bull, 1983; Schmid, 1983; Hagele, 1985; Steinmann-Zwicky et al., 1990; Traut and Willhoeft 1990; Villeneuve and Meyer, 1990; Bownes, 1992). The parallel evolution of sex determination systems in different groups of animals strongly suggest that although a variety of mechanisms are used for determination of sex in different species, a relatively simple evolutionary force have been involved in it. However, sex determination often involves the differentiation of the structure of the sex chromosomes. During evolution, structural changes in the Y chromosome is associated with stepwise reduction of the Y chromosome activity. The evolution of genetic inertness of the Y chromosome cause severe imbalance in gene dosage between sexes—a functional aneuploidy. The deleterious effects associated with X chromosome aneuploidy between two sexes produce a strong selection pressure to develop a regulatory mechanism for compensation. In consequence, compensatory mechanisms are adopted to restore the balance between autosomal and X chromosomal genes products. A comparative study of the mechanism of dosage compensation systems in different group of animals further suggests that it is the product of a complex evolutionary process. A seenario can be developed to explain the compensation system in different animals without greatly involving molecular mechanisms of this system. To date, the data suggest that a single principle of dosage compensation system is operative in all taxa, but there is no resemblance in terms of molecular biology (Section III). As natural selection is opportunistic and always utilise common mechanisms in different taxa, it is considered that somatic dosage compensation and X chromosome inactivation in germ line of the heterogametic sex may have evolved as independent solutions of degeneration and/or absence of a chromosome (i.e., in case of XO male) in different animals. This may lead us to suggest that different systems of dosage compensation found today may be the refinement of different biochemical processes of X chromosome regulation in the whole animal kingdom. Although the imminent understanding of the mechanisms of dosage compensation in different animal groups has yielded some insights into the evolution of dosage compensation (Muller, 1950; Charlseworth, 1978, 1991; Lucchesi, 1978, 1989; Villeneuve and Meyer, 1990; Lyon, 1992, 1993; Baker et al., 1994) to make further progress more evidence is needed on the comparative genetics and moleculer biology of sex determination and dosage compensation systems, particularly for the sex chromosomes that have originated recently.

Acknowledgments

The studies presented here have been supported in part by Council of Scientific and Industrial Research [37(0850)/94/EMR-II, dt. 28.11.1994] and an UGC grant [F. 3–10/95 (SR-II) dt. 23.12.1995]. The author also wishes to thank Dr. Rajiv Raman, Banaras Hindu University, Varanasi, and Dr. Lucas Sánchez, Centro de Investigaciones Biologicas, Madrid, Spain, for many useful discussion and criticisms on various aspects during the preparation of the manuscript.

REFERENCES

Abraham, I. and Lucchesi, J.C. (1974). Dosage compensation of genes on the left and right arms of the X chromosome of *Drosophila pseudoobscura* and *Drosophila willistoni*. *Genetics*, **78**, 1119–1126.

Ali, S. and Rao, S.R.V. (1982). Insect sex chromosome VII. Negative heteropycnosis and transcriptional activity of the X chromosome in the spermatogonia of *Acheta domesticus* (L). *Chromosoma*, **86**, 571–576.

Arora, P. and Rao, S.R.V. (1979). Insect sex chromosomes IV. DNA replication in the chromosomes of *Gryllotalpa fossor. Cytobios.*, **26**, 45–55.

Baker, B.S.; Gorman, M. and Martin, I. (1994). Dosage compensation in *Drosophila. Annu. Rev. Genet.,* **28**, 491–521.

Bardoni, B.; Zanaria, E.; Guioli, S.; Floridia, G.; Worley, K.C.; Tonini, G.; Ferrante, E.; Chiumello. G.; McCabe, E.R.B. and Fraccaro, M.I. *et al.* (1994). A dosage sensitive locus at chromosome Xp21 is involved in male-to-female sex reversal. *Nal. Genet.,* **7**, 497–501.

Barr, M.L. and Bertram, L.G. (1949). A morphological distinction between neurones of the male and female, and the behavior of the molecular statellite during accelerated nucleo-protein synthesis. *Nature*, **163**, 676–677.

Baverstock, P.R.; Adams, M.; Polkinghorne, R.W. and Gelder, M. (1982). A sex linked enzyme in bird-Z-chromosome conservation but no dosage compensation. *Nature,* **296**, 763–766.

Bell, L.R.; Horabin, J.I.; Schedl, P. and Cline, T.W. (1991). Positive autoregulation of *Sex lethal* by alternative splicing maintains the female determined state in *Drosophila. Cell*, **65**, 229–239.

Belote, J.M. and Lucchesi, J.C. (1980). Male specific lethal mutaions of *Drosophila melanogaster. Genetics,* **96**, 165–186.

Bernstein, M. and Cline, T.W. (1994). Differential effects of *Sex lethal* mutations on dosage compensation early in *Drosophila* development. *Genetics*, **136**, 1051–1061.

Berstein, R.; Koo, G.C. and Wachtel, S.S. (1980). Abnormality of the X-chromosome in the human 46XY female with dysgenic ovaries. *Science*, **207**, 768–769.

Bhadra, U. and Chatterjee, R.N. (1986). Dosage compensation and template organisation in *Drosophila: In situ* transcriptional analysis of the chromatin template activity of the X and autosomes of *Drosophila melanogaster* strains trisomic for the left arm of the second and third chromosomes. *Chromosoma*, **94**, 285–292.

Birchler, J.A.; Hiebert, J.C. and Paigen, K. (1990). Genetic basis of autosomal dosage compensation in *Drosophila melanogaster. Genetics*, **124**, 677–686.

Bone, J.R.; Lavender, J.; Richman, R.; Palmer, M.J.; Turner, B.M. and Kuroda, M.I. (1994). Acetylated histone H_4 on the male X chromosome is associated with dosage compensation in *Drosophila. Genes Dev.,* **8**, 96–104.

Bopp, D.; Bell, L.R.; Cline, T.W. and Schedl, P. (1991). Developmental distribution of female-specific *Sex-lethal* proteins in *Drosophila melanogaster. Genes Dev.*, **5**, 403–415.

Bownes, M. (1992). Molecular aspects of sex determination in insects. In *'Insect Molecular Science'* (J.M. Crampton and P. Eggleston eds.) pp 76–100. Academic Press, London.

Bridges, C.B. (1921). Triploid intersexes in *Drosophila melanogaster. Science*, **54**, 252–254.

Brockdorff, N.; Ashworth, A.; Kay, G.F.; McCabe, V.M.; Norris, D.P.; Cooper, P.J.; Swift, S. and Rastan, S. (1992). The product of the mouse *Xist* gene is a 15 kb inactive X specific transcript containing no conserved ORF and located in the nucleus. *Cell,* **71**, 515–526.

Brown, C.J.; Hendrich, B.D.; Rupert, J.L.; Lafrenierge, R.G.; Xing, Y.; Lawrance, J. and Willard, H.F. (1992). The human *Xist* gene: analysis of a 17 kb inactive X specific RNA that contains conserved repeats and is highly localized within the nucleus. *Cell,* **71**, 527–542.

Bull, J.J. (1983). Evolution of sex determining mechanisms. Benjamin/Cummings, California.

Burla, H.; Cunha, A.B.D.A.; Cordeiro, A.R.; Dobzhansky, T.; Malagolowkin, C. and Pavan, C. (1949). The *willistoni* group of sibling species of *Drosophila. Evolution,* **3**, 300–314.

Carson, H.L. and Stalker, H.D. (1968). Polytene chromosome relationships in Hawaiian species of *Drosophila* I. The *Drosophila grimshawi* subgroup. *Univ. Texas Publ.,* **6818**, 335–354.

Chandra, H.S. (1985). Is human X chromosome inactivation a sex determining device? *Proc. Natl. Acad. Sci* (USA), **82**, 6947–6949.

Chandra, H.S. (1991). How do heterogametic females survive without gene dosage compensation? *J. Genet.*, **70**, 137–146.

Chandra, H.S. and Brown, S.W. (1975). Chromosome imprinting and the mammalian X chromosome. *Nature,* **253**, 165–168.

Charlesworth. B. (1978). Model for evolution of Y chromosomes and dosage compensation. *Proc. Natl. Sci. Acad.* (USA), **75**, 5618–5622.

Charlesworth, B. (1991). The evolution of sex chromosomes. *Science*, **251**, 1030–1033.

Charlesworth, B. (1996). The evolution of chromosomal sex determination and dosage compensation. *Current Biol* **6**, 149–162.

Chatterjee, R.N. (1985). X chromosomal organisation and dosage compensation: *In situ* transcription of chromatin template activity of X chromosome hyperploids of *Drosophila melanogaster.* Chromosoma, **91**, 259–266.

Chatterjee, R.N. (1992). Mechanisms of X chromosome regulation in *Drosophila melanogaster. The nucleus,* **35**, 31–44.

Chatterjee, R.N. and Mukherjee, A.S. (1977). Chromosomal basis of dosage compensation in *Drosophila* IX. Cellular autonomy of the faster replication of the X chromosome in haplo X cells of *Drosophila melanogaster* and synchronous initiation. *J. Cell Biol.,* **74**, 168–180.

Chatterjee, R.N. and Mukherjee, A.S. (1981). Chromosomal basis of dosage compensation in *Drosophila* X. Assessment of hyperactivity of the male X *in situ. J. Cell Sci.,* **47**, 295–309.

Chatterjee, R.N.; Mukherjee, A.S.; Derksen, J. and Ploeg, M.V.D. (1980). Role of non-histone chromosomal protein in the attainment of hyperactivity of the X chromosome of male *Drosophila*: A quantitative cytochemical study. *Ind. J. Exptl. Biol.,* **18**, 574–576.

Chatterjee, R.N.; Bunick, D.; Manning, J.E. and Lucchesi, J.C. (1992). Control of LSP

1 α gene expression of *Drosophila melanogaster* and ectopic sites. *Pers. Cytol. Genet.,* **7** (G.K. Manna and S.C. Ray, eds.) pp 323–336.

Chino, M. (1936). A case of inversion of the fifth chromosome of *Drosophila virilis. Jap. J. Genet.,* **12,** 63–65.

Chuang, P.; Albertson, D.G. and Meyer, B.J. (1994). DPY-27: a chromosome condensation protein homolog that regulates *C. elegans* dosage compensation through association with X. chromosome. *Cell,* **79,** 459–474.

Clarke, C.A.; Mittwoch, U. and Traut, W. (1977). Linkage and cytogenetic studies in the swallow tail butterflies, *Papilio polyxenes* Fab and *Papilio machaon* L. and their hybrids. *Proc. R. Soc. Lond. B. (Biol. Sci.),* **198,** 385–399.

Cline, T.W. (1984). Autoregulatory functioning of a *Drosophila* gene product that establishes and maintains the sexually determined state. *Genetics,* **107,** 231–277.

Cline, T.W. (1985). Primary events in the determination of sex in *Drosophila melanogaster.* In *'Origin and Evolution of Sex'* (H.O. Halvorson and A. Monroy eds.) pp 301–327, Liss, New York.

Cline. T.W. (1986). A female specific lethal lesion in an X linked positive regulator of the *Drosophila* sex determination gene, *Sex lethal. Genetics,* **113,** 641–663.

Cline, T.W. (1988). Evidence that *sisterless*-a and *sisterless*-b are two of several discrete 'Numerator Elements' of the X/A sex determination signal in *Drosophila* that switch *Sxl* between two alternative stable expression states. *Genetics,* **119,** 829–862.

Cline, T.W. (1993). The *Drosophila* sex determination signal: how do flies count to two? *Trends Genet.,* **9,** 385–390.

Cock, A.G. (1964). Dosage compensation and sex-chromatin in non-mammals. *Genet. Res.,* **5,** 354–365.

Cock, J.M. (1993). Sex determination in the Hymenoptera: A review of models and evidence. *Heredity,* **71,** 421–435.

Cooper, D.W.; Vandeberg, J.L.; Sharman, G.B. and Poole, W.E. (1971). Phosphoglycerate-kinase polymorphism in kangaroos provides further evidence for paternal X-inactivation. *Nature New Biol.,* **230,** 155–157.

Cornmiller, C.; Schedl, P. and Cline, T.W. (1988). Molecular characterization of *daughterless, Drosophila* sex determination gene with multiple roles in development. *Genes Dev.,* **2,** 1666–1676.

Crouse, H.V. (1960). The controlling element in sex chromosome behaviour in Sciara. *Genetics,* **45,** 1429–1443.

Danilevskaya, O.; Lofsky, A.; Kuronova, E.L. and Pardue, M.L. (1993). The Y chromosome of *Drosophila melanogaster* contains a distinctive subclass of Het-A related repeats. *Genetics,* **134,** 531–543.

Delong, L.; Plenefisch, J.D.; Klein, R.D. and Meyer, B.J. (1993). Feedback control of sex determination by dosage compensation revealed through *Caenorhabdities elegans, sdc*-3 mutations. *Genetics,* **133,** 875–896.

DePamphilis, M.L. (1993). Eukaryotic DNA replication: Anatomy of an origin. *Annu. Rev. Biochem.,* **62,** 29–63.

Devlin, R.H.; Holm, D.G. and Grigliatti, T.A. (1982). Autosomal dosage compensation in *Drosophila melanogaster* strains trisomic for the left arm of chromosome 2. *Proc. Natl. Acad. Sci* (USA), **79,** 1200–1204.

Devlin, R.H.; Holm, D.G. and Grigliatti, T.A. (1988). The influence of whole arm trisomy on gene expression in *Drosophila. Genetics,* **118,** 87–101.

DiBartolomeis, S.M.; Tartof, K.D. and Jackson, F.R. (1992). A superfamily of *Drosophila* satellite related (SR) DNA repeats restricted to the X chromosome euchromatin. *Nucl. Acid. Res.,* **20,** 1113–1116.

Dobzhansky, T. (1957). The X chromosome in the larval salivary glands of hybrids *Drosophila insularis* X *Drosophila tropicalis. Chromosoma,* **8,** 691–698.

Dobzhansky, T. and Socolov, D. (1939). Structure and variation of the chromosomes in *Drosophila azteca. J. Hered.*, **30**, 3–19.

Dobzhansky, T. and Tan, C.C. (1936). Studies in hybrid sterility III. A comparison of the gene arrangement in two species, *Drosophila pseudoobscura* and *Drosophila miranda. Z i A V*, **72**, 88–114.

Dobzhansky, T.; Ehrman, L. and Pavlovsky, O. (1957). *Drosophila insularis*, a new sibling species of the *willistoni* group. *Univ. texas Publ.*, **5714**, 39–47.

Donahue, R.P. (1972). Cytogenetic analysis of the first cleavage division in mouse embryos. *Proc. Natl. Acad. Sci* (USA), **69**, 74–77.

Dorbic, T. and Witting, B (1987). Chromatin from transcribed genes contain HMG 17 only downstream from the starting point of transcription. *EMBO J.*, **6**, 2373–2379.

Duffy, J.B. and Gergen, J.P. (1991). The *Drosophila* segmentation gene *runt* acts as a position specific numerator element necessary for the uniform expression of the sex determining gene *Sex lethal. Genes Dev.,* **5**, 2176–2187.

Ecklund, P.S. and Levine, L. (1975). Mouse sperm basic nuclear protein. Electrophoretic characterization and fate after fertilization. *J. Cell Biol.,* **66**, 251–262.

Ellis, H.M.; Spann, D.R. and Posakony, J.W. (1990). *extramacrochaetae*, a negative regulator of sensory organ development in *Drosophila,* defines a new class of Helix-Loop-Helix proteins. *Cell*, **61**, 27–38.

Erickson, J.W. and Cline, T.W. (1991). The molecular nature of the *Drosophila* sex determination signal and its link to neurogenesis. *Science,* **251**, 1071–1074.

Erickson, J.W. and Cline, T.W. (1993). A bZIP protein, *sisterless*-a collaborates with bHLH transcription factors early in *Drosophila* development to determine sex. *Genes Dev.*, **7**, 1688–1702.

Foster, G.G.; Whitten, M.J.; Konvalov, C.; Arnold, J.T.A. and Maffi, G. (1981). Autosomal genetic maps of the Australian sheep blowfly: *Lucilia Cuprina dorsalis* (Diptera: Calliphoridae) and possible correlations with the linkage maps of *Musca domestica* and *Drosophila melanogaster. Genet. Res.,* **37**, 55–69.

Fredga, K. (1970). Unusual sex chromosome inheritance in mammals. *Philos Trans. Roy. Soc. (Lond.) B. (Biol. Sci.)*, **259**, 15–36.

Fredga, K. (1983). Aberrant sex chromosome mechanisms in mammals: evolutionary aspects. *Differentiation,* **23**, S23–S30.

Gartler, S.M. and Riggs, A.D. (1983). Mammalian X chromosome inactivation. *Annu. Rev. Genet.*, **17**, 155–190.

Ghosh, S.; Chatterjee, R.N.; Bunick, D.; Manning, J.E. and Lucchesi, J.C. (1989). The LSP 1α gene of *Drosophila melanogaster* exhibits dosage compensation when it is relocated to a different site on the X chromosome. *EMBO J.,* **8**, 1191–1196.

Gorman, M and Baker, B.S. (1994). How flies make one equal two: dosage compensation in *Drosophila. Trends Genet.*, **16**, 376–380.

Granadino, B.; Campuzano, S. and Sánchez, L. (1990). The *Drosophila molanogastor fl (2)d* gene is needed for the female specific splicing of *Sex lethal* RNA. *EMBO J.*, **9**, 2597–2602.

Granadino, B.; Sanjuan, A.; Santamaria, P. and Sanchez, L. (1992). Evidence of a dual function in *fl(2)d a* gene needed for *Sex lethal* expression in *Drosophila melanogaster. Genetics,* **130**, 597–612.

Grant, S.G. and Chapman, V.M. (1988). Mechanisms of X chromosome regulation. *Annu. Rev. Genet.*, **22**, 199–233.

Graves, J.A.M. (1987). The evolution of mammalian sex chromosomes and dosage compensation: clues from marsupials and monotremes. *Trends Genet.*, **3**, 252–256.

Graves, J.A.M. (1995). The origin and function of the mammalian Y chromosome and Y borne genes—an evolving understanding. *BioEssays,* **17**, 311–321.

Graves, J.A.M. and Watson, J.M. (1991). Mammalian sex chromosomes: evolution of organisation and function. *Chromosoma, 101*, 63–68.

Graves, J.A.M. and Foster, J.W. (1994). Evolution of mammalian sex chromosomes and sex determining genes. *Inter. Rev. Cytol., 154*, 191–259.

Grula, J.W. and Taylor, O.R. Jr. (1980a). Some characteristics of hybrids derived from the sulfur butterflies, *C. eurytheme* and *E. philodice*. *Evolution, 34*, 673–687.

Grula, J.W. and Taylor, O.R. Jr. (1980b). The effect of X chromosome inheritance on mate-selection behaviour in the sulfur butterflies, *Colias eurytheme* and *Colias philodice*. *Evolution, 34*, 688–895.

Hagele, K. (1985). Identification of a polytene chromosome band containing a male sex determiner of *Chironomus thummi thummi*. *Chromosoma. 89*, 37–41.

Hodgkin, J. (1990). Sex determination compared in *Drosophila* and *Caenorhabditis*. *Nature, 344*, 721–728.

Hofmann, A. and Korge, G. (1987). Upstream sequences of dosage compensated and non-dosage compensated alleles of the larval secretion protein gene sgs-4 in *Drosophila*. *Chromosoma, 96*, 1–7.

Huijser, P.; Hennig, W. and Dijkhof, R. (1987). Poly (dC-dA/dG-dT) repeats in the *Drosophila* geneome: a key function for dosage compensation and position effects? *Chromosoma, 95*, 209–215.

Hughes, R.D. (1939). An analysis of the chromosomes of two sub-species *Drosophila virilis* and *Drosophila virilis americana*. *Genetics, 24*, 811–834.

Hsu, D.R. and Meyer, B.J. (1993). X chromosome dosage compensation and its relationship to sex determination in *C. elegans*. *Semin. Dev. Biol., 4*, 93–106.

Hsu, T.C. (1952). Chromosomal variation and evolution in the *virilis* group of *Drosophila*. *Univ. Texas Publ., 5204*, 35–72.

Jaffe, E. and Laird, C. (1986). Dosage compensation in *Drosophila*. *Trends Genet., 2*, 316–321.

Jeppesen, P. and Turner, B.M. (1993). The inactive X chromosome in female mammals is distinguished by a lack of histone H_4 acetylation, a cytogenetic marker for gene expression. *Cell, 74*, 281–289.

John, B. (1988). The biology of heterochromatin. In 'Heterochromatin: Molecular and Structural aspects (R.S. Verma ed.) p 1–147. Cambridge University Press, Cambridge.

Johnson, M.S. and Turner, J.R.G. (1979). Absence of dosage compensation for a sex linked enzyme in butterflies. (*Heliconius*). *Heredity, 43*, 71–77.

Jones, K.W. (1984). The evolution of sex chromosomes and their consequeces for the evolutionary process. In *'Chromosomes Today'*. Vol 8 (M.D. Bennett, A. Gropp and U. Wolf eds.). pp 241–255. Allen and Unwin, London.

Jones, K.W. (1989). Inactivation phenomena in the evolution and function of sex chromosomes. In *'Evolutionary Mechanisms in Sex Determination'* (S.S. Wachtel ed.) pp. 69–78 CRC Press Florida.

Jones, K.W. and Singh, L. (1985). Snakes and the evolution of sex chromosomes. *Trends Genet., 1*, 55–61.

Kay, G.F.; Barton, S.C.; Surani, M.A. and Rastan, S. (1994). Imprinting and X chromosome counting mechanisms determine *Xist* expression in early mouse development. *Cell, 77*, 639–650.

Kelley, R.L.; Solovyeva, I.; Lyman, L.M.; Richman, R.; Solovyev, V. and Kuroda, M.I. (1995). Expression of *msl-2* causes assembly of dosage compensation regulators on the X chromosomes and female lethality in *Drosophila*. *Cell, 81*, 867–872.

Kerr, W.E. (1962). Genetics of sex determination . *Annu. Rev. Ent., 7*, 157–176.

Khesin, R.B. (1973). Binding of thymus histone F1 and *E. coli* RNA polymerase to DNA of polytene chromosomes of *Drosophila*. *Chromosoma, 44*, 255–265.

Kikkawa, H. (1938). Studies on the genetics and cytology of *Drosophila ananassae*. *Genetica*, **20**, 458–516.

King, V.; Korn, R.; Kwok, C.; Ramkissoon, Y.; Wanaderle, V. and GoodFellow, P. (1995). One for a boy, two for a girl? *Current Biol.*, **5**, 37–39.

Kistler, W.S.; Geroch, M.E. and Williams-Ashman, H.G. (1973). Specific basic proteins from mammalian testes. Isolation and properties of small basic proteins from rat testes and epididymal spermatozoa. *J. Biol. Chem.*, **248**, 4532–4543.

Klein, R.D. and Meyer, B.J. (1993). Independent domains of the *sdc-3* protein control sex determination and dosage compensation in *C. elegans*. *Cell*, **72**, 349–364.

Korge, G. (1981). Genetic analysis of the larval secretion gene *sgs*-4 and its regulatory chromosome sites in *Drosophila melanogaster*. *Chromosoma*, **84**, 373–390.

Kuroda, M.I.; Kernan, M.J.; Kreber, R.; Ganetzky, B. and Baker B.S. (1991). The *maleless* protein associates with the X chromosome to regulate dosage compensation in *Drosophila*. *Cell*, **66**, 935–947.

Kuroda, M.I.; Plamer, M.J. and Lucchesi, J.C. (1993). X chromosome dosage compensation in *Drosophila*. *Semin, Dev. Biol.*, **4**, 107–116.

Lancefield, D.E. (1922). Linkage relations of sex linked characters in *Drosophila obscura*. *Genetics*, **7**, 335–384.

Lemeunier, F. and Ashburner, M. (1976). Relationship within the *melanogaster* species subgroup of the genus *Drosophila* (Sophophora) II. Phylogenetic relationships between six species based upon polytene chromosome banding sequences. *Proc. R. Soc. (Lond. B.)*, **193**, 275–294.

Lifschytz, E. (1972). X chromosome inactivation: an essential feature of normal spermiogenesis in male heterogametic organisms. *Proc. Int. Symp. Genet. Spermatozoon.* (RA Beatty and S. Gluecksohn Waelsch ed.), pp 223–232.

Lifschytz, E. and Lindsley, D.L. (1972). The role of X chromosome inactivation during spermatogenesis. *Proc. Natl. Acad. Sci. (USA)*, **69**, 182–186.

Lowenhaupt, K.; Rich, A. and Pardue, M.L. (1989). Non-random distribution of mono and dinucleotide repeats in *Drosophila* chromosomes: correlations with dosage compensation, heterochromatin and recombination. *Mol. Cell Biol.*, **9**, 1173–1182.

Lucchesi, J.C. (1978). Gene dosage compensation and the evolution of sex chromosomes. *Science*, **202**, 711–716.

Lucchesi, J.C. (1989). On the origin of the mechanism compensating for gene-dosage differences in *Drosophila*. *Amm. Nat.*, **134**, 474–485.

Lucchesi, J.C. and Manning, J.E. (1987). Gene dosage compensation in *Drosophila melanogaster*. *Adv. Genet.*, **24**, 371–429.

Lyon, M.F. (1961). Gene action in the X chromosome of the mouse (*Mus musculus* L). *Nature*, **190**, 372–373.

Lyon, M.F. (1974). Mechanisms and evolutionary origins of variable X chromosome activity in mammals. *Proc. R. Soc. (Lond. B.)*, **187**, 243–268.

Lyon, M.F. (1988). X chromosome inactivation and the location and expression of X linked genes. *Amer. J. Human Genet.*, **42**, 8–16.

Lyon, M.F. (1992). Some milestones in the history of X chromosome inactivation. *Annu. Rev. Genet.* **26**, 15–27.

Lyon, M.F. (1993). Controlling the X chromosome. *Current Biol.*, **3**, 242–244.

Mackenson, O. (1955). Further studies on a lethal series in the honey bee. *J. Hered.*, **46**, 72–74.

Malacrida, A.; Gasperi, G.; Biscaldi, G.F. and Milani, R. (1984). Functional significance of gene clusters in the housefly, *Musca domestica* and in other Diptera. Atti, 2′ Congr. Soc. Ital. E.

Mainx, F.; Koske, Th. and Smital, E. (1953). Untersuchungen Uber dic chromosomale struktur europaischer vertreterder *Drosophila obscura. Gruppe Z Vererbungsl,* **85**, 354–372.

Mattei, J.F.; Mattei, M.G.; Baeteman, M.A. and Giraud, F. (1981). Trisomy 21 for the region 21q223: Identification by high resolution R-banding patterns. *Human Genet,* **56**, 409–411.

McBurney, M.W. (1988). X chromosome inactivation: A hypothesis. *BioEssays.,* **9**, 85–88.

McCarrey, J.R. and Dilworth, D.D. (1992). Expression of *Xist* in mouse germ cells correlates with X chromosome inactivation. *Nat. Gent.,* **2**, 202–203.

McKay, L.M.: Wrigley, J.M. and Graves, J.A.M. (1987). Evolution of mammalian X inactivation: Sex chromatin in monotremes and marsupials. *Aus. J. Biol. Sci.,* **40**, 397–404.

Metz, C.W. (1916). Additional types of chromosome groups in the Drosophilidae. *Amer. Nat.,* **50**, 587–599.

Migeon, B.R. (1994). X chromosome inactivation: molecular mechanisms and genetic consequences. *Trends. Genet.,* **10**, 230–234.

Miller, D.D. (1939). Structure and variation of the chromosomes in *Drosophila algonquin. Genetics,* **24**, 699–708.

Monk, M. (1986). Methylation and the X chromosome. *BioEssays,* **4**, 294–208.

Monk, M. (1988). Genomic imprinting. *Genes Dev.,* **2**, 921–925.

Moore, T. and Haig, D. (1991). Genomic imprinting in mammalian development: a paternal tug-of-war. *Trends Genet,* **7**, 45–49.

Mukherjee, A.S. and Beermann, W. (1965). Synthesis of ribonucleic acid by the X chromosomes of *Drosophila melanogaster* and the problem of dosage compensation. *Nature,* **207**, 785–786.

Mukherjee, A.S. and Chatterjee, S.N. (1975). Chromosomal basis of dosage compensation in *Drosophila* VIII. Faster replication and hyperactivity of both arms of the X chromosome in males of *Drosophila pseudoobscura* and their possible significance. *Chromosoma,* **53**, 91–105.

Muller, H.J. (1932). Further studies on the nature and causes of gene mutation. Proc 6th Int. Cong. Genet. 1. 213–255.

Muller, H.J. (1940). Bearing the *Drosophila* work on systematics. The New systematics. pp 185–268. Oxford University Press, Oxford.

Muller, H.J. (1950). Evidence of the precision of genetic adaptation. *Harvey Lect.,* **43**, 165–229.

Muller, H.J. (1964). The relation of recombination to mutational advance. *Mutant. Res.,* **1**, 2–9.

Muller, H.J.; Laegue, B.B. and Offerman, C.A. (1931). Effects of dosage changes of sex linked genes and the compensatory effect of other gene difference between male and female. *Anat. Rec.,* **51** (Suppl.), 110.

Murtagh. C.E. (1977). A unique cytogenetic system in monotremes. *Chromosoma,* **65**, 37–57.

Nonet, M.I. and Meyer, B.J. (1991). Early aspects of *Caenorhabdities elegans* sex determination and dosage compensation are regulated by a Zinc finger protein. *Nature,* **351**, 65–68.

Norris, D.P.; Patel, D; Kay, G.F.; Penny, G.D.; Brockdorff, N.; Sheardown, S.A. and Rastan, S. (1994). Evidence that random and imprinted *Xist* expression is controlled by presumptive methylation. *Cell,* **77**, 41–51.

Nur, U. (1982). Evolution of unusal chromosome systems in scale insects (Coccoidea: Homoptera). In 'Insect cytogenetics, symposia of the Royal Entomological Soc. Lond. (RL Blackman GM Hewitt and M. Ashburner ed) pp 97–117. Blackwell Scientific Publication, Oxford.

Nur, U.; Werren, J.H.; Eickbush, D.G.; Burke, W.D. and Eickbush, T.H. (1988). A 'selfish' B chromosome that enhances its transmission by eliminating the paternal genome. *Science,* **240**, 512–514.

Ohno, S. (1967). Sex chromosomes and sex linked genes. Springer Verlag. Berlin.

Ohno, S. (1983). Phylogeny of the X chrmosome of man. In 'Cytogenetics of the Mammalian X chromosome. Part A.' Basic mechanisms of X chromosome behaviour (A.A. Sandberg ed) pp 1–19 Alan R. Liss Inc. New York.

Oliver, B.; Perriman, N. and Mahawald A.P. (1988). Genetic evidence that the *sans-fille* locus is involved in *Drosophila* sex determination. *Genetics,* **125**, 535–550.

Palmer, M.J; Mergner, V.A.; Richman, R.; Manning, J.E.; Kuroda, M.I. and Lucchesi, J.C. (1993). The male-specific lethal one (*msl-1*) gene of *Drosophila melanogaster* encodes a noval protein that associates with the X chromosome in males. *Genetics,* **134**, 545–557.

Palmer, M.J.; Richman, R.; Richter, L. and Kuroda, M.I. (1994). Sex specific regulation of the *male specific lethal-1* dosage compensation gene in *Drosophila. Genes Dev.,* **8**, 698–706.

Pardue, M.L.; Lowenhaupt, K.; Rich, A. and Nordheim, A. (1987). (dC-dA)n (dG-dT)n sequences have evolutionarily conserved chromosomal locations in *Drosophila* with implications for roles in chromosome structure and function. *EMBO J.,* **6**, 1781–1789.

Parkhurst, S.M. and Meneely, P.M. (1994). Sex determination and dosage compensation: lessions from flies and warms. *Science,* **264**, 924–932.

Patterson, J.T. (1944). A new member of *virilis* group. *Univ. Tex. Pub.,* **4445**, 102–103.

Patterson, J.T. and Stone, W.S. (1952). Evolution in the genous *Drosophila.* Macmillan, New York.

Patterson, J.T.; Stone, W.S. and Griffen, A.B. (1942). Genetic and cytological analysis of the *virilis* species group. *Univ. Tex. Pub.,* **4228**, 162–200.

Penny, G.D.; Kay, G.F.; Sheardown, S.A.; Rastan, S. and Brockdorff, N. (1996). Requirement for *Xist* in X-chromosome inactivation. *Nature,* **379**, 131–137.

Rao, P.M. and Ranganath, H.A. (1991). Karyotype differentiation among members of the *immigrans* species group of *Drosophila. Genetica,* **83**, 145–152.

Rao, S.R.V. and All, S. (1982). Insect sex chromosomes VI. A presumptive hyperactivation of the male X chromosome in *Acheta domensticus* (L). *Chromosoma,* **86**, 325–339.

Rastan, S. (1994). X chromosome inactivation and the *Xist* gene. *Current Biol.,* 4, 292–295.

Rice, W.R. (1987). Genetic hitchhiking and the evolution of reduced genetic activity of the Y sex chromosome. *Genetics,* **116**, 161–167.

Richardson, B.J.; Czuppon, A.B. and Sharman, G.B. (1971). Inheritance of glucose-6-phosphate dehydrogenase variation in kangaroos. *Nature New Biol.,* **230**, 154–155.

Roberts, D.B. and Evans-Roberts, S. (1979). The X linked α-chain gene of *Drosophila* LSP 1α does not show dosage compensation. *Nature,* **280**, 691–692.

Rothenbuhler, W.C. (1957). Diploid male tissue as new evidence on sex determination in honey bees. *J. Hered.,* **48**, 160–168.

Russell, L.B. (1963). Mammalian X chromosome action: Inactivation limited in spread and in region of origin. *Science,* **140**, 976–978.

Schmid, M. (1983). Evolution of sex chromosomes and heterogametic systems in Amphibia. *Differentiation,* **23**, (Suppl.) S13–S22.

Sharman, G.B. (1971). Late DNA replication in the paternally derived X chromosome in female kangaroos. *Nature,* **230**, 231–232.

Smithies, O. (1982). The control of globin and other eukaryotic genes. *J.Cell Phys. (Suppl),* **1:** 137–143.

Solter, D. (1988). Differential imprinting and expression of maternal and paternal genomes. *Annu. rev. Genet.,* **22**, 127–146.

Speicher, B.R. (1935). Cell size and chromosomal types in *Habrobracon. Amer. Nat.,* **69**, 79–80.

Spencer, W.P. (1949). Gene homologies and mutants in *Drosophila hydei,* pp 23–44. In *'Genetics Paleontology and Evolution'.* Princeton, Univ. Press, Princeton.

Spradling, A. and Orr-Weaver, T. (1987). Regulation of DNA replication during *Drosophila* development. *Annu. Rev. Genet.,* **21**, 373–403.

Steinemann, M. (1982). Multiple sex chromosomes in *Drosophila miranda:* a system for study the degeneration of a chromosome. *Chromosoma,* **86**, 59–76.

Steinemann, M. and Steinemann, S. (1991). Preferential Y chromosomal location of TRIM, a noval transposable element of *Drosophila miranda,* obscura group. *Chromosoma,* **101**, 169–179.

Steinemann, M. and Steinemann, S. (1992). Degenerating Y chromosome of *Drosophila miranda*: A trap for retrotransposons. *Proc. Natl Acad. Sci. (USA),* **89**, 7591–7595.

Steinemann, M. and Steinemann, S. (1993). A duplication including the Y allele of Lcp^2 and the TRIM retrotransposon at the *Lcp* locus on the degenerating neo-Y chromosome of *Drosophila miranda*: Molecular structure and mechanisms by which it may have arisen. *Genetics,* **134**, 497–505.

Steinemann, M., Steinemann, S. and Turner, B.M. (1996). Evolution of dosage compensation. *Chromosome Res.,* **4**, 185–190.

Steinmann–Zwicky, M. (1988). Sex determination in *Drosophila;* the X chromosomal gene *liz* is required for *Sxl* activity. *EMBO J.,* **7**, 3889–3898.

Steinmann-Zwicky, M.; Amrein, H. and Nothiger, R. (1990). Genetic control of sex determination in *Drosophila. Adv. Genet.,* **27**, 189-237.

Stern, C. (1960). Dosage compensation-development of a concept and new facts. *Cand. J. Genet. Cytol.,* **2**, 105–118.

Stone, W.S. and Patterson, J.T. (1947). The species relationships in the *virilis* group. *Univ. Texas Pub.,* **4720**, 157–160.

Stone, W.S.; Griffen, A.B. and Patterson, J.T. (1942). *Drosophila montana* a new species of the *virilis* group. *Genetics,* **27**, 172.

Sturtevant, A.H. (1921). The North American species of *Drosohilla. Carne. Inst. Publ.,* **301**, 1–150.

Sturtevant, A.H. (1940). Genetic data on *Drosophila affinis* with discussion of the relationships in the subgenus *Sophophora. Genetics,* **25**, 337–353.

Sturtevant, A.H. and Dobzhansky, T. (1936). Observations on the species related to *Drosophila affinis,* with descriptions of seven new forms. *Am. Nat.,* **70**, 574–584.

Sturtevant, A.H. and Novitski, E. (1941). The homologies of the chromosome elements in the genus *Drosophila. Genetics,* **26**, 517–547.

Tan, C.C. (1935). Salivary gland chromosomes in the two races of *Drosophila pseudoobscura. Genetics,* **20**, 392–402.

Therman, E.; Sarto, G.E. and Patau, K. (1974). Centre for Barr body condensation on the proximal part of the human Xq : A hypothesis. *Chromosoma,* **44**, 361–366.

Throckmorton, L.H. (1975). The phylogeny, ecology and geography of *Drosophila.* In *'Hand Book of Genetics'* Vol. III. R.C. King ed., pp 421–469. Planum Press, New York.

Torres, M. and Sánchez, L. (1989). The scute (T4) gene acts as a numerator element of the X: A signal that determines the state of activity of *Sex lethal* in *Drosophila. EMBO J.,* **8**, 3079–3086.

Torres, M. and Sánchez, L. (1992). The segmentation gene *runt* is needed to activated *Sex lethal,* a gene that controls sex determination and dosage compensation in *Drosophila. Genet. Res.,* **59**, 189–198.

Traut, W. and Willhoeft, U. (1990). A jumping sex determining factor in the fly *Megaselia scalans. Chromosoma,* **99**, 407–412.

Trempe, J.P.; Lindstrom, Y.I. and Leftak, M. (1988). Opposite replication polarities of transcribed and non-transcribed histone H5 genes. *Mol. Cell Biol.,* **8**, 1657–1663.

Turner, B.M.; Birley, A.J. and Lavender, J. (1992). Histone H_4 isoforms acetylated at specific lysine residues define individual chromosomes and chromatin domains in *Drosophila* polytene nuclei. *Cell,* **69**, 375–384.

Uchida, S.; Uenoyams, T. and Oishi, K. (1981). Studies on the sex specific lethals of *Drosophila melanogaster:* III. A third chromosome male specific lethal mutant. *Jap. J. Genet.,* **56**, 523–527.

Villeneuve, A.M. and Meyer, B.J. (1990). Regulatory hierarchy controlling sex determination and dosage compensation in *Caenorhabdities elegans. Adv. Genet.,* **27**, 177–188.

Wakhoma, K. and Kitagawa, O. (1972). Evolutionary and genetical studies of *Drosophila nasuta* subgroup II. Karyotypes of *Drosophila nasuta* collected from the Scychelles Islands. *Jap. J. Genet.,* **47**, 129–131.

Waring, G.L. and Pollack, J.C. (1987). Cloning and characterization of a dispersed, multicopy, X chromosome sequence in *Drosophila melanogaster. Proc. Natl. Acad. Sci. (USA),* **84**, 2843–2847.

Wasserman, M. (1962). Cytological studies of the *repleta* group of the geneus *Drosophila* IV. The *mulleri* subgroup. Stud. Genet. II. *Univ. Texas Publ.,* **6205**, 73–84.

White, M.J.D. (1973). Animal cytology and evolution. 3rd Ed. Cambridge University Press, cambridge.

Whiting, P.W. (1943). Multiple alleles in complementary sex determination in *Habrobracon. Genetics,* **28**, 365–382.

Whiting, P.W. (1960). Polyploidy in *Mormoniella. Genetics,* **45**, 949–970.

Williams, N. (1995). How males and females achieve X equality? *Science,* **269**, 1826–1827.

Wolf, S.F.; Jolly, D.J.; Lunnen, K.D.; Axelman, J. and Migeon, B.R. (1984). Methylation of the hypoxanthine phosphoribosyltransferase locus on the human inactive X chromosome: implications for X chromosome inactivation. *Proc. Natl. Acad. Sci. (USA),* **81**, 2806–2810.

Wrigley, J.M. and Graves, J.A.M. (1988). Sex chromosome homology and incomplete, tissue specific X inactivation suggest that monotremes represent an intermediate stage of mammalian sex chromosome evolution. *J. Heredity.,* **79**, 115–118.

Younger-Shepherd, S.; Vaessin, H.; Bier, E.; Jan, L.Y. and Jan, Y.M. (1992). *deadpan* an essential pan-neural gene encoding an HLH protein, acts as a denomitor in *Drosophila* sex determination. *Cell,* **70**, 911–922.

Zawel, L. and Reinberg, D. (1995). Common themes in assembly and function of eukaryotic transcription complexes. *Annu. Rev. Biochem.,* **64**, 533–561.

Zhou, S.; Yang, Y.; Scott, M.J.; Pannuti, A.; Fehr, K.C.; Eisen, A.; Koonin, E.V.; Fouts, D.L.; Wrightsman, R.; Manning, J.E. and Lucchesi, J.C. (1995). *Male specific lethal-2,* a dosage compensation gene of *Drosophila,* undergoes sex specific regulation and encodes a protein with a RING finger and a metallothionein-like cysteine cluster. *EMBO J.,* **14**, 2884–2895.

Zuccotti, M. and Monk, M. (1995). Methylation of the mouse *Xist* gene in sperm and eggs correlates with imprinted *Xist* expression and paternal X inactivation. *Nat. Genet.,* **9**, 316–320.

Genome Analysis in Eukaryotes: Developmental and evolutionary aspects
R.N. Chatterjee and L. Sánchez (Eds)
Copyright © 1998 Narosa Publishing House, New Delhi, India

9. On How the Memory of Determination is Kept, and What May Happen to Forgetful Cells

Pedro Santamaria and Neel B. Randsholt

Centre de Génétique Moléculaire du C.N.R.S.
F-91198 Gif sur Yvette, Cedex, France

I. Introduction

The development of eukaryotes is as much controlled by the activity of specific genes in particular tissues as by the perpetuation of such specific activity throughout cell divisions. Maintenance of the identity of a tissue once the initial "on" or "off" input has disappeared is conceptually equivalent to its determination, a transient state that increases in definition as development progresses. Much is known about the scalar events that in *Drosophila* lead from the fertilized egg to the segmented embryo, in which the identity of each unit is specified by a combination of selector or master genes. As the factors that first provoke a given pattern of gene expression are transient, other factors must propagate this pattern through cell divisions. Indeed, the state of selector gene activity, once established, is accurately propagated autonomously through cell lineage, and the simple solution of autoregulatory loops in which selector gene proteins bind to their own promoters does not seem to be sufficient, at least not to explain repression.

Two groups of genes are considered to be instrumental on preserving the state of other genes (i) active—*trithorax group (trx-G)* (Kennison and Tamkun, 1988) and (ii) repressed—the *Polycomb group (Pc-G)* (Jürgens 1985). Many excellent recent reviews exist about the *Polycomb group* of genes (Moehrle and Paro 1994, Bienz and Müller 1995, Orlando and Paro 1995, Paro 1995, Simon 1995). Here, we would first like to summarize briefly a number of data concerning the *Pc-G*, and then to underline and to discuss some open questions about the multi-faceted functions of these genes.

II. The *Polycomb group* of genes

Presently, about a dozen loci make up the best known of the *Pc-G* genes (see Table 1). They include *Additional sex combs (Asx), cramped (crm), extra sex combs (esc), multi sex combs (mxc), pleiohomeotic (pho), Polycomb (Pc), Enhancer of zeste/polycombeotic [E(z)/pco], Polycomblike (Pcl), polyhomeotic (ph), Posterior sex combs (Psc), Sex comb extra (Sce), Sex comb on midleg (Scm), super sex combs (sxc) and Suppressor of zeste 2 [Su(z)2].* The *Pc-G* genes are required in most segments of *Drosophila* for correct differentiation. The criterion to include them in the *Pc-G* is the common and almost coincident pleiotropic phenotype of their mutants, that corresponds to apparent gain of function mutations of primary homeotic genes of the Bithorax and Antennapedia gene complexes (BX-C and ANT-C). ANT-C and BX-C genes specify segmental identity in *Drosophila* embryos and adults. In severe *Pc-G* gene mutant embryos, all segments are transformed towards the eighth abdominal segment. Viable hypomorphic *Pc-G* alleles induce more or less penetrant transformations of antennae towards legs, wings towards halters, meso- and metathoracic legs towards prothoracic legs, and abdominal segments towards more posterior ones (first towards second, fourth towards fifth, sixth and seventh towards eighth). These transformations are synergistically enhanced in *Pc-G* double mutants. The actual ectopic expression of BX-C and ANT-C genes has been shown by molecular analysis of most of the *Pc-G* mutants (Struhl and Akam 1985, Weeden et al 1986, Riley et al 1987, McKeon and Brock 1991, Simon et al 1992). After an initial correct expression pattern, transcripts of selecetor genes are ectopically expressed in *Pc-G* mutant embryos and imaginal discs (Dura and Ingham 1988, Busturia and Morata 1988, McKeon and Brock 1991, Simon et al 1992), suggesting that normal *Pc-G* products are needed for fixing the repressed state of the genes that they regulate. Paro and Hogness (1991) proposed that the *Pc-G* genes cooperate by organizing the chromatin structure of their targets in such a way that these genes are maintained repressed throughout development, a phenomenon functionally similar to imprinting (Paro 1995). Alternatively, *Pc-G* products could leave the chromatin fibre unaltered, but help assign the genes to be repressed to a nuclear compartment whereto different transcription factors do not have equal access (Schlossherr et al 1994). The products of *polyhomeotic* and *Polycomb* associate in a large protein complex (Franke et al 1992) and they bind to about one hundred sites on polytene chromosomes, all of which are common (Zink and Paro 1989, De Camillis et al 1992, Franke et al 1992). The products of *Posterior sex combs, Suppressor (2) of zeste* and *Polycomblike* share many binding sites on polytene chromosomes with PH and PC (Martin and Adler 1993, Rastelli et al 1993, Lonie et al 1994). No individual *Pc-G* protein has to date been found to be clearly able to mediate specific binding to DNA.

By indirect calculations Jürgens (1985) suggested that there could be around forty *Pc-G* genes in the *Drosophila* genome. Using two different calculations, Landecker et al (1994) also agreed to consider that their actual number is somewhere between thirty and fourty. The multimeric complex analyzed by Franke et al (1992) seems to be formed by 10-15 proteins. It is thus probable that all *Pc-G* gene products do not play identical roles on their common targets, and also that they have uncommon targets (Cheng et al 1994, Campbell et al 1995, Pirrotta 1995, Soto et al 1995). Certain homeotic transformations observed in different *Pc-G* mutants, especially those not easily ascribed to primary homeotic gene deregulation, are only shown by subsets of *Pc-G* mutant alleles. In terms of expression patterns, the *Pc-G* is not homogeneous either. While some, like *Pc* and *Psc,* are uniformly expressed in the embryo (Paro and Zink 1992, Martin and Adler 1993) the distribution of the *polyhomeotic* protein or mRNA in the embryo or imaginal discs is rarely uniform (Deatrick 1992, De Camillis and Brock 1994), and *ph* activation in germ-band elongated embryos is first mediated by the localized product of the segmentation gene *engrailed* (Serrano et al 1995).

III. Are there targets for *Pc-G* genes other than HOM (*Hox*) genes?

The pleiotropic phenotypes of *Pc-G* mutants cannot all be explained by ectopic expression of primary homeotic BX-C and ANT-C genes. On the salivary gland chromosomes, *ph, Pc, Psc* and *Pcl* antibodies share numerous binding sites that include *Pc-G* genes, along with the BX-C and ANT-C, and possibly also many other targets. Embryonic and imaginal disc stainings have shown that *Pc-G* genes can regulate loci such as posterior group genes and gap genes (Pelegri and Lehmann 1994), or pair rule and segment polarity genes (Dura and Ingham 1988, Busturia and Morata 1988, Smouse et al 1988, Moazed and O'Farrell 1992, McKeon et al 1994). Smouse et al (1988) showed that neuronal development is impeded in *ph* mutant embryos, due to aberrant axonal guidance, *pho* and *Psc* products are also required for central and peripheral neural development (Girton and Jeon 1994, Wu and Howe 1995). Thus these important trans-regulators cannot be confined to BX-C and ANT-C regulation, but seem needed in many different, yet specific processes of development.

IV. When *Pc-G* genes are required during *Drosophila* life?

Prima facies, the *trx* and *Pc* groups have often been considered as antagonistic (Kennison and Russell 1987, Kennison and Tamkun 1988, 1992, Locke et al 1988, Tamkun 1995) and competing for maintaining the activity of the same homeotic targets, yet different levels of regulation can be ascribed to the two groups (Kennison 1993, Chinwalla et al 1995, Gutjahr et al 1995, Pirrota 1995, Reijnen et al 1995, Sathe and Harte 1995). Even if activation and repression can alternate as development progresses (Johnson 1995), to start out development by relieving generalized repression is not equivalent to repressing already activated genes. Hartl et al (1993) have provided *in vitro* proof of the suggestion of Edgar and Schubiger (1986) that transcription is actively repressed in the embryo until the apparition of a G2 phase at the fourteenth cellular division. Many selector gene products must be locally restricted during early embryonic development, therefore some mechanism could turn them "off" during oogenesis to prevent their ectopic products from aborting the embryo (Paro and Zink 1992, Lawrence and Morata 1994).

Deregulation of development only becomes apparent in *Pc-G* mutants after initiation of BX-C and ANT-C activity (Struhl and Akam 1985, Kuziora and Mc Ginnis 1988), yet most *Pc-G* genes have maternal effects (Lawrence et al 1983, Breen and Duncan 1986, Dura et al 1988, Jones and Gelbart 1990, Phillips and Shearn 1990, Martin and Adler 1993, De Camillis and Brock 1994). Even if a majority of *Pc-G gene* products do not zygotically interfere with the set-up of initial homeotic gene repression, the perdurance of some of their maternal products do cover early embryonic needs for *Pc-G* products, prior to homeotic gene regulation (Paro and Zink 1992). Indeed, the early embryonic mutant phenotype of *Psc* can be rescued by maternal products (Martin and Adler 1993). Early requirement for *Pc-G* activity is also shown by the fact that total lack of *ph* product in the oocyte cannot be rescued by a wild-type *ph* copy carried by the sperm (Dura et al 1988).

Other data support the idea that *Pc-G* genes play an important role in early development. The posterior identity in the young *D. melanogaster* embryo is specified by maternally provided products of the genes forming the posterior group. Zygotical gap proteins then normally specify regions corresponding to several metameric units of the embryo. Pelegri and Lehmann (1994) have shown that certain *Pc-G* mutants [*E(z)*, *Su(z)2*, *Psc* and *pho*] have a maternal dominant effect that suppresses the posterior group gene *nanos (nos)*, and thus allows to rescue the lethality of embryos born from *hb nos/hb⁺ nos* mothers. The authors show that the normal products of these *Pc-G* genes are required for maintaining the expression domains of the gap genes *knirps* and *giant*, controlled by the maternal *hunchback (hb)* protein gradient. Wild-type *mxc* has probably a similar role as some mutant *mxc* alleles also suppress the lethality of embryos born from *hb nos/hb⁺ nos* mothers (Santamaria and Forquignon, unpublished results). The consequences of simultaneous zygotic and maternal deprival of certain *Pc-G* gene products on the expression of primary homeotic genes show spatial and temporal effects different from those due to zygotic deprival alone (Soto et al 1995). The *multi sex combs* product is needed both for normal gametogenesis and normal somatic development (Santamaria and Randsholt 1995, Docquier et al 1996), suggesting that *mxc* is also required for the silencing of other genes than the known selector genes. Finally, the *extra sex combs* protein product seems to play a transient role during early steps of embryogenesis in transcriptional inhibition, and in directing other *Pc-G* proteins to assembly with specific targets (Gutjahr et al 1995, Sathe and Harte 1995, Simon et al 1995). Together these data clearly indicate that the *Pc-G* genes have (either singly or together) played an important role in development well before transcriptional regulation of homeotic selectors is initiated in the embryo.

V. What triggers DNA binding of the *Pc-G* complex?

The specificity of *Pc-G* product targets has been shown by immunocytological analysis on polytene chromosomes (Zink and Paro 1989, De Camillis et al 1992, Messmer et al 1992, Lonie et al 1994). *Pc-G* protein binding is rendered impossible by *Pc-G* mutations, not only of the gene whose product is being tested but also by mutations of other *Pc-G* loci (Messmer et al 1992, Franke et al 1995). The cis-regulatory elements of the genes that are maintained in repress condition by *Pc-G* products are called *Pc-G* Response Elements (PRE) (Chan et al 1994, Chiang et al 1995). They exert their switch-off effect on reporter genes in a *Pc-G* gene dependent manner (Chiang et al 1995, Gindhart and Kaufman 1995), and function as orientation dependent silencers (Zink and Paro 1995). M. Bienz and her collaborators (Müller and Bienz 1991, Zhang and Bienz 1992) have proposed that the early repression of homeotic genes is mediated by the product of *hunchback (hb)* and by other gap gene proteins. These authors consider the early repression as independent of most *Pc-G* proteins. Müller (1995) suggested that *hb* protein acts as the natural "tether" to recruit *Pc-G* proteins. However, Pirrotta et al (1995) have shown that the PRE can be functionally separated from other enhancers and does not contain HB binding sites, indicating that the exact role of *hunchback* protein in this process remains to be determined.

The *trithorax* proteins also bind to specific chromosomal sites and co-localize with *Pc* and *ph* proteins at many of these, suggesting a complex antagonistic effect (Chinwalla et al 1995). Repressors do not infrequently co-occupy DNA with activators, and can still prevent the activator from functioning. Infact, certain activators and repressors cooperate with each other to bind to DNA, yet in most cases the repressor dominates the outcome (Johnson 1995). *trx* binding has been shown to depend on the presence of *E(z)* and *Psc* (Rastelli et al 1993, Kuzin et al 1994, Paro 1995), suggesting that the antagonistic genetic interaction between *trx-G* and *Pc-G* gene products is at a different level than that of competition for exclusive binding to their target sequences.

VI. Have the *Pc-G* genes been conserved during evolution?

Any multicellular organism undergoing development requires a mechanism to maintain the process of cellular determination. Once established, such a mechanism could have been conserved during evolution. Many targets of the *Pc-G* genes are, as far as we know, conserved through evolution (Burke et al 1995). This is true not only for the primary homeotic selector genes but also for other selectors such as *engrailed* and *evenskipped.* It could thus be expected that at least part of the *Pc-G* genes would have homologues in vertebrates, mammals included, possibly with similar functional roles.

Indeed, *Posterior sex combs (Psc)* has a murine homologue called *bmi*-1 (van Lohuizen et al 1991, Brunk et al 1991, van der Lugt et al 1994). Animals lacking *bmi*-1 show pleiotropic posteriorly directed homeotic transformations that parallel the *Drosophila Pc-G* mutant phenotypes, probably due to ectopic expression of *Hox* genes (Alkema et al 1995). Mutant *bmi-1* proteins induce lymphomas and collaborate with *myc* in tumorigenesis (Haupt et al 1993). A second mouse homologue of *Psc, mel-18* acts as a transcriptional repressor of *Hox* genes (Kanno et al 1995). *Polycomb* has also a mouse homologue, *M33* (Pearce et al 1992), that can substitute for the lack of Pc^+ activity in transgenic flies (Müller et al 1995). In *Xenopus laevis,* the *Pc* and *Psc* homologous proteins have overlapping expression patterns, and they interact by forming multimeric complexes, like their *D. melanogaster* counterparts (Reijnen et al 1995).

The mouse homologue of *polyhomeotic, rae-*28 (*retinoic acid early*) (Nomura et al 1994) is expressed in cell cultures in response to treatment with retinoic acid (RA), a product that controls the expression of *Hox* genes in mammals (Tabin 1995), as these genes can be induced by RA. The product of *E(z)/pco* shares the SET protein domain with the human mixed-lineage leukemia *MLL/HRX/ALL-1* gene (Lawrence and Largman 1992, Jones and Gelbart 1993), and with the mouse *Mll* gene whose mutation induces homeotic transformations associated with *Hox* gene deregulation (Yu et al 1995). Finally, the *extra sex combs* protein contains multiple copies of the WD40 repeat, which probably mediates protein-protein contacts (Gutjahr et al 1995, Simon et al 1995).

Clearly, almost all *Pc-G* genes that have been cloned to date have turned out to code for proteins that share at least certain functional domains with vertebrate protein products. Some of these function in a similar manner, and are required for comparable functions in vertebrates as their *Drosophila* counterparts. In addition, the proteins encoded by *bmi-1, Pc, Psc* and *Su(z)2* can all mediate repression of transcription in mammalian cells (Bunker and Kingston 1994), indicating that the structure and functions of *Pc-G* genes have been conserved to a large extent during evolution

VII. Could *Pc-G* genes control the appendicular axes?

Each segment of the *Drosophila* embryo is patterned essentially according to two body axes, the antero/posterior (A/P) and the dorso/ventral (D/V) ones. To these, a third proximo/distal (P/D) axis must be added for correct patterning of the limbs (for reviews see Blair 1995, Campbell and Tomlinson 1995, Held 1995). A/P identity within each segment depends on posterior compartment expression of the homeodomain protein encoded by *engrailed*. In the appendage primordia, *engrailed* induces and is co-expressed with the trans-membrane protein *hedgehog (hh)*. At the A/P compartment border, *hedgehog* triggers a signal transduction pathway that will lead to expression of *decapentaplegic (dpp)* along the A/P border of dorsal appendages, and to expression of *wingless (wg)* and of *dpp* along this border in ventral appendages. *dpp,* a member of the transforming growth factor β (TGF-β) family then functions as an organizer of the future appendage. In the ventral limb primordia, the secreted *wingless* protein, homologous to *int* oncogenes, also acts as an organizer molecule (Vincent 1994).

Engrailed is ectopically expressed in *Pc-G* mutant embryos (Dura and Ingham 1988, Moazed and O'Farrell 1992, McKeon et al 1994). The *engrailed-invected (en-inv)* gene complex is not only used in the set-up of embryonic compartments; it also defines the A/P compartments of the appendages of the imago (Guillen et al 1995, Tabata et al 1995). *engailed,* if ectopically expressed in the anterior wing compartment, can activate *hedgehog* anteriorly; if ectopic *en* expression is near the dorso/ventral boundary, it can lead to ectopic activation of *decapentaplegic* and can initiate cell growth associated with pattern duplications (Guillen et al 1995). Conversely, the ectopic expression of *hedgehog* can also activate *engrailed* (de Celis and Ruiz-Gomez 1995), *trithorax* is required for normal expression of *en* in the posterior compartment (Breen et al 1995). Busturia and Morata (1988) have shown that *Polycomb* mutant cells derepress *en* in the anterior compartment of the developing wing. The need for *Pc-G* gene repressive action on *en* in the anterior compartment extends to *ph* as we have found that clones of *ph⁻* cells induced by mitotic recombination also express *engrailed* ectopically in the anterior appendage primordia (N.B. Randsholt, unpublished). We have also shown that such clones sort-out, and that they can occasionally induced overgrowths in the neighbouring cells resulting in pattern duplications (Santamaria et al 1989). These duplications are similar to those produced by ectopic expression of *en-inv*, (Tabata et al 1995), or by ectopic expression of *hedgehog or* of *decapentaplegic* (Struhl and Basler 1993, Basler and Struhl 1994, Capdevila and Guerrero 1994), arguing for a role of *Pc-G* genes in the establishment or the maintenance of the A/P compartments.

Dorso/ventral compartmentalisation of the future *Drosophila* wing is controlled by the spatially localized expression of the LIM-type homeodomain protein product of the *apterous (ap)* gene (Cohen et al 1992). *apterous* is expressed in the primordial dorsal wing cells and specify their dorsal identity (Diaz-Benjumea and Cohen 1993). Dorso/ventral identity of appendages can also be affected by many *Pc-G* mutants such as *ph* (Dura et al 1988), *Pc* (Denell and Fredericks 1983, Tiong and Russell 1990) or *Psc* (Adler et al 1991), manifested by a homeotic transformation of the ventral surface of the wing into the dorsal one (Santamaria 1993). This transformation is the opposite of the dorsal into ventral one caused by loss of the *apterous* protein in the dorsal wing cells. A second gene, *Dorsal wing (Dlw),* can induce ventral into dorsal transformation in heterozygous mutants, while somatic homozygous clones of *Dlw* cells induce the dorsal surface to differentiate as ventral (Tiong et al 1995). Clones of cells for lethal alleles of *trx* can also induce dorsal into ventral transformations of the wing margin. As *Pc* and *ph* proteins bind to the region of the *apterous* locus on salivary gland polytene chromosomes, and as *Pc-G* genes act as negative regulators of Dlw^+ (Tiong et al 1995), it is tempting to propose that either *ap,* or *Dlw,* or both could, in *Pc-G* mutants, be ectopically expressed in the ventral surface of the wing, where at least *apterous* is normally repressed. This gain of function of yet another homeodomain-encoding gene could then account for the wing transformations observed.

Both A/P and D/V identities seem to require *Pc-G* activity. More difficult to explain are the effects of some *Pc-G* mutants on the proximo/distal axis. The P/D axis, at least of the legs, does not seem to be related to clear-cut compartments, in contrast to the other two axes (Steiner 1976). The future distal-most point has been defined as the site where cells expressing *wingless* meet cells that express *decapentaplegic* strongly (Struhl and Basler 1993, Buratovich and Bryant 1995, Campbell and Tomlinson 1995). It has been suggested that cell identity along the proximo/distal axis could be an intrinsic consequence of the fixation of the A/P and D/V ones (Diaz Benjumea et al 1994, Held 1995). Yet, the action of at least two homeobox genes, *Distal-less* and *aristaless,* is also needed for formation of specific pattern elements along the P/D axis of *Drosophila* appendages (Cohen and Jürgens 1989, Campbell et al 1993).

Mutants of *Pc* (Capdevila et al 1986), *E(z)/pco* (Jones and Gelbart 1990), *mxc* (Santamaria and Randsholt 1995) or *crm* (Lindsley and Zimm 1992) show a transformation of the prothoracic second tarsus towards the sex comb-carrying basitarsus. Other pleiotropic mutations can induce the same transformation, among which are *bric à brac* (Godt et al 1993) and *Montium-like* (Docquier, Santamaria and Randsholt, unpublished). Both *trx-G* and *Pc-G* genes appear to be necessary for correct P/D identity. Felsenfeld and Kennison (1995) have suggested that *Moonrat (Mrt),* a gain of function allele of *hh,* transforms the proximal part of the wing into the distal one. These authors also show that a number of *trx-G* mutants can suppress the *Mrt* phenotype. We have established (Randsholt and Santamaria, unpublished results) that the *Mrt* mutant phenotype is synergistically enhanced by hypomorphic mutant alleles of *ph* or *Psc.* Yet the targets through which the *Pc-G* may act on the P/D axis are not obvious to define. If *hh* regulation depends on *Pc-G* products, then the proximo/distal axis could be indirectly affected through the transforming growth factor *decapentaplegic.* TGF-βs form a family of secreted growth factors, and *dpp* may not only provide a signal for the establishment of proximo/distal positional information (Campbell et al 1993), but also control growth and pattern, acting as an organizer (Capdevila and Guerrero, 1994). *dpp* can exert a long range organizing influence on surrounding imaginal tissues, specifying anterior or posterior patterns depending on the state of *en* activity of the neighbour cells (Zecca et al 1995).

Axis determination has been studied in many other organisms. Vertebrate axial limb morphology is controlled by *Hox* genes (Morgan and Tabin 1993, Tickle and Eichele 1994, Burke et al 1995, Tickle 1995) and by the homologue of *hh, sonic hedgehog (shh)* (Johnson and Tabin 1995). In these animals, growth and patterning are strongly implicated, and retinoic acid plays an important role (Conlon 1995, Tabin 1995) as some *Hox* genes appear to be directly regulated by RA (Langston and Gudas 1994, Marshall et al 1994, Ogura and Evans 1995a,b), and as the treatment with RA can induce expression of *sonic hedgehog* in the chick wing bud (Helms et al 1994). As the actors of this inductive mechanism have been conserved during evolution (Ingham 1995), it remains now to be determined whether the type of regulation-involving *Pc-G* gene products-could also be conserved.

VIII. Tumorigenicity of *Pc-G* mutants

Several lines of evidence suggest that *Pc-G* genes may have tumour suppressor activity. Genetic alterations of developmental processes such as cell communication, signal transduction, regulation of gene expression, cytoskeletal organization and regulation of the cell-cycle have been found at the origin of tumorous cells. Normally these genes are classified according to the assumed effect of their wild-type allele on cell proliferation as proto-oncogenes for those with positive effects, and as tumour suppressors for those with negative effects. However, the most frequently applied criterion for classification is the dominance or recessivity of mutants, assuming a gain of function for the proto-oncogenes and a loss of function for the tumour suppressors. It is interesting that in *Drosophila*, most genes related with tumorigenicity have been identified as tumour suppressors (Bryant 1993, Gateff 1994, Watson et al 1994). Yet, without the knowledge of direct or indirect relation between a given gene and its hypothetical target, one must keep in mind that the loss of function of a repressor is concomitant with the gain of function of its targets. The loss of function of a repressor can be haplo-insufficient and produce a dominant phenotype, as some *Pc-G* gene mutations do, and the haplo-insufficiency of a gene has a contingent dominance in a stochastic manner in some cells that is undetected in others, as it happens for most *Pc-G* mutants. These arguments could explain the "predisposition" or "susceptibility" to tumours of individuals heterozygous for hypomorphic mutants of haplo-insufficient loci. In flies, 10-30% of individuals will show at least one homozygous clone for a given heterozygous marker (Garcia-Bellido 1972), showing that spontaneous mitotic recombination can also make a homozygous mutant clone from an originally heterozygous cell. The fact that many processes are affected in tumorous cells has led to the general assumption that cancer cells suffer from multiple mutational events. This has been proved for some types of cancer, but the possibility remains that in others a unique mutation of a repressor with multiple targets (about one hundred for *Pc-G* genes) can provoke the apparent misregulation of all of its downstream genes. Weinberg (1983) has already suggested that cancer is probably due to a simple genetic alteration of developmental mechanisms with pleiotropic effects.

In Table 1, we show the connections between certain *Pc-G* loci and other genes related with tumorigenicity. *Enhancer of zeste-polycombeotic* shares the SET protein domain (Jones an Gelbart 1993) not only with *trx* genes but also with the human protein mixed lineage leukemia *Mll/HRX/All-1* (Lawrence and Largman 1992) implied in acute leukemia, and with the mouse Mll locus that regulates *Hox* gene expression (Yu et al 1995). The *Sex comb on midleg* protein appears to present structural homologies with the *Drosophila* gene *l(3) malignant brain tumour* implied in brain tumorigenesis (Borneman et al 1996). The best documented case concerns *Psc*. *Psc* proteins contain a RING zinc finger motif that is present in other organisms from yeast to mammals (Brunk et al 1991, van Lohuizen et al 1991) and, as mentioned previously, *Psc* is homologous to the mouse genes *bmi-1* and *mel-18* involved in control of cell proliferation, of tumorigenesis and of correct pattern formation (Brunk et al 1991, Goebl 1991, van Lohuizen et al 1991, van der Lugt et al 1994, Alkema et al 1995).

Table 1. Members of the *Polycomb* group

Known *Polycomb* group genes, and their connections with tumour genes

Pc-G genes	Connection with tumorigenesis
Additional sex comb (Asx)[c]	
cramped (crm)	
Enhancer of zeste/polycombeotic	Mixed lineage leukemia Mll/All/Hrx[a]
[E(z)pco][c] *extra sex combs (esc)*[c]	
multi sex combs (mxc)	*lethal (1) malignant blood neoplasm*[b]
pleiohomeotic (pho)	
Polycomb (Pc)[c]	
Polycomblike (Pcl)[c]	
polyhomeotic (ph)[c]	retionic acid early *(rae-28)*[a]
Posterior sex combs (Psc)[c]	B-limphoma Mo-M/V insertion-region-1[a]
Sex combs extra (Sce)	(bmi-1, and mel-18)
Sex combs on midleg (Scm)[c]	*lethal (3) maliganant brain tumour*[a]
super sex combs (sxc)	B-limphoma Mo-M/V insertion-region-1[a]
Suppressor 2 of zeste [Su(z)2][c]	(bmi-1, and mel-18)

[a]sequence holomogy; [b]allelism; [c]*Drosophila* genes that have been cloned.

Orlando and Paro (1995), in a clever parallel suggest the similarity between the dotted nuclear staining pattern of *Pc-G* proteins that disappears in *Pc-G* mutant cells (Franke et al 1995), and the behaviour of a human nuclear organite called POD for Promyelocyte oncogenic domains, formed by the PML (promyelocytic) RING zinc finger product and several other proteins. PODs are revealed as nuclear dots by immuno-histo-chemical staining of the PML proteins. When a translocation fuses the PML gene to the Retinoic Acid Receptor α (RARα), a PML-RAR protein is produced which forms a complex with the available PML, the POD nuclear staining pattern disappears, and acute promyelocytic leukemias (APL) are induced. RA treatment restores the ability of PML to associate with its normal partners, and regenerates normal appearing PODs (Dyck et al 1994, Weis et al 1994), suggesting that PODs-and other nuclear organites could play important roles, not only in transcriptional regulation but also in control of tumorigenesis.

The *ph* murine homologue, *rae-28*, is induced in cell cultures by RA treatment and provides another case possibly connecting a *Pc-G* gene to growth control or tumorigenesis (Nomura et al 1994). Retinoic acid originally attracted the attention of developmental biologists because of its teratogenic effects. RA has an efficient therapeutic action in acute promyelocytic leukemias (Huang et al 1988), but can also provoke abnormal pattern formation, since local application of RA to the anterior margin of the chick limb bud results in pattern duplications reminiscent of those that develop after anterior grafting of posterior cells from the zone of polarizing activity (Helms et al 1994). The effects of RA on growth are probably a consequence of cells changing positional identity, since RA administration to limb cells seems to convert anterior cells to posterior identity via *Hox* gene ectopic expression (Helms et al 1994). The induction of cells with posterior cell-identity among cells with an anterior cell-identity could trigger the observed effect on cell growth. The direct and indirect links of RA with *Hox* genes and with patterning activity seem strengthened by recent analysis (Conlon 1995). The pattern deformations provoked by RA are also reminiscent of the duplications that can be induced—mostly in the anterior compartment of the *Drosophila* wing—by clones of *ph⁻* cells (Santamaria et al 1989). As mentioned above, this phenomenon follows sorting out of *ph* deficient cells—indicating that cell identity and cell-cell recognition have been affected and is associated with ectopic *engrailed* expression in *ph⁻* cells. Similar duplications of imaginal discs can be induced by a mutation of the *Drosophila* tumour suppressor gene *l(2)giant larvae [l(2)gl]* (Agrawal et al 1995), and *engrailed* expression is affected in *l(2) gl* clones (Muhkerjee et al 1995). As treatment with RA promotes the expression not only of *Hox* genes (Helms et al 1994) but also of *sonic hedgehog*, and as *hh* expression can be induced by *engrailed* protein that is expressed ectopically in *ph⁻* cells, both cases of pattern duplications could possibly pass through the ectopic expression of *hh* (Hidalgo 1996, Joyner 1996), and finally *dpp*.

Two alleles of the *Pc-G* gene *multi sex combs* (Santamaria and Randsholt 1995, Docquier et al 1996) were isolated by E. Gateff as *malignant blood neoplasms* (Gateff 1978a, b, 1982, 1994, Gateff and Mechler 1989). In fact, all *mxc* alleles can manifest, in heteroallelic combinations, both syndroms: blood tumours and homeotic transformation. Small pieces of *mxc^{mbn}* hematopoietic organs implanted into the body cavity of wild-type flies grow in an autonomous lethal invasive fashion. The blood cells encapsulate and melanize parts of the posterior fat body and of the gut. The imaginal disc epithelia are invaded and partially destroyed by the neoplasic blood cells. These tumours fulfill the criteria for malignacy defined by: (a) rapid autonomous growth, (b) growth and host lethality after transplantation, and (c) invasiveness (Gateff 1978a). Gateff (1982) suggested that the hematopoietic organs of *mxc* mutants overproduce new blood cells due to the loss of the mechanism that controls hemocyte production, but leave blood cell differentiation intact. The plasmatocyte cells that are normally capable of controlled division (Rizki 1978) multiply in an unrestrained manner in the mutant. Gateff (1994) has also shown that mutant mxc^{mbn} larvae constitutively express the antibacterial peptide "diptericine", suggesting that this gene has, directly or indirctly, lost its normal regulation in *mxc* mutants. It is also interesting to note here that nuclei of mutant mxc^{mbn} plasmatocytes exhibit a high number of Virus like particles compared to a wild-type control (Gateff 1978b). These particles are constant components of *Drosophila* cells, but their number reflects the physiological state of the cells, and increases in aging or in neoplastic cells (Gateff 1978b). We would suggest the possibility that mobile elements normally present in the genome (some of which are retroviruses; Kim et al 1994) could be repressed because they are included in non-active eu- or heterochromatic chromosome regions. These elements could become active because of the derepression of their insertion sites, caused by *Pc-G* gene mutations.

It seems clear that in *Drosophila*, the critical step in the generation of a malignant tumour is an impediment on the normal development and not a direct stimulus of cell multiplication (Harris 1990). Mutation of a *Pc-G* gene does provoke a disturbance of development, so if these loci act like tumour suppressors, *Pc-G*'s indirect regulation of TGF-β/*dpp*, that in turn regulates the cell-cycle (Kamb 1995), could be the last step of the process. This would explain the role of developmental genes in tumorigenesis, as TGF-β growth regulators can inhibit replication of certain cells and stimulate the growth of others (Klein 1987). Aberrant expression of homeobox-containing transcription factors induced by *Pc-G* mutations might also be a cause for tumour generation. The deregulation of *Hox* genes, or of other homeobox-coding loci such as the *Pax* genes, has already been proved to be able to induce tumourigenesis in mammals (Lawrence and Largmen 1992, Maulbecker and Gruss 1993, Stuart et al 1995), and the injection of *Hox* expressing cells into athymic mice can result in tumour formation. Many cases of over-expressing *Hox* genes have been related to leukemia (Blatt 1990, Aberdam et al 1991, Lu et al 1991, Castronovo et al 1994, Stuart et al 1995), while Lawrence and Largmen (1992) have shown that aberrant expression of *Hox* genes is related to leukemia, and that *Hox* products can bind hematopoietic control genes.

The available data on *Pc-G* gene protein products show a number of them to be homologuous to vertebrate proto-oncogenes or to tumour suppressors. Some of these homologues appear to form multimeric protein complexes, and they can also play a role in regulation of vertebrate homeotic gene expression. In parallel to the situation in vertebrates, and to explain how *Pc-G* genes and their homologues could induce tumorigenesis, we suggest that *Pc-G* mutants, via deregulation of selector genes, could be at the origin of tumours. As *Hox* gene products regulate the *Drosophila* homologues of *dpp, hh* and *wingless* (homologues, respectively of TGF β1, *sonic hedgehog* and *int* oncogenes), these three products might then be among the possible effectors of tumorigenicity. Indeed, misexpression of *dpp* in *Drosophila* has already been related with tumour formation (Buratovich and Bryant 1995). Alternatively, *hh* or *wg* could be direct *Pc-G* regulatory targets, as binding sites for the *Pc-G* complex have been observed in the chromosomal regions of these loci (De Camillis et al 1992, Paro and Zink 1992).

Using *Drosophila* as a paradigm for the study of tumorigenesis has already been proposed by Gateff (1994) and by Watson et al (1994). Molecular analysis of more *Drosophila Pc-G* genes will hopefully in the near future show whether these loci do play the same tumour suppressor or proto-oncogenic roles as their vertebrate homologues do. If this assumption is correct, *Pc-G* of *Drosophila* might then provide an adventageous new field for cancer research.

IX. Role of *Pc-G* genes in evolution

It is probably more than an amusing coincidence that mutant phenotypes of *Pc-G* genes frequently coincide with systematic characteristics of Drosophilids and other Dipterans (Santamaria 1993). To quote some examples: the three spermatecae seen in *Polycomblike* and *pleihomeotic* (Duncan 1982, Girton and Jeon 1994) are the characteristic of most Muscomorpha (McAlpine 1989); and the transformation of the second tarsus towards the first, observed for certain *Pc-G* mutant alleles is also a sexual secondary character, and so might be at the origin of speciation of the subgroups *D. montium. D. subobscura, D. algonkin, D. athabasca* among others (McAlpine 1989, Santamaria 1993). The lack of post-pronotal bristles (humerals), presented by *ph, Asx* and other *Pc-G* mutants, is a characteristic that differentiates Aeschiza from Schizophora (McAlpine 1989).

Gain of function of BX-C genes could explain some of these phenotypes. For instance, transformation of the first abdominal segment towards the second, seen in many *Pc-G* mutants, and supposed to be caused by ectopic BX-C protein expression, is a characteristic of *D. victorina, coracina* or *pattersonii* (Garcia-Bellido 1983). Reduction of the 6th tergite, due to a transformation towards the 7th that normally is inexistent in *D. melanogaster* males is a common phenotype of most *Pc-G* mutants, and a general trend throughout the Dipteran order (Calyptratae versus Acalyptratae 44).

Several authors have proposed that modification of *Hox* gene expression, mostly by gain of function, could account for certain evolutionary trends (Ahlberg 1992, Slack et al 1993, Holland et al 1994, Averof and Akam 1995, Tabin 1995, Gibson and Hogness 1996, Tautz 1996). It is also plausible that variations in the regulation of HOM genes could be due to a combinatorial difference in the multimers formed by the *Pc-G* proteins on specific targets, that would lead to a new spatial distribution of a selector product. In this model, to change the identity of one or several body compartments, it would not be necessary to alter the HOM-*Hox* genes themselves, nor the *Pc-G* genes as such. Rather it would be sufficient to change topologically the kind or the number of elements that associate as aggregated proteins and form the regulatory complexes responsible for the repression of specific targets.

X. Conclusion

A large body of knowledge about *Polycomb* group genes has accumulated since Zink and Paro (1989) first showed that the *Polycomb* protein binds to *Drosophila* chromatin at specific targets. In the present review, we chose not to focus upon how *Pc-G*-mediated repression is molecularly imposed on the chromatin structure, although many elegant experiments have investigated, and are investigating this important aspect of *Pc-G* function. We believe (as the reader may have gathered) that our understanding of the many tasks accomplished by the *Pc-G* gene control of their numerous targets is only at its beginning. Here we have only tried to speculate about, or to suggest, what some of these roles might be, and now look forward to the data that will tell us how many of these assumptions correspond to biological reality.

Acknowledgments

We thank Françoise Forquignon, Olivier Saget and France Docquier for many useful and fruitful discussions. This work was supported by grants from the Association pour la Recherche contre le Cancer.

REFERENCES

Aberdam, D., Negreanu, V., Sach, L., and Blatt, C., (1991). The oncogenic potential of an activated Hox-2.4 homeobox gene in mouse fibroblasts. *Mol. Cell. Biol.* **11**: 554–557

Adler, P.N., Martin, E.C., Charlton, J., and Jones, K., (1991). Phenotypic consequences and genetic interactions of a null mutation in the *Drosophila Posterior sex combs* gene. *Dev. Genet.* **12**: 349–361

Agrawal, N., Kango, M., Mishra, A., and Sinha, P., (1995). Neoplastic transformation and aberrant cell-cell-interaction in genetic mosaics of *lethal(2) giant larvae (lgl)*, a tumour suppressor gene of *Drosophila*. *Dev. Biol,* **127**: 218–229

Ahlberg, P.E., (1992) Coelacanth fins and evolution. *Nature* 358: 459

Alkema, M.J., van der Lugt, N., Bobelijk, R.C., Berns, A., and van Lohuizen, H., (1995). Transformation of axial skeleton due to over expression of *bmi-1* in transgenic mice. *Nature* **374**: 724–727

Averof, M., and Akam, M., (1995) Hox genes and the diversification of insect and crustacean body plans. *Nature* **376**: 420–423

Basler, K., Struhl, G., (1994) Compartment boundaries and the control of *Drosophila* limb pattern by *hedgehog* protein. *Nature* **368**: 208–214

Bienz, M., and Müller, J., (1995) Transcriptional silencing of homeotic genes in *Drosophila*. *BioEssays* **17**: 775–784

Blair, S., (1995) Compartments and appendage development in *Drosophila*. *BioEssays* **17**: 299–309

Blatt, C., (1990) The betrayal of homeobox genes in normal development: The link to cancer. *Cancer Cells* **6**: 186–189

Bornemann, D., Miller, E. and Simon, J., (1996) The *Drosophila Polycomb* group gene *Sex comb on midleg (Scm)* encodes a zinc finger protein with similarity to *polyhomeotic* protein. *Development* **122**: 1621-1630

Breen, T.R. and Duncan, I.M., (1986) Maternal expression of genes that regulate the expression of the *bithorax* complex of *Drosophila melanogaster*. *Dev. Biol.* **118**: 442–456

Breen, T.R., Chinwalla, V., and Harte, P., (1995) *Trithorax* is required to maintain *engrailed* expression in a subset of *engrailed*–expressing cells. *Mech. Dev.* **52**: 89–98

Brunk, B.P., Martin, I.M., and Adler, P.N., (1991) *Drosophila* genes *Posterior Sex Combs* and *Suppressor two of zeste* encode proteins with homology to the murine *bmi-1* oncogene. *Nature* **353**: 351–353

Bryant, P., (1993) Towards the cellular functions of tumour suppressors. *Trends Cell Biol.* **3**: 31–39

Bunker, C.A., and Kingston, R.E., (1994) Transcriptional repression by *Drosophila* and mammalian *Polycomb* group proteins in transfected mammalian cells *Mol. Cell. Biol.* **14**: 1721–1732

Buratovich, M.E., and Bryant, P.T., (1995) Duplication of *l(2)gd* imaginal discs in *Drosophila* is mediated by ectopic expression of *wg* and *dpp*. *Dev. Biol.* **168**: 452–463

Burke, A.C., Nelson, C.E., Morgan, B.A., and Tabin, C., (1995) Hox genes and the evolution of vertebrate axial morphology. *Development* **121**: 333–346

Busturia, A., and Morata, G., (1988) Ectopic expression of homeotic genes by the elimination of the *Polycomb* gene in *Drosophila* imaginal epidermis. *Development* **104**: 713–720

Campbell, G., and Tomlinson, A., (1995) Initiation of the proximo distal axis in insect legs. *Development* **121**: 619–628

Campbell, G., Weaver, T., and Tomlinson, A., (1993) Axis specification in the developing *Drosophila* appendage: the role of *wingless, decapentaplegic* and the homeobox gene *aristaless*. *Cell* **74**: 1113–1123

Campbell, R.B., Sinclair, D.A.R., Couling, M., and Brock, H.W., (1995) Genetic interactions and dosage effects of *Polycomb* group genes of *Drosophila*. *Mol. Gen. Genet.* **246**: 291–300

Capdevila, J., and Guerrero, I., (1994) Targeted expression of the signaling molecule *decapentaplegic* induces pattern duplications and growth alterations in *Drosophila* wings. *EMBO J* **13**: 4459–4468

Capdevila, M.P., Botas, J., and Garcia–Bellido, A., (1986) Genetic interactions between the *polycomb* locus and the *Antennapedia* and *Bithorax* complexes of *Drosophila*. *Roux's Arch. Dev. Biol.* **195**: 417–432

Castronovo, V., Kusata, M., Chariot, A., Gielen, T., and Sobel, M., (1994) Homeobox genes: Potential candidates of the transcriptional control of the transformed phenotype and invasive phenotype. *Biochem. Pharmacol.* **47**: 137–143

Chan, C.S., Rastelli, L., and Pirrotta, V., (1994) A *Polycomb* response element in the

Ubx gene that determines an epigenetically inherited state of repression. *EMBO J.* **13**: 2553–2564

Cheng, N.N., Sinclair, D.A.R., Campbell, R.B., and Brock, H.W., (1994) Interactions of *polyhomeotic* with *Polycomb* Group genes of *Drosophila melanogaster. Genetics* **138**: 1151–1162

Chiang, A., O'Connor, M.B., Paro, R., Simon, T., and Bender, W., (1995) Discrete *Polycomb* binding sites in each parasegmental domain of the *bithorax* complex. *Development* **121**: 1681–1689

Chinwalla, V., Jane, E.P., and Harte, P.J., (1995) The *Drosophila trithorax* protein binds to specific chromosomal sites and colocalizes with *Polycomb* at many sites. *EMBO J.* **14**: 2056–2065

Cohen, B., McGuffin, M.E., Pfeifle, C., Segal, D., and Cohen, S.M., (1992) *apterous,* a gene required for imaginal disc development in *Drosophila* encodes a member of the LIM family of developmental regulatory proteins. *Genes Dev.* **6**: 715–729

Cohen, S.M., and Jürgens, G., (1989) Proximal-distal pattern formation in *Drosophila*: graded requirement for *Distal-less* gene activity during limb development. *Roux's Arch. Dev. Biol.* **198**: 157–169

Conlon, R.A., (1995) Retinoic acid and pattern formation in vertebrates. *Trends Genet.* **11**: 314–319

Deatrick, J., (1992) Localization *in situ* of *polyhomeotic* transcripts in *Drosophila* embryos reveals spatially retricted expression beginning at the blastodem stage. *Dev. Genet.* **13**: 326–330

De Camillis, M., and Brock, H.W., (1994) Expression of the *polyhomeotic* locus in development of *Drosophila melanogaster. Roux's Arch. Dev. Biol.* **203**: 429–438

De Camillis, M., Cheng, N., Pierre, D., and Brock, H.W., (1992) The *polyhomeotic* gene of *Drosophila* encodes a chromatin protein that shares polytene chromosome binding sites with *Polycomb. Genes Dev.* **6**: 223–232

de Celis, J.F., and Ruiz-Gomez, M., (1995) *groucho* and *hedgehog* regulated *engrailed* expression in the anterior compartment of the *Drosophila* wing. *Development* **121**: 3467–3476

Denell, R.E., and Fredericks, R.D., (1983) Homeosis in *Drosophila*: a description of the *Polycomb* lethal syndrome. *Dev. Biol.* **97**: 34–47

Diaz-Benjumea, F., Cohen, B. and Cohen, S.M., (1994) Cell interaction between compartments establishes the proximal-distal axis of *Drosophila* legs. *Nature* **372**: 175–178

Diaz-Benjumea, F., and Cohen, S.M., (1993) Interaction between dorsal and ventral cells in the imaginal discs directs wing development in *Drosophila. Cell* **75**: 741–752

Docquier, F., Saget, O., Forquignon, F., Randsholt, N.B., and Santamaria, P., (1996) *multi sex combs* a *Polycomb*-group gene of *Drosophila melanogaster* is required for proliferation of the germline. *Roux's Arch. Dev. Biol.* **205**: 203–214

Duncan, I.M., (1982) *Polycomblike:* a gene that appears to be required for the normal expression of the *Bithorax* and *Antennapedia* gene complexes of *Drosophila melanogaster. Genetics* **102**: 49–70

Dura, J.M., Deatrick, J., Randsholt, N.B., Brock, H.W., and Santamaria, P., (1988) Maternal and zygotic requirement for the *polyhomeotic* complex genetic locus in *Drosophila. Roux's Arch. Dev. Biol.* **197**: 239–246

Dura, J.M., and Ingham, P., (1988) Tissue and stage-specific control of homeotic and segmentation gene expression in *Drosophila* embryos by *polyhomeotic* gene. *Development* **103**: 733–741

Dyck, J.A., Maul, G.G., Müller, W.H., Cohen, J.D., Kakizuka, A., and Evans, R.M.,

(1994) A novel macromolecular structure is a target of the Promyelocyte-Retinoic acid receptor oncoprotein. *Cell* **76**: 333-343

Edgar, B.A., and Schubiger, G., (1986) Parameters controlling transcriptional activation during early *Drosophila* development. *Cell* **44**: 871–877

Felsenfeld, A.L., and Kennison, J.A., (1995) Positional signaling by *hedgehog* in *Drosophila* imaginal disc development. *Development* **121**: 1–10

Franke, A., De Camillis, M., Zink, D., Cheng, N., Brock, H.W., and Paro, R., (1992) *Polycomb* and *polyhomeotic* are constituants of a multimeric protein complex in chromatin of *Drosophila melanogaster*. *EMBO J.* **11**: 2941–2950

Franke, A., Messmer, S., and Paro, R., (1995) Maping functional domains of the *Polycomb* protein of *Drosophila melanogaster*. *Chromosome Res.* **3**: 351–360

Garcia-Bellido, A., (1972) Some parameters of mitotic recombination in *Drosophila melanogaster*. *Mol. Gen. Genet.* **115**: 54–72

Garcia-Bellido, A., (1983) Comparative anatomy of cuticular patterns in the genus *Drosophila*. *In* "Development and evolution." (B.C. Goodwing, N. Holder and C.C. Wylie, eds.), pp. 227–255. Cambridge University Press.

Gateff, E., (1978a) Malignant neoplasms of genetic origin in *Drosophila melanogaster*. *Science* **200**: 1448–1459

Gateff, E., (1978b) Malignant and benign neoplasms. *In* "The Genetics and Biology of *Drosophila*." (M. Ashburner and T.R.F. Wright, eds.), Vol. **2B**, pp. 181–275. Academic Press, New York.

Gateff, E., (1982) Cancers, genes and development: *The Drosophila* case. *In* "Advances in Cancer Research." Vol 37, pp. 33–74. Academic press.

Gateff, E., (1994) Tumor suppressor and overgrowth suppressor genes of *Drosophila melanogaster*. Developmental aspects. *Int. J. Dev. Biol.* **38**: 565–590

Gateff, E., and Mechler, B.M., (1989) Tumor-Suppressor genes of *Drosophila melanogaster*. *In* "Critical Reviews in Oncogenesis." Vol. 1, pp. 221–245. CRC Press.

Gibson, G., and Hogness, D.S., (1996) Effect of polymorphism in the *Drosophila* regulatory gene *Ultrabithorax* on homeotic stability. *Science* **271**: 200–203

Gindhart, T.G., and Kaufman, T.C., (1995) Identification of *Polycomb* and *trithorax* group responsive elements in the regulatory region of the *Drosophila* homeotic gene *Sex comb reduced*. *Genetics* **139**: 797–814

Girton, J.R., and Jeon, S.H., (1994) Novel embryonic and adult homeotic phenotypes are produced by *pleiohomeotic* mutations in *Drosophila*. *Dev. Biol.* **161**: 93–407

Godt, D., Couderc, J.L., Cramton, S.E., and Laski, F., (1993) Pattern formation in the limbs of *Drosophila*: *bric à brac* is expressed in both a gradient and a wave-like pattern and is required for specification and proper segmentation of the tarsus. *Development* 119: 799–812

Goebl, M.G., (1991) The *bmi-1* and *mel-18* gene products define a new family of DNA-binding proteins involved in cell proliferation and tumorigenesis. *Cell* **66**: 623.

Guillen, I., Mullor, J.L., Capdevila, J., Sanchez-Herrero, E., Morata, G., and Guerrero, I., (1995) The function of *engrailed* and the specification of *Drosophila* wing pattern. *Development* **121**: 3447–3456

Gutjahr, T., Frei, E., Spicer, C., Baumgartner, S., White, R.A. H., and Noll, M., (1995) The *Polycomb* group gene, *extra sex combs* encodes a nuclear member of the WD-40 repeat family. *EMBO J.* **14**: 4296–4306

Harris, H., (1990) The role of differentiation in the suppression of malignacy. *J. Cell Sci.* **97**: 5–10

Hartl, P., Gottesfeld, T., and Douglas, J.F., (1993) Mitotic repression of transcription *in vitro*. *J. Cell. Biol.* **120**: 613–624

Haupt, Y., Bath, M.L., Harris, A.W., and Adams, J.M., (1993) *bmi-1* transgene induces lymphomas and collaborates with *myc* in tumorigenesis. *Oncogene* **8**: 3161–3164

Held, L.I., (1995) Axis boundaries and coordinates: the ABC of the fly leg development. *BioEssays* **17**: 721–732

Helms, J., Thaller, C., and Eichele, G., (1994) Relationships between retinoic acid and *sonic-hedgehog,* two polarizing signals in the chick wing bud. *Development* **120**: 3267–3274

Hidalgo, A., (1996) The roles of *engrailed. Trends Genet.* **12**: 1–4

Holland, P.W.H., Garcia-Fernandes, J., Williams, N.A., and Sidow, A., (1994) Gene duplications and the origins of vertebrate development. *Development Supl:* 125–133

Huang, M.F., Ye, Y.C., Chen, R.S., Chai, J.R., Zhoa, L., Gu, L.J., and Wang, Z.Y., (1988) Use of all trans retinoic acid in the treatment of acute promyelocytic leukemia. *Blood* **72**: 567–572

Ingham, P.W., (1995) Signalling by *hedgehog* family proteins in *Drosophila* and vertebrate development. *Curr. Opin. Genet. Dev.* **5**: 492–498

Johnson, A.D., (1995) The price of repression. *Cell* **81**: 655–658

Johnson, R.L., and Tabin, C., (1995) The long and the short of *hedgehog* signalling. *Cell* **81**: 313–316

Jones, R.S., and Gelbart, W.M., (1990) Genetic analysis of the *Enhancer of zeste* locus and its role in gene regulation in *Drosophila melanogaster. Genetics* **126**: 185–199

Jones, R.S., and Gelbart, W.M., (1993) The *Drosophila* Polycomb group gene *Enhancer of zeste* contains a region with sequence similarity with *trithorax. Mol. Cell. Biol.* **13**: 6357-6366

Joyner, A.L., (1996) *Engrailed, Wnt* and *Pax* genes regulate midbratin-hindbrain development. *Trends Genet.* **12**: 15–20

Jürgens, G., (1985) A group of genes controlling the spatial expression of the *bithorax* complex in *Drosophila. Nature* **316**: 153–155.

Kamb, A., (1995) Cell-cycle regulators and cancer. *Trends Genet.* **11**: 136–140

Kanno, M., Hasegawa, M., Ishida, A., Isono, K., and Taniguchi, M., (1995) *mel-18* a *Polycomb* group related mammalian gene, encodes a transcriptional negative regulator with tumor suppressive activity. *EMBO J.* **14**: 5672–5678

Kennison, J.A., (1993) Transcriptional activation of *Drosophila* homeotic genes from distant regulatory elements. *Trends Genet.* **9**: 75–79

Kennison, J.A., and Russell, M.A., (1987) Dosage dependant modifiers of homeotic mutations in *Drosophila melanogaster. Genetics* **116**: 75–86

Kennison, J.A., and Tamkun, J.W., (1988) Dosage dependant modifiers of *Polycomb* and *Antenapedia* mutations in *Drosophila. Proc. Natl. Acad. Sci.* USA **85**: 8136–8140

Kennison, J.A., and Tamkun, J.W., (1992) Transregulation of genes in *Drosophila. The New Biologist* **4**: 91–96

Kim, A., Terzian, C., Santamaria, P., Pelisson, A., Prud'homme, N., and Bucheton, A., (1994) Retroviruses in invertebrates: The gypsy retrotransposon is apparently an infectious retrovirus of *Drosophila melanogaster. Proc. Natl. Acad. Sci.* USA **91**: 1285–1289

Klein, G., (1987) The approaching era of tumor suppressor genes. *Science* **238**: 1539–1545

Kuzin, B., Tillib, S., Sedkov, Y., Mizrochi, Z., and Mazo, A., (1994) The *Drosophila trithorax* gene encodes a chromosomal protein and directly regulates the region specific homeotic gene *fork-head. Genes Dev.* **8**: 2478–2440

Kuziora, M.A., and Mc Ginnis, W., (1988) Different transcripts of the *Drosophila AbdB* gene correlate with distinct genetic subfunctions. *EMBO J.* **7**: 3233–3244

Landecker, H.L., Sinclair, D.A.R., and Brock, H.W., (1994) Screen for enhancers of *Polycomb* and *Polycomblike* in *Drosophila melanogaster. Dev. Genet.* **15**: 425–434

Langston, A.W., and Gudas, L.J., (1994) Retinoic acid and homeobox gene regulation. *Curr. Opin. Genet. Dev.* **4**: 550–555

Lawrence, J., and Largman, C., (1992) Homeobox genes in normal hematopoiesis and leukemia. *Blood* **80**: 2445–2453

Lawrence, P., Johnston, P., and Struhl, G., (1983) Different requirements for Homeobox genes in the soma and germ line of *Drosophila. Cell* **35**: 27–34

Lawrence, P., and Morata, G., (1994) Homeobox genes: their function in *Drosophila* segmentation and pattern formation. *Cell* **78**: 181–189

Lindsley, D., and Zimm, G., (1992) The genome of *Drosophila melanogaster.* Academic Press, Harcourt Brace Javanovich. San Diego.

Locke, J., Kotarski, M.A., and Tartof, K.D., (1988) Dosage dependent modifiers of position effect variegation in *Drosophila* and a mass action model that explains their effect. *Genetics* **120**: 181–198

Lonie, A., D'Andrea, R., Paro, R., and Saint, R., (1994) Molecular characterization of the *Polycomblike* gene of *Drosophila melanogaster,* a transacting negative regulator of homeotic gene expression. *Development* **120**: 2629–2636

Lu, M., Gong, Z.Y., Shen, W.F., and Ho, A.D., (1991) The TCL–3 protooncogene altered by chromosomal translocation in T-Cell leukemia codes for a *Homeobox* protein *EMBO J.* **10**: 2905–2910

Marshall, H., Studer, M., Pöpperl, H., Aparicio, S., Kuroiwa, A., Brenner, S., and Krumlauf, R., (1994) A conserved retinoic acid response element required for early expression of the *homeobox* gene Hoxb-1. *Nature* **370**: 567–571

Martin, E.C., and Adler, P.N., (1993) The *Polycomb* group gene *Posterior sex combs* encodes a chromosomal protein. *Development* **117**: 641–655

Maulbecker, C.C., and Gruss, P., (1993) The oncogenic potential of Pax genes. *EMBO J.* **12**: 2361–2367

McAlpine, J.F., (1989) Phylogeny and classification of the Muscomorpha. *In* "Manual of Neartic Diptera." (J.F., McAlpine, ed.), Vol. 3, Monograph **32**, pp. 1397–1502. Research Branch, Agriculture Canada.

McKeon, J., and Brock, H.W., (1991) Interactions of the *Polycomb* group of genes with Homeotic loci of *Drosophila. Roux's Arch. Dev. Biol.* **199**: 387–396

McKeon, J., Slade, E., Sinclair, D.A.R., Cheng, N., Couling, M., and Brock, H.W., (1994) Mutations in some *Polycomb* group genes of *Drosophila* interfere with regulation of segmentation genes. *Mol. Gen. Genet.* **244**: 474–483

Messmer, S., Franke, A., and Paro, R., (1992) Analysis of the functional role of the *Polycomb* chromo–domain in *Drosophila melanogaster. Genes Dev.* **6**: 1241–1254

Moazed, D., and O'Farrell P.H., (1992) Maintenance of the *engrailed* expression pattern by *Polycomb* group genes in *Drosophila. Development* **116**: 805–810

Moehrle, A., and Paro, R., (1994) Spreading the silence: Epigenetic transcriptional regulation during *Drosophila* development. *Dev. Genet.* **15**: 478–484

Morgan, B.A., and Tabin, C.J., (1993) The role of homeobox genes in limb development. *Curr. Opin. Genet. Dev.* **3**: 668–674

Mukherjee, A., Lakhotia, S.C., and Roy, J.K., (1995) *l(2) gl* gene regulates late expression of segment polarity genes in *Drosophila. Mech. Dev.* **51**: 227–234

Müller, J., (1995) Transcriptional silencing by the *Polycomb* protein in *Drosophila* embryos. *EMBO J.* **14**: 1209–1220

Müller, J., and Bienz, M., (1991) Long range repression: comparing boundaries of *Ultrabithorax expression* in *Drosophila* embryos. *EMBO J.* **10**: 3147–3156

Müller, J., Gaunt, S., and Lawrence, P., (1995) Function of the *Polycomb* protein is conserved in mice and flies. *Development* **121**: 2847–2852

Nomura, M., Takihara, Y., and Shimada, K., (1994) Isolation and characterization of retinoic acid-inducible cDNA clones in F9 cells: One of the early inducible clones encodes a novel protein sharing several highly homologous regions with a *Drosophila polyhomeotic* protein. *Differentiation* **57**: 39–50

Ogura, T., and Evans, R.M., (1995a) A retinoic acid-triggered cascade of Hox B1 gene activity. *Proc. Natl. Acad. Sci. USA* **92**: 387–391

Ogura, T., and Evens, R.M., (1995b) Evidence for two distinct retinoic acid response pathways for HoxBI gene regulation. *Proc. Natl. Acad. Sci. USA* **92**: 392–396

Orlando, V., and Paro, R., (1995) Chromatin multiprotein complexes involved in the maintenance of transcription patterns. *Curr. Opin. Genet. Dev.* **5**: 174–179

Paro, R., (1995) Propagating memory of transcriptional states. *Trends Genet.* **11**: 295–297

Paro, R., and Hogness, D.S., (1991) The *Polycomb* protein shares a homologous domain with a heterochromatin associated protein of *Drosophila*. *Proc. Natl. Acad. Sci. USA* **88**: 263–267

Paro, R., and Zink, B., (1992) The *Polycomb* gene is differentially regulated during oogenesis and embryogenesis of *Drosophila melanogaster*. *Mech. Dev.* **40**: 37–46

Peare, J.J., Sing, P. B., and Gaunt, S., (1992) The mouse has a *Polycomb* like chromobox gene. *Cell* **76**: 345–356

Pelegri, F., and Lehmann, R., (1994) A role of *Polycomb* group genes in the regulation of *gap* gene expression in *Drosophila*. *Genetics* **136**: 1341–1353

Phillips, M.D., and Shearn, A., (1990) Mutations in *polycombeotic, a Drosophila* Polycomb group gene, cause a wide range of maternal and zygotic phenotypes. *Genetics* **125**: 91–101

Pirrotta, V., (1995) Chromatin complexes regulating gene expression in *Drosophila*. *Curr. Opin. Genet. Dev.* **5**: 466–472

Pirotta, V., Chan, C.S, McCabe, D., and Quian, S., (1995) Distinct parasegmental and imaginal enhancers and the establishment of the expression pattern of the *Ubx* gene. *Genetics* **141**: 1439–1450

Rastelli, L., Chan, C.S., and Pirrotta, V., (1993) Related chromosome binding sites for *zeste, Suppressor of zeste* and *Polycomb* group proteins in *Drosophila* and their dependance on *Enhancer of zeste* function. *EMBO J.* **12**: 1513–1522

Reijnen, M.J., Hamer, K.M., den Blaauwin, J.L., Lambrechts, C., Schoneveld, I., van Driel, R., and Otte, A.P., (1995) *Polycomb* and *bmi-1* homologs are expressed in overlapping patterns in *Xenopus* embryos and are able to interact with each other. *Mech. Dev.* **53**: 35–46

Riley, P. D., Caroll, S.B., and Scott, M.P., (1987) The expression and regulation of *Sex comb reduced* protein in *Drosophila* embryos. *Genes Dev.* **1**: 716–730

Rizki, T.M., (1978) The circulatory system and associated cells and tissues. *In* "The genetics and biology of *Drosophila*" (M. Ashburner and T.M.F., Wright eds.) Vol 2B, pp. 397–444. Academic Press London, New York.

Santamaria, P., (1993) Evolution and aggregulates: Role of the *Polycomb* group genes of *Drosophila*. *C.R. Acad. Sci. Paris* **316**: 1200–1206

Santamaria, P., Deatrick, J., and Randsholt, N.B., (1989) Pattern triplications following genetic ablation on the wing of *Drosophila*. Effect of eliminating the *polyhomeotic* gene. *Roux's Arch. Dev. Biol.* **198**: 65–77

Santamaria, P., and Randsholt, N.B., (1995) Characterization of a region of the X chromosome of *Drosophila* including *multi sex combs (mxc)* a *Polycomb* group gene which also functions as a tumour suppressor. *Mol. Gen. Genet.* **246**: 282–290

Sathe, S.S., and Harte, P., (1995) The *Drosophila extra sex combs* protein contains WD motifs essential for its function as a respressor of homeotic genes. *Mech. Dev.* **52**: 77–87

Schlossherr, J., Eggert, H., Paro, R., Cremer, S., and Jack, R.S., (1994) Gene inactivation in *Drosophila* mediated by the *Polycomb* gene product or by position-effect variegation does not involve major changes in the accessibility of the chromatin fibre. *Mol. Gen. Genet.* **243**: 453–462

Serrano, N., Brock, H.W., Demeret, C., Dura, J.M., Randsholt, N.B., Kornberg, T.B., and Maschat, F., (1995) *polyhomeotic* appears to be a target of *engrailed* regulation in *Drosophila. Development* **121**: 1691–1703

Simon, J., (1995) Locking in stable states of gene expression: transcriptional control during *Drosophila* development. *Curr. Opin. Cell. Biol.* **7**: 376–385

Simon, J., Chiang, A., and Bender, W., (1992) Ten different *Polycomb* group genes are required for spatial control of the abdA and AbdB homeotic products. *Development* **114**: 493–405

Simon, J., Bornemann, D., Lunde, K., and Schwartz, C., (1995) The *extra sex combs* product contains WD40 repeats and its time of action implies a role distinct from other *Polycomb* group products. *Mech. Dev.* **53**: 197–208

Slack, J.M.W., Holland, P.W.H., and Graham, C.F., (1993) The zootype and the phylotypic stage. *Nature* **361**: 490–492

Smouse, D., Goodman, C., Mahowald, A., and Perrimon, N., (1988) *polyhomeotic* a gene required for the embryonic development of axon pathways in the central nervous system of *Drosophila. Genes Dev.* **2**: 830–840

Soto, M.C., Chou, T.B., and Bender, W., (1995) Comparison of germline mosaics of genes in the *Polycomb* group of *Drosophila melanogaster. Genetics* **140**: 231–243

Steiner, G., (1976) Establishment of compartments in the developing leg imaginal discs of *Drosophila melanogaster. Roux's Arch. Dev. Biol.* **180**: 9–30

Struhl, G., and Akam, M., (1985) Altered distribution of Ultrabithorax transcripts in *extra sex combs* mutant embryos of *Drosophila. EMBO J.* **4**: 3259–3264

Struhl, G., and Basler, K., (1993) Organizing activity of *wingless* protein in *Drosophila. Cell* **72**: 527–540

Stuart, E.T., Yokota, Y., and Gruss, P., (1995) *PAX* and *HOX* in Neoplasia. *Advances in Genetics* **33**: 255–274

Tabata, T., Schwartz, C., Gustavson, I., Ali, Z., and Kornberg, T.B., (1995) Creating a *Drosophila* wing the novo, the role of *engrailed* and the compartment border hypothesis. *Development* **121**: 3359–3369

Tabin, C., (1995) The initiation of the limb bud: growth factors, *Hox* Genes, and Retinoids. *Cell* **80**: 671–674

Tamkun, J., (1995) The role of *brama* and related proteins in transcription and development *Curr. Opin. Genet. Dev.* **5**: 473–477

Tautz, D., (1996) Selector genes, polymorphism and evolution. *Science* **271**:160–161

Tickle, C., (1995) Vertebrate limb development. *Curr. Opin. Genet. Dev.* **5**: 478–484

Tickle, C., and Eichele, G., (1994) Vertebrate limb development. *Ann. Rev. Cell. Biol.* **10**: 121–152

Tiong, S.Y.K., and Russel, M.A., (1990) Clonal analysis of segmental and compartmental homeotic transformations in *Polycomb* mutants of *Drosophila melanogaster. Dev. Biol.* **141**: 306–318

Tiong, S.Y.K., Nash, D., and Bender, W., (1995) *Dorsal wing* a locus that affects dorsoventral wing patterning in *Drosophila. Development* **121**: 1649–1656

van der Lugt, N.M.T., Domen, J., Linders, K., van Roon, M., Robanus–Maandag, E., Te Riele, H., van der Valk, M., Deschamps, J., Sofroniew, M., van Lohuizen, M., and Berns, A., (1994) Posterior transformation, neurological abnormalities and severe hematopoietic defects in mice with a targeted deletion of the *bmi-l* protooncogene. *Genes Dev.* **8**: 750–769 van Lohuizen, M., Frasch, M., Wientjens, E., and Berns, A., (1991) Sequence similarity between the mammalian *bmi-l* protooncogene and the *Drosophila* regulatory genes *Psc* and *Su(z)2*. *Nature* **353**: 353–355

Vincent, J.P., (1994) Morphogens dropping like flies? *Trends Genet.* **10**: 383–385

Watson, K.L., Justice, R.W., and Bryant, P.J., (1994) *Drosophila* in cancer research: the first fifty tumor suppressor genes. *J. Cell Sci. Supl.***18**: 19–33

Weden, C., Harding, K., and Levine, M., (1986) Spatial regulation of *Antennapedia* and *bithorax* gene expression by *Polycomb* locus. *Cell* **44**: 739–748

Weinberg, R.A., (1983) A molecular basis of cancer. *Scientific American* **249**: 102–106

Weis, K., Rambaud, S., Lavau, C., Jansen, J., Carvallo, T., Carmo–Fonseca, M., Lamond, A., and Dejean, A., (1994) Retinoic acid regulates aberrant nuclear localization of PML-RAR in acute promyelocytic leukemia cells. *Cell* **76**: 345–356

Wu, C.-T., and Howe, M., (1995) A genetic analysis of the *Suppressor 2 of zeste* complex of *Drosophila melanogaster. Genetics* **140**: 139–181

Yu, B.D., Hess, J.L., Horning. J.L., Horning, S.E., Brown, G.A.J., and Korsmeyer, S.J., (1995) Altered *Hox* expression and segmental identity in *Mll*-mutant mice. *Nature* **378**: 505–508

Zecca, M., Basler, K., and Struhl, G., (1995) Sequential organizing activities of *engrailed,* *hedgehog* and *decapentaplegic* in the *Drosophila* wing. *Development* **121**: 2265–2278

Zhang, C.C., and Bienz, M., (1992) Segmental determination in *Drosophila* conferred by *hunchback* a direct repressor of the *homeotic* gene *Ultrabithorax. Proc. Natl. Acad. Sci.* USA **89**: 7511–7515

Zink, B., and Paro, R., (1989) *In vivo* binding pattern of a trans-regulator of *homeotic* genes in *Drosophila melanogaster. Nature* **337**: 468–471

Zink, B., and Paro, R., (1995) *Drosophila Polycomb* group regulated chromatin inhibits the accessibility of a trans-activator to its target DNA. *EMBO J.* **14**: 5660–5671

Genome Analysis in Eukaryotes: Developmental and evolutionary aspects
R.N. Chatterjee and L. Sánchez (Eds)
Copyright © 1998 Narosa Publishing House, New Delhi, India

10. Operational Redundancy: An evolutionary link between replication and the establishment of repressive chromatin structures

Francesco De Rubertis and Pierre Spierer

Department of Zoology and Animal Biology, University of Geneva,
30, quai Ernest-Ansermet, CH-1211 Geneva 4, Switzerland

I. Introduction

Molecular and cell biologists are innate reductionists. They adore the clean-cut dissection of biological functions as distinct and separable events led by distinct and separable macromolecules. Evolution, unfortunately, sometimes borrows more tortuous ways. It tries hard to make many uses for each of the macromolecules that it has painstakingly assembled. As organisms were built by chance and necessity, and not engineering, there is plenty of added complexity for us to decipher. It is therefore not so surprising to unravel common features of distinct cellular processes, and to discover different functions for one protein. In its simplest expression, this implies that some proteins are able to function in different contexts. Their specificity might depend on several factors: interaction and recruitment by specialized factors, different schedule of action, or place of action, in the cell cycle or organism development. This is true not only in prokaryotes, where the level of structural organization is low and the dialogue with other cells limited, but also in eukaryotes, where examples of operational redundancy are numerous.

II. DNA replication and heterochromatin formation:
An overview

A good example of operational redundancy is the machinery of DNA replication, as it also controls cell cycle progression and repair of genetic damage. This factory seems related to another phenomenon, heterochromatin formation. This emerging link is the subject of this review. To have all the actors on the scene, let us also mention that heterochromatin seems to provide structural scaffold to chromosomes enabling centromeres to be correctly segregated and telomeres to be replicated and maintained without loss of substance at each cell cycle. Last but not least, heterochromatinization is associated with epigenetic gene silencing, as revealed by centromeric and telomeric position effects resulting from chromosomal rearrangements. Heterochromatinization also seems implicated in the regulation of homeotic gene complexes.

To make this intricacy even worse, the principle of economy of the cell at this biochemical level is illustrated by the several proteins that are able to act both as transcriptional activators and repressors depending on metabolic conditions. In evolutionary terms, it could be a risk for the cell to employ few proteins in multiple process, as a mutation in a single gene could result in several phenotypes. Analysis shows however that chance and necessity are making a wide use of this mode of organization of the cell, and that both genetic specificity and redundance are widespread in eukaryotic organisms. Recently, a number of laboratories working on a variety of organisms have started to collect evidences of the above-mentioned operational redundancy in two categories of phenomena: the processes of DNA replication and heterochromatin formation. Evidences originate from the study of gene silencing in yeast at the mating type locus, and from the phenomenon of position-effect variegation in *Drosophila melanogaster.* We shall present here different ideas and models that could explain, at least partially, this correlation.

III. Mating type silencing in yeast

Gene silencing has been extensively studied in yeast, particularly at the mating type locus (MAT), and these investigations have produced a wealth of information. In short, two different alleles are naturally occuring at the MAT locus, MAT*a* and MATα, each one determining one of the two mating types of yeast. Haploid yeast cells can mate only if expressing the two opposite mating types. The resulting diploid *a*/α is itself incompetent for mating (Herskowitz, 1988). The mating type switch in *Saccharomyces cerevisiae* strains relies on the expression of a DNAse, the HO endonuclease. Mechanically, the mating type switch, depends on the replacement of the expressed allele present at the MAT locus, by the other one which, as recently demonstrated, is present on the same chromosome but at a distinct locus (HM locus). It is kept there in a silenced transcriptional state by a mechanism of local repression that has features analogous to chromosomal position-effect silencing. There are two such donor loci, namely HML, lying 300 kb on the left of the MAT locus and usually carrying a copy of the α allele and HMR, located on the other side of MAT, with a silent copy of the *a* allele (Herskowitz, 1988; Laurenson and Rine, 1992). These silent copies are hence named HMR*a* and HMLα. Their transcriptional silencing requires four cis-acting elements: HMR-E, HMR-I, HML-E and HML-I. A number· of trans-acting factors are also involved, including the proteins NAT1, ARD1, RAP1, the histone H4 and, most importantly, the SIR proteins (Silencing Information Regulator) and the Origin Recognition complex (ORC)

(Bell *et al*, 1993; Foss *et al*, 1993; Micklem *et al*, 1993). Originally, the evidences that mutations in histone H4 lead to HM loci derepression (Kayne *et al*, 1988) were the first indications of a link between chromatin structure and silencing in yeast. Yeast geneticists found then that several strains defective at HM loci repression carried mutations in genes encoding proteins which intervene directly in the structuring of heterochromatin, both at telomeres and at the mating type loci. This has been the most compelling evidence of the establishment of heterochromatic structures as a necessary step in gene transcriptional silencing (see below).

Recent studies of the functioning of the mating type locus switch and of its genetic and structural requirements have opened new perspectives: the particular nature of the already cited cis-acting elements and of the trans-acting factors which were demonstrated to play a key role in yeast gene silencing strongly suggests a tight functional link between the formation of heterochromatic structures and the progression of cell cycle. We shall focus on this particular relationship.

IV. Control elements of the mating type and replication

The cis-acting silencing elements (silencers) lie next to the HM loci and, with the exception of the different action exerted on the regulation of gene transcription, present some features similar to those of enhancer elements: they can work in both orientation, and they can repress the transcription of their target gene even if placed at distance. Finally, another characteristics outlining a functional similarity with enhancers is the ability of silencers to act on unrelated genes, transcribed either by RNA pol II or pol III (Laurenson and Rine, 1992 and ref. therein). Nevertheless, it is not yet clear whether chromosomal location can affect the repressive function of a silencer. The best studied silencer is HMR-E. Its genetic dissection shows that it is composed of three different elements: a pRAP1 binding site (E element), a pABF1 binding site (B element) and, most intriguingly, the conserved core motif of Autonomous Replicating Sequence (ARS), known as ARS consensus sequences (ACS). This element binds *in vitro* to the origin recognition complex (ORC) and can work as an autonomous replicating sequence in plasmids (A element). Among the four silencers, however, only HMR-E functions as an ARS at the native chromosomal position (Bell *et al*, 1993). The three elements present a certain degree of redundancy, as indicated by the fact that mutations in at least two of them are necessary to derepress the adjacent silenced gene (Brand *et al*, 1987).

The protein pRAP1 was found to play a direct role in the regulation of the mating type locus. Particularly, a series of *rap1* mutant strains are defective in the transcriptional activation of the MATα locus and some of them show even silencing defects at HMR. The allele specificity of these phenotypes suggests that repression at the silenced locus and activation of the expressed one could be mediated by different domains of the same protein (Kurtz and Shore, 1991). The protein pABF1 was found to bind to HMR-I and HML-I. The interest for this protein resides in the fact that it can bind a subset of autonomous replicating sequences in *S. cerevisiae*. The function of this protein is not yet clarified, but there are strong indications that it acts in silencing: the construction of synthetic silencers clearly showed that each one of the three constitutive elements (A, B, E) is necessary to have a perfectly functional silencer (Laurenson and Rine, 1992).

The A-element is undoubtly the most intriguing among the three, as it has been shown to be necessary both for origin of replication and gene silencing. This provides support to the idea of a coordination between the two phenomena. A protein which binds to ARS and also to HMR-E has been identified (pACBP). A mutation in the target site prevents the binding of pACBP, and also causes derepression of the HM locus. This suggests an involvement of the protein is silencing, but, at the moment, just like in the case of ABFI, *abcp* mutants presenting silencing defects have not been isolated (McNally and Rine, 1991).

V.   Replication and heterochromatin formation in yeast

The first indication of a link between replication and heterochromatin formation was given by the evidence that mutations in HMR-E affect its ability to function both in replication and in silencing. This finding left open the possibility that the sequence interacts at different times of the cell cycle with unrelated proteins, each one acting in one of the two pathways. The proof of a true functional link between the two cellular events has come with the finding that yeast strains carrying mutations in the genes ORC2 and ORC5, that code for two of the subunits of ORC, are defective in both DNA replication and transcriptional repression at the HMR locus (Bell *et al*, 1993; Foss *et al*, 1993). In particular, *orc2* mutants bring to full derepression and HMR partially derepressed by a point mutation in binding site for RAP1, thus blocking the yeast mating ability. The HM locus derepression is only partial in the case of the single point mutation at the pRAP1 binding site because of functional redundancy. As expected, mutation in ORC2 cause full derepression of the silenced loci flanked by a synthetic silencer lacking the functional redundancy (Bell *et al*, 1993). This demonstrates a functional involvement of the protein in the mechanisms of gene silencing and as the other phenotypes of *orc*2 mutants are replication and segregation defects, a link is established between the two cellular processes. The fact that also *orc5* mutants (another subunit of ORC) have been shown to be defective at silencing has ruled out the possibility that the silencing defects observed in four independent *orc2* alleles were specifically linked to this protein.

However it is not known whether silencing needs simply the assembly of a precise multisubunit protein complex. It would include ORC, which, being available only at a restricted moment during cell cycle, would give an indirect cell cycle-dependence to silencing. The passage of a fork of replication might also be necessary to establish repression as a prerequisite to assemble a protein complex. This is suggested by the bidirectional spreading of silencing. The replication fork *per se* is not sufficient to induce repression of the adjacent genes: the particular origins of replication associated with the silencers could be programmed to be active only at a particular moment of cell cycle, exactly when the heterochromatin components are available to build up repressive structures.

A second line of evidence, strongly suggesting a tight functional relationship between DNA replication and hetrochromatin formation, comes from experiments conducted with strains carrying temperature sensitive mutations in genes coding for proteins playing an important role in silencing. The SIR proteins are directly involved in the establishment of the repressive heterochromatic structures at the HM loci, as demonstrated by the fact that the HO endonuclease, responsible for the double-strand cut at a target sequence in the MAT locus, can produce cuts at the HM loci only in *sir* strains (Nasmyth, 1982). Moreover, other proteins, like DNA repair enzymes, methyltransferases, and different restriction endonucleases seem to be prevented from acting directly on DNA at HM loci in a SIR dependent manner, strengthening the proofs for the function of SIR proteins in inducing a closed chromatin conformation (Nasmyth, 1982). *sir³*, thermosensitive mutants, for example, show derepression of the HM locus when shifted to the non permissive temperature regardless of the moment at which the shift occurs during the cells cycle. Conversely, it is necessary to pass through an S-phase, most probably hence through DNA replication, to reestablish transcriptional repression. This evidence can be looked at in different ways: either the ARS at HMR is very important in silencing and the elements of the silencer can be bound by the compacting factors only if chromatin is undergoing replication or, alternatively, when bound to the HMR locus, an activator protein can be displaced only by the passage of a replication fork.

Intriguingly, a mutation in CDC7, a gene important for the progression of cell cycle at the G1/S boundary, was isolated in an unrelated screen for suppressors of silencing defects due to mutations at HMR-E silencer, which abolished the *in vitro* binding of pRAP1 and pABF1 (Axelrod and Rine, 1991). Various loss of function alleles were able to suppress silencing defects at HMR, but did not repress transcription when the mating type allele resided at the MAT locus, thus indicating a specific effect in silencing. Overexpression of CDC7, similarly, interferes with the normal establishment of silencing: it enhances the derepression threshold level of the HM locus (Axelrod and Rine, 1991). Finally, in experiments with double mutants, mutations in CDC7 are unable to suppress the nonmating phenotype (absence of repression) imposed by mutations in *sir⁴*, but they are able to restore repression, at least at HMR, in *sir³* mutants, indicating that the ability of CDC7 to repress HMR depends solely on *sir⁴* function.

How does CDC7 affect transcriptional repression at the silenced mating type locus? At the moment, there is not clear cut answer to this question, but there are some indicative experiments. First of all, *cdc7* mutations do not restore binding of pRAP1 and pABF1 to their target site in the silencer (Axelrod and Rine, 1991). Secondly, the finding that *cdc7* mutants cannot suppress the nonmating phenotype of diploids if the MA*Ta* locus is carried on a plasmid rather than on the chromosome argues against the hypothesis of non-specific phenotypic suppression of the defect by *cdc7* mutations (Axelrod and Rine, 1991). Another hypothesis could be that the restoring of silencing results from a sort of bypass of the requirement of HMR-E, through the activation of HMR-I. This idea is in contrast with the known interaction between CDC7 and SIR4, that argues against the appearance of a completely new mechanisms through which gene silencing would be accomplished. Since CDC7 is implicated in the initiation of DNA replication, a possible implication of cell cycle in the formation of heterochromatin has to be taken into account. It is known that euchromatic regions are replicated relatively early during cell cycle and that heterochromatin, in contrast, it replicated quite late. HMR is normally replicated near the end of the S-phase and HMR-E mutations could be responsible of an early replication of the locus, causing its transcriptional derepression. *cdc* mutations could then restore silencing by simply delaying again the replication of HMR to the correct moment during cell cycle and hence establishing a tight relationship between late replication and transcriptional repression. Alternatively, it is possible to imagine that initiation of DNA replication at the ARS sequence of HMR-E is a necessary step to establish repression and that *cdc* mutations restore silencing by increasing the firing of this origin of replication. This idea awaits a proof of the ability of silencers to work as origins of replication at their native sites on the chromosomes.

VI.   Position-effect variegation (PEV) in *Drosophila* and DNA replication

Old and new evidences suggesting a link between cell cycle and formation of repressive chromatin structures arise from work with *Drosophila melanogaster,* and more particularly from the phenomenon of position-effect variegation (PEV). When an euchromatic gene is placed by a chromosomal rearrangement next to heterochromatin, it can become permanently inactive in some cells and not in others, producing thus a mosaic phenotype. This transcriptional silencing, clonally inherited, is currently explained by the spreading of adjacent heterochromatin over the locus, which undergoes a general chromatin compaction. The most fashionable rearrangement is called *white mottled*. It is a chromosomal inversion within the X chromosome placing the *white* gene at proximity of a block of centromeric heterochromatin. The *white*+ gene is responsible for the red color of the eye. In *white mottled* rearrangements, the neighbouring heterochromatin seems to spread toward the *white* gene and inactivates it permanently in some cells, and thus results in a mottled phenotype of clones of active (red) and inactive (white) cells (Reuter and Spiere, 1992).

This rearrangement and its easily scorable phenotype has been used to isolate dominant extragenic mutations affecting the ratio between active and inactive clones in flies carrying the rearrangement. These modifiers of variegation fall into two major groups, enhancers and suppressors, the former increasing the relative number of clones inactivated by the heterochromatin spreading and the latter increasing, conversely, the number of clones in which heterochromatin does not "diffuse". It can be expected that these genes encode structural components of chromatin (suppressors), or proteins acting to decondense it (enhancers), or proteins which control or modify these factors. Several enhancers and suppressors have been cloned and the results fit quite well with these expectations (Reuter and Spierer, 1992). An unpredicted result, however, is the cloning of different proteins which are, directly or indirectly, related to cell cycle and to the progression through S-phase. Before discussing in more details the enhancers and suppressors of variegation, let us consider the structural basis of position-effect variegation and the different models for heterochromatin formation.

The most popular model, the "spreading of heterochromatin" has been proposed long ago, and was reformulated by Locke *et al* (1988). On the chromosomes, there would be boundaries which delimit the domains of formation of heterochromatin, itself assembled as a multimeric protein complex. In a large chromosomal rearrangement, like the *white mottled* inversion, a boundary element is displaced and heterochromatin is not any longer confined to its natural domain. It is free to assemble cooperatively along the chromosome, as long as structural subunits aggregating to form the multimeric protein complex are in sufficient supply. This spreaded heterochromatic structure can form a compact and inaccessible chromatin fiber over an adjacent euchromatic gene and consequently be able to inactivate it. The extent of spreading could vary in length from cell to cell and results in a variegated transcriptional profile of that euchromatic locus.

Another model argues that heterochromatin and euchromatin are not structurally different and present a similar pattern of accessibility to restriction endonucleases, methylases and to DNAase digestions (Schlossherr *et al*, 1994). Karpen (1994) believed that a different subnuclear localization could be the cause of different transcriptional activity. Heterochromatin is located next to the nuclear membrane, where the structural determinants and transcription factors required respectively for the structuring and the genetic activity of this type of chromatin would also be located. These factors would differ from those present in the central compartment of the nucleus, the nest of euchromatin. According to this "compartimentalization" model, a rearrangement is seen as variegating when it puts an euchromatic gene in the heterochromatic compartment, exposing it to the determinants that would induce its heterochromatin-like compaction. The interest of this model resides in the fact that, unlike the spreading model, it explains very easily also the variegation observed in the case of heterochromatic gene (i.e. the *light* gene, Devlin *et al*, 1990).

These two models explain variegation with a stochastic transcriptional repression. Other observations, namely the fact that heterochromatin is late replicating in the cell cycle and underrepresented in polytene chromosomes, have lent support to other propositions. In polytene tissues, variegation could result from underreplication (reviewed in Karpen, 1994). The programmed delay of replication of heterochromatin could also correspond to a specific window of availability of transcription factors. The limit of the first idea is that it cannot be applied to diploid tissues and therefore is unsufficient to elucidate the structural basis of position-effect variegation, as we know that it occurs also in diploid tissues. Nevertheless, these are stimulating prospects in our search for links between gene silencing and DNA replication, especially now that PEV modifier mutations were shown to occur in genes encoding proteins directly involved in cell cycle progression (see section VII).

VII.   Modifiers of position effect variegation and replication

What could be the basis of a relationship between modifies of position effect variegation and replication? Let us first consider the different models. The model implying that an euchromatic gene variegates when present in low copy number in a polytene tissue directly links to the control of DNA replication. Any mutation in genes encoding proteins intervening in the mechanism of polytenization, therefore, would be predicted to affect the DNA copy number and consequently to modify the level of gene silencing. Theoretically, the partial or full inactivation of a positive regulator of DNA replication could result in a PEV enhancer phenotype and, conversely, a mutation in a protein negatively regulating cell cycle progression or inducing the somatic loss of a certain number of the copies could result in a suppressor phenotype. This prediction is consistent with the recent demonstration that loss of a dose of the gene encoding dE2F, the *Drosophila* homologue of a mammalian transcriptional activator, necessary for the progression of cell cycle at the boundary G1/S, is an enhancer of position-effect variegation (Seum *et al*, 1996). It is known that during G1, E2F is bound to the retinoblastoma protein (RB). A series of sequential cyclin-dependent phosphorylations induce the separation of the two proteins, allowing E2F to activate its target genes at the onset of the S-phase. Interestingly, RB interacts *in vivo* with the protein phosphatase type 1 catalytic subunit, PP1 (Durfee *et al*, 1993): loss of function mutations in the *Drosophila* PP1 have been shown to suppress PEV (Baksa *et al*, 1993). Other completely different hypotheses also are consistent with the data. *dE2F* could act either by a direct action on chromatin or by being a regulator of a gene affecting itself PEV. Even as a cell cycle regulator, it could act indirectly at the time of euchromatin and/or heterochromatin assembly.

An involvement of DNA replication appears to be different, probably less direct, in the "spreading" model. In the case of heterochromatin as a multimeric protein complex, its formation would depend on the availability of structuring factors at a very precise moment during cell cycle, for example, at the moment of replication fork progression. If a deregulation at the level of cycle progression occurs, the consequent uncoordination between the replication of the part of the chromosome to be condensed and the presence of the heterochromatic factors would modify the final level of heterochromatin formed in the cell. Alternatively, it is possible to imagine that in mutants with an extended S-phase, heterochromatin, usually formed late during cell cycle, can be formed in larger amounts, simply because there is a longer time window available for its assembling. Consequently, this would lead to an increase in gene silencing and to PEV enhancer phenotype. Any mutation inducing an extension of the S-phase could, in principle, be an enhancer of position-effect variegation. In this respect, a new enhancer of PEV, which encodes a deubiquitinating enzyme, *D-Ubp-64E*, is of interest (Henchoz *et al*, 1996). Ubiquitin is among many other functions, directly involved in the control of cell cycle, inducing its completion through the degradation of some of its positive regulators (cyclins, for example). A delay in this step of cell cycle would extend the S-phase, thus possibly causing an increase in the amount of heterochromatin formed. In support of this idea, a mouse line presenting defects at the boundary S/G2 of cell cycle carries a mutation in the ubiquitin activating enzyme (Yasuda *et al*, 1981). Here again, other hypotheses cannot be disqualified.

Finally, in the model of "compartements", the correct progression along cell cycle would be important if the peripheric localization of heterochromatin and the more internal one of euchromatin are synchronized with DNA replication. The late replication of an euchromatic region would simply put it in the peripheral compartment of the nucleus, changing the biology of the chromatin fibre. This could happen for example as a consequence of unspecific steric hindrances, distribution of specific anchoring factors, or local density of factors involved in chromatin assembly.

The proliferative cell nuclear antigen (PCNA) is an essential protein in eukaryotic DNA replication, as established by biochemical criteria. It works as an auxiliary factor for DNA polymerase δ and ε, and is continuously expressed throughout cell cycle, though more abundantly during S-phase. It localizes to replication foci and seems to be important for the coordination of the replication of the leading and of the lagging strands. Most interestingly PCNA has been genetically linked to two other important aspects of cell biology: DNA repair and heterochromatin formation. The involvement in the DNA repair machinery can, of course, be explained with its involvement in the synthesis of the new DNA strand, but the fact that PCNA is also a suppressor of PEV suggests a role at the level of chromatin condensation, mediating the action of the enzymes directly involved in repair (Henderson *et al*, 1994).

The gene *cramped (crm)* is a newly cloned gene belonging to the Polycomb-group genes, the negative regulators of homeotic genes (Yamamoto *et al*, 1995). Involvement of *crm* in chromatin condensation is strongly suggested by the fact that this protein is also suppressor of position-effect variegation, confirming previous evidences of the relationship existing between negative regulation of the bithorax complex and PEV suppression. Antibody staining of embryos shows a cell cycle dependent variation in the protein distribution which is extremely similar to that of the PCNA protein. It has been shown that the *cramped* protein is distributed in the nucleus during interphase, but goes to cytoplasm when chromosomes condensate. The two proteins, *crm* and PCNA have also been colocalized on polytene chromosomes. It is not know at the moment if *crm* has a role in DNA replication. However, if this will turn out to be the case, a clear link between cell cycle progression and heterochromatin formation will be definitively established (Yamamoto *et al*, 1995).

The "compaction" model, predicting that gene inactivation derives from the establishment of a more condensed chromatic fiber, can explain more easily the variegating phenotype of PCNA mutant flies: PCNA could be a structural protein and its absence would affect chromatin organization. This would lead to activation of previously inactive genes, thus resulting in a suppression of variegation. But, based on its biochemical function, it is really impossible to judge the role of PCNA. It is possible that PCNA affects the onset of replication, during S-phase, of the heterochromatic regions of chromosomes. A decrease in the amount of the factor could trigger an early replication of heterochromatin, leading to the loss of its normal physical and chemical state, and would consequently lead to failure in repression of adjacent genes.

VIII.   As a temptative conclusion

At this time, it is impossible to assess the functional meaning of the relationship between DNA replication and heterochromatin formation. What is known is that replication can be a necessary step for gene silencing, but not for derepression, and that silencer sequences can work as origins of replication. When defective, their functions in both transcriptional control and replication are affected. The finding that two proteins active at the G1/ S boundary, in different organisms, namely CDC7 in yeast, and *dE2F* in *Drosophila*, seem both able to promote hetrochromatin formation is rather intriguing. The first restores HM silencing, while the second suppresses position-effect variegation.

This striking result is to be confronted with the recent data obtained by

Laman *et al.* (1995). In a screen designed to clone suppressors of silencing defects at HMR in *rap* 1 mutants, these authors found that mutations causing a delay in the completion of cell cycle were able to restore silencing. The possibility that a specific period of the cycle is connected to silencing has been ruled out by the fact that they were able to find mutations in genes coding for proteins acting at different steps of cell-cycle: the transcriptional regulators SW16, SW14, RNR1 and MBP1, the cyclins CLB2 and CLB5, important for the progression through S-phase, the cyclin CLN3, acting at late G1 and the protein CIN8, required at the G2/M boundary. What Laman *et al* (1995) observed, is that the lengthening of any phase of cell cycle can restore gene silencing. They could prove that this effect is specific because a similar lengthening induced by lowering the incubation temperature or by growing cells on alternative carbon sources did not affect gene inactivation. They could also prove that the restoration of silencing that they observe in the mutants with an extended S-phase is not due to the shortening of the G1 phase. They used a particular *cln3* allele, in which a deletion of the C-terminus enhances the stability of the protein, thus fastening the passage to S-phase: silencing is not restored. Laman and coworkers concluded that a general disturbance in cell cycle enhances gene silencing in *S. cerevisiae.* The validity of the observation is strengthened by the evidence that silencing can be restored also by chemicals delaying the completion of cell cycle, as hydroxyurea. All these data seem to orient us in one direction: the necessity of a correct cell-cycling to correctly form heterochromatin.

Among hypotheses to be tested, the following comes first: is the functional link between DNA replication and gene silencing based on the necessity to disrupt the existing chromatin structure to establish subsequently a heterochromatic silencing complex? Chromatin disruption is well documented during the passage of a replication fork. In another form: does a new set of structural proteins become available to build additional heterochromatin around genes during S-phase, most probably during late S-phase? In respect to this second possibility, the ARS activity associated with silencers would have the particular meaning of determining the precise moment in cell cycle during which the locus to be silenced should be replicated. It is then obvious that any disturbance in the progression of the cycle would lead to a modification in the temporal availability of heterochromatin components. Another and completely unrelated hypothesis is put forward by Laman *et al.* (1995). It takes advantage of the observation that a delay in cell cycle causes an increase in cell volume. The consequent increase in overall biosynthesis could actually cause variations in the nuclear concentration of silencing factors, favoring heterochromatin formation. In this case the link between cell cycle and silencing would be indirect. In any case, the answers to all these open questions require the characterization of other proteins acting in both processes. It will be of interest, for example, to identify the targets of *D-Ubp 64E* encoded ubiquitin-deconjugating enzyme in *Drosophila* or to understand its mode of action. The targets of the *crm* encoded protein and of PCNA need also be determined.

The parallels presented in this chapter between yeast and *Drosophila* open a wide and potentially productive field of investigation, as both systems allow combined molecular and genetical analysis. We might soon understand the molecular basis of the relationship between the establishment of repressive chromatin structures and the correct progression of cell cycle.

Acknowledgments

The authors thank S. Henchoz, D. Pauli, and C. Seum for discussions of unpublished work, and acknowledge support by the Swiss National Science Foundation and the State of Geneva.

REFERENCES

Axelrod, A. and Rine, J. (1991). A role for CDC7 in repression of transcription at the silent mating-type locus HMR in *S. cerevisiae*. *Mol. Cell. Biol.* **11**: 1080–1091

Baksa, K., Morawietz, H., Dombradi, V., Axton, M., Taubert, H., Szabo, G., Torok, I., Udvardy, A., Gyurkovics, H., Szoor, B., Glover, D., Reuter, G. and Gausz, J. (1993). Mutations in protein phosphatase 1 gene at 87B can differently affect suppression of Position Effect Variegation and mitosis in *Drosophila melanogaster*. *Genetics* **135**: 117–125

Bell, S.P., Kobayashi, R. and Stillman, B. (1993). Yeast Origin Recognition Complex functions in transcription silencing and DNA replication. *Science* **262**: 1844–1848

Brand, A.H., Micklem, G. and Nasmyth, K. (1987). A yeast silencer contains sequences that can promote autonomous plasmid replication and transcriptional activation. *Cell* **51**: 709–719

Devlin, R.H., Bingham, B.B., and Wakimoto, B.T. (1990). The organization and expression of the *light* gene, a heterochromatin gene of *Drosophila melanogaster*. *Genetics* **125**: 129–140

Durfee, T., Becherer, K., Chen, P.L., Yeh, S.H., Yang, Y., Kilburn, A.E., Lee, W.H., and Elledge, S.J. (1993). The retinoblastoma protein associates with the protein phosphatase type 1 catalytic subunit. *Genes Dev.* **7**: 555–569

Foss, M., McNally, F.J., Laurenson, P. and Rine, J. (1993). Origin Recognition Complex in transcriptional silencing and DNA replication in *Saccharomyces cerevisiae*. *Science* **262**: 1838–1843

Henchoz, S., De Rubertis, F., Pauli, D. and Spierer, P. (1996). The dose of a putative ubiquitin-specific protease affects position-effect variegation in *Drosophila melanogaster*. *Mol. Cell. Biol.* **16**: 5717–5725

Henderson, D.S, Banga, S.S., Grigliatti, T.A. and Boyd, J.D. (1994). Mutagen sensitivity and suppression of Position Effect Variegation result from mutations in *mus 209*, the *Drosophila* gene encoding PCNA. *EMBO J* **13**: 1450–1459

Herskowitz, I. (1988). Life cycle of the budding yeast *Saccharomyces cerevisiae*. *Microb. Rev.* **52**: 536–553

Karpen, G.H. (1994). Position Effect Variegation and the new biology of heterochromatin. *Curr. Opin. Genet. Dev.* **4**: 281–291

Kayne, P.S., Kim, U.J., Han, M., Mullen, J.R., Yoshizaki, F. and Grunstein, M. (1988). Extremely conserved histone H4 N terminus is dispensable for growth but essential for repression the silent mating loci in yeast. *Cell* **55**: 27–39

Kurtz, S. and Shore, D. (1991). Rap1 protein activates and silences transcription of mating type genes in yeast. *Genes Dev.* **5**: 616–628

Laman, H., Balderes, D. and Shore, D. (1995). Disturbance of normal cell-cycle progression enhances the establishment of transcriptional silencing in *Saccharomyces cerevisiae*. *Mol. Cell. Biol.* **15**: 3608–3617

Laurenson, P. and J. Rine, J. (1992). Silencers, silencing and heritable transcriptional states. *Microb. Rev.* **56**: 543–560

Locke, J., Kotarski, M.A., and Tartof, K.D. (1988). Dosage-dependent modifiers of Position Effect Variegation in *Drosophila* and a mass action model that explains their effects. *Genetics* **120**: 181–198

McNally, F.J. and Rine, J. (1991). A synthetic silencer mediates SIR-dependent functions in *Saccharomyces cerevisiae*. *Mol. Cell. Biol.* **11**: 5648–5659

Micklem, G., Rowby, A., Harwood, I., Nasmyth, K. and Diffley, J.F. (1993). Yeast Origin Recognition Complex is involved in DNA replication and transcriptional silencing. *Nature* **366**: 87–89

Nasmyth, K. (1982). The regulation of yeast mating-type structure by SIR: an action at a distance affecting both transcription and transposition. *Cell* **30**: 567–578

Reuter, G. and Spierer, P. (1992). Position effect variegation and chromatin proteins. *Bioessays* **14**: 605–612

Schlossherr, J., Eggert, H., Paro, R., Cremer, S. and Jack, R.S. (1994). Gene inactivation in *Drosophila* mediated by the Polycomb gene product or by position-effect variegation does not involve major changes in the accessibility of the chromatin fibre. *Mol. Gen. Genet.* **243**: 453–462

Seum, C., Spierer, A., Pauli, D., Szidonya, J., Reuter, G. and Spierer, P. (1996). Position-effect variegation in *Drosophila* depends on dose of the gene encoding the E2F transcriptional activator and cell cycle regulator. *Development* **122**: 1949–1956

Yamamoto, Y., Affolter, M. and Gehring, W.J. (1995). Linkage between stable repression and DNA replication: the gene *cramped of Drosophila melanogaster. Experientia* **51**: A27/S05–S55

Yasuda, H., Matsumoto, Y., Mita, S., Marunouchi, T. and Yamada, M. (1981). A mouse temperature-sensitive mutant defective in H1 histone phosphorylation defective in deoxyribonucleic acid synthesis and chromosome condensation. *Biochemistry* **20**: 4414–4419

Subject Index

REVOLUTION AND CONTINUITY

STUDIES IN PHILOSOPHY
AND THE HISTORY OF PHILOSOPHY

General Editor: Jude P. Dougherty

Studies in Philosophy
and the History of Philosophy   Volume 24

Revolution
and Continuity
Essays in the History
and Philosophy of
Early Modern Science

edited by Peter Barker and Roger Ariew

THE CATHOLIC UNIVERSITY OF AMERICA PRESS
Washington, D.C.

LIBRARY OF CONGRESS CATALOGING-IN-PUBLICATION DATA

Revolution and continuity : essays in the history and philosophy of
early modern science /edited by Peter Barker and Roger Ariew.
    p.   cm.—(Studies in philosophy and the history of
philosophy : v. 24)
    Includes bibliographical references and index.
    ISBN 0-8132-0738-X (permanent paper)
    1. Science—History. 2. Science—History—17th century.
3. Science—Philosophy—History. 4. Continuity. I. Barker, Peter,
1949– . II. Ariew, Roger. III. Series.
B21.S78 vol. 24
[Q175.3]
100 s—dc20
[501]   90-19633

Contents

Introduction

PETER BARKER AND ROGER ARIEW

REVOLUTION

When Leopold Von Rancke began his research in diplomatic history in the early nineteenth century, the field was dominated by myths. Many myths failed to survive even a brief exposure to evidence available in the diplomatic archives he opened. But this evidence was not self-authenticating. Introducing archival sources required a carefully developed method for weighing and validating evidence. Rancke's great legacy was therefore twofold. He provided the first modern solutions to the twin problems of sources and of method. His achievement became the foundation for the modern discipline of history.

If the prevalence of myth is a measure of the maturity of a field, then the history and philosophy of science is scarcely beyond its infancy. This is especially true of the period sometimes called the scientific revolution, the early modern period that witnessed the demise of the relatively stable medieval pattern of learning and the founding of the disciplines that became modern science. Here also, many myths persist because of ignorance of primary sources. But these sources are neither widely available nor easily intelligible to the increasingly broad audience for historical, philosophical, and sociological studies of science. Hence the need for a volume of this type, presenting new material based on historical research and treating the methodological problems that naturally arise in this research. Our introduction will examine several myths surrounding the supposedly revolutionary nature of Copernicus's work, and then review the historiographical debate on the continuity of medieval and modern science. Historical and philosophical concerns run through both questions.

The Originality of Copernicus

One cycle of myth that should long ago have yielded to evidence from contemporary sources surrounds the role of Copernicus in the scientific revolution. This myth has many dimensions. Part of the myth is Copernicus's heroic stature, bestriding history like a colossus, rejecting ancient science and founding modern science. Connected myths distort the status of Copernican astronomy and misrepresent the grounds for adoption of Copernicanism by later scientists. An important supporting myth presents astronomy as the engine that drives the scientific revolution. The complexity of these historical fables probably deserves a volume in itself, but enough may be said in a short space to dispel the plausibility of some typical examples.

Let us first consider the myth of Copernicus's originality. Heliocentrism of course was not original. It had been proposed in antiquity by the Pythagoreans and Aristarchus, and was known throughout the Middle Ages in the Latin West. It might be thought, then, that Copernicus's originality lay in his embodiment of heliocentrism in a technically proficient mathematical astronomy. Certainly Wittenberg astronomers such as Erasmus Reinhold admired the technical devices used in *De revolutionibus*—for example, the compounding of two circular motions to produce harmonic motion along a straight line in Copernicus's model for the precession of the equinoxes. The Wittenberg astronomers also approved Copernicus's elimination of the equant. Reinhold is in fact an early exponent of the myth of Copernicus's originality:

God in his goodness kindled a great light in him [Copernicus] so that he discovered a great host of things which until our day, had not been known or had been veiled in darkness.[1]

If we are to take this as an estimation of the originality of Copernicus's work, and not rhetorical embellishment, we must conclude that Northern European astronomers like Reinhold never encountered the Arabic astronomy available in Southern Europe at the time of Copernicus's education. Neither Copernicus's objections to the equant nor the technical devices he employed to avoid it were original.

Objections to Ptolemaic astronomy based on the equant are at least as old as the eleventh century, when Arab critics of Ptolemy proposed a return to the strict homocentrism of Aristotle and, later, Averroes.[2] The new homocentric schemes had little success. But the same objec-

1. Erasmus Reinhold, *Logistice scrupulorum astronomicorum* (Wittenberg, 1551), p. 21. This work precedes the Prutenic Tables. The translation is that of Doland and Maschler from Pierre Duhem, *To Save the Phenomena* (Chicago: University of Chicago Press, 1969), p. 73.

2. A. I. Sabra, "An Eleventh Century Refutation of Ptolemy's Planetary Theory," *Studia Copernicana* 16 (1978): 117–31; "The Andalusian Revolt Against Ptolemaic Astronomy," in E. Mendelsohn (ed.), *Transformation and Tradition in the Sciences* (Cambridge: Cambridge University Press, 1984), pp. 133–53.

tions, based on the equant, motivated later Arab astronomers, who succeeded in reconciling essentially Ptolemaic planetary models with the Aristotelian restriction that heavenly motions must consist of circles traversed at constant speed. They were able to achieve this through the use of a new mathematical device, known today as a Tusi-couple. Both of these ideas—the criticisms of Ptolemy and the Tusi-couple—reached Europe in time to appear in new attempts to revive homocentric astronomy written in Padua during the Averroist resurgence at the beginning of the sixteenth century.[3] Significantly, this revival coincided with Copernicus's education in Padua.

The most important Arabic astronomers so far identified by historical research are Nasir al-Din al-Tusi (1201–74), who founded the observatory at Maragha in Persia; his student Qutab al-Din al-Shirazi (1236–1311); Mu'ayyad al-Din al-Urdi (d. 1226), who assisted in constructing the observatory's instruments; and the Damascus astronomer Ibn al-Shatir (c. 1304–c. 1375). The new mathematical device provided by al-Tusi produced harmonic motion along a straight line by compounding two circular motions. Physically, this construction could be produced in two ways. In one version a small sphere rolls (without slipping) inside a large sphere twice its diameter. A second version compounds small circumpolar circles on two homocentric spheres with inclined axes.[4] Al-Tusi and later members of the Maragha school applied the device to produce successively more sophisticated solutions to the problems of planetary motions in longitude and latitude, and the problems posed by precession of the equinoxes.[5]

3. N. Swerdlow, "Aristotelian Planetary Theory in the Renaissance: Giovanni Battista Amico's Homocentric Spheres," *Journal for the History of Astronomy* 3 (1972): 36–48. On Nifo, see Duhem, *To Save the Phenomena*, p. 48ff.

4. F. Jamil Ragep, "Two Versions of the Tusi Couple," in D. A. King and G. Saliba (eds.), *From the Deferent to the Equant* (New York: New York Academy of Sciences, 1987), pp. 329–56.

5. For a general view of the development of Arabic astronomy see D. Pingree, "Greek Influence On Early Islamic Mathematical Astronomy," *Journal of the American Oriental Society* 93 (1973): 32–43, and Bernard R. Goldstein, "The Making of Astronomy in Early Islam," *Nuncius: Annali di Storia della Scienza* 1 (1986): 79–92. Some main sources on the Maragha school are W. Hartner, "Nasir Al-Din Al-Tusi's lunar theory," *Physis* 11 (1969): 287–304; "Ptolemy, Azarquiel, Ibn Al-Shatir, and Copernicus on Mercury," *Archives internationales d'histoire des sciences* 24 (1974): 5–25, "Ptolemaische Astronomie im Islam und zur Zeit des Regiomontanus," in G. Hamann (ed.), *Regiomontanus Studien* (Vienna: Osterreicher Akademie der Wissenschaft, 1980), pp. 109–24; E. S. Kennedy and I. Ghanem (eds.), *The Life and Work of Ibn al-Shatir* (Aleppo: Institute for the History of Arabic Science, 1976); F. Jamil Ragep, "Cosmography in the 'Tadhkira' of Nasir al-Din al-Tusi," 2 vols. (Ann Arbor, MI: University Microfilms, 1982), Harvard Ph.D. thesis; G. Saliba, "The Original Source of Qutb Al-Din Al-Shirazi's Planetary Model," *Journal for the History of Astronomy* 3 (1979): 4–18; "The First Non-Ptolemaic Astronomy" at the Maragha School," *Isis* 38 (1979): 571–76; "Arabic Astronomy and Copernicus," *Zeitschrift*

The mathematical device admired by the Wittenberg astronomers when Copernicus used it to explain precession of the equinoxes is, of course, a Tusi-couple. Copernicus used the same device to account for the variation of the obliquity of the ecliptic and also to generate motions in latitude by rocking the planes of planetary orbits. The details of his models for the main planetary motions—the motion in longitude represented by Ptolemy using the famous deferent, epicycle, eccentric, and equant construction—also follow lines established by Maragha astronomers. Copernicus's models for the superior planets correspond to the models developed by al-Urdi and al-Shirazi, while the model for the inferior planets corresponds to al-Shatir's model. Not only is Copernicus's double epicycle model for the moon the same as that proposed by al-Shatir; the values of the three key parameters—the radii of the deferent and epicyclic circles—are identical.[6]

The correspondences between Copernicus's mathematical astronomy and the Maragha models are so overwhelming that as early as 1968 Otto Neugebauer wrote: "The mathematical logic of these methods is such that the purely historical problem of contact or transmission, as opposed to independent discovery, becomes a rather minor one."[7] In his 1973 analysis of the *Commentariolus,* Noel Swerdlow argued for the existence of an intermediary, perhaps a Byzantine manuscript, that supplied Copernicus and his Paduan contemporaries with versions of the Maragha models.[8] More recently these authors have said: "The question . . . is not whether, but when, where and in what form [Copernicus] learned of Maragha theory."[9] What is left of Copernicus's originality? Well, he did revive the "Pythagorean hypothesis" that the sun, not the earth, was the body closest to the common center of planetary motions, but in all other respects he must be regarded as a European

für Geschichte der Arabisch-Islamischen Wissenschaften 1 (1984): 73–87; "The Role of Maragha in the Development of Islamic Astronomy: a Scientific Revolution before the Renaissance," *Revue de Synthèse* 108 (1987): 361–73.

6. For summaries of these correspondences see O. Pedersen and M. Pihl, *Early Physics and Astronomy* (New York: Elzevier, 1974), and the general introduction to N. Swerdlow and O. Neugebauer, *Mathematical Astronomy in Copernicus's De Revolutionibus* (New York: Springer-Verlag, 1984), which may also be recommended as a general introduction to Copernicus scholarship. For more detailed information, see the sources listed in the previous note.

7. O. Neugebauer, "On the Planetary Theory of Copernicus," *Vistas in Astronomy* 10 (1968): 90.

8. N. Swerdlow, "The Derivation and First Draft of Copernicus Planetary Theory: A Translation of the Commentariolus with Commentary," *Proceedings of the American Philosophical Society* 117 (1973): 423–512.

9. N. Swerdlow and O. Neugebauer, *Mathematical Astronomy in . . . De Revolutionibus,* p. 47.

satellite of the Maragha astronomers, whose work he either copied or repeated.

We may now dispense quickly with the connected myths that Copernicus's originality lay in his realism, and that the Ptolemaic astronomical tradition was instrumentalist. Inescapable evidence that Ptolemy himself was a realist (and proposed elaborate solid models of the heavens to produce motions corresponding to the mathematical models of the *Almagest*) perhaps came to the attention of historians and philosophers of science after the recovery of Book I, Part II of the *Planetary Hypotheses*.[10] But the content of this part of Ptolemy's work had been preserved and transmitted through the Ptolemaic tradition in Islam, which continued to employ both mathematical and solid models.[11] A decisive argument for treating the whole tradition as realist is the motivation for Arabic improvements on the original Ptolemaic forms—the nature of the motion required by the equant. The objection to the equant has no force unless one is describing real spheres. Copernicus, of course, also treated the equant problem as a motive for reform. It must be conceded that the reality of eccentrics and epicycles was disputed throughout the Middle Ages, in both Islam and the Latin West. Sixteenth-century astronomers also took a variety of positions on the reality of heavenly motions. But a primary goal of the Maragha research program was to supply physically admissible models, and the technical tradition in Western astronomy before Copernicus, the *Theorica* tradition, also treated celestial spheres as real objects.[12] There can be little doubt that the traditions that supplied Copernicus with his mathematical methods regarded heavenly motions and the spheres that produced them as physically real.

A number of other myths surround the evidence for Copernicanism. Beginning with Copernicus's own work, considered strictly, the claim is still made that Copernicus for the first time provided a means to calculate planetary distances. Now the *Planetary Hypotheses* provides a scheme for calculating planetary distances; furthermore, the calculated values agree with the independently derived solar distance used by Ptolemy. Similar schemes were developed by the Ptolemaic tradition in Islam, and were so well known in the Latin West that a recent historical account suggests that figures for planetary distances and sizes

10. Bernard R. Goldstein, "The Arabic Version of Ptolemy's Planetary Hypotheses," *Transactions of the American Philosophical Society* 57, part 4, (1967): 3–55.

11. Ragep, *Cosmography in the "Tadhkira" al-Tusi.*

12. E. J. Aiton, "Peurbach's *Theorica novae planetarum*," *Osiris* 3, (1987): 5–44. O. Pedersen "Decline and Fall of the Theorica Planetarum," in *Science and History: Studia Copernicana XVI* (Warsaw: Polish Academy of Sciences, 1978), pp. 157–85.

formed part of the intellectual background of any educated European during the later Middle Ages.[13] The Copernican values for planetary distances are an alternative to an established set of values. The issue between Copernicanism and geocentrism is emphatically not one of resources provided by heliocentrism and unavailable in the rival tradition.

Galileo and the Phases of Venus

A similar myth surrounds another alleged piece of evidence for the Copernican system—Galileo's telescopic observations of the phases of Venus. Such historically oriented philosophers of science as Imre Lakatos and Thomas Kuhn both treat the phases of Venus as a uniquely strong argument for the Copernican position, specifically a phenomenon correctly predicted by Copernicus about which his predecessors are silent.[14] More recently this myth has resurfaced in accusations that Galileo stole the idea of phases and their deployment in support of Copernicus from Benedetto Castelli.[15] These opinions are based on neglect of readily available historical sources—or perhaps on lack of attention to philosophical distinctions.

First, it is important to distinguish between Aristotelian cosmology (as found, for example, in the tradition of commentators on *De caelo*) and Ptolemaic astronomy (as found, for example, in the *Theorica* tradition). Galileo's observation of a full range of phases for Venus may be an objection to some versions of Ptolemaic astronomy, but their discovery is, if anything, congenial to Aristotelian cosmology. From the time Avicenna became readily available in the West, commentators such as Albert of Saxony were in agreement that Aristotelian cosmology required the phases of Venus. There was general agreement that planets were not self-luminous. As extended bodies between the earth and the sun, both Venus and Mercury should therefore show phases. But no such thing is evident to the unaided eye. Albert avoided this embarrassment to Aristotle by postulating that the inferior planets absorbed and retransmitted the light of the sun, and that Venus itself was too thin to retransmit it properly.[16]

13. A. Van Helden, *Measuring the Universe* (Chicago: University of Chicago Press, 1985).

14. I. Lakatos and E. Zahar, "Why Did Copernicus' Research Programme Supercede Ptolemy's?" in R. Westman (ed.), *The Copernican Achievement* (Berkeley: University of California Press, 1975), pp. 374–75. T. S. Kuhn, *The Structure of Scientific Revolutions*, 2d ed. (Chicago: University of Chicago Press, 1970), p. 154.

15. R. Westfall, "Science and Patronage: Galileo and the Telescope," *Isis* 76 (1985): 11–30.

16. R. Ariew, "The Phases of Venus before 1610," *Studies in History and Philosophy of Science* 18 (1987): 81–92.

The *Theorica* tradition based Ptolemaic astronomy upon Aristotelian cosmology. Any astronomer who had studied the commentators on Aristotle would therefore have expected Venus to show phases. Convention placed the whole epicycle of Venus between the earth and the moving sun. This would lead to a sequence of phases different from those Galileo saw. But it might still be possible to defend Ptolemaic astronomy by the simple expedient of shifting the center of the epicycle to the mean sun, and a precedent for this configuration of the planets had existed from antiquity in the work of Martianus Capella.[17]

So the phases of Venus are not a unique prediction of Copernican astronomy; they are required also by Aristotelian cosmology, and may perhaps be reconciled with Ptolemaic astronomy. Working in a milieu that understood these points, it is hardly surprising that Galileo's contemporaries failed to attribute the significance to the phases of Venus imagined by modern commentators.[18] Moreover, Galileo can hardly be guilty of plagiarizing an idea that had been in circulation for two centuries; it is very probable that Galileo had read Albert of Saxony (perhaps in George Lokert's compendium edition of Albert, Themon Judaeus, and Buridan, considered in the next section of this essay), and it is quite certain that Galileo had access to the work of Collegio Romano professors who used Albert's work.[19]

The case of the phases of Venus shows the importance of appraising claims about the content of science through a knowledge of its context. Here the immediate context includes the Collegio Romano, and the wider context includes at least the cosmological commentators and the *Theorica* tradition. A final example suggests the need for even wider consideration of the milieu in which early modern scientists worked.[20]

17. Ibid.; also, Bruce S. Eastwood, " 'The Chaster Path of Venus' in the Astronomy of Martianus Capella," *Archive Internationale d'Histoire des Sciences* 32 (1982): 145–58.

18. See for example the quotation from Kepler's 1625 *Appendix to the Hyperaspistes*, reproduced in Ariew, "Phases of Venus before 1610," pp. 87–88.

19. Ariew, "Phases of Venus before 1610," pp. 87–88; W. Wallace, *Galileo and His Sources* (Princeton: Princeton University Press, 1984).

20. The most important attempts to provide a context for Copernicus appear in the groundbreaking work of Robert Westman. The interested reader is particularly directed to "The Astronomer's Role in the Sixteenth Century: A Preliminary Survey," *History of Science* 18 (1980): 105–47; and "Proof, Poetics and Patronage: Copernicus's Preface to *De Revolutionibus*," in D. C. Lindberg and R. S. Westman (eds.), *Reappraisals of the Scientific Revolution* (Cambridge: Cambridge University Press, 1990), the latter bringing to prominence Copernicus's use of Horatian aesthetics and his place within the humanist clerical bureaucracy.

For important new insights on the reception of Copernicanism, see O. Gingerich and R. S. Westman, *The Wittich Connection: Conflict and Priority in Late Sixteenth Century Cosmology*, Transactions of the American Philosophical Society, Vol. 78, part 7, 1988. See also the papers on the Wittenberg interpretation of Copernicus, including that referred to in note 26 below.

Tycho, Observational Astronomy, and Comets

Accounts of the period from Copernicus to Newton often assign a key role to Tycho Brahe's evidence that comets were celestial, contrary to the Aristotelian doctrine that they were meteorological phenomena in the upper regions of the terrestrial realm. Tycho brought technical tools to bear on the problem of comets: a mathematical method for determining parallax, supported by unusually large and accurate instruments. But a study of the development of these tools shows that they are part of continuous traditions stretching from the middle ages through the time of Newton, rather than being new and unique resources of an emerging "modern" science. Most surprisingly for the orthodox account, Tycho's work produced no decisive change in theories of the nature of comets.

Astrological interest in comets was the motivation for increasingly accurate observations from the thirteenth century onward. Practical astrology was primarily a concern of the medical profession—indeed in sixteenth-century France the medical profession enjoyed a monopoly on the writing of books of prognostications. The persuit of increasingly accurate positional observation required innovation in scientific instruments. Here the medical profession again proved singularly well equipped to pursue the study of comets. A pioneer in the history of technology—Lynn White—went so far as to conjecture that every early modern scientist who actually built a new instrument had medical training.[21] Although there are prominent exceptions to White's generalization, the role of medical astrologers in the early modern period is conspicuous, and underappraised by historians.

One exception to White's thesis is the Provençal natural philosopher Levi ben Gerson (1288–1344). Levi was responsible for the invention of the Jacob's staff or *radius astronomicus*. He also proposed an adaptation of Ptolemy's method for determining lunar parallax that would yield distances for comets, and his conclusions suggest that he actually made such observations with the aid of a Jacob's staff.[22]

By the end of the fifteenth century both the technique of parallax determination for comets and the Jacob's staff were in general circulation among Puerbach and his students. The instrument is described in a work by Regiomontanus devoted to this particular application. But

21. Lynn White, "Medical Astrologers and Medieval Technology," in his *Medieval Religion and Technology* (Berkeley, University of California Press, 1978), pp. 297–315.
22. Bernard R. Goldstein, *The Astronomy of Levi Ben Gerson (1288–1344)* (New York: Springer-Verlag, 1985). Peter Barker and Bernard R. Goldstein, "The Role of Comets in the Copernican Revolution," *Studies in History and Philsophy of Science* 19 (1988): 299–319.

the estimates of cometary distances produced by all these workers seem to have been consistent with the Aristotelian theory that comets were below the moon.[23] The first serious incursion on the Aristotelian theory took the form of an attack on the constitution of comets rather than on their location.

Early sixteenth-century makers of astronomical instruments were also interested in comets. No later than 1532 Peter Apian recognized that the direction of a comet's tail always pointed radially away from the sun. He concluded that the tail was an effect of the sun's rays shining through the head of the comet. Both Apian's observation and his explanation were repeated by Gemma Frisius in his book on the *radius astronomicus*. In Southern Europe, Fracastoro and Cardano arrived at similar ideas at about the same time. With the appearance of Girolamo Cardano's *De Subtilitate* in 1550, and Jean Pena's *De Usu Optices* in 1557, the implications of these new developments for Aristotle's theory of comets were explicitly drawn.[24]

Aristotle had described comets as burning gases, variously positioned in the upper atmosphere. But the anti-solar direction of their tails suggested that comets were lenses focusing the rays of the sun. Burning vapors simply would not produce this effect, as Pena pointed out. Both Pena and Cardano rejected Aristotle's account of the constitution of comets, and substituted the view that the body of a comet was a spherical lens. Contemporaries of Tycho (for example, Christopher Rothmann, and the young Johann Kepler) accepted the new theory. Tycho himself also adopted this ready-made theory, which was not significantly criticized until the later career of Kepler. Thus, in an important sense, Aristotle's theory of comets had already been abandoned by the time Tycho located them in the heavens. It was not his observations that proved that comets could not have the constitution required by Aristotle.[25]

Understanding the role comets played in the Copernican revolution and the nature of the evidence they provided for the various world views in contention at the turn of the seventeenth century requires that we examine at least the traditions of cosmological commentators and technical astronomers already identified, and in addition the mathe-

---

23. Jane Jervis, *Cometary Theory in Fifteenth-Century Europe:* Studia Copernicana XXVI (Warsaw: Polish Academy of Sciences, 1985).

24. Barker and Goldstein, "Role of Comets in the Copernican Revolution," pp. 315–17.

25. The optical theory was not, of course, universally accepted, nor were Tycho's observations immediately taken as conclusive evidence for the position of comets. Galileo, for example, resisted the relocation of comets. Barker and Goldstein, "Role of Comets in the Copernican Revolution."

matical practitioners and instrument makers such as Regiomontanus, Apian, Pena, Tycho, and Kepler. Overlapping all these groups are the medical astrologers, including Toscanelli, Gemma Frisius, Fracastoro, and Cardano, who may represent the largest single group of proto-scientists throughout the early modern period. It is worth recalling that Copernicus practiced medicine for most of his life; even Kepler was advised to go away and qualify as a doctor at a low point in his career.

CONTINUITY

So far we have examined a variety of myths, still prevalent in the historical and philosophical literature, that have been used to present modern science as discontinuous from earlier science. Several of these myths (for example, the myth about Tycho refuting Aristotle's theory of comets) may be partly excused due to the inaccessibility of historical sources. But in other cases (for example, the uniqueness of Copernicus's mathematical astronomy or the phases of Venus) the historical evidence undermining the myths has been available for a considerable time. In these cases we must look beyond ignorance of the sources to reasons for ignoring the sources. Here we enter the realm of historiography. A common feature of these myths is the application to a discipline's formative period of standards and arguments that became available only after the consolidation of the discipline under study. This problem of anachronism has been noted by many authors, but its current importance in the history of science is underlined by the attention drawn to it in recent discussions—those of Robert Westman, Noel Swerdlow, Bruno Latour, and Peter Dear, to name a few.[26]

The motives for anachronism may be various. One is the desire to present a founding figure as possessing truths that were revealed by others elaborating upon the founding figure's work.[27] Anachronism is also used to sharpen the contrast between a figure, whose work is still admired, and his contemporaries, who conspicuously lacked some modern virtue. A related motive may be the philosophical commitment to the idea that there is a single, specifiable scientific method and that anything of value anywhere in science's history is valuable because it

26. R. S. Westman, "The Melanchton Circle, Rheticus and the Wittenberg Interpretation of the Copernican Theory," *Isis* 66 (1975): 165–93, esp. pp. 165–68. N. Swerdlow, "Pseudodoxia Copernicana," *Archives Internationales d'Histoire des Sciences* 26, (1976): 108–58, esp. pp. 111–14. Bruno Latour, *Science in Action* (Cambridge: Harvard University Press), Rule 3 p. 258. Peter Dear, *Mersenne and the Learning of the Schools* (Ithaca: Cornell University Press, 1988).

27. N. Swerdlow, "Pseudodoxia Copernicana," *Archives Internationales d'Histoire des Sciences* 26, (1976): 108–58.

conforms to that standard. Another philosophical commitment takes the view that the history of science may be divided into periods marked by distinct methods. Both of these views sharply distinguish the method supposedly exemplified by modern science from the science of a previous period. The former view treats the origin of modern science as the initial point for anything worthy of the name "scientific method." The latter—popular among historically oriented philosophers of science—allows the existence of something worthy of the name "science" before the origin of modern science, but proposes a radical discontinuity between modern science and its antecedents. These themes appear, at the very beginning of modern history of science, in the works of Duhem; they reappear in the most popular philosophical accounts of science in the second half of the twentieth century, as the debate between revolution and continuity. Historical and philosophical themes are inextricably linked in this debate.

Duhem on Continuity

At the turn of the twentieth century there was a consensus that nothing from the middle ages was worthy of the name "science"; no history of medieval science could even be written: "The two great periods of School Philosophy . . . were that of the Greeks and that of the Middle Ages;—the period of the first waking of science, and that of its mid-day slumber."[28] Modern science was thought to begin with the seventeenth-century rejection of the first waking, that is, of Greek science, and its mid-day slumber, called medieval mysticism. The following passage of Whewell's *History of the Inductive Sciences,* from the chapter entitled "Of the Mysticism of the Middle Ages," was a typical treatment:

A new and peculiar element was introduced into the Greek philosophy which . . . tinged a large portion of the speculations of succeeding ages. We may speak of this peculiar element as *Mysticism.* . . . Thus, instead of referring the events of the external world to space and time, to sensible connexion and causation, men attempted to reduce such occurrences under spiritual and supersensual relations and dependencies; they referred them to superior intelligences, to theological conditions, to past and future events in the moral world, to states of mind and feelings, to the creatures of an imaginary mythology or demonology. And thus their physical Science became Magic, their Astronomy became Astrology, the study of the Composition of bodies became Alchemy, Mathematics became the contemplation of the Spiritual Relations of number and figure, and Philosophy became Theosophy. . . . This tendency materially affected both men's speculations and their labours in the pursuit of knowledge. . . . And by calling into a prominent place astrology, alchemy, and magic, it long occupied

28. William Whewell, *History of the Inductive Sciences,* 3d ed. (London, 1857), Book I, Introduction.

most of the real observers of the material world. In this manner it delayed and impeded the progress of true science; for we shall see reason to believe that human knowledge lost more by the perversion of men's minds and the misdirection of their efforts, than it gained by any increase of zeal arising from the peculiar hopes and objects of the mystics.[29]

Given this intellectual context, it is understandable that Pierre Duhem's *L'évolution de la mécanique*, published in 1903, dismissed the middle ages as a scientifically sterile period. Similarly, in 1900, when Duhem completed a history of chemical combination, *Le mixte et la combinaison chimique*, his discussion jumped from Aristotle's concept of *mixtio* to the modern concepts. It was only in 1904, while writing a history of statics, *Les origines de la statique*, that he came across an unusual reference in a sixteenth-century book to an unknown medieval thinker, Jordanus de Nemore, and decided to follow it up. From this the history of medieval science was born. Unlike Duhem's previous attempts, *Les origines de la statique* contained a number of chapters on medieval thought: one on Jordanus de Nemore; another on the Jordanus' followers; one on their influence on Leonardo da Vinci. As Duhem later told us, what he discovered was:

The Christian Middle Ages had known the Greek writings on statics; some of these writings had come directly and others through Arab commentaries. But the Latins who read those works were not at all the uninventive and servile commentators that people depicted. The remnants of Greek thought that they received from Byzantium or from Islamic science did not stagnate in their minds; these relics were sufficient to arouse their attention, to fertilize their intellect. And, from the thirteenth century on, perhaps even before then, the school of Jordanus opened to students of mechanics some paths unknown to the ancients.[30]

In 1905–6, when writing volume two of *Les origines de la statique*, Duhem went into full swing: he wrote chapters on seventeenth-century statics—on Galileo, Stevin, and Roberval—as expected; but he also returned to medieval statics, with four chapters on geostatics. In these, Duhem continued the story of the development of statics from the Greeks to the moderns, adding another strand depicting the development of geostatics from Albert of Saxony in the fourteenth century to Torricelli in the seventeenth.[31]

29. Ibid., Book IV, chap. 3.

30. Pierre Duhem, "Notice sur les titres et travaux scientifiques de Pierre Duhem, rédigée par lui-même lors de sa candidature à l'académie des sciences (mai 1913)," in *Mémoires de la société des sciences physiques et naturelles de Bordeaux*, series 7, vol. 1 (1917), p. 160.

31. "Leonardo da Vinci, that indefatigable reader, perused and pondered endlessly the writings of the school of Jordanus, on the one hand, and the scholastic questions of

Duhem did not rest content with his demonstration that the history of modern statics was continuous with medieval statics. From 1906 to 1913, he delved deeply into his primary source for the recovery of the past, the scientific notebooks of Leonardo da Vinci. He wrote a series of essays published as the three-volume *Etudes sur Léonard de Vinci: ceux qu'il a lus et ceux qui l'ont lu,* which, as its subtitle indicates, was devoted to uncovering da Vinci's medieval sources and their influence on the moderns. The third volume gained a new subtitle, *Les précurseurs parisiens de Galilée,* announcing Duhem's bold new thesis that even the works of Galileo had a medieval heritage:

When we see the science of Galileo triumph over the stubborn Peripatetism of Cremonini, we believe, since we are ill-informed about the history of human thought, that we are witness to the victory of modern, young science, over medieval philosophy, so obstinate in its mechanical repetition. In truth we are contemplating the well-prepared triumph of the science born at Paris during the fourteenth century over the doctrines of Aristotle and Averroes brought back into fashion by the Italian Renaissance.[32]

Duhem proceeded to detail Galilean dynamics as continuous with medieval dynamics. He discovered medieval impetus theory and traced its development from an early criticism of Aristotle by John Philoponus to its mature statements in the fourteenth-century works of John Buridan and Nicole Oresme:

The role played by *impetus* in Buridan's dynamics is exactly the one that Galileo attributed to *impeto* or *momento,* Descartes to the *quantity of motion,* and finally Leibniz to the *vis viva.* So exact is this correspondence that, in order to exhibit Galileo's dynamics, Torricelli, in his *Lezioni accademiche,* often took up Buridan's reasons and almost his exact words.[33]

Duhem then sketched the extension of impetus theory from earthly dynamics to a theory about the motion of the heavens and earth:

Nicole Oresme attributed to the earth a natural *impetus* similar to the one Buridan attributed to the celestial orbs. In order to account for the vertical fall of weights, he allowed that one must compose this *impetus* by which the mobile rotates around the earth with the *impetus* engendered by weight. The principle he distinctly formulated was only obscurely indicated by Copernicus and

---

Albert of Saxony, on the other. The former, by acquainting him with the law of the equilibrium of the bent lever, led him to the following memorable law, which governs the composition of concurrent forces: with respect to a point taken on one of the components or on the resultant, the two other forces have equal moments. Moreover, Albert of Saxony's ideas on the role of the center of gravity allowed da Vinci to discover the rule of the subtended polygon, which Villalpand plagiarized. Thus, we find the origins of several basic principles of statics in the writings composed during the 13th and 14th centuries." Duhem, "Notice," pp. 160–61.

32. Duhem, "Notice," p. 162.   33. Duhem, "Notice," p. 164.

merely repeated by Giordano Bruno. Galileo used geometry to derive the consequences of that principle, but without correcting the incorrect form of the law of inertia implied in it.[34]

Duhem continued his work by describing further developments in the medieval science of weights, the law of free fall, analytic geometry, hydrostatics, and geology. He ended the essays on Leonardo da Vinci by speculating about the actual mechanism of transmission for medieval science. Since most of the works of Buridan and Oresme had remained in manuscript form, Duhem assumed that Albert of Saxony, whose works were printed and reprinted during the sixteenth century, was the likely link to Galileo. Duhem thought he had found the key to understanding the transmission of medieval science in some of Galileo's manuscripts that made references to the *Doctores Parisienses,* that is, the Parisian masters, or Buridan and Oresme; he even thought he could tell, because of the unusual doctrines and the particular order of the questions referred to by Galileo, which of the sixteenth-century editions of Albert of Saxony was read by Galileo (namely, the compilation by George Lokert of the works of Albert of Saxony, Themo Judaeus, and others).[35]

In the three years from 1913 to his death in 1916, Duhem wrote the ten-volume *Le système du monde,* published in fits and starts over the next four decades, from 1913 to 1959. He died before finishing the project, so we do not have his final thoughts on the continuity thesis; on the other hand, we have an overwhelming amount of data about medieval astronomy, astrology, tidal theory, geostatics, and the developments in doctrines associated with such concepts as infinity, place, time, void, and the unity and singularity of the world.

Continuity's Critics: Favaro and Dear

From the time it was first announced until the present, Duhem's continuity thesis has been the topic of numerous discussions, both as a historical thesis and a historiographical thesis.[36] The thesis is perhaps most compelling as a contingent claim about the history of science: continuity just happens to be the case; it could have been otherwise. As such, it was rejected by most of the leading scholars in the history and philosophy of science, from the 1920s to the present. Almost immediately, it was rejected by Antonio Favaro, the editor of Galileo's *Opera Omnia.*[37]

---

34. Duhem, "Notice," p. 166.
35. Pierre Duhem, *Etudes sur Léonard de Vinci* 3 (Paris, 1913), pp. 582–83.
36. See, for example, Joseph Agassi, *Towards an Historiography of Science, History and Theory,* Beiheft 2 (1963), pp. 1–117.
37. Antonio Favaro, "Galileo Galilei e i Doctores Parisienses," *Rencondita della Accade-*

It was then rejected in the 1940s and 1950s by such historians of science as Alexandre Koyré and Annaliese Maier.[38] It was also rejected by Imre Lakatos and Thomas Kuhn, the leading historians and philosophers of science in the 1960s and 1970s.[39] And its rejection continues to the present, as can be seen in the final chapter of Peter Dear's *Mersenne and the Learning of the Schools,* published in 1988.[40] In appraising the present state of the debate, it will be instructive to consider the first and the last of Duhem's detractors, in the hope that the process will shed some light on the intermediaries.[41]

Favaro objected that the Galilean manuscripts upon which Duhem based his continuity thesis were mere juvenilia, Galileo's lecture notes from his student days, which did not represent Galileo's own thoughts. Favaro even refused to include one of the three manuscripts in his edition of the works of Galileo. He did publish the other two, but gave one of them the title *Juvenilia.*

However, as the recent work of A. C. Crombie, Adriano Carugo, and William Wallace has shown, these allegedly youthful Galileo manuscripts date from a later period.[42] Wallace has argued cogently that they

---

mia dei Lincei 27 (1918): 3–14 and "Léonard de Vinci a-t-il exercé une influence sur Galilée et son école?" *Scientia* 20 (1916): 257–65.

38. For example, Alexandre Koyré, *Etudes Galiléennes* (Paris, 1939) and the collection of Annaliese Maier's writings translated into English as *On the Threshold of Exact Science,* S. D. Sargent (ed. and tr.) (Philadelphia: University of Pennsylvania Press, 1982).

39. T. Kuhn, *Structure of Scientific Revolutions;* I. Lakatos, *Philosophical Papers, Volume 1* (Cambridge: Cambridge University Press, 1978).

40. P. Dear, *Mersenne and the Learning of the Schools* (Ithaca: Cornell University Press, 1988).

41. Here we pass over—as well known—the period in the debate corresponding to the works of Maier and Koyré. Noteworthy recent work includes Stephen Sargent's edition and translation of a number of Maier's most important essays, published as A. Maier *On the Threshold of Exact Science* (Philadelphia: University of Pennsylvania Press, 1982), and the September 1987 issue of *Science in Context* (vol. 1, no. 2) devoted to Duhem's continuity thesis in historical context. In a forthcoming article, Stephen Menn makes a case that Maier's argument against Duhem's continuity thesis is cogent only in the restricted context of the continuity between medieval terrestrial physics and modern physics, but not in the context of the continuity between medieval celestial mechanics and modern physics. In the same collection, William Wallace details the fortunes of Koyré's attempt to argue for discontinuity between medieval and early modern mechanics. See, respectively, S. Menn, "Descartes and some predecessors on the divine conservation of motion," and W. Wallace, "Duhem and Koyré on Domingo de Soto," pp. 215–38 and 239–60 in *Pierre Duhem: Historian and Philosopher of Science,* R. Ariew and P. Barker (eds.), *Synthese* 83 (1990).

42. W. A. Wallace, *Galileo's Early Notebooks: The Physical Questions. A Translation from the Latin, with Historical and Paleographical Commentary* (Notre Dame: Notra Dame University Press, 1977); *Prelude to Galileo. Essays on Medieval and Sixteenth-Century Sources of Galileo's Thought* (Dordrecht: Reidel, 1981); A. C Crombie and A. Carugo, "The Jesuits and Galileo's Ideas of Science and of Nature," *Annali dell'Istituto e Museo di Storia della Scienza di Firenze* 8, fasc. 2 (1983): S. Drake "Galileo's Pre-Paduan Writings: Years, Sources, Motivations," *Studies in History and Philosophy of Science* 17 (1986): 429–88.

represent Galileo's notes for his own lectures at the University of Pisa; Crombie and Carugo give them an even later date. Since Galileo's manuscripts do contain many fourteenth-century doctrines, Duhem's continuity thesis gains credibility.[43] Interestingly, one of the manuscripts is a commentary on late medieval interpretations of Aristotle's *Posterior Analytics*, so that one can add a thesis of continuity of scientific methodology to the continuity thesis. Moreover, the actual mechanism of transmission is now better understood, given the works of Crombie, Carugo, and Wallace. Duhem had speculated that Buridan's and Oresme's doctrines arrived in the sixteenth century through Albert of Saxony's writings made popular by such figures as George Lokert and the Dominican Domingo de Soto. That speculation is substantially correct, but one has to add that between the early sixteenth-century scholastics and Galileo lie the Jesuits, and, in particular, de Soto's brilliant student, Franciscus Toletus, who became a Jesuit and taught natural philosophy in that order's Collegio Romano. As Wallace has shown, Galileo's manuscripts were copied from the lecture notes and published books of Collegio Romano professors.[44] On the other hand, some of Duhem's claims for Galileo's knowledge of medieval doctrines must be weakened somewhat. As Wallace has also shown, the peculiarities of reference that Duhem was relying upon for his linkages can be found in the Jesuit notes and texts Galileo was copying. Therefore, one cannot conclude that Galileo read Albert of Saxony directly, but only that he was aware of fourteenth-century doctrines through the agency of the Jesuits.[45]

Peter Dear's implicit rejection of the continuity thesis is more subtle and seems to be the articulation of a widely shared view. First, Dear distinguishes between two continuity theses, continuity of theory and continuity of method. He dismisses continuity of theory, claiming that although elements of medieval science did find "their way to Galileo, his enterprise differed sufficiently to render them of only secondary value in understanding the development of his mechanics."[46] Thus, lack of continuity is due to changes in the context of science, its goals, and perhaps even its methods, although the theories at stake might remain outwardly unchanged. Dear accepts continuity of method, but dismisses method itself as insignificant to science:

The strongest [of the objections to the methodological varieties of the continuity thesis] is that "science" differs radically from "applied scientific method." As Paolo Rossi has written, one cannot explain the Scientific Revolution in terms

43. Cf. Wallace, *Galileo and His Sources*.   44. Ibid.
45. W. A. Wallace, "Galileo Galilei and the *Doctores Parisienses*, in R. E. Butts and J. C. Pitt (eds.), *New Perspectives on Galileo* (Dordrecht: Reidel, 1978), pp. 87–138.
46. Dear, *Mersenne*, p. 232.

of the development or application of correct scientific method, because such a method—taken as a regulative, determinate procedure generating objective knowledge of nature—does not exist; science is not that simple.[47]

Even if the importance or unimportance of method to science could be definitively settled, its significance for the historian of science would still need clarification. Dear speculates that Wallace and other proponents of methodological continuity are motivated by an image of science producing true results, given a correct method. That may or may not be true; whatever their motivations, their work stands or falls independently. On the other hand, Dear correctly argues, in his attack on anachronism, that the truth or falsity of a theory and correctness or incorrectness of a method cannot legitimately be accepted by historians as a basis for their explanations:

> People do not believe things *because* they are true. Consequently, quite apart from Rossi's surely correct observation that there is no "scientific method" or set of rules for producing true knowledge, we may say that even if there were, it could never, even in principle, assist in producing accounts or explanations of Galileo's work in particular or the Scientific Revolution in general.[48]

Ultimately, Dear asserts that "true continuity would reside in a continuity of natural philosophical agendas, presupposing a continuity not only of specific ideas but also of goals."[49]

Dear's implicit rejection of the continuity thesis is premature. Discontinuities of context, method, or goals do not entail that the continuity thesis is false. Continuity does not require identical contexts, methods, or goals.[50] This point may be supported by considering the case of Duhem, who cannot be accused of thinking that science produces true results. Duhem was well aware that scientific methods and goals were often discontinuous; he understood that peripatetics, atomists, Cartesians, Newtonians, and so on, disagreed about both methods and goals. In the *Aim and Structure of Physical Theory,* Duhem argued for instrumentalism, or the doctrine that physical theory is classificatory as opposed to explanatory, in an attempt to safeguard physical theory against radical changes of methods and goals;[51] physical theory as clas-

---

47. Ibid., pp. 234–35.   48. Ibid., p. 235.
49. Ibid., p. 238.
50. The demand for continuity of large-scale philosophical agendas as "true continuity" is too stringent a demand; given that criterion we would have to conclude that there is no continuity between Galileo and Descartes, or between Mersenne and Descartes, or between early and late Galileo, for example.
51. P. Duhem, *Aim and Structure of Physical Theory,* chaps. 1–2. Duhem distinguishes sharply between classification, which occurs in science, and explanation, which he confines to metaphysics. Explanation can proceed only from the one correct ontology. Science cannot make ontological claims, although the categories introduced by science

sificatory need not be wedded to any particular metaphysics or philosophical agenda.

The only continuity Duhem ever claimed was continuity of phenomenal laws, not that of large-scale theories, methods, ontology, or goals and agendas. Perhaps he was aware of the possibility of other kinds of continuities, but their discovery would have been a surprise to him.

In chapter 3 of the *Aim and Structure of Physical Theory*, Duhem announced an interesting thesis, that "no metaphysical system suffices in constructing a physical theory." Duhem's thesis is the precursor to Dear's point of view, that no scientific method suffices in constructing a physical theory; both are extensions of a thesis often repeated by Duhem, that observations do not suffice in determining particular laws— or that hypotheses, laws, theories, and so forth are underdetermined by observation. Duhem's favorite example was the observational equivalency of epicycles and eccentrics.[52]

Underdetermination can be extended to include underdetermination at all levels of science—that is, between physical and metaphysical theories, between methods and theories, between theories and goals. Just as one set of observations can be compatible with two different hypotheses, one method can be compatible with two different theories, and one theory can be compatible with two different goals. Moreover, it is also clear that the reverse holds: two theories can be compatible with a single method or a single goal, and so on. Thus, Dear's own point of view should allow him to conclude that the same theory might be found in different contexts or in two different philosophical agendas. We may conclude that neither continuity nor discontinuity should be legislated a priori. In some historical cases, hypotheses, laws, or theories may be continuous despite changes in goals. The issue of continuity or discontinuity can be resolved only by historical investigation.

DESCRIPTION OF PAPERS

The papers in the present volume form four groups. David Lux and Mordechai Feingold address historiographical questions concerning the role of institutions such as universities and scientific academies in the founding of modern science. Lux examines the pervasive influence of Martha Ornstein's work concerning seventeenth-century scientific societies. He concludes that despite the absence of a single work of equivalent influence rejecting Ornstein's position, the historiographical tide has turned decisively against her work in the past decade. Fein-

---

constitute a classification of nature that approximates more and more closely to the ultimate metaphysical categories as science progresses.

52. See, for example, Duhem's *To Save the Phenomena*.

gold substantially qualifies the thesis that universities were centers of resistance to the new science. Both authors move toward accounts that mute the sharp intellectual and institutional discontinuities of earlier histories of science.

Harold Cook and Roger Ariew extend the range of historical studies into generally neglected areas of science. The overemphasis on the role of astronomy and physics in the history of science is still evident in the disproportionate attention paid to these disciplines by historians and philosophers of science. We have already referred to the central role played by physicians in early modern science. Cook's essay provides an entry into a range of issues connecting medicine and other sciences. Ariew's paper reminds us that, during the same period, astronomy and physics were by no means the exclusive concerns of those now remembered as founders of modern physics.

Alan Gabbey, Joseph Pitt, and Bernard R. Goldstein provide papers in more traditional areas of interest to historians and philosophers of science, but their conclusions are distinctly novel. Gabbey offers the first modern treatment of a central problem in heliocentric astronomy and cosmology—the nature of the moon's motion. Pitt offers a distinctly different picture of the relationship between Bellarmine and Galileo. Goldstein punctures one of the great myths of the Copernican revolution—Alfonso of Castile's supposed dissatisfaction with the Ptolemaic tradition. The myth of Alfonso X is used very frequently to support a thesis of discontinuity between medieval and early modern astronomy. Alfonso is often given as an example of the degeneracy of astronomy in the Middle Ages—epicycles upon epicycles, etc.—to justify the modern break with the past. Goldstein shows the myth to be the invention of those historians most wedded to the novelty and superiority of early modern astronomy; as such the myth is incapable of providing support for a thesis of discontinuity.

Finally, François De Gandt and Emily Grosholz offer papers on early modern mathematics. Although it is a historiographic commonplace that early modern science gave a new importance to mathematics in the study of nature, astonishingly little historical attention has been paid to the development of mathematics itself in this period. De Gandt examines Cavalieri's mathematical practice during the period of the development of the calculus. Grosholz considers the mathematical practice of Descartes. Both De Gandt and Grosholz are concerned with constraints, with breaking ties to the past. De Gandt examines Cavalieri's attempts to break free of Greek mathematical thinking. Grosholz shows that Descartes' own mathematical ideals were an impediment to his full utilization of the mathematical resources exploited by his successors.

PART I

SCIENTIFIC INSTITUTIONS

1   Societies, Circles, Academies, and Organizations: A Historiographic Essay on Seventeenth-Century Science

DAVID S. LUX

Numerous works published during the 1980s broke with a long-established historiographic interpretation of the social and institutional history of seventeenth-century science. These new works began to challenge accepted notions of dramatic conceptual and methodological shifts underlying the new scientific organizations appearing in the second half of the seventeenth century. This is an important historiographic development, especially given the significance of the past half century of debates over issues such as Puritanism and science. The literature of the 1980s actually began to recast the terms of such controversies and suggested that the long-standing arguments for social influences on science added little to our basic understanding of the social dynamics of seventeenth-century science. The revisionism running through this new literature constituted a profound historiographic shift, one that warrants close attention. It appears the rejection of a historiographic tradition began in these works.

Martha Ornstein's *Rôle of Scientific Societies in the Seventeenth Century* stands as the classic in the modern literature on scientific institutions.[1] Not only does her work remain in print after seventy-five years, but Ornstein's volume also continues to exercise deep influence over historical understanding. The work is still cited as the only major comparative

---

1. Martha Ornstein, *The Rôle of Scientific Societies in the Seventeenth Century* (Chicago: The University of Chicago Press, 1928 [1913]). Originally published in 1913 as a private printing of Ornstein's doctoral dissertation, the work was revived, copyrighted, and published commercially for the first time in 1928 at James Harvey Robinson's suggestion. That and subsequent editions have all followed the dissertation text faithfully. Robinson had been the dissertation director who suggested the topic for the *Rôle* to Ornstein, and after her untimely death in an accident at age 36 (1915) Robinson felt the work deserved wider dissemination. Despite the separate edition of 1928 and the production of a third edition in 1938, then, the *Rôle* remains fundamentally unchanged from the 1913 dissertation.

study treating the full complement of seventeenth-century scientific institutions.[2] More importantly, Ornstein's work continues to shape the conceptual and methodological frameworks governing research on scientific organizations—both the organizations of the seventeenth century and those of other periods. Indeed, because of the seminal influence of Ornstein's *Rôle* in shaping research, parsing her work can still yield fundamental insights into the basic questions driving research on the seventeenth century's scientific organizations.

As with any "classic" that drives a literature,[3] Ornstein's *Rôle* did several things: It formalized a historical research program; it surveyed the subject's dimensions; then it argued a strong historical interpretation explaining its self-defined issues. As with other such classic interpretations, Ornstein's work also established the benchmark from which subsequent works have been gauged—either as corroboration and amplification, or as rebuttal and redirection. More than any other work, Ornstein's *Rôle of Scientific Societies* has defined the modern research problem for the institutional history of science: to explain why the first autonomous, legally protected scientific organizations appeared during the seventeenth century. Indeed, that is the Ornstein question: Why did seventeenth-century science require separate and distinct scientific organizations? The major issue to address with that formulation is whether it might not constitute *une question mal posée.*

The Ornstein question followed a straightforward rationale: The oldest continuously operating "modern" scientific research organizations trace their origins to the mid-seventeenth century. Also, those first research societies were founded outside the universities.[4] Finally, from

2. See, for example, the bibliographic essay in Michael Hunter's *Science and Society in Restoration England* (Cambridge: Cambridge University Press, 1981), p. 204.

3. The most widely known of such classics in the history of science is Robert Merton's *Science, Technology and Society in Seventeenth-Century England* (New York: Howard Fertig/Harper Torchbooks, 1970 [originally published in *Osiris* 4 (1938): 360–632]). With the possible exception of analysis concerning the Duhem thesis, however, the history of science has not the same fertile levels of scholarship based on historiographic debate as those seen, for example, with economic and political history (the Pirenne thesis or the Beard thesis), intellectual history (Burckhardt), or medieval social history (Bloch). For an overview of research inspired by the Pirenne thesis, see Alfred F. Havighust (ed.), *The Pirenne Thesis: Analysis, Criticism, and Revision*, 3d ed. (Lexington: D. C. Heath, 1976). For Burckhardt and the problem of the Renaissance, see Wallace K. Ferguson, *The Renaissance in Historical Thought: Five Centuries of Interpretation* (Boston: Houghton Mifflin Company, 1948). For the impact of Charles Beard and Frederick Jackson Turner on American historiography, see Richard Hofstader, *The Progressive Historians: Turner, Beard, Parrington.* (New York: Alfred Knopf, 1968). For the legacy of Marc Bloch as currently practiced by a leading medievalist among the *Annalistes*, see Jacques Le Goff, *Time, Work, and Culture in the Middle Ages*, Arthur Goldhammer (tr.) (Chicago: The University of Chicago Press, 1980).

4. See the Feingold essay in this volume.

Ornstein's vantage point in 1913, the "best" modern scientific research was being done in universities—institutions tracing their roots to the twelfth and thirteenth centuries. In other words, the universities predated the seventeenth-century societies; they subsequently superseded them as centers for scientific research. Why then did the seventeenth-century scientists need the new institutions?[5]

Ornstein's answer derived from one of the oldest postulates underlying the history of science. Following a logic progressively elaborated through the writings of Fontenelle, Voltaire, Condorcet, William Whewell, John William Draper, and Andrew Dickson White,[6] Ornstein asserted that scholasticism's intellectual tyranny and the generally benighted state of university learning forced mathematical and experimental practitioners to create new institutional forms during the seventeenth century. But, to solidify that traditional formulation, Ornstein also made an original departure in her thesis. The outline for that departure appears clearly in the *Rôle's* Table of Contents.

Obviously, Ornstein was concerned with more than just the seventeenth-century scientific organizations; she was more specifically interested in university hostility toward science. As the Table of Contents reveals, Ornstein laid out the book in three parts. In the first she treated selected practitioners from the early century.[7] In part two, she turned to the "Learned Societies and Journals."[8] Then, in part three, Ornstein presented her capstone argument on science—or rather, its absence—in the universities.[9]

Ornstein laid the foundation for her thesis as she asserted, "[A] divid-

---

5. Ornstein posed her research agenda as follows: "On the one hand, . . . science obtained its most valuable, nay indispensable, aid from the scientific societies of the day; and, on the other hand, . . . the universities failed to supply such aid. Yet no existing work—so far as the writer knows—tries to show how this aid was given and why the societies were indispensable; nor is there any treatment which follows the work of the universities during this period and points out wherein it was inadequate" (ibid., p. xi.).

6. Fontenelle, *Histoire de l'Académie Royale des Sciences (1666–1699)* (Paris, 1733), furnished a general statement on the secularization of science. Voltaire's corpus of writings added considerably to that theme, and developed the added theme of traditional Christian institutions exercising an obscurantist influence on entire course of human progress. William Whewell's *History of the Inductive Sciences* (3 vols., London, 1837) carried the theme even further, especially in his dismissal of any possibility of science in the "Dark Ages" dominated by the Church. John William Draper, with *The Conflict between Religion and Science* (7th ed., London, 1876), and Andrew Dickson White, with *A History of the Warfare of Science with Theology in Christendom* (2 vols., New York, 1896), carried the theme of hostility between Christianity and science to its most radical extreme. In large part, the professional history of science is still struggling to shed the last vestiges of this insidious historiographic tradition.

7. Gilbert, Galilei, Torricelli, Pascal, Harvey, van Helmont, Bacon, Descartes, von Guericke, and the Amateurs; *Rôle*, pp. 21–69.

8. Ibid., pp. 73–216.   9. Ibid., pp. 213–263.

ing line may be drawn at about the middle of the century. . . . [The] first half seems more like a 'mutation' than a normal, gradual evolution from previous times. . . . [The societies of] the second half of the century elaborated [these results]."[10]According to Ornstein's thesis, individuals created entirely new scientific concepts and practices during the first half of the seventeenth century; during the second half of the century, newly founded societies consolidated and elaborated those intellectual breakthroughs. She then produced the rationale for her thesis as she concluded that, "with the exception of the medical faculties, universities contributed little to the advancement of science. . . . It is true that several of the greatest scientists occupied professorial chairs, but in many such cases it can be shown that their efficiency and prominence were due to forces external to the universities."[11]

Taken out of context, such an argument lacks the force of logic. In context, however, Ornstein's formulation persuaded several generations of graduate students and historians (including myself) to follow her argument, which asserts: first, an intellectual "mutation" occurred in science during the early years of the seventeenth century;[12] second, the universities proved incompatible with that new science;[13] and third, eager practitioners called for (and then created) new societies.[14] Thus, her argument postulated a tension between creative, vibrant scientific activity and hostile institutional frameworks. For Ornstein, that tension forced self-conscious groups of practitioners toward formalization of new scientific organizations in the second half of the seventeenth century.

Several generations of twentieth-century historians have researched and written about seventeenth-century scientific organizations under the Ornstein influence. No work matching the scope of her Europe-wide survey has yet appeared; nor has any work explicitly challenged the overall coherence of her thesis.[15] Only within the past ten years have any significant numbers of historians broken fundamentally with specific components to the Ornstein thesis. These breaks with Ornstein are significant. Before characterizing this recent trend, however, it is important to emphasize just how deeply the Ornstein influence pervades the twentieth-century literature on the institutional, organizational, and social history of science.

---

10. Ibid., p. 21.
11. Ibid., p. 257.
12. Ibid., p. 21.
13. Ibid., p. 257.
14. Ibid., p. 259.
15. At least in part, this reluctance to challenge the totality of the thesis must be attributed simply to the fact that no comparative work of similar scope has appeared in the historical literature.

The Ornstein argument has proven itself supple in stretching across ideological, conceptual, and methodological barriers in the history of science. Perhaps the best way to illustrate this influence is to note that even in the well-worn debates between the internalists and externalists—in either the earliest versions or the 1980s reprises[16]—the Ornstein influence permeates both sides. On the issue of the depth to the Ornstein influence, there is little difference between the internalists and the externalists, except in the ways the two sides bend Ornstein's framework to accommodate specific arguments.

The old-line, traditional internalists and intellectual historians such as Harcourt Brown, Marie Boas Hall, Margery Purver, or W. E. Knowles Middleton may appear to gain the most from the Ornstein framework; nevertheless, even such avowed externalists and social historians as Robert Merton, Roger Hahn, Charles Webster, or Margaret Jacob have produced works showing strong Ornstein influences. These strongly social formulations accept the four basic components to the Ornstein thesis just as readily as the internalists: (1) two discontinuities in the seventeenth century—one intellectual, the other institutional; (2) the central importance of self-conscious communities of practitioners; (3) the reality of anti-scientific hostility within the entrenched institutions—especially the universities; and (4) the necessity for focusing research on the histories of the special purpose, politically chartered societies.

The literature's clearest articulation of an argument carrying the Ornstein thesis appears in Margery Purver's *The Royal Society: Concept and Creation.*[17] Purver's work stands as a prime example of the fully developed internalist and intellectual history approach to the seventeenth-century scientific organizations. Fundamentally, her argument portrays the Royal Society as the product of a small community of practitioners who were determined to create a society codifying the Baconian method. Intellectual and institutional discontinuities, self-conscious group activity, entrenched institutional obscurantism, and a main-line focus on a successful society—every component in the Ornstein thesis appears. Indeed, those constructs furnish the driving motors for Purver's argument, an argument that casts the Royal Society as the cornerstone for any historical explanation of seventeenth-century science.

W. E. Knowles Middleton's *The Experimenters: A Study of the Accademia*

16. Charles E. Rosenberg's "Woods or Trees: Ideas and Actors in the History of Science" (*Isis* 79 (1988): 565–70) offers perceptive commentary on the continued viability of the debate, despite professional unwillingness to acknowledge its reality.
17. London: Routledge and Kegan Paul, 1967.

del Cimento[18] uses the same explanatory constructs, but for a purpose very different from the one found in Purver's *Royal Society*. In treating the Cimento, Middleton employed mirror-image reversals of the Ornstein thesis to explain why the Cimento failed in 1667. The intellectual and organizational discontinuities were too great: the membership chafed under the dull and repetitious routines that the Medici brothers established.[19] Self-conscious group identity failed to solidify, at least in part because the Medici brothers lacked the time necessary to reinforce their personal commitment to the Cimento.[20] Institutionalized obscurantism and hostility appeared in the "delicate religious situation in which the Italians found themselves after the condemnation of Galileo,"[21] in the failure of the *Saggi* (the Academy's only publication) to gain acceptance in the larger scientific community,[22] and (paradoxically) in the membership's distaste for their own work.[23] Indeed, in his insistence that the Cimento's exercised negligible influence on the larger scientific community of the 1660s, Middleton furnished a powerful corroborative statement reinforcing the importance of the Ornstein focus on main-line societies.[24]

18. Baltimore: The Johns Hopkins University Press, 1971.
19. Ibid., pp. 309–29, esp. p. 327.   20. Ibid., pp. 27–40; 56–64.
21. Ibid., p. 6; also see p. 333.
22. Ibid., pp. 333–38. In fact, the evidence Middleton presented on this point (considerable transalpine correspondence in the early 1660s [pp. 282–308] and the seven Italian, one English, two Latin, and two French editions of the *Saggi* before 1761 [pp. 337; 347–54]) seems to argue against his own stated conclusion of minimal impact. As explanation for this seeming contradiction of his own evidence, Middleton claimed these many editions were "of no use to the professionals," but served instead the "greedy" appetites of educated laymen who were gullible enough even to purchase the "trivia" produced at Theophraste Renaudot's academy (p. 337).
23. Ibid., pp. 27–40; 57–65; 310–19; 327–328.
24. Ibid., p. 346. This point is brought home most strongly in Middleton's extensive discussion of the influence of the "Transalpine Relations of the Academy," pp. 281–308. As with Middleton's arguments on the minimal impact of the *Saggi*, this chapter too seems marshaled in contradiction of the evidence presented. Opening the chapter with the statement that the academy had "no transalpine relations whatsoever" (p. 281), Middleton proceeded to document extensive correspondence between its members and the larger European intellectual community. The resolution to that apparent paradox comes in Middleton's explanation that there was only one "official" letter addressed to the academy (p. 281). As for the members' correspondence, Middleton found a "surprising" number having to do with the book trade (p. 282), and later found it a "far-fetched association of ideas that all this miscellaneous correspondence of Leopold's can be related to the Accademia del Cimento" (p. 308). On balance, Middleton's argument ignores the organizational and social significance of "intelligencer networks" in seventeenth-century science. For Middleton's argument, scientific correspondence must be straightforward, direct, and uncluttered with any extraneous discussion. For a very different treatment of such correspondences, including some originating from the Cimento, see David S. Lux, *Patronage and Royal Science in Seventeenth-Century France: The Académie de Physique in Caen* (Ithaca: Cornell University Press, 1989).

In a paradoxical attempt to revise the Ornstein thesis, Harcourt Brown's *Scientific Organizations in Seventeenth Century France*[25] produced a similarly skewed argument. Brown openly acknowledged that his study stood in the shadow of Ornstein's book, even as he announced his goal of producing a more fundamentally historical work.[26] Brown then set out to explore the entire complex of organized French scientific activity between 1620 and 1680. In the process he found four significant anomalies inexplicable under the Ornstein thesis: (1) general trends of intellectual and organizational continuity and development throughout the period;[27] (2) difficulty in tracing self-conscious, coherent communities of practitioners;[28] (3) general openness in the ways individuals moved across disciplinary boundaries prior to 1666;[29] and (4) the existence of vibrant scientific activity throughout France during the period before Colbert founded the Académie Royale des Sciences.[30] Yet, in the end, Brown's *Scientific Organizations* turned these anomalies around as he concluded that after 1685 the social, political, and economic climate of France destroyed an open scientific community that had reached its apogee about 1675.[31] In that final argument, the Ornstein thesis found corroboration: Intellectual and organizational discontinuities, formation of a self-conscious community, institutional opposition to science, and the need to focus research on the Académie Royale.

Even after recognizing Brown's failure to follow through on the significance of the anomalies his research uncovered, his *Scientific Organizations* still stands as one of the most important twentieth-century works on early-modern scientific organizations. Despite the initial disclaimer, Brown's *Scientific Organizations* gave sympathetic treatment to the Ornstein thesis. Nevertheless, the work carried important material and conveyed real insight into the institutional history of seventeenth-century science. Brown's account of the events leading to Colbert's founding of the Académie Royale, for example, implies a research agenda still worth pursuing.[32] Likewise, his chapter on provincial science calls for a much-needed analysis of provincial relations with Paris.[33] More-

25. Baltimore: The Johns Hopkins University Press, 1934.
26. Ibid., pp. x.   27. Ibid., pp. 161–253.
28. Ibid., pp. 1–16; 161–84.   29. Ibid., pp. 1–134.
30. Ibid., pp. 262–264.   31. Ibid., pp. 254–65.
32. Ibid., pp. 135–60. More specifically, Brown's account is the only to appear in the literature to date pursuing the origins of the Académie Royale primarily through the contemporary private correspondences. His attempt in that direction produced significant results, and offered the promise of additional insights available to anyone who pursues such research further.
33. Ibid., pp. 208–30. Brown provided a virtual catalogue of unexplored centers of provincial activity, bringing his discussion to a close by saying: "The few examples cited

over, his material on the changes in French science after 1675 and on the absence of discussions on science in the Parisian salons of the 1690s furnishes important evidence on the consequences of the Académie Royale's foundation.[34]

Overall, the intellectual and internalist histories of seventeenth-century scientific organizations contain material sufficient to vindicate their place in the canon of standard works on seventeenth-century science. With Middleton and Brown's works especially, the quality of the archival research and the presentation of documents guarantees their importance. Within the limits of their research design and arguments, such works are extremely valuable.

Within the literature attempting to apply the techniques of social history to early-modern scientific organizations, the possibility of such scholarly contributions appears less clear-cut. Perhaps time will prove the critics wrong, but it now appears unlikely that the trend toward social (or externalist) treatments of seventeenth-century scientific organizations can yield the same level of benefits to general scholarship.[35] In this historical genre, Boris Hessen's work, especially "The Social and

---

will perhaps convey some idea of the ferment that was working in many parts of France, as in Europe generally, to lead men of taste towards active research among the ways of nature" (p. 215). In his extensive case history of the academy in Caen, Brown suffered from a lack of documentation (a problem he partially overcame in 1939 with his "L'Académie de Physique de Caen (1666–1675) d'après les lettres d'André de Graindorge" in the *Mémoires de l'Académie Nationale des Sciences, Arts et Belles-Lettres de Caen* 9 (1939): 117–208. For a full case study of this academy, see Lux, *Patronage and Royal Science.*

34. As presented in Brown's final chapter (pp. 254–65) the closing down of the amateur scientific community in Paris occurred after 1675, as the result of changes in the broader political, economic, and social climates of France. Brown's argument stands as an important counterweight to simplistic assumptions about the death of private patronage in Paris prior to 1665, but ignores any possible influence of the opening of the Académie Royale. Despite the chronology he gives (Académie Royale opening in 1666 and the continuation of amateur activity into the 1670s), Brown ultimately subscribes to the notion that the Académie Royale resulted from an inevitable organizational shift brought about by the end of private patronage. For a very different view, in which the Académie Royale itself played a causal role in bringing about the end of private patronage, see Lux, *Patronage and Royal Science.*

35. While critics such as Michael Hunter, Mordechai Feingold, or Harold Cook focus on conceptual and methodological biases, this comparison with the older literature raises a different point. The type of intellectual history practiced by Harcourt Brown and W. E. Knowles Middleton allowed the presentation of significant bodies of documents as part of finished work. Their work could, and did, inspire derivative studies based on the documentation presented. With the social history as it is currently practiced, each new monograph requires immersion in the archives for the production of a unique research base. This requirement for extensive archival work significantly reduces the possibility of elaborative studies following from the publication of a seminal work. Such a methodological limitation, in fact, imposes a general restriction on the rate at which historians can produce high-quality social history for any period or any nation.

Economic Roots of Newton's *Principia*,"[36] and Robert Merton's *Science, Technology and Society in Seventeenth-Century England*[37] stand as the seminal works, but the historiographic controversies surrounding those works actually may have delayed acceptance of social approaches to seventeenth-century science.

In 1971 the publication of Roger Hahn's *The Anatomy of a Scientific Institution: The Paris Academy of Sciences, 1666–1803*[38] marked the true emergence of social history into the mainstream of scholarship on seventeenth-century scientific organizations. Since the publication of Hahn's work, a host of works carrying social explanations of seventeenth-century science have appeared. The list of such works now multiplies annually, and the number of practitioners applying the techniques of social history to seventeenth-century science is impressive. Among these social historians, however, Roger Hahn, Margaret Jacob, Simon Schaffer, Steven Shapin, and Charles Webster hold a special place. It is in the corpus of works these historians have produced that the full range of possible social explanations for seventeenth-century science finds its best expression. At one end of this range (Hahn and Webster), social forces act as conditioning factors; at the other end (Jacob, Schaffer, and Shapin), society becomes a determinant "constructing" scientific reality.

Critics claim to find weaknesses throughout this literature. Indeed, these social accounts appear to suffer from what the historiographer David Hackett Fischer identifies as "fallacies of narration."[39] More particularly, Fischer's "fallacy of presentism," in which explanations for manifold consequences rest on narrowly defined causality,[40] and that fallacy's opposite, the "antiquarian fallacy," in which a myriad of factors all come to bear in producing a unitary event,[41] both appear. And, with both, the ties to the Ornstein thesis are strong.

The publication of Roger Hahn's *Anatomy of a Scientific Institution* marked a turning point in the institutional literature on seventeenth-century science, but the work itself holds a curious place in that literature. Besides the fact that Hahn's book treats French science while most of the subsequent literature has dealt with English science, his work is also anomalous in that it takes up its account in the mid-seventeenth century and then carries out the bulk of analysis on the eighteenth and

---

36. In *Science at the Cross Roads* (London: Frank Cass, 1971 [1931]).

37. New York: Howard Fertig/Harper Torchbooks, 1970; originally published in *Osiris* 4 (1938): 360–632.

38. Berkeley: University of California Press.

39. *Historians' Fallacies: Toward a Logic of Historical Thought* (New York: Harper Torchbooks, 1970), pp. 131–63.

40. Ibid., pp. 135–40.   41. Ibid., pp. 140–42.

early nineteenth centuries.[42] In the *Anatomy of a Scientific Institution* the account of seventeenth-century institutionalization actually serves only as an extended introduction to the work's real task: to explain the eighteenth-century Académie Royale in the context of French society and politics. An extended commentary on that central theme in the *Anatomy of a Scientific Institution* lies beyond the scope of this essay, but it is important to note that through its heavy reliance on Parsonian social theory[43] Hahn's *Anatomy of a Scientific Institution* carries a structural-functional argument in which the Académie Royale's eighteenth-century professionalism and elitist character increasingly brought it into conflict with strong social forces, forces that erupted in revolution in 1789. Caught between the monarchy (its creator) and the inexorable forces of social change, the Académie Royale had become an anachronism by 1793, when it was swept away by the Revolution.[44] In establishing the basis for that argument, Hahn's account of the Académie Royale's seventeenth-century institutionalization furnishes one of the literature's clearest examples of the pervasive Ornstein influence on the history of science.

The opening chapter, "Initiating a Tradition," in *The Anatomy of a Scientific Institution* set the stage for Hahn's argument on eighteenth-century French science. Emerging professionalism, failing patronage, new forms of practice, opposition from entrenched institutional interests—all supported Hahn's main-line historical focus on the Académie Royale. Significantly, this chapter presents very little new or original in its treatment of the origins, founding, and first thirty years of the Académie Royale's history. Indeed, the chapter's value lies in its synthesis of the literature.[45] It is in this synthesis that the Ornstein thesis receives one of its strongest expressions.

42. A sense of Hahn's priorities can be gained from a comparison of the treatments given to the founding and the Revolutionary period. In *The Anatomy of a Scientific Institution,* the account of the Académie Royale's founding (chap. 1, "Initiating a Tradition") takes the academy from its prehistory through 1699 and occupies just 34 pages (pp. 1–34). On the other hand, the account of the academy between 1789 and 1795 (the establishment of the Institut) receives three chapters and 127 pages (pp. 159–285). The book's final chapter, on the Institut, carries the account forward through the Napoleonic reforms and adds another 27 pages (pp. 286–312) to the treatment of the period of less than twenty years following 1789.

43. Ibid., pp. 328–29.

44. Hahn's work then traces the academy's history as it was recreated as part of the institute (1795), and then caught up in the Napoleonic reforms (1803).

45. Of the 60 separate primary and secondary sources cited in the chapter's 71 footnotes, 57 are published works. Of the three citations to unpublished works, one is to Condorcet's "Sur les Académies" (n. 18), one to the Académie Royale's Registre des procès-verbaux des séances for 1699 (n. 40), and one to Mlle. Suzanne Delorme's unpublished study (1965) of publication of the academy's work in the *Journal des Savants* (n. 55). In short, the chapter carries just one citation to an unpublished archival document dating from the first thirty years of the academy's history.

Although Ornstein's *Rôle of Scientific Societies* does appear in Hahn's "Bibliography," the work itself is cited neither in the footnotes to the text nor in the *Anatomy's* "Bibliographical Note." The Ornstein influence remains evident, however.[46] In characterizing his own historiographic tradition, Hahn correctly traced his specific interpretation of the Académie Royale's origins and founding to the *Histoire de l'Académie Royale des Sciences (1666–1699)* authored anonymously by Bernard de Fontenelle.[47] Nevertheless, the interpretation in this chapter goes far beyond anything found in Fontenelle's work, which concentrates simply on the assertion of intellectual and organizational discontinuities of the mid-seventeenth century. Hahn's emphasis on new "professional norms,"[48] the "expression of academic practices already in use,"[49] the failure of private patronage and the rise of government interests in science,[50] and the opposition of corporate interests[51] all speak more directly to the Ornstein influence than to anything found in Fontenelle. The source for those constructs lay in the twentieth-century literature from which Hahn drew his synthesis.

The overall effect of Hahn's use of that literature was to limit the narrative and argument possible in the *Anatomy of a Scientific Institution*. Built on the narrow interpretative basis offered in the literature, the sweeping argument on the social dynamics of science in eighteenth-century France must stand ultimately on the strength of the linkages forged between the monarchy and the seventeenth-century community. By accepting the twentieth-century historical literature based on the Ornstein thesis as the cornerstone for the account of the Académie Royale's origin, Hahn limited his ability to explicate those linkages. As presented, they rest on the two Ornstein discontinuities, the coalescence of group identity (professionalization), government support in overcoming obscurantism and institutional barriers, and a narrow focus on the Académie Royale as constituting the only significant line of institutional development in French science. Overall, the argument possesses coherence, but suffers from a narrowing of historical causality and a subsequent broadening of consequences in the eighteenth

46. Ibid., pp. 321–29, esp. p. 322. This is especially evident in Hahn's reliance on works by Guillaume Bigourdan, James King, Harcourt Brown, Albert J. George, Pierre Gauja, René Taton, and John Milton Hirschfield—works featured prominently in both the footnotes to chap. 1 and the *Anatomy's* "Bibliographical Note."

47. Ibid., p. 1. Hahn's claim to follow in the specifically Fontenelle tradition is supported by his reference in the "Bibliographical Note" to the French tradition of scholarship on the academy as expressed in the works of L. F.-Alfred Maury, Joseph Bertrand, Ernest Maindrou, Léon Aucoc, Pierre Gauja, and Lucien Plantefol (p. 321).

48. Ibid., p. 2, 4, and passim.   49. Ibid., p. 5.
50. Ibid., pp. 5–12.   51. Ibid., pp. 13–14.

century. In short, reliance on the Ornstein constructs leads the argument in *The Anatomy of a Scientific Institution* to develop along lines Fischer categorizes under the rubric of the "fallacy of presentism."[52]

While Hahn's *Anatomy of a Scientific Institution* carries the Ornstein thesis almost as a distilled essence derived from the literature on French science, the works on English science by Margaret Jacob, Simon Schaffer, Steven Shapin, and Charles Webster constitute an entirely different type of social history. No less theory-driven than Hahn's *Anatomy*, the works of these historians nonetheless offer accounts of the organizational history of seventeenth-century science that are based on broad-ranging (often voracious) archival and primary source research. Here there is no danger that reliance on the Ornstein thesis can produce narratives based on the fallacy of presentism. On the contrary, where intensive research in archival sources, manuscripts, and published primary sources is used in supporting the interpretative conclusions of the Ornstein thesis, the danger for these social histories of English science lies in Fischer's "antiquarian fallacy," a fallacy of narration in which the historian draws on a multitude of causal events to explain a narrow consequence.[53]

The publication of Charles Webster's, *The Great Instauration: Science, Medicine and Reform, 1626–1660*[54] marked a second turning point in the development of a social history of science. Webster's solid erudition gives his *Great Instauration* an almost encyclopedic quality and the appearance of an exhaustive treatment of English science during a critical period. Yet, despite such appearances, Webster's work is less than comprehensive, as he clearly acknowledged.[55] Throughout its survey of scientific culture, organizations, education, and institutions, *The Great Instauration* pursues a close argument on the centrality of puritan millenarianism in the development of English science. More specifically, Webster identified Puritanism as the "dominant element in English society in the middle of the seventeenth century"[56] and argued the thesis that "the rise of the English scientific movement correlates extremely closely with the growth in strength of the puritan party."[57]

In pursuing this argument Webster attacked a historiographic tradition he traced to Merton, one in which the real flowering of English science occurred in the Restoration—following a lull in activity during the Civil War and Interregnum. Within the terms posed, Webster succeeded in pressing that argument forcefully and well. Nevertheless, the

---

52. Fischer, *Historians' Fallacies*, pp. 135–40.
53. Ibid., pp. 140–42.   54. London: Duckworth, 1975.
55. Ibid., pp. xiii–xvi.   56. Ibid., p. xiv.
57. Ibid., p. 503.

terms Webster used were highly selective. First, his use of "Puritan" can be challenged, especially as it applies to any coherent, self-defined group of the seventeenth century. Second, the choice of representative individuals, organizations, and institutions was highly selective. And third, even as Webster broke with the specific historiographic conclusions derived from Merton, he accepted the standard categorization contrasting the "good" science of the Restoration to the "less good" science of the previous twenty years.[58]

By accepting that dichotomy, Webster opened the door to Ornstein's discontinuities: The "mutations" of the early century found their "elaboration" in the organizational productivity of the new organizations in the second half of the seventeenth century. His clearly articulated focus on Puritanism brought a significant expansion to Ornstein's description of self-conscious practitioners championing new scientific approaches, but the basic explanatory construct remained firm: a clearly defined, self-conscious group carried the heaviest burden in the creation of the century's new organizations; thus, that group carried the burden of the social and institutional history of English science. With Webster, institutional hostility and entrenched obscurantism certainly play their parts, but *The Great Instauration* gave these a new twist in its use of puritan millenarianism as the causal motor driving the new science. Finally, even as Webster showed puritan influences on English science during the 1640s and 1650s, his acceptance of the real flowering of English science during the 1660s strongly supported the Ornstein emphasis on special-purpose organizations such as the Royal Society. Webster's only real challenge to that emphasis came from substituting puritan influences for Ornstein's individual geniuses as the basis of the "elaborations" of the later century.

Overall, Charles Webster's *Great Instauration* offers a vast survey of the social and intellectual history of early seventeenth-century English Puritanism for the purpose of arguing its significance to English science in the Restoration. Moving from broad causal events, the argument concentrates on a narrow consequence in explaining receptivity to new scientific ideas. In effect, as presented in *The Great Instauration,* all English Puritanism seemed to drive toward Restoration science. Hence, the argument falls under the rubric of Fischer's antiquarian fallacy, the movement from broad causal events to a specific consequence.

Margaret C. Jacob's *The Newtonians and the English Revolution, 1689–1720*[59] pursues a similar, but more sharply defined, argument.

---

58. Webster's argument holds that the science of the Interregnum furnished a necessary basis for the flourishing science of the 1660s (pp. 484–520, esp. p. 491).

59. Hassocks [Eng.]: The Harvester Press, 1976.

Whereas Webster's *Great Instauration* presents English Puritans as a social group whose ideology gave them a favorable disposition toward the new science during the mid-seventeenth century, Jacob's latitudinarians of the later century chose their specific scientific formulations precisely because they expressed a social and political ideology: "This book has argued that the Newtonians articulated their natural philosophy with constant reference to their social and political context."[60] With this argument, Jacob made Newtonian science a dependent variable, the precise value of which was determined by the course of political and religious debates. Once again, with the focus firmly fixed on defining the causal motor driving the creation of English science, such an argument produced an antiquarian fallacy of narration. As described in *The Newtonians and the English Revolution*, the entire sweep of seventeenth-century English social, intellectual, and political history came to bear in producing English Newtonianism.

Steven Shapin and Simon Schaffer's *Leviathan and the Air-Pump: Hobbes, Boyle, and the Experimental Life*[61] takes this type of explanation based on the social construction of knowledge considerably further. Shapin and Schaffer correlate the philosophical debates over the practice of science with the course of political history, ultimately arriving at the conclusion that it is possible to describe the entire history of modern scientific practices through a "relationship between our knowledge and our polity that has, in its fundamentals, lasted for three centuries."[62] With this work, the significant innovation for the social and institutional history of science appears in the use of a case-study approach comparing the scientific practices and the politics of Robert Boyle and the experimentalists with those of Thomas Hobbes. Quite consciously, Simon and Schaffer avoided the question of why Boyle won the scientific debate,[63] but in establishing the notion that Restoration polity and experimental science shared "a form of life," *Levianthan and the Air Pump* clearly argues that the validation of scientific knowledge derives ultimately from the polity.[64] Indeed, Simon and Schaffer argue the impossibility of any truly meaningful separation between scientific and political discourse—first in Restoration England, then in any larger contexts.[65]

In the elaboration of the social history of seventeenth-century science between Hahn's *Anatomy of a Scientific Institution* (1971) and Simon and Schaffer's *Leviathan and the Air Pump* (1985) there were clear progres-

60. Ibid., p. 271.
61. Princeton: Princeton University Press, 1985.
62. Ibid., p. 343.   63. Ibid., p. 341.
64. Ibid., pp. 342–43.   65. Ibid., pp. 283–344.

sions toward a more strongly defined social construction of knowledge and toward a fuller analysis of the social and ideological contexts of scientific knowledge. Nevertheless, the fundamental constructs of intellectual and organizational discontinuity, the importance of self-conscious groups championing a new science, the conflict of scientific practitioners with the representatives of traditional institutional interests, and the organizational flowering of science in the 1660s and 1670s remained as firmly fixed as the cornerstone for this genre as it had been for the older internalists and intellectual historians.

Throughout the articulation of social approaches to seventeenth century organizational and institutional history the four basic constructs of the Ornstein thesis remained unchallenged. Consequently, this radical departure from the traditional approaches to seventeenth-century organizational and institutional history furnished implicit corroboration for their validity. Indeed, by 1980, discontinuities, the analysis of selected groups, the assumption of entrenched hostility toward science, and the central role of the major organizations of the 1650s and 1660s had become so thoroughly integrated into historical explanations that they had taken on a canonical status. No work had challenged any of those constructs. Usage appeared to validate their empirical status. Yet, no one had ever really questioned their validity.[66]

Michael Hunter's *Science and Society in Restoration England*[67] furnished what proved to be only the opening salvo in a barrage of works that have collectively destroyed the foundations of the Ornstein thesis as it applies specifically to seventeenth-century science. Indeed, the 1980s brought forth numerous historical studies implicitly challenging the various components of the Ornstein thesis. In addition to Michael Hunter's work, the more important of these studies include works by Barbara Shapiro, Mordechai Feingold, Harold Cook, L. W. B. Brockliss, and Peter Dear.[68]

In these historians' works, the relationships between "new" and "traditional" science are far more complex than ever conceived previously.

---

66. In this sense, the Ornstein thesis became the basis for the "fallacy of hypostatized proof," which Fischer describes as a substitution of theory for empirical investigation in constructing historical interpretations (Fischer, *Historians' Fallacies*, pp. 55–56).

67. Cambridge: Cambridge University Press, 1981.

68. Barbara J. Shapiro, *Probability and Certainty in Seventeenth-Century England* (Princeton: Princeton University Press, 1983); Mordechai Feingold, *The Mathematicians' Apprenticeship: Science, Universities and Society in England, 1560–1640* (Cambridge: Cambridge University Press, 1984); Harold Cook, *The Decline of the Old Medical Regime in Stuart London*, Ithaca: Cornell University Press, 1986); L. W. B. Brockliss, *French Higher Education in the Seventeenth and Eighteenth Centuries* (Oxford: Oxford University Press, 1987); Peter Dear, *Mersenne and the Learning of the Schools* (Ithaca: Cornell University Press, 1988).

Practitioners become far more difficult to sort out into clearly defined groups than ever before. The institutional frameworks (both those supporting science and those nominally opposed) are much more overlapped, nuanced, and ambiguous than ever conceived under the Ornstein influence. And finally, the continuities and discontinuities (both intellectual and organization) begin to appear far more subtle than ever imagined possible within the terms of the Ornstein thesis. In this literature, even the universities boast considerable intellectual and scientific vitality.

The implicit repudiation of Ornstein's sharp intellectual and institutional discontinuities has become the hallmark of recent historical work on seventeenth-century scientific institutions. Only slightly less central are the denial of too easy an identification of any unique "scientific community," the repudiation of clear-cut institutional obscurantism, and the shift away from exclusive focus on the major special-purpose institutions. Nothing suggests, however, that the authors of this literature are united in any anti-Ornstein movement or that they are ready to articulate a full set of generalizations challenging the Ornstein thesis. Indeed, there is no evidence suggesting a unified or coordinated program among these works at all, except insofar as the works on English science all address the conceptual and methodological weaknesses of predecessors such as Webster and Jacob. The rationale behind this lack of coordinated effort is not difficult to discern.

The strongest common theme uniting the recent literature flows from the methodological insistence on context-specific historical analysis. Most of these recent works depend on micro-level, source-specific research woven into case studies or into synthetic treatments of particular historical themes. The only notable exception to that general rule appears in Michael Hunter's *Science and Society in Restoration England.* Indeed, Hunter's work is unique with respect to the clarity with which he defined his study as a synthesis filling a chronological gap in the available scholarship. Referring directly to Hessen, Merton, and Webster[69] and indirectly to Jacob,[70] Michael Hunter focused his study precisely in the period framed by their works. Whereas these others had assumed the Ornstein scientific "elaboration" occurred during the Restoration, Hunter made the investigation of Restoration science itself the object of his research—with significant results.

Hunter's most surprising findings appeared in the intellectual continuities and the relative low priority science held throughout the Resto-

---

69. Cambridge: Cambridge University Press, 1981; p. 2.
70. Ibid., p. 188.

ration. Much of the vaunted progress of science turned out to be rhetorical rather than substantial, and although the audiences for science began changing, the practitioners remained fixed in older patterns. Overall, the actual record of scientific achievements was disappointing, as "science also remained far from dominant in the intellectual realm."[71] In Hunter's survey, even the Royal Society became something less than the lodestar of English science during the second half of the seventeenth century.[72]

Michael Hunter's *Science and Society in Restoration England* still stands alone as a synthetic, contextual survey of any period, albeit a relatively short period, in seventeenth-century science. Nevertheless, other historians have recently provided a convincing series of case studies validating their pursuit of context and corroborating or amplifying themes similar to those found in Hunter's work. Barbara J. Shapiro's *Probability and Certainty in Seventeenth-Century England: A Study of the Relationships between Natural Science, Religion, History, Law, and Literature* (1983) offers a truly impressive display of scholarship in pursuing the single issue of the relationships between various forms of knowledge. Her conclusions on continuity, on the relative lateness of major conceptual shifts in science, and on the interrelations between forms of knowledge offer powerful antidotes to simplistic notions of intellectual "mutation" and "elaboration" in science.[73] Likewise, by treating university teaching in historical context, Mordechai Feingold's *The Mathematicians' Apprenticeship: Science, Universities and Society in England, 1560–1640* debunks the myth of sharply divided university and London intellectual communities. Moreover, Feingold clearly demonstrates the centrality of universities in furnishing both an educational and a social locus for English mathematicians. Overall, Feingold's work shows definite continuity in English scientific activity from the middle of the sixteenth century to the middle of the seventeenth.[74]

Approaching many of these same general issues from an entirely different direction, Harold J. Cook's *The Decline of the Old Medical Regime in Stuart London* traces the declining fortunes of the Royal College of Physicians through the seventeenth century. In the process, Cook perceptively treats both the professional contexts of medicine and the more general context of natural philosophy across the entire seventeenth century. His work is unique in treating that chronological span in its entirety, and the results of the effort are significant: the illustration of divisiveness and continual reformulation within the communi-

71. Ibid., p. 189.   72. Ibid., pp. 32–86.
73. Shapiro, *Probability and Certainty*, pp. 267–72.
74. Feingold, *The Mathematicians Apprenticeship*, pp. 214–16.

ties of practitioners; a contextualized account of the practical (as opposed to ideological) implications of science establishing political ties; and an entirely new picture of the cycles of rise and decline characterizing seventeenth-century scientific organizations—including the Royal Society.[75]

Overall, the scope of this recent scholarship on English science in the seventeenth century has outstripped anything available for continental science, but such an imbalance seems merely to reflect the traditional weight of relative scholarly effort. Despite the relatively slower emergence of new interpretations, research on seventeenth-century French science shows every sign of proceeding along lines similar to those that have emerged in scholarship on England. L. W. B. Brockliss's *French Higher Education in the Seventeenth and Eighteenth Centuries* (1987) delivers for France much of what Shapiro and Feingold have produced on English intellectual life and education. Built narrowly on intensive treatment of statutes, the curriculum actually studied, and the social contexts of university knowledge, Brockliss's work demonstrates the same interlocking systems of knowledge and practitioners Shapiro and Feingold found in England.

Among more focused case studies, Peter Dear's *Mersenne and the Learning of the Schools* (1988) debunks the notion of sharp intellectual discontinuities in the early seventeenth century. Lynn S. Joy's *Gassendi the Atomist: Advocate of History in an Age of Science*[76] shows the complexities of intellectual life early in the century and makes a first effort (since Brown's *Scientific Organizations*) at delineating the actual formation of communities around patrons and new intellectual programs. David S. Lux's *Patronage and Royal Science in Seventeenth-Century France: The Académie de Physique in Caen*[77] follows similar themes while tracing the fortunes of private-patronage science through the 1660s. Overall, the collective impact of these case studies is to challenge the notions of sharp discontinuities, simplistic group identification, overemphasis on the antiscientific obscurantism of French institutions, and the blind assumption that Paris and the Académie Royale were the culmination of seventeenth-century French science.

Taken individually, none of these recent works challenges the totality of the Ornstein thesis directly, or even indirectly. Indeed, in many of these works the *Rôle of Scientific Societies* fails to find a place even in the

75. On this last point, Cook's chap. 4, "Political Weakness and Intellectual Threats, 1660–1672" (pp. 132–82) constitutes an especially valuable contribution to the literature on the Royal Society.
76. New York: Cambridge University Press, 1988.
77. Ithaca: Cornell University Press, 1989.

bibliography. At one level, this is a curious phenomenon. A fundamental historiographic rejection of Ornstein's thesis is underway, but those historians who are most centrally involved seem unaware, or only slightly concerned, with the fundamental historiographic nature of their task. At another level, however, this tacit dismissal of Ornstein is entirely comprehensible. Each of these works does situate itself within the immediate historiographic tradition. And these historians seem to look for their models in work produced on periods other than the seventeenth century. Indeed, the reassessment of the institutional history of seventeenth-century science has not sprung from any general historiographic debate or focused internal questioning of available constructs. Rather, the revisionist movement among historians of seventeenth-century science seems to be inspired by a broader historiographic shift affecting the entire history of science. Here, we find the final irony in the strength of the hold Ornstein constructs have held over seventeenth-century scholarship.

In the history of science, a revision of scholarship on the entire corpus of medieval and early modern literature has been underway for some time. Contextual studies have appeared regularly since the early 1970s. There is really nothing extraordinary about their emergence in the literature on the social and institutional history of seventeenth-century science—nothing extraordinary, that is, except for their timing. Despite the pivotal role historians of seventeenth-century science tend to claim for their studies, the impetus for contextualized studies of science has come from scholars of other centuries—centuries on both sides of the seventeenth.[78]

78. In the medieval history of science, works from the 1970s by Brian Stock, *Myth and Science in the Twelfth Century: A Study of Bernard Silvester* (Princeton: Princeton University Press, 1972), William Wallace, *Causality and Scientific Explanation*, vol. 1 (Ann Arbor: University of Michigan Press, 1972), and Nicholas H. Steneck, *Science and Creation in the Middle Ages* (Notre Dame: Notre Dame University Press, 1976) offer particularly strong examples of contextual analysis of science within intellectual history. More broadly, C. S. Lewis's *The Discarded Image: An Introduction to Medieval and Renaissance Literature* (Cambridge: Cambridge University Press, 1964) and Charles Trinkaus's "*In Our Image and Likeness*," 2 vols. (Chicago: University of Chicago Press, 1970) offer even earlier examples of intellectual history treating medieval and renaissance science in relation with other forms of knowledge—in ways unknown for seventeenth-century science until the work of Shapiro, Feingold, and Brockliss.

For the sixteenth century, there is an equally impressive body of literature significantly antedating the historiographic shift for the seventeenth. Although not given over totally to contextual studies, *The Copernican Achievement*, edited by Robert S. Westman, contains numerous noteworthy contributions. Likewise, shorter pieces by Bruce Moran (e.g., "German Prince-Practitioners: Aspects in the Development of Courtly Science, Technology, and Procedures in the Renaissance," *Technology and Culture* 22 (1981): 253–74) provide important work on sixteenth-century patronage. Also noteworthy is R. J. W. Evans' *Rudolf II and his World: A Study in Intellectual History, 1576–1612* (Oxford:

The earlier appearance of such studies for periods bracketing the seventeenth century (a full ten years before any real break with the Ornstein thesis) suggests how the revisionist literature on seventeenth-century social and institutional history of science could have developed with so little direct reference to Ornstein or the ways the standard literature has been dominated by her hypostatized explanatory constructs. Quite simply, the new studies on seventeenth-century science seem to flow naturally out of the broader literature rather than from any direct reassessment of the specific historiographic tradition governing seventeenth-century science. Indeed, the tacit silence maintained in the new literature concerning the overall historiographic traditon governing seventeenth-century science furnishes what is perhaps the strongest evidence on the pervasiveness of the Ornstein influence. With the corroboration nominally provided by the social (externalist) historians, the explanatory constructs spelled out in her work had indeed come to imbue the literature so pervasively by the late 1970s as to pass uncited as empirical reality.[79] Thus, the revisionist movement of the 1980s was launched almost with the notion that what was being displaced was an incomplete historical record.

Actually, what has been revised was an interpretative historiographic tradition, one based very strongly in the progressive "new history" of

---

The Clarendon Press, 1973). R. J. W. Evans has also produced a seminal article on seventeenth-century intellectual organizations, "Learned Societies in Germany in the Seventeenth Century," *European Studies Review* 7 (1977): 129–51, which conveys an impressive intellectual taxonomy placing science in a context very different from anything seen elsewhere. In a similar vein, Keith Thomas's *Religion and the Decline of Magic: Studies in Popular Beliefs* (New York: Charles Scribner's Sons, 1971) offers important contextual background for understanding early-modern science, as does Robert Muchembled's *Popular Culture and Elite Culture in France, 1400–1750* (trans. Lydia Cochrane. Baton Rouge: Louisiana State University Press, 1985). Finally, among the specialized studies of sixteenth-century science, Owen Hannaway's *The Chemists & the Word: The Didactic Origins of Chemistry* (Baltimore: The Johns Hopkins University Press, 1975) offers a particularly striking example of a type of contextual analysis of discipline formation that has yet to appear for any seventeenth-century activity.

For eighteenth-century science, Karl Hufbauer's *The Formation of the German Chemical Community (1720–1795)* (Berkeley: University of California Press, 1982) presents another such disciplinary study. Eighteenth-century science also furnishes one of the best examples of contextualized history of science in the literature with Keith Baker's *Condorcet: From Natural Philosophy to Social Mathematics* (Chicago: The University of Chicago Press, 1975). In Baker's work, biography moves beyond the normal limits of intellectual history to provide treatments of institutional history, discipline formation, and social history.

79. While the internalists and older generation of intellectual historians regularly cited Ornstein, this does not hold true for the literature of the 1970s and 1980s. Of the social history and revisionist works on seventeenth-century science cited in this essay, only three, Roger Hahn's *Anatomy of a Scientific Institution,* Michael Hunter's *Science and Society in Restoration England,* and David Lux's *Patronage and Science in Seventeenth-Century France,* carry citations to Ornstein's *Rôle* in their bibliographies.

the early twentieth-century American scholarship. Indeed, historians of science who have relied on Ornstein's constructs have perhaps overlooked or forgotten the fact that she wrote her work under the guidance and supervision of James Harvey Robinson, the progenitor of the "new history's" avowed presentism: "The present has hitherto been the willing victim of the past; the time has now come when it should turn on the past and exploit it in the interests of advance." Robinson wrote those words in *The New History* published in 1912, at the very time he was supervising Ornstein's completion of *The Rôle of Scientific Societies* in essentially the form it appears today. As with the corpus of Robinson's work, and the work of his other followers in the "new history," Ornstein's *Rôle* stressed a broad, synthetic approach to the intellectual and social trends of an age, especially those that can be foreshortened to elucidate the origins of the modern world.[80] Such a historiographic formula may have served its purposes in the early twentieth century, but in the last years of the century the time has surely come to reject its influence on the history of science.

80. For a succinct summary of the "New History" and its rejection in the broader historical profession during the mid-1930s, see John Higham, *History: The Development of Historical Studies in the United States* (Englewood Cliffs, NJ: Prentice Hall, 1965), pp. 108–31. For Ornstein's own programmatic statement on history and Robinson's inspiration, see the *Rôle*, pp. xi–xii.

2   Tradition versus Novelty: Universities and Scientific Societies in the Early Modern Period

MORDECHAI FEINGOLD

"Not only were the universities of Europe not the foci of scientific activity, not only did natural science have to develop its own centers of activity independent of the universities, but the universities were the principal centers of opposition to the new conception of nature which modern science constructed."[1]

Thus decreed a distinguished historian of science a few years ago. But by no means was he either the first or the last to lash out against the traditional citadels of learning. Indeed, the chorus of critics to berate the universities is almost as numerous as the authors to deal with the subject. Again and again it has been claimed that the ultra-conservativism of the universities and their overt hostility to the "new science" drove out all those interested in such topics, thereby causing them to gravitate to new centers of scientific support. For most of the Continent, such centers grew up around the princely courts;[2] in England and Hol-

---

1. Richard S. Westfall, *The Construction of Modern Science* (New York, 1977), p. 105. Even more blunt is Westfall's "Isaac Newton in Cambridge: The Restoration University and Scientific Creativity," in P. Zagorin (ed.), *Culture and Politics from Puritanism to the Enlightenment* (Berkeley and Los Angeles, 1980), pp. 135–64; for a more balanced view, see Barbara J. Shapiro, "The Universities and Science in Seventeenth Century England," *Journal of British Studies* 10 (1971): 47–82; Robert G. Frank, Jr., "Science, Medicine and the Universities of Early Modern England: Background and Sources," *History of Science* 11 (1973): 194–216, 239–69; Nicholas Tyacke, "Science and Religion at Oxford before the Civil War," in D. Pennington and K. Thomas (eds.), *Puritans and Revolutionaries* (Oxford, 1978), pp. 73–93; John Gascoigne, "The Universities and the Scientific Revolution: The Case of Newton and Restoration Cambridge," *History of Science* 23 (1985): 391–434; See also note 4 below.

2. R. S. Westman, "The Astronomer's Role in the Sixteenth Century: A Preliminary Study," *History of Science* 18 (1980): 105–47; Bruce T. Moran, "German Prince-Practitioners: Aspects in the Development of Courtly Science, Technology and Procedures in the Renaissance," *Technology and Culture* 22 (1981): 253–74; Idem, *The Hermetic World of the German Court: Medicine and Alchemy in the Circle of Maurice of Hesen-Kassel* (forthcoming).

land, they established themselves in the emerging mercantile, bourgeois cities. As one historian put it: "The science of Elizabeth's reign was the work of merchants and craftsmen, not dons; carried on in London, not in Oxford and Cambridge."[3]

Such a line of reasoning was frequently extended to include the scientific academies that from roughly the middle of the seventeenth century began to emerge. These new "progressive" institutions, it was claimed, were intended as centers for scientific study and hence alternatives to the backward, traditional universities. In view of this new function, it is not surprising—so the argument goes—that the academies became the focus of attack by the universities and victims of their hostility toward the new science.

Certainly, there existed a debate over the nature and boundaries of science in the second half of the seventeenth century. More certain still, there existed a real and acute antagonism between the universities and the new institutions. Thus far I endorse the received view. However, I differ in the attribution of the origins and development of such antagonism. It is my contention that such rancor was not the direct result of a conflict between modern and old-fashioned forms of science, as is often assumed; rather, it emerged for reasons only tangentially concerned with science—either old or new—and only as the debate evolved and grew increasingly heated were such weighty issues as the new science injected into it. But even during this later stage, it seems to me, there was never any official persecution by the universities of that intellectual pursuit we call science.

Needless to say, a comprehensive treatment of the relations between the universities and the scientific academies would necessitate an extensive study of the state of scientific studies at the universities during the seventeenth century. Considerations of space will prevent me from attempting such a task here, except for reiterating at the outset that the received image of the universities as institutions clinging to their beloved Aristotle and opposing any new knowledge, the new experimental science in particular, cannot be substantiated.[4] My aim here will be to deal instead with the mechanism that animated the conflict between the universities and the new institutions, and thus to present a more complex—and I hope a more accurate—picture. For the purpose of

3. Christopher Hill, *Intellectual Origins of the English Revolution* (Oxford, 1965), p. 15. See also, most recently, J. A. Bennett, "The Mechanics' Philosophy and the Mechanical Philosophy," *History of Science* 24 (1986): 1–28.

4. Mordechai Feingold, *The Mathematicians' Apprenticeship: Science, Universities and Society in England, 1560–1640* (Cambridge, 1984); "Universities and the Scientific Revolution: The Case of Oxford," in H. A. M. Snelders and R. P. W. Visser, et. al. (eds.), *New Trends in the History of Science* (Amsterdam, 1989), pp. 29–48.

expedience, I have divided the components of this conflict into three categories: institutional, intellectual, and theological.

The universities of the early modern period regarded themselves as heirs to a monopoly on higher education. In fact, their centuries-old hegemony had invested their strength, position, and importance in society with mythic proportions, so that they looked upon themselves—and were looked upon by society—as the only legitimate guardians of learning. Not surprisingly, therefore, their initial opposition to new institutions was based not on whether a particular academy—owing to its curriculum, academic body or intended purpose—posed a specific threat, but on whether there existed any place at all for a non-university institution of higher learning. For just this reason, the repeated assurances of the new institutions that their foundations would not confer degrees or in any way interfere with, or rival, a university education—all of which were intended to mitigate the anticipated hostility of the universities—were irrelevant. The real threat was not the particular institution, but the very concept of non-university education; the real fear was that the success of one such institution would serve as a dangerous precedent for future foundations.

An important aspect of the universities' intention to remain the sole guardians of higher education was financial and political. Leaving aside the inevitable loss of income that would result from a declining enrollment, it must be remembered that the universities were highly dependent upon private benefactors—almost all of whom were university alumni—for revenues. Equally important, the deflection of both students and benefactors to new institutions would weaken the traditional bond that had evolved between the educated elite and the universities, a bond that until now had been a source of financial strength and political leverage. Whether in court or in church, university alumni traditionally served as advocates for their former institutions. In view of such anticipated financial and political reverberations, it mattered little, therefore, whether a projected institution aimed at becoming another university or a theological seminary, a finishing school for the nobility or a scientific academy. The corporate academic bodies were bound to oppose them. A few examples should suffice to illustrate the essential characteristics of such opposition.

In March 1575 the authorities at Cambridge learned of Sir Thomas Gresham's intention to rescind his pledge to bestow on them £500 for the purpose of founding a new college. Instead, Gresham resolved now to establish a college in London that would teach the "seven liberal sciences"—divinity, law, medicine, rhetoric, music, geometry and astronomy—in the vernacular, to the citizens of London, free of charge. The

Cambridge orator was promptly ordered to discharge a letter to Sir Thomas, urging him to abandon such a project, which, it was claimed, would be "to the detriment or almost ruin of either university." Simultaneously, the Cambridge authorities sought to dissuade Gresham from undertaking such a project by employing the services of other persons as well. Two decades later, when the time finally came for the establishment of Gresham College, the college trustees, wishing to reassure Oxford and Cambridge that no threat to their monopoly was intended, sent them extremely cautious and respectful letters, asking them to assist the new enterprise and even to nominate their own candidates for the seven chairs. Thus, every attempt was made to pacify the universities. The Cambridge vice-chancellor, however, still fearful that the new institution "in time it may be greatly prejudicial to our universitys" would have preferred to combat the college. But the chancellor of the university, William Cecil, Lord Burghley, had been a close friend of Gresham and refused to oppose the last will of the founder. Thus, the college was allowed to open without a fight. The loyal Oxford and Cambridge men who filled the seven professorships, however, ensured that the new institution would never pose a threat to the universities.[5]

The example of Gresham College was not very different from that of the Collège de France, founded earlier in the century. The French institution, too, was intended to provide public lectures, free of charge, this time to the citizens of Paris. Naturally, the University of Paris duly put up a fight, but to no avail.[6] In a similar manner, when the burghers of Amsterdam decided in 1629 that they, too, deserved an institution that would provide public lectures for their citizens—for the better preparation of the youth for the universities, as they justified their goal—and hence proceeded to found the Athenaeum Illustre, the neighboring University of Leiden fiercely combatted the scheme, but with no more success than either the French or English universities had had the previous century.[7]

5. John Ward, *The Lives of the Professors of Gresham College* (London, 1740), p. 38; Ian Adamson, "The Foundation and Early History of Gresham College, London, 1596–1704," Ph.D. dissertation, Cambridge University, 1975, pp. 38–40 and passim; idem, "The Administration of Gresham College and Its Fluctuating Fortunes as a Scientific Institution in the Seventeenth Century," *History of Education* 9 (1980): 13–25; Feingold, *The Mathematicians' Apprenticeship*, chap. 5.

6. Abel Lefranc, *Histoire du Collège de France* (Paris, 1893, rep. Geneva, 1970), passim, especially pp. 119–23, 143–49; Arthur Tilley, *Studies in the French Renaissance* (Cambridge, 1922), pp. 123–48.

7. Chris L. Heersakkers, "Foundation and Early Development of the Athenaeum Illustre at Amsterdam," *Lias* 9 (1982), 4–55. The Leiden authorities even sued the Amsterdam magistrate for the "secretive" manner in which G. J. Vossius was enticed to leave Leiden for Amsterdam. The Court of Holland not only found in Amsterdam's

Indeed, from the point of view of the universities it seemed more than sensible to oppose all such schemes indiscriminately, for once the plans of a new foundation were set in motion, there was no telling what shape it would eventually take; all past promises and declarations would count for nought. In fact, of the new institutions just cited, this is more or less what eventually happened. The Collège de France moved quite rapidly beyond the stage of providing lectures in Greek and Hebrew as originally stipulated, and in rapid succession chairs in mathematics, medicine, philosophy, eloquence, botany, Arabic, and Canon Law were added. In Amsterdam, from the very beginning the burghers plotted to go beyond the stipulated two professorships—in history and philosophy—and within a few years we encounter professors in mathematics, law, Oriental languages, and, later still, medicine. Gresham College, we already noted, was prevented by its incumbent professors, all university men, from proceeding further into becoming perhaps, London University. And everyone in Europe appreciated and feared the quick and efficient methods employed by the Jesuits to encroach upon existing universities.

Consequently, the crude laws of survival and the politics of patronage dictated much of the initial reaction of the universities to any and all schemes of higher learning. Promises not to meddle with subjects taught by the universities or to confer degrees could not allay the fears of the universities. Changing circumstances or shifts in the balance of power might nullify the initial assurances and benefit such new institutions. The universities were not about to countenance what one day might prove to be their bane. Once we realize the sensitivity of the universities to any proposal for new institutions of higher learning, we can better appreciate their initial reaction to the new, up-and-coming scientific academies. Thus, the same corporate bodies of the Sorbonne, the Medical Faculty, and the Parlement of Paris that conspired in 1635 to prevent the chartering of the Académie Française,[8] joined forces again in 1666, but this time with their old foe, the Académie Française. The purpose of this new alliance was to thwart the efforts to bring to fruit the "General Academy" envisaged by Charles Perrault and apparently increasingly favored by Jean-Baptiste Colbert, who ventured the estab-

---

favor, but also stated that the foundation of the Athenaeum Illustre "did not conflict with the privilege of Leiden, provided that in Amsterdam promotions and official exams were foregone." C. S. M. Rademaker, *Life and Work of Gerardus Joannes Vossius* (Assen, 1981), p. 241.

8. D. Maclaren Robertson, *A History of the French Academy* (New York, 1910), pp. 24–28. Fr. Olivier-Martin's important study, *l'Organization corporative de l'ancien régime* (Paris, 1938), should be consulted for the background of French corporate reaction to external threats.

lishment of a sort of "Ministry of Culture" under his authority, and one that would best further the deification of Louis XIV.[9] The combined front managed to win this particular battle, but ironically it led to Colbert's endorsement of the more modest scheme that resulted in the establishment of the Académie Royale des Sciences—a body that managed to successfully weather the sundry intrigues of the various corporate institutions.[10] A similar story can be told of the Académie de Physique at Caen, which managed to operate relatively peacefully under the private patronage of Pierre-Daniel Huet until it was learned that the academy was about to obtain the rank of Royal Institution. Such a change in status immediately resulted in forceful opposition by the local university.[11]

Events in the Holy Roman Empire followed a similar pattern. The foundation of the Collegium Naturae Curiosorum in 1651 was a relatively uneventful occasion. However, once the members of this medical body acquired the direct patronage of Emperor Leopold I, eventually changing the academy's name to Academia Caesarea Leopoldina, it became a different matter; not only was the academy granted a charter, but its official status equalled that of the University of Vienna, and it was also granted elaborate printing privileges. The "success" of the academy, in fact, encouraged quite a few other individuals to look to Leopold for patronage of similar designs.[12] Small wonder, then, that the Jesuits and the University of Vienna were subsequently well prepared to oppose the attempt made by Leibniz—founder and president of the Berlin Society (later Academy) of Sciences—to found a comprehensive Imperial Academy in the city in 1712–14.[13] Perhaps only in

9. Pierre Clément, *Lettres, Instructions et Mémoires de Colbert* 8 vols. (Paris, 1861–82), vol. 5, pp. 512–13; John M. Hirschfield, *The Académie Royale des Sciences 1666–1683* (New York, 1981), pp. 14–16; Roger Hahn, *The Anatomy of a Scientific Institution* (Berkeley, Los Angeles, 1971), pp. 12–13.

10. See Harcourt Brown, *Scientific Organizations in Seventeenth-Century France* (New York, 1934, repr. 1967), pp. 147–48; and more fully in David S. Lux, "Colbert's plan for La Grande Académie: Royal Policy toward Science, 1663–1667," paper delivered at the 51st annual meeting of the Southern Historical Association, Houston, 14–16 December 1985.

11. David Lux, *Patronage and Royal Science in Seventeenth-Century France: The Académie de Physique in Caen* (Ithaca, 1989).

12. For the academy, see Erwin Reichenbach and Georg Uschmann (eds.), *Nunquam otiosus. Beiträg zur Geschichte der Präsidenten der Deutsche Akademie der Naturforscher Leopoldina* (Leipzig, 1970).

13. The insinuation that the Jesuits objected to Leibniz's proposal because of the latter's failure to convert to Catholicism should not be given excessive credence; their claims concerning the damage such a Royal Academy could inflict upon corporate interests were, in fact, sincere. See John M. Mackie, *Life of Godfrey William Leibnitz* (Boston, 1845), pp. 248–51; Ornstein, *Rôle of Scientific Societies*, pp. 196–97; E. J. Aiton, *Leibniz* (Bristol and Boston, 1985), pp. 319–20. See in general, O. Klopp, "Leibniz Plan

Italy, where the function of the universities had always been confined to "the instruction of the youth," could "the academies [grow] up along side them without any evidence of rivalry or discord."[14] These examples illustrate how jealously the universities guarded their privileges from any threat, real or imagined. Never far was the fear that any new foundation might expand into a degree-granting institution and eventually bring about a loss of revenue as well as of potential allies in church and state. But the greatest threat posed by these new academies was a direct result of the special relations they enjoyed with the local rulers; the new academies were seen as potentially powerful and dangerous rivals.

The pattern in England was similar. Charles II returned to London from exile on 25 May 1660. On 15 July 1662 he granted the charter that officially created the Royal Society, and a generous charter it was. It permitted the president and members of the society to determine the range of topics to be investigated (which, according to Sprat, included "all the objects of mens thoughts");[15] to accept and manipulate to the society's advantage gifts of money and land; to correspond freely with persons of all nationalities; to found a college—or colleges—for its use at will; to employ printers and engravers, and to publish or sanction publication of all works concerning natural knowledge.[16] Given such a munificent charter, how could Oxford and Cambridge afford not to be alarmed?

This was only the beginning. Within three years of obtaining their first charter, the Royal Society totaled 202 registered members, nearly a third of whom were titled, including four Dukes, fourteen Earls, four Viscounts, one Marquis, nine Lords, forty-one Knights, not to omit the king himself.[17] To add insult to injury, from early on the Royal Society adopted the irritating practice of boasting of its new "acquisitions" by regularly publishing lists of members. Surely, if such were the accom-

---

der Gründung einer Societät der Wissenschaften in Wien," *Archiv für Osterreichische Geschichte* 40 (1869): 159–255; A. Harnack, *Geschichte der Königlichen preussischen Akademie der Wissenschaften zu Berlin*, 3 vols. (Berlin, 1900), vol. 1; and, for the German background, Robert J. W. Evans, "Learned Societies in Germany in the Seventeenth Century," *European Studies Review* 7 (1977): 129–51.

14. Eric W. Cochrane, *Tradition and Enlightenment in the Tuscan Academies 1690–1800* (Chicago, 1961), pp. 41–42.

15. Thomas Sprat, *History of the Royal Society*, eds., Jackson I. Cope and Harold Whitmore Jones (St. Louis and London, 1959), p. 81.

16. For the charters of the Royal Society, see *The Record of the Royal Society of London*, 4th ed. (London, 1940), pp. 215–86.

17. The figures are based on information gathered from *The Record of the Royal Society*, and Michael Hunter, *The Royal Society and Its Fellows 1660–1700*, British Society for the History of Science, Monograph 4, corrected reprint (Chalfont St. Giles, 1985).

plishments of the Society within such a short period, the future promised great things. In view of this manifestly instant success, not many university men needed to be reminded that the Society's meetings took place in Gresham College, the founding of which had given some headaches to the universities a few decades earlier.

What further aggravated the situation was that England was unaccustomed to such munificent displays of royal patronage. The last time something of this magnitude had occurred was during the carefree days of Henry VIII, a century and a half earlier, when the College of Physicians had received its charter, or, a few years later, when the five Regius professorships at both Oxford and Cambridge were founded. But Henry's munificence proved to be the closest England would ever come to imitating Continental Renaissance courts. Far more familiar was the parsimonious stance of Elizabeth I, who preferred to have noblemen and private individuals promote learning, as well as most everything else. Thus, jealousy of the Royal Society undoubtedly blended with the universities' concern for their monopoly, or even their future existence. By all accounts, chief among the initial concerns of both universities following the foundation of the Royal Society was the likelihood that the new foundation might evolve into an arch-rival institution of higher learning. Such a conclusion can easily be drawn from the rhetoric used by both proponents and adversaries of the Royal Society throughout the 1660s, and from the indignant response of Thomas Sprat to the insinuations incorporated into Samuel Sorbière's account of his impressions of the Society.[18]

This institutional conflict between the universities and the Royal Society was the background against which the intellectual and theological aspects of the conflict were debated. Despite pronouncements to the contrary, advocates and apologists of the Royal Society knew that in the course of legitimizing their new institution, they purposely misrepresented the state of science—and of knowledge in general—in the universities. Similarly, their attempts to portray themselves as doing something new and utterly different from what was being done at Oxford and Cambridge were equally exaggerated. But such caricature was necessary if they were to generate the support they needed. At the same time, in projecting a distorted state of affairs, they also encountered that same dilemma that had faced all projectors of schemes of higher learning: On the one hand, if a new proposal was to generate wide support, it had to differentiate itself from existing institutions and to

18. Thomas Sprat, *Observations on Monsieur Sorbière's Voyage* (London, 1665).

demonstrate in what way the learning it offered would be distinct from that offered by the universities. On the other hand, by singling out studies that the universities ostensibly were unable, or unwilling, to provide, such institutions jeopardized their repeated endeavors to appease the universities and allay their fears. Simply put, the problem was how to demonstrate that they were simply filling an existing void within the educational structure without insinuating that Oxford and Cambridge had failed to address themselves to such a void. A difficult problem, and one that would prove the stumbling block to most schemes for higher learning during the early modern period. Most projectors, fearful of antagonizing the universities, became trapped by the very moderation they espoused. Afraid of making the universities appear redundant, most schemes for higher learning were themselves made redundant.

The apologists of the Royal Society were also among those who attempted the exacting acrobatics of balancing a boastful pride in the new age their society was about to inaugurate and the circumspection necessary to appease the universities. But by claiming they could not possibly harm Oxford or Cambridge for the simple reason that the Royal Society engaged only in studies neither of the two universities engaged in, they opened a Pandora's box. For if the universities were truly guardians of all knowledge—science included—the insinuation that they had repudiated one of its constituents amounted to a direct challenge. If Oxford and Cambridge acquiesced to such reasoning, it would be an admission of failure to live up to the very principle of "comprehensive education" upon which their monopoly rested. Hence, what began as an institutional/monopolistic conflict over corporate interests blossomed into a war over intellectual issues.

What, then, were the manifestations of the intellectual dimension of the conflict? As I see it, there were three principal issues involved, each of which worked to entrench the positions taken by university men on the one hand and the champions of the Royal Society on the other. To begin with, as a result of the aspiration of many men of science to acquire an autonomous status for the scientific disciplines, there ensued a breakdown of the traditional framework of knowledge and a subsequent reordering of the disciplines. By this I mean that the medieval concept of the unity of knowledge, according to which every educated man was instructed in the entire arts and sciences curriculum—and more important still, was deemed proficient to contribute to any of its constituents—began to disintegrate. Owing largely to the appropriation of science from the once all-encompassing domain of the "general scholar," this unity of knowledge was no longer valid. Instead, there emerged a new type of man of science: someone not just talented in the sciences, but dedicated to its pursuit as his vocation. Moreover, this new breed of intellectual ventured not only to secure distinguishable characteristics for himself, but also to exclude from the domain of science those "general scholars" who lacked his scientific sophistication and aptitude. What occurred, then, was a distinct rupture between the new men of science—who strove to legitimize their activities, found a new profession, and generate mandatory support and patronage—and the "general scholars." The latter, either unaware of the extent of the breach, or unwilling to accept the new ordering of the disciplines and the ensuing implications concerning their competency in the sciences, chose to fight back.

Closely related to this fracture is the phenomenon often referred to as the "Ancients and Moderns" controversy. However, in view of what has been said thus far it is more accurate, I believe, to view the dispute as a far more serious polemic between two competing images or priorities of knowledge. Much more study needs to be devoted to this painful and sincere confrontation between the "humanistic" and "scientific" world views. I would just like to reiterate here that it was not predominantly a clash between progress and reaction; between speculative, book-and-authority adoring dons who clung tenaciously to antiquity and the modern man who used experimental methods and the new way of reasoning in his quest for "true" knowledge. However, the fact that some of the slogans tossed about by the apologists of the new science— in particular those directed against "book learning" which made the apologists sound alarmingly reminiscent of the enthusiasts of the 1650s—is relevant to our discussion. Such an antibookish stance and its implied denigration of an important function of the universities only fueled the grievances of Oxford and Cambridge men.[19] It must be remembered that the universities considered themselves to be the custodians of all knowledge and that such rhetoric, clearly the expression of a far too narrow view of learning, provoked their animosity. And this brings us to the third manifestation of the intellectual aspect of the debate: the issue of the alleged forfeiture of all standards of learning in the wake of the scientific vogue.

To attract rich and influential patrons, the campaign of the Royal

19. See Michael R. G. Spiller, *"Concerning Natural Experimental Philosophy": Meric Casaubon and the Royal Society* (The Hague, 1980); Michael Hunter, "Ancients, Moderns, Philogists, and Scientists," *Annals of Science* 39 (1982), pp. 187–92; B. C. Southgate, " 'Forgotten and Lost': Some Reactions to Autonomous Science in the Seventeenth Century," *Journal of the History of Ideas* 50 (1989), pp. 249–68; Idem, " 'No Other Wisdom?' Humanist Reactions to Science and Scientism in the Seventeenth Century," *The Seventeenth Century* 5 (1990), pp. 71–92.

Society, as well as of other scientific societies, was based, in large part, on a careful marketing of the "new science." Its champions promised the rich and powerful that the science advocated by the society could be easily procured, without long laborious study, and better yet, could be procured in an extremely pleasant manner. Robert Boyle certainly served as a revered model:

There are many ingenious persons, especially among the nobility and gentry, who, having been first drawn to like this new way of philosophy by the sight of some experiments, which for their novelty or prettiness they were much pleased with, or for their strangeness they admired, have afterwards delighted themselves to make or see variety of experiments, without having ever had the opportunity to be instructed in the rudiments of fundamental notions of that philosophy whose pleasing or amazing productions have enamored them of it.[20]

Hence, the focus on experiments by the Royal Society and the overall effort to involve patrons in the scientific process by asking them to re-cord the weather, forward astronomical observations of all kinds or relate to the society events of an unusual nature. Again and again these patrons were reassured that they were contributing significantly to the growth of knowledge; that they were active participants in a field that hitherto they had been only passively footing the bill for. The envisaged rewards of such a "marketing" approach were too tempting, and almost all men of science succumbed to such practices.

Yet the critics of the Royal Society were aware, at least in part, of the inherent dangers of such an approach that distorted knowledge in general, and science in particular. Moreover, by seeming to promise profundity in science attained exclusively via experiments, the new men of science suggested, by implication, to those rich and powerful patrons the possibility of easily attained proficiency in such other disci-plines as philosophy and theology. Most dangerous, however, were the ramifications of such an approach for other domains of learning that were dependent upon munificent patronage. For if patrons were con-verted to the promotion of scientific studies by virtue of some juvenile fascination with its more entertaining aspects and a somewhat exagger-ated conviction of its benefits to society, would they be willing to perse-vere in their support of the more "speculative," and less tangible do-mains of learning? Would "usefulness" and "enjoyment" become the prerequisites for patronage? And if so, what would happen to the lion's share of human knowledge? As Hobbes once presented the need for

20. Robert Boyle, "The Origin of Forms and Qualities according to the Corpuscular Philosophy," in M. A. Stewart (ed.), *Selected Philosophical Papers of Robert Boyle* (Manches-ter, 1979), p. 4.

disinterested support of learning to his patron the Earl of Newcastle, why not simply "suffer the liberall sciences to be liberall"?[21]

I mentioned earlier that the intellectual dimension of the conflict originated in the need to differentiate the Royal Society from the universities. Unlike the projectors of most other schemes, who were so moderate that they moderated themselves out of existence, the early members of the Royal Society were frequently brash. Struck by what in the early 1660s seemed to be the swift materialization of their aspirations, the early members of the Society became over-confident, almost intoxicated, with a newly acquired sense of power following their procurement of the royal charter. Believing their success guaranteed, the support of king and nobility permanent, and the new Society invincible, they threw caution to the wind. The anonymous "Ballad of Gresham Colledge"—perhaps written by William Godolphin circa 1663—is an excellent example of this hubris. Among other verses, we find the following:

> Thy Colledge, Gresham, shall hereafter
> Be the whole world's Universitie.
> Oxford and Cambridge are our laughter;
> Their learning is but Pedantry.
> These new Collegiates doe assure us
> Aristotle's an Asse to Epicurus.[22]

Similarly, the prefatory ode of Abraham Cowley to Sprat's *History of the Royal Society* could also be read as an attack on priests and scholars, books and letters, in contrast with the triumphant liberation of philosophy,[23] and even the routinely cautious Henry Oldenburg could occasionally be carried away with anti-university rhetoric and grandiose dreams—particularly in his correspondence with foreign scholars. Congratulating Pierre de Carcavy, for example, on the recent foundation of the Académie des Sciences, the secretary of the Royal Society continued:

> I am thoroughly persuaded that the societies newly established here, in France, and in Italy will serve as a stimulus, within a few years, to incite all the other nations of Europe to take up the same studies and to oblige them wholly to desert the quodlibetic learning of the schools which serve no other purpose than to befog the spirit and drag out disputes which are not only useless but often very pernicious."[24]

21. *Historical Manuscripts Commission. The Manuscripts of His Grace the Earl of Portland preserved at Welbeck Abbey* (London, 1893), vol. 2, pp. 126.

22. Dorothy Stimson, *Ancients and Moderns* (London, 1949), p. 58.

23. Sprat, *History of the Royal Society*, sig. B-b3.

24. *The Correspondence of Henry Oldenburg*, A. Rupert and Marie Boas Hall (eds.), 13 vols. (Madison, London, 1965–86), vol. 4, p. 101.

Worst of all were the almost fanatic writings of Joseph Glanvill, who abused real and imaginary enemies alike, obscuring issues that it would have been in the best interests of the Society to elaborate on, or hammering on others that perhaps should have been left alone. As a result, "It may be the fact that Glanvill's efforts on behalf of the Royal Society did as much to generate critics of the new science as it did to silence them."[25] Small wonder, then, that even such a staunch supporter of the Society as John Beale would rather have done without the help of such a champion or that Oldenburg wrote Boyle in 1664 of his fear that Glanvill's propaganda "may be of more prejudice, yn advantage to ym, if they be not competently endowed wth a revenue, to carry on their Undertakings." Four years later, though lamenting that he was prohibited from including a poem praising the Royal Society in a book he had published in Cambridge, Peter Du Moulin proceeded nonetheless to tell Boyle that it grieved him "to see a feud between that Noble Society and the Universities, to which Mr. Glanvill's books have much contributed."[26] Hence, what clearly happened was that cardinal issues pertaining to knowledge, its aims and its application, got mixed up with personal and institutional slanders, thereby resulting in the entrenchment of the combatants in ever more uncompromising positions.

The third component of the conflict between the Royal Society and the universities is the theological one. Here, too, underlying tensions existed even before the foundation of the Royal Society, and the new institution merely brought them to the surface. Like philosophy and the arts curriculum, theology also suffered from the breakdown of the traditional unity of knowledge. In fact, since the discipline had always commanded a position at the very pinnacle of the pyramid, subsuming to its own ends all other disciplines, the appropriation of science from the general body of learning was even more ominous. The desire for autonomy and legitimacy on the part of the new breed of men of science resulted, therefore, in some dangerous repercussions. Christian intellectuals, who in the past had concentrated their energies and aspirations on God, salvation, and primary causes, now demanded the right to pursue "secular" topics and concentrate on secondary causes alone. In addition, the Church, like the universities, was alarmed by the explicit demand of all scientific academies for printing and licensing privileges and even more, by the granting of such requests. Their alarm was rooted not only in their desire to guard their monopoly, but also by a

25. Nicholas H. Steneck, " 'The Ballad of Robert Crosse and Joseph Glanvill' and the Background to Plus Ultra," *The British Journal for the History of Science* 14 (1981): 60.

26. *The Correspondence of Henry Oldenburg* 2, p. 332; *The Works of the Honourable Robert Boyle*, ed. Thomas Birch (London, 1772), vol. 6, p. 579.

genuine fear of the consequences of a policy that would allow publication of anything and everything pertaining to philosophy—the traditional handmaiden of divinity—by men who had not been trained as theologians.

Such tensions, then, had existed all along and would have erupted in due course. However, the natural instability was aggravated and intensified by the early apologists of the Royal Society. Fearing religious prejudice, they attempted to anticipate it just as they made some attempts to anticipate corporate opposition by the universities. Hence, they boisterously proclaimed both their own piety and the great impetus the new science would give to religion. Clearly, they were too vocal in their enthusiasm, for instead of allaying fears, they awakened dormant prejudices and created new resentments. By offering their peculiar services to religion, not only were these "virtuosi" proposing to enter a field in which they were not qualified—and en route to teach theologians how to go about their business—but at the same time they attacked or denigrated those humanistic disciplines upon which theology was grounded. As far as the theologians were concerned, not only were the members of these new societies toying with dangerous ideas, but they appeared inclined to deploy such ideas in religion as well.

The theologians were not alone in their perception of the inherent perils of the new science. The virtuosi themselves understood all too well how their advocacy of the new science and the reordering of the disciplines would directly affect theology. However, the men of science had an axe to grind, and in the course of fighting for the formation of new, legitimate disciplines, they blinded themselves with their own rhetoric. The "divines"—many of whom were precisely those "general scholars" who were rapidly being excluded from the earnest pursuit of science—opposed the breaking down of the prevailing framework of knowledge, and they were not necessarily interested in seeing the establishment of an autonomous discipline of science. Such divines were well aware of the inherent dangers posed by the new science to theology as they understood it, and would spare little effort in spelling these out.[27]

Such are, all too briefly, some of my notions concerning the nature of the conflict between the Royal Society—and other scientific academies—and the universities in the early modern period. Regardless of

27. I intend to discuss these issues in a separate study. See, in general, Michael Hunter, *Science and Society in Restoration England* (Cambridge, 1981), esp. chapter 7; John Gascoigne, *Cambridge in the Age of the Enlightenment* (Cambridge, 1989); John Morgan, *Godly Learning* (Cambridge, 1986); idem, "Puritanism and Science: A Reinterpretation," *The Historical Journal* 22 (1979): 535–60. See also the sources listed in n. 19 above.

the objectives of the new societies, the universities were bound to oppose them, as any proposal could pose a threat to their monopoly on higher learning. Equally threatening was the possible loss of patronage, upon which they were so dependent, to the scientific societies. For their part, the new societies could not help but differentiate between their role and that of the universities. Although such differentiation was intended to generate patronage and not necessarily to assail the universities, it fueled the underlying tensions inherent in the very concept of a non-university institution. An interesting illustration of such attitudes can be found in the opposition of Leiden University as late as 1759 to the attempt of the first scientific society in the Netherlands—that of Haarlem, organized seven years earlier—to "acquire the official support of the States General." The letter written by the Leiden Senate is illuminating:

Although the Haarlem Society, in so far as nothing pertaining to the sciences is taught there orally, nor doctoral degrees granted, etc., is of a different nature [than the university], experience has nonetheless shown how much the glory of more than one university has been beclouded and darkened when public authority was also bestowed upon other societies and when illustrious persons consented to become members of the same. This has been the experience in England, for the universities of Cambridge and Oxford have lost much of their previous brilliance and luster since the Royal Society in London was established by public authority. The university of Paris, so famous in former times, is scarcely mentioned anymore since the Royal Academy of Sciences there was brought to bloom by the personal protection of the king and the accession of illustrious persons as honorary members.[28]

While these were the "objective" sources for the conflict—those that would have come into play no matter what scheme was involved—the progression of the strife took a new, and different, course. The seemingly unprecedented victory of the Royal Society made some of its protagonists imprudent and arrogant; some of their arguments became insulting, tending even toward intemperate caricaturization, while certain of their claims and propositions became downright offensive. Both the universities and many churchmen—fearful, jealous, and irritated—reacted angrily, thus bringing about, prematurely, a heated general debate over science which, I believe, also contributed to the relative decline of English science by the end of the seventeenth century.

28. *Bronnen tot de Geschiedenis der Leidsche Universiteit,* ed. P. C. Molhuysen, 7 vols. (s'Gravenhage, 1913–24), vol. 5, p. 209*, quoted in Edward G. Ruestow, *Physics at Seventeenth and Eighteenth-Century Leiden* (The Hague, 1973), p. 150n33.

PART II

MEDICINE AND GEOLOGY

3   Physick and Natural History
in Seventeenth-Century England

HAROLD J. COOK

Among the many people affected by the "scientific revolution" were
the physicians. Exploring some of the responses of physicians to the
intellectual changes of early modern Europe can be quite telling, for
they were well-educated men with a large stake in the intellectual estab-
lishments of their day. Since the one and only mark universally distin-
guishing physicians from all other kinds of medical practitioners was
their higher academic degrees, their medical doctorates, any alterations in the framework of knowledge were felt keenly among these
learned men. The physician's M.D. was supposed to certify, and ordinarily did certify, that he was well educated in "physic," a branch of
university study that demanded much philosophical sophistication
from its devotees. Because the physicians were well educated, and because that education alone separated them from other practitioners,
physicians were very well aware of the implications of almost any significant change in the intellectual currents of the day. Their reactions
to the development of the new philosophy, then, are a bellwether that
can lead to a better understanding of what was important about the
new philosophy to contemporaries, throwing light on some of the implications of the new philosophy commonly overlooked in studying the
philosophical innovators alone.

Among the difficulties of the physicians in the seventeenth century
were those arising from the competition of medical rivals. There had
always been only a handful of university-educated physicians in comparison with the large number of medical practitioners. But by the early
modern period, new forms of advertising and the growth of the market
economy encouraged the commoditization of medical products and
services.[1] Many of the people involved in the rapidly growing medical

---

1. For more on the growth of a consumer economy, see Margaret Spufford, *The
Great Reclothing of Rural England: Petty Chapmen and Their Wares in the Seventeenth Century*
(London: Hambleton Press, 1984); and Joan Thirsk, *Economic Policy and Projects: The*

marketplace were practitioners who offered explanations for disease and treatment that were at variance with the academic explanations given by physicians. Paracelsianism and other varieties of chemical medicine certainly posed a threat to the preeminence of learned physic.[2] But so too did the more mundane and more accessible vernacular medical tracts, the variety of which grew tremendously in the seventeenth century.[3] The ordinary, nonacademic practitioners who offered their services or medicines to the public commonly also offered rationales for their practices that either assumed a certain medical outlook on the part of the public or developed a novel viewpoint of their own.

The many controversies within the medical community of the period are often reduced to the somewhat simplistic terms of rationalism vs. empiricism. Many of the assumptions of the historical literature can be formulated in the following line of reasoning: (a) the physicians were well educated in a rational natural philosophy; (b) their medical rivals, some well educated, some self-educated, some reading only the vernacular or reading not at all, were called "empirics" because, like the ancient philosophical school, they relied almost entirely upon experience as their medical guide, declining to look closely into the causes of health and disease by the use of reason; (c) therefore, it seems sensible to view the physicians as supporters of intellectual traditions placing primacy on the ability to reason through to causes, and to view their rivals as advocates of new approaches that placed primacy on the ability to find out new things through experience. The intellectual battles between physicians and their rivals were therefore battles between old and new, public authority and individual liberty, university learning and craft tradition, reason and experience.[4]

Development of a Consumer Society in Early Modern England (Oxford: Clarendon Press, 1978); for the medical marketplace, see Harold J. Cook, *The Decline of the Old Medical Regime in Stuart London* (Ithaca, NY: Cornell University Press, 1986) pp. 28–69; R. Porter, *Health for Sale: Quackery in England 1660–1850* (Manchester: Manchester University Press, 1989).

2. Allen G. Debus, *The Chemical Philosophy: Paracelsian Science and Medicine in the Sixteenth and Seventeenth Centuries*, 2 vols. (New York: Science History Publication, 1977); P. M. Rattansi, "Paracelsus and the Puritan Revolution," *Ambix* 11 (1963): 24–32; Charles Webster, "English Medical Reformer of the Puritan Revolution: A Background to the 'Society of Chymical Physitians,' " *Ambix* 14 (1967): 16–41; and "Alchemical and Paracelsian Medicine," in C. Webster (ed.) *Health, Medicine and Mortality* (Cambridge: Cambridge University Press), pp. 301–34.

3. I am currently surveying the medical literature of seventeenth-century England; for a study of that literature in the sixteenth century, see Paul Slack, "Mirrors of Health and Treasures of Poor Men: The Uses of the Vernacular Medical Literature of Tudor England," in C. Webster (ed.), *Health, Medicine and Mortality*, pp. 237–73.

4. See, for example, Christopher Hill, *Intellectual Origins of the English Revolution* (Oxford: Oxford University Press, 1965); Theodore M. Brown, "The College of Physicians and the Acceptance of Iatromechanism in England, 1665–1695," *Bulletin of the*

Indeed, just these issues have often been raised in interpreting the causes of the scientific revolution;[5] and too, there is enough sense in this dialectical argument to use it as a staging post from which to push on. But there are also many reasons to think that while this dialectic captures much of the debate over ideas between the physicians and their rivals, it does not do so with enough nuance to be true to the contemporary content of the debate. We are all aware that there are difficulties in using any one term like "reason," "experiment," or the "experimental method," as the key determinate of the scientific revolution: for one thing, many "scientists" were hardly experimentalists by our lights. So, too, there are problems with speaking about the scientific revolution as a consequence of the "inductive method," or of Baconianism or Cartesianism, much less Platonism, Aristotelianism, or any other "isms." All these attempts to characterize the intellectual issues at stake in the intellectual struggles of the day are ultimately reductionistic, presuming that there is "a" scientific revolution that has a particular essence, outlook, world view, philosophy, or approach at its root, a kind of Hegelian spirit of the age into which historians can with effort finally, if incompletely, gain insight.

One basic problem with this manner of thinking about the changes characterized as the new philosophy is the tendency to pose the problem in terms of dialectical entities: reason vs. experience or physician vs. empiric. This fails to capture the nuances of the contemporary struggles for many reasons. Partly the failure is due to using terms that are abstractions rather than tangible entities. "The physicians," for instance, were deeply divided on many significant intellectual (as well as social and political) issues; so, too, were "the apothecaries," empirics, chemists, Helmontians, and so forth, not to mention the philosophers and virtuosi. Therefore, before making any larger statements about the *causes* of the scientific revolution, it is worth once again trying to sharpen the *description* of a part of it, leaving explanations for another time.

The description that follows, then, tries to give an overview of what some learned physicians, who as a group were particularly sensitive to shifts in the intellectual winds, thought to be new about some of the

History of Medicine 44 (1979): 12–30; and "Medicine in the Shadow of the 'Principia,'" *Journal of the History of Ideas* 48 (1987): 629–48.

5. For example, Paolo Rossi, *Philosophy, Technology and the Arts in the Early Modern Era*, S. Attanasio (tr.), B. Nelson (ed.) (New York: Harper and Row, 1970); P. M. Rattansi, "Early Modern Art, 'Practical' Mathematics and Matter Theory," in Rom Harré (ed.), *The Physical Sciences since Antiquity* (London: Croom Helm, 1986), pp. 63–77; J. A. Bennett, "The Mechanics' Philosophy and the Mechanical Philosophy," *History of Science* 24 (1986): 1–28.

changes of the day. These physicians clearly saw that many elements of the new philosophy were indeed far more empirical than the old. At the same time, however, although the new philosophy was more empirical, it was not without philosophical content. The particular combination of empiricism and philosophy that characterized much of the new philosophy is best captured in the phrase "natural history."

1. NATURAL PHILOSOPHY AND PHYSIC

To understand the responses of the physicians to various parts of the new philosophy, it is well to begin by trying to understand their intellectual traditions. That is, what was "physic," that profession practiced by physicians?

The distinction between "medicine" and "physic" was an important one in late medieval and Early modern English.[6] Like all linguistic distinctions, it was somewhat messy and ambiguous, with different authors using the words in different ways.[7] But like all linguistic distinctions, too, it deserves our attention, for it provides a useful analytical device. "Medicine," we can say, is the art of administering therapies to the sick (derived from the Latin *medico,* to apply drugs—or dyes). "Physic," on the other hand, is a word derived from the Greek *phusis,* or "nature." It was the art of counseling people to live their lives so as to live in accordance with nature (to retain health), or, if health had already been lost, to help them regain health by counseling them about how to reharmonize their lives with nature.

The ancient Greeks, who invented physic, had emphasized soundness of body along with soundness of mind as a fundamental constituent of living the good life. Soundness involved not just strength but harmony and balance with nature. Living in tune with nature meant retaining health and living a long life, while disharmonies and imbalances meant illness, possibly even death. In order to live harmoniously, people had to regulate their lives so that each person's unique "constitution" (or "temperament") might remain in tune with an ever-changing nature. The physician's task, then, was to help individual people regulate their lives according to universal principles of nature so as to help

6. Cook, *Decline of the Old Medical Regime,* pp. 62–66; Harold J. Cook, "Physicians and the New Philosophy: Henry Stubbe and the Virtuosi-Physicians," in R. French and A. Wear (eds.), *The Medical Revolution of the Seventeenth Century* (Cambridge: Cambridge University Press, 1989).

7. Jerome J. Bylebyl is working on a study of the Latin uses of *"physica"* and *"medica"* in late medieval Europe. See Bylebyl, "The Medical Meaning of 'Physica'," *Osiris,* 2d series, 6 (1990): 16–41.

them retain health and prolong life: that is, to advise on regimen.[8] In the early modern period, because the ideal of the best physic continued to stress the regulation of life according to nature, it was sometimes called "preventive" medicine or "dietetic" medicine, the Greek word *diaita* meaning a way of living or a mode of life. The principles of physic were therefore intended for the use of the healthy as well as the sick. Clearly, then, the art of physic entailed more than mere medical treatment by drugs or surgery.

Like its fellow professions of law and theology, early modern physic was a learned science obtained through university study and intended to affect the behavior of the person seeking professional guidance; in the case of physic, the end of the physician's advice was to retain health and prolong life. Since the client's behavior would, ideally, be reinforced or changed by the physician's counsel, the physician needed to be persuasive: that is, he had to be good at logic and rhetoric. The ultimate purpose of the physician was to keep his client a good or to make him a better person. *Care* rather than *cure* was the learned physician's first duty. To preserve health and prolong life required something more than mere skill in curing diseases: it meant acquiring the ability to give advice on how a person might regulate his or her life in order to remain in balance with the environment. Like his learned counterparts in law and church, then, the primary goal of the physician was to provide pastoral advice and care that would prevent difficulties, although a secondary end was to correct problems that had already occurred.

In order to maintain the health of clients and to cure the ills of patients, the physician had to be able to probe the interior natures of each uniquely tuned human being so as to maintain or re-establish a harmonious balance with nature. To do this, the physician needed to know how nature worked: to know the general principles of nature, or natural philosophy. He also needed to know how to apply those universals to particulars: to apply general principles to unique individuals. Physic was therefore applied natural philosophy, and the physician needed to be very well grounded in the knowledge of nature as well as in the logical and rhetorical arts.

Consequently, academic books on learned physic had long divided the subject into *theoria* and *practica*, at least since the time of the ninth-

8. See esp. Owsei Temkin, "Greek Medicine as Science and Craft," *Isis* 44 (1953): 213–25; Ludwig Edelstein, "The Dietetics of Antiquity," and "The Relation of Ancient Philosophy to Medicine," reprinted in O. Temkin and C. L. Temkin (eds.), C. L. Temkin (trans.), *Ancient Medicine: Selected Papers of Ludwig Edelstein* (Baltimore: Johns Hopkins University Press, 1967), pp. 303–16 & pp. 349–66, respectively.

century translators at Baghdad. One of the most important authors of scholastic physic, known in Europe by his Latin name Joannitius (usually identified as Hunain ibn Ishaq), wrote a book that remained the best short summary of physic until the seventeenth century, the *Isagoge*. He wrote:

> Medicine is divided into two parts, namely theoretic and practical. And of these two the theoretic is further divided into three, that is to say, the consideration of the naturals, the non-naturals, and the contra-naturals. From the consideration of these arises the knowledge of sickness, of health, and of the mean state, and their causes and significations.[9]

The same division of the knowledge of physic into the naturals, non-naturals, and contra-naturals became standard in the scholastic curricula, built not only on Joannitius but on the works of Avicenna, Isaac Judeaus, a few of Galen's and Hippocrates' works, and so forth.[10]

But, by *theoria* and *practica* the learned physicians did not have in mind the kind of differences we commonly do. If one turns to Avicenna, one finds a typically lucid explanation of this point.[11] Physic, like philosophy, he says, has both theoretical and practical parts, but the difference needs explaining in regard to medicine because people often have the wrong idea about medical practice:

> Thus, when in regard to medicine, we say that practice proceeds from theory, we do not mean that there is one division of medicine by which we know, and another, distinct therefrom, by which we act—as many examining this problem suppose. We mean instead that these two aspects are both sciences—but one dealing with the basic problems of knowledge, the other with the mode of operation of these principles. The former is theory; the latter is practice.[12]

In other words, Avicenna determined to show that, in both "theory" and "practice," physic is a science rather than an art. Avicenna contin-

---

9. Joannitius, *Isagoge*, in E. Grant (ed.), *A Source Book of Medieval Science* (Cambridge: Harvard University Press, 1974), p. 705. The "naturals" were those principles of nature that made up the human body; the elements, temperaments, humors, faculties, spirits, and so on. The six "non-naturals" were those things that affected the naturals; the air, food and drink, labor and rest, sleeping and waking, evacuation and retention, and passions and perturbations of the mind. The "contra-naturals" were the host of things that operated against nature, including accidents and remedies.

10. On scholastic physic, see Charles C. Talbot, "Medicine," in David C. Lindberg (ed.), *Science in the Middle Ages* (Chicago: University of Chicago Press, 1978), pp. 391–428; Faye M. Getz, "Medicine at Medieval Oxford University," in J. Catto (ed.), *The History of Oxford University* 2 (Oxford: Clarendon Press, forthcoming).

11. The following point was first suggested to me by Faye M. Getz. Since this paper was delivered, Nancy Siraisi's excellent book *Avicenna in Renaissance Italy: The Canon and Medical Teaching in Italian Universities after 1500* (Princeton: Princeton University Press, 1987) has appeared. Siraisi has a fine discussion of the meaning of theoria and practica to Avicenna and his sixteenth-century commentators, esp. pp. 97–100, 226–38.

12. Avicenna, *Canon*, in E. Grant (ed.), *A Source Book of Medieval Science*, p. 716.

ues, in this important passage, to argue yet further that both parts of physic are sciences, i.e., based upon principles of reasoning rooted in fundamental truths about nature. "Theory" is utterly certain; "practice," the intellectual elaboration of true principles, results in somewhat less certainty.

Theory is that which, when mastered, gives us a certain knowledge, apart from any question of treatment. . . . The *practice* of medicine is *not* the *work* which the physician carries out, but is that branch of *medical knowledge* which, when acquired, enables one to *form an opinion* upon which to base the proper plan of treatment. . . . Here the *theory guides an opinion,* and the opinion is the basis of treatment. Once the purpose of each aspect of medicine is understood, you can become skilled in both theoretical and applied knowledge, *even though there should never come a call for you to exercise your knowledge.*[13]

The practice of physic, then, concerned the ability to move intellectually from certain knowledge to opinion based upon that certainty; to associate the universal and the particular. The physician ideally did this based upon his skill in philosophy rather than upon his clinical experience, so that he could practice even if he saw no patients.

This tradition of the "practice" of physic, like a lawyer's or a cleric's practice of their sciences, was far from being an art rooted in mere clinical experience, much less the empirical skill of curing; and so the English university faculties made formal provision for study and debate, but none for clinical study.[14]

2. NATURAL HISTORY AND MEDICINE

But the idea was changing in the sixteenth and seventeenth centuries that academic physic was a science based on the established principles of natural philosophy. More and more learned men were arguing that the science of physic was at root rather the art of medicine, the *techne* of treating the sick with medicaments. Others argued that while physic ought to emphasize therapy more than it had, it could still be a science if it built upon new philosophical principles rather than the old. Naturally, there were disagreements among those who shared this viewpoint about which principles could establish the true foundations for a therapeutic science: chemical, "mechanical," or other principles. Among the various positions in favor of a renewed science of physic was a line of

13. Ibid., p. 716; my emphasis.
14. For more, see Phyllis Allen, "Medical Education in Seventeenth-Century England," *Journal for History of Medicine* 1 (1946): 115–43; Robert G. Frank, Jr., "Science, Medicine and the Universities of Early Modern England: Background and Sources," *History of Science* 2 (1973): 194–216, 239–69.

argument advanced by many of the English virtuosi and of the physicians most associated with the virtuosi. This proposition was that the new physic ought to be rooted in natural historical endeavors.

Sir Francis Bacon had been among those Englishmen who had declared that the "advancement" of science had to be rooted in new natural historical endeavors;[15] Robert Hooke penned a tract explaining the foundations of the new science as a natural historical endeavor;[16] and Robert Boyle wrote many treatises on "physiology" (or natural investigation) that focused on physic and "specific" medicines.[17] But it was a physician who authored one of the most popular English natural historical books of the century. A physician of Norwich, Thomas Browne (knighted in 1671), wrote *Pseudodoxia Epidemica*, which broke important ground by publishing an account of many things in nature presumed to be true but that were in fact false. He told the reader in his Preface that the model for his work was James Primrose's *De vulgi in medicina erroribus*,[18] a book that examined and dismissed many popular misapprehensions concerning medicine, while just a few pages earlier his first remark about mistaken beliefs concerned the "fruitlesse importunity of Uroscopy" burdening his time:[19] that is, the public coming to him for prognostications based upon the inspection of urines.[20]

It would certainly be wrong on our part to think that Browne's work was directly concerned with medical matters, when in fact it went far toward avoiding the subject covered so well by Primrose. Rather, Browne's book exhibited his approach to learning more than any immediate medical utility: he meant the book to be edifying rather than remedial.[21] His effort to pay close attention to detail, and to correct

15. For example, Francis Bacon, *The Plan of the Great Instauration,* prefixed to his *New Organon* (London, 1620). Of Bacon's published work, the largest amount (if the least read) is of a natural historical nature.

16. D. R. Oldroyd, "Some Writings of Robert Hooke on Procedures for the Prosecution of Scientific Inquiry, Including His 'Lectures of things Requisite to a Natural History,'" *Notes and Records of the Royal Society* 41 (1987): 145–67.

17. See esp. Robert Boyle, *Some Considerations Touching the Usefulness of Experimental Natural Philosophy, Propos'd in Familiar Discourses to a Friend, by way of Invitation to the Study of it* (Oxford, 1663), which is addressed to the five parts of the "physical" institutes, and *Of the Reconcileableness of Specific Medicines to the Corpuscular Philosophy* (London, 1685).

18. James Primrose, *De vulgi in medicina Erroribus Libri quatuor* (London, 1638); and *Popular Errors, or the Errours of the People in Physick,* Robert Wittie (tr.) (London, 1651).

19. Thomas Browne, *Pseudodoxia Epidemica: or, Enquiries into Commonly Presumed Truths* (London, 1646), "To the Reader."

20. For other physicians attacking the inspection of urines without the patient being present, see Peter Forrest, *The Arraignment of Urines* (London, 1625); Thomas Brian, *The Pisse-Prophet* (London, 1637).

21. Browne's views were probably formed during his medical studies at Montpellier, Padua, and Leiden (where he took his M.D. in 1633), places that promoted a natural historical approach to nature; he did not refer to Bacon as an inspiration for his work.

error, was deeply affected by a view of learning spelled out in the opening sentences. If truth were an active principle,[22] "we could be content, with Plato, that knowledge were but Remembrance. . . . [But] to purchase a clear and warrantable body of Truth, we must forget and part with much we know."[23] That is, we have to give up our common assumptions and inquire into everything anew. To do so demands far more than deep philosophical discourse; it demands labors in nature's own garden, that is, examining the particulars closely.

Browne's project suggested not only an attention to detail but a task that would be open-ended for many years. Such a sentiment was clearly voiced by that other great mid-century English natural historian and virtuoso, Izaak Walton. His book is still warmly regarded for its sentiments about fishing. But his intention was not only to speak of the edification and spiritual peace brought by the "experience" of angling, but to convey a wealth of information about the sport and fish themselves:

I undertake to acquaint the Reader with many things that are not usually known to every Angler; and I shall leave gleanings and observations enough to be made out of the experience of all that love and practise this recreation, to which I shall encourage them. For Angling many be said to be so much like Mathematicks, that it can ne'r be fully learnt; at least not so fully, but that there will still be more new experiments left for the tryal of other men that succeed us.[24]

For some physicians, the new emphasis on natural history provided a foundation for a new kind of learning that would have direct utility for physic. The utility of natural history was contained both in the therapeutical improvements growing from a better knowledge of natural detail and in the way that it provided the foundations for a new kind of certainty in medical knowledge. One of Browne's correspondents, Christopher Merrett, is an excellent example of an English virtuoso-physician who argued for the central importance of natural history for physic.[25] Merrett was a stubborn defender of the rights of the academically trained physicians in London, throwing himself into the campaign during the Restoration to re-establish the prestige and authority of academically trained physicians over other practitioners, and into both the contemporary disputes between apothecaries and physicians and the

22. Browne's phrase was, "Would Truth dispense. . . ."
23. Browne, *Pseudodoxia Epidemica*, "To the Reader."
24. Isaak Walton, *The Compleat Angler* (1653) (London: Oxford University Press, 1935), p. 7, "To the Reader."
25. Much of Browne's correspondence with Merrett, begun after Merrett published his *Pinax Rerum Naturalium Britannicarum, Continens Vegetabilia, Animalia, et Fossilia, In haec Insula reperta inchoatus* (London, 1667) is printed in Thomas Browne, *Notes and Letters on the Natural History of Norfolk . . . from the MSS. of Sir Thomas Browne, with notes by Thomas Southwell* (London: Jarrold and Sons, 1902), pp. 57–89.

controversy between Henry Stubbe and the virtuosi.[26] He was also a strong proponent of the new philosophy. At Oxford he had been part of William Harvey's circle.[27] When he moved to London he took part in the "1645 group" of natural philosophers.[28] And from its founding until the end of the 1670s, he actively participated in the Royal Society.[29] Merrett's own book on natural history was among the "exceptional books by English authors," according to the Italian visitor Lorenzo Magalotti.[30] Merrett also published a translation of Antonio Neri's *The Art of Glass, how to colour Glass* in 1662, to which he added his own *An Account of the Glass-drops* (or Prince Rupert's Drops). He had his work on cold published as an appendix to Robert Boyle's *New Experiments touching cold* (1665). He presented at least six formal papers on natural history to the Royal Society.[31] And he headed up the Society's committee on the history of trades.[32]

This is how Merrett put the physician's task in a work of about 1680:

The word Physician, derived from the Greek *pusikos*, is plainly and fully rendred by the word Naturalist, (that is) one well vers'd in the full extent of Nature, and Natural things; hereunto add the due, and skilful preparation and application of them to Mens Bodies, in order to their Health, and prolongation of Life, and you have a comprehensive Definition of a Physician.[33]

26. Cook, *Decline of the Old Medical Regime*, pp. 162–80.

27. Robert G. Frank, Jr., *Harvey and the Oxford Physiologists: Scientific Ideas and Social Interactions* (Berkeley and Los Angeles: University of California Press, 1980), pp. 74–75.

28. See the letters of John Wallis on the background to the Royal Society, reprinted in Sir Henry Lyons, *The Royal Society 1660–1940: A History of Its Administration under Its Charters* (Cambridge: Cambridge University Press, 1946), pp. 8, 11.

29. Thomas Birch, *The History of the Royal Society of London* (London, 1756), vols. 1, 2; Michael Hunter, *The Royal Society and Its Fellows 1660–1700: The Morphology of an Early Scientific Institution* (Chalfont St. Giles, Buck., 1982), pp. 162–163.

30. Merrett, *Pinax Rerum Naturalium Britannicarum;* W. E. Knowles Middleton (ed. & tr.), *Lorenzo Magalotti at the Court of Charles II: His* Relazione d'Inghilterra *of 1668* (Waterloo, Ontario: Wilfrid Laurier University Press, 1980), p. 149. For a modern evaluation of Merrett's book, see Charles E. Raven, *English Naturalists from Neckam to Ray: A Study in the Making of the Modern World* (Cambridge: Cambridge University Press, 1947), pp. 305–38.

31. Christopher Merrett, "A Paper Concerning the Mineral Called Zaffora by Dr. Merrett found amongst Dr. Hook's papers by Mr. Waller" (Royal Society, RBO.RBC.9.360); "The Arts of Refining Lead" (RS, Cl.P.IX[ix]1); "Some Observations Concerning the Ordering of Wines" (RS, RBO, RBC.1.278; later published at the end of Walter Charleton's Discourses on the Wits of Men [1692]); "An Account of the Tynn Mines and working of Tinn in the County of Cornewall" (RS, RBO.RBC.2.119); "Observations concerning the Uniting of the Barks of Trees cut, to the tree itself" (RS, RBO.RBC.2.301). These works appeared between 1660 and 1675.

32. Birch, *History of the Royal Society* I, p. 439.

33. Christopher Merrett, *The Character of a Compleat Physician or Naturalist* (London, 1680?), pp. 2–3.

Merrett believed, then, that the physician should become what he called a "naturalist." If the physician did so, Merrett argued, he would know the foundations of physic with certainty and also find many new cures for diseases.[34] Being a naturalist instead of merely a philosopher would end the strong criticisms of physic being made by many people outside the profession who claimed to be superior to the physicians because they worked with things themselves instead of ideas.[35]

But perhaps the most famous mid-century application of natural history to medicine originated with neither Browne nor Merrett, but with Thomas Sydenham.[36] Sydenham became known as the "English Hippocrates"; the Hippocrates represented in the phrase was thought to be the epitome of the natural historian.[37] Sydenham's publications emphasized case histories (carefully noting the changing symptoms a patient experienced over the whole course of the disease), together with a discussion organized by season of the weather and diseases prevailing in a locality and what kind of constitutions were most affected over the course of a year. In fact, this method is perfectly exemplified by the Hippocratic work *Epidemics I*, with *Epidemics II* and *III* being further collections of case histories. Sydenham also became known for his support of the idea of specific diseases: that is, that each disease had a precise set of distinct symptoms, and could be classified according to these outward signs in the same way that plants could be known and arranged. According to one recent commentator on Hippocratic medicine, "At the centre of Hippocratic pathology is the concept of specific disease."[38]

The physician's task, then, according to Sydenhamian medicine, was to be a natural historian of disease: to examine the clinical cases carefully and exactly, to identify and classify the specific disease entity as one would any other natural object, to describe that species in its precise

34. The connection between Merrett's stress on natural history and his discovery of new treatments is nicely made by the fact that he later publicly advertised his cures: Barbara Simons kindly alerted me to Merrett's single sheet folio advertisement in the Bodleian Library, Rawlinson MS C. 419, fol. 17.

35. This is the argument of the preface to Merrett's *Pinax Rerum Naturalium Britannicarum*, as well as the thrust of his arguments against the apothecaries and Henry Stubbe: see Cook, *Decline of the Old Medical Regime*, p. 169; Cook, "Physicians and the New Philosophy."

36. David Reisman, *Thomas Syndenham Clinician* (New York: Paul B. Hoeber, 1926); Kenneth Dewhurst, *Dr. Thomas Sydenham (1624–1689): His Life and Original Writings* (Berkeley: University of California Press, 1966); Donald G. Bates, "Thomas Syndenham: The Development of His Thought, 1666–1676," Ph.D. dissertation (Baltimore, MD: The Johns Hopkins University, 1975).

37. See Wesley D. Smith, *The Hippocratic Tradition* (Ithaca, NY: Cornell University Press, 1979), pp. 13–60.

38. Paul Potter, *Short Handbook of Hippocratic Medicine* (Quebec: Les Editions du Sphinx, 1988), p. 40.

surroundings (or what we would call environment), and to carefully note the treatments given and their various effects. He wrote that there were two ways to improve physic: first by "a History, or Description of all Diseases, as graphically and naturally as possibly may be, and, secondly, by a perfect and stable Practice or Method respecting them."[39] He went on to quote Francis Bacon on the difficulties of natural history. Then, in describing how actually to carry out a natural historical program in physic, he listed four points:

> It is necessary that all Diseases should be reduced to certain and definite Species, with the same diligence we see it is done by Botanick writers in their Herbals. . . . [I]n writing a History of Diseases, every Philosophical Hypothesis that has . . . inveighed the Writers Mind, ought to be set aside, and then the clear and natural Phaenomena of Diseases, how small soever they are, should be exactly marked, as Painters express the smallest Spots or Moles in the Face. . . . It is necessary in describing any Disease to mention the peculiar and perpetual Phaenomena apart from those which are accidental and adventitious. . . . Lastly, the Seasons of the Year, which chiefly favour any kind of Diseases, are carefully to be observed.[40]

Such a natural historical approach had great utility "with respect to practice," especially "in comparison wherewith the nice Discourses, which nauseously stuff the Books of modern Authors are of no value."[41]

For a generation or more Sydenham's "clinical" teachings became the foremost example for many physicians of how to establish a new certainty in physic in both its principles and its therapies, based upon the application of natural historical methods. Many of Sydenham's followers stressed the "practical" knowledge of diseases and remedies following upon natural history over the study of natural philosophy. For instance, John Pechy, the translator of Sydenham's works, made several typical comments in the preface to one of his own books intended for practical physicians. "Romancing on the Nature or the Causes of Diseases" has obstructed the art of physic, "so that in some [authors] scarce a Page can be spared for the Cure, that which is the main of the Business being huddled up or touch'd on by the by."[42] That is, natural philosophy and its attendant dietetics were not the point of the "art of physic"; curing specific diseases was.

Another translator of Sydenham, William Salmon, argued that while some authors divided physic into five parts (physiology, pathology, se-

39. Thomas Sydenham, *The Whole Works of that Excellent Practical Physician Dr. Thomas Sydenham,* John Pechy (trans.) (London, 1696) sig. Av.

40. Sydenham, ibid., sigs, A2-A2v.   41. Sydenham, ibid., sig. A3.

42. John Pechy, *The Store-house of physical practice* (London, 1695) sig. A2.

miotics, hygiene, and therapy), he preferred a division into three parts: physiology, pathology, and therapy. That is, semiotics and hygiene were of no use to the modern "practical" physician who simply wanted to cure diseases. As Salmon went on to explain, the practical physician had to know a bit of physiology (in which category of natural knowledge he included not any theory but rather a bit of human anatomy and a greater knowledge of *materia medica,* the preparation of medicines, and pharmacology), and a lot more about specific diseases and therapies.[43] While the books of Pechy and Salmon were not intended for academic audiences, similar changes in the direction of natural history rather than natural philosophy are seen even in contemporary Latin texts rooted in the scholastic tradition of Avicenna. One such text, the *Fundamenta medicinae physico-anatomica,* had its origin in the curriculum of Louvain university.[44] Originally written by François van den Zype, or Zypaeus,[45] and first published in 1683,[46] it was altered and republished in London by Joannes Groenevelt.[47] So successful was this introduction to the science of physic that it had two entirely separate English translations in the eighteenth century.[48] As the second English translator put

43. John Dolaeus, *Systema Medicinale, A Compleat System of Physick, Theoretical and Practical,* William Salmon (tr.) (London, 1686): Salmon's preface.

44. The seventeenth-century medical statutes of Louvain required the teaching of the five medical institutes "iuxta seriem doctrinarum Avicennae": L. van der Essen, *L'Université de Louvain (1425–1940)* (Bruxelles: Editions Universitaires, 1945), pp. 253–54.

45. On Zypaeus, a teacher at Louvain, see Joannis Jacobi Mangeti, *Bibliotheca Scriptoribus Medicorum Veterum et Recentiorum* (Geneva: Perachon & Cramer, 1731), vol. 2, p. 699; CC. Broeckx, *Essai sur l'Histoire de la Medicine Belge Avant le XIX siecle* (Zaventem: Sequoia, 1981), pp. 114–15.

46. Francois Zypaeus, *Fundamenta medicinae physico-anatomica* (Brussels, 1683); his *Fundamenta* was republished at Brussels in 1687 and 1693, went through a fourth edition at Lyons in 1692, and yet a fifth (at Brussels) in 1737.

47. Johannes Groenevelt, *Fundamenta Medicinae Scriptoribus, tam inter Antiquos quam Recentiores, Praestantioribus deprompta, Quorum Nomina Pagina sequens exhibet* (London, 1714); *The Grounds of Physick, Containing so much of Philosophy, Anatomy, Chimistry, and the Mechanical Construction of the Humane Body, as is necessary to the Accomplishment of a Physitian: with the Method of Practice in Common Distempers* (London, 1715); and *Fundamenta Medicinae Scriptoribus . . . editio noviss* (Venetiis, 1743). A comparision of the versions of Zypaeus and Groenevelt shows that Groenevelt introduced only a few significant changes in the course of making Zypaeus's discourse into a dialogue between a teacher and pupil; but for the sake of brevity, what follows is from the 1753 translation of the second Groenevelt edition (see next note).

48. Johannes Groenevelt, *Fundamenta medicinae Scriptoribus, Tam inter Antiquos quam Recentiores, Praestantioribus . . . Secundum Dictata D. Zypaei, M.D. et Medicinae Professoris Eruditissimi in Academia Lutetiana.* Editio Seconda (London, 1715); *The Rudiments of Physick Clearly and Accurately Describ'd and Explain'd, in the most easy and familiar Manner, by Way of Dialogues between a Physician and his Pupil . . . First collected from the instructions of a celebrated Professor of Medicine in the Royal Academy of Paris: And since Improv'd from the Authors, Ancient and Modern by John Groenevelt* (Sherborne and London, 1753). The fact that the second translation was done without apparent knowledge of the first suggests

it, "Dr. Groenevelt by a most happy Genius, has contracted the whole Substance of Physick into so small a Compendium, that he hath rendered the Study of it both easy and pleasant." In doing so, he had written something far more than a book of medical receipts. It was because the book introduced the student of medicine to theory that "there has been nothing yet of this kind in our language."[49]

On a basic point, Avicenna and the *Fundamenta* are in complete agreement: both argued that the end of physic was two-fold. To quote the author of the seventeenth century text: "Physick is the Art of preserving Health, and restoring it, when lost; or it is that Science . . . by the knowledge of which Life and Health are preserved, or lost Heath Restored."[50] But in other respects, the *Fundamenta* differs from Avicenna. In the first place, like many of the textbooks of the seventeenth century, in place of what Avicenna had called *theoria* it put the "Institutes," or the five parts of what had been *practica*.[51] That is, what had been "theory" (the description of the elements, qualities, four causes, form and matter, naturals, non-naturals, and contra-naturals) is jettisoned after a few general remarks, while the parts of scholastic medicine that had been "practice" become the new "theory." As the translator explained, all the "Systems" of the ancients were "rigidly accomodated to the particular Problems of Philosophy then in Vogue," problems that were of no concern to modern readers.[52] Groenevelt wrote that previous doctrine had been changed by a revival of "the Doctrine of Hippocrates . . . in the Academies of France,[53] [and] by the Experiments of the Chymists." Physic was further "improved with the greatest Pains, by Observations made in Mechanics, Natural Philosophy, and Chymistry, without Regard to any particular Sect."[54]

But this generous and eclectic view meant that no particular theory on the frame of nature was offered. Instead, Groenevelt immediately remarked that the art of physic is acquired by means of "Observation and Reasoning." Observation must be of "all Things in the human

---

the continuing value of the Latin edition and the disappearance of the first English translation into private hands.

49. Groenevelt, *Rudiments of Physick*, 1753, pp. vii, viii.

50. Ibid., p. 17.

51. The five institutes were ordinarily taken to be physiology (which included a discussion of the elements), hygiene, pathology, semiotics, and therapeutics: Siraisi, *Avicenna In Renaissance Italy;* Avicenna, *Canon*, p. 101.

52. Groenevelt, *Rudiments of Physick*, 1753, p. vi.

53. See I. M. Lonie, "The 'Paris Hippocratics': Teaching and Research in Paris in the Second Half of the Sixteenth Century," in A. Wear, R. K. French, and I. M. Lonie (eds.), *The Medical Renaissance of the Sixteenth Century* (Cambridge: Cambridge University Press, 1985), pp. 155–74, 318–26.

54. Groenevelt, *Rudiments of Physick*, 1753, pp. 22–23.

Body, either well, sick, dying, or dead," while reasoning is "an accurate Observation, by which those Things which pass in the human Body, unobservable by the Senses, are discovered and demonstrated."[55] Such a view of observation and reasoning is only very slightly more "rational" than the less academic views of Pechy and Salmon. Observation and "experience" provide a foundation for practice; physic is now rooted not in philosophical principles but in natural historical investigation. Instead of philosophy, then, the book began with a discussion of the five institutes, which had been Avicenna's *practica*.

Then, where Avicenna had placed *practica*, Groenevelt moved immediately to a description of various therapies, which Avicenna had not considered part of the science but the art of physic.[56] In other words, the new academic account of learned medicine dropped any discussion of foundational natural philosophy (the old "theory"), began with what had been the old "practice" (the five institutes), and elevated the knowledge of the empirical details of disease and drug lore to the rank of *practica*. The principles by which one could preserve health (the old *practica*) had become the new *theoria*, while mere empirical details of therapy had become the new *practica*: a division of theory and practice more like that we would expect today.

Apparently, then, even academic textbooks were beginning to treat physic more like medicine. Academic physic still placed weight on *theoria*, for the five institutes remained as subjects to be mastered by study and discourse. Among the five institutes, hygiene (understanding how one should live in order to prevent illness) still preceded therapy. But such doubt had been cast on the principles of natural philosophy that they were no longer taught as the necessary propaedeutic to understanding that part of nature concerning the physician. The five institutes remained the last bastion of scholastic *theoria* in physic. More and more, even universities began to teach what Avicenna had considered to be the art of medicine rather than the science of physic, the knowledge derived from experience rather than the philosophical search for causes that had formerly carried the presumption of certainty, an experience oriented far more toward therapeutic management than preventive advice.

The attack on physic by nonacademic practitioners clearly picked out

55. Ibid., pp. 22–23.
56. Siraisi finds Santorio Santorio asserting in 1625 that the task of the medicus "was not to treat individuals but to treat diseases; hence an effective medicine should be understood as one that cured the same disease in any number of people, an idea that gave therapy a universal [and hence "scientific"] aspect" (Siraisi, *Avicenna in Renaissance Italy*, p. 237).

preventive physic as the last place where theory was still supposed to give the learned physician an advantage over his rivals. These "practical" men stressed, instead, the importance of an experienced mastery of the details of medical therapy alone. One author simply stated that "Preventive Physick [is] a cheat, and a trick to get Money by."[57] Another argued that according to the new manner, medicine was divided into two parts; but instead of *theoria* and *practica*, both of which are to be mastered by study, he divided medicine into the prophylactical and the therapeutical. While prophylaxis, which depended on dietetics, might "in theory" prevent disease, "in practice" regulating dietetics exactly enough to prevent disease was impossible, he wrote. Therefore, this author's first chapters were devoted to a criticism of five of the six non-naturals (all except exercise); the rest of the book argued that disease was caused by an improper fermentation of the blood, and that the author had two sovereign remedies to promote fermentation, red coral and steel. He ended with this advertisement: "The true prepared coral and sugar of steel to be sold by Mr. Nathaniel Brook at the Angel in Cornhil and Mr. Simon Miller, Stationer, at the Star and Bible at the West end of St. Paul's Church."[58] The match between contemporary medical "empiricism" as both an attack on the last vestiges of academic physic (an understanding of individual hygiene via the non-naturals) and salesmanship for specific drugs could hardly be clearer.

Most seventeenth-century medical books in vernacular English stressed therapy, especially a knowledge of curative remedies. A great many of them, Pechy's and Salmon's included, also promoted the skills or the remedies of the practitioner who published the book. The intellectual assault of the new philosophy on scholasticism gave them added cachet: like Sydenham, many took to quoting from Bacon or Boyle.[59] The nonacademic authors tended to stress the empirical details of curative therapies, most commonly the drugs that they recommended. While regimen remained an important element in one important vernacular genre,[60] in it the new advice about regimen was good for everyone rather than tailored for the individual's unique temperament. The

57. Robert Godfrey, *Various injuries and abuses in chymical and Galenical physick; committed both by physicians and apothecaries detected* (London, 1674), p. 199.

58. Richard Browne, *Coral and Steel: A most Compendious Method of Preserving and Restoring Health. Or, a Rational Discourse, grounded upon Experience* (London, 1660).

59. For example, see Marchamont Nedham, *Medela Medicina. A Plea for the free Profession, and a Renovation of the Art of Physick* (London, 1665).

60. Virginia Smith, "Physical Puritanism and Sanitary Science: Material and Immaterial Beliefs in Popular Physiology, 1650–1840," in W. F. Bynum and R. Porter (eds.), *Medical Fringe and Medical Orthodoxy 1750–1850,* (London: Croom Hel, 1987), pp. 174–97.

older connection between the universals of nature and the particulars of the specific constitution were not necessary anymore, because the rules were general enough to apply to everyone.

In this new medical literature, unlike the literature on physic, the authors tried to derive universal principles from the groundwork of their experiences. The empirical "facts" had become more certain than the principles of natural philosophy—hence the privileged place of the natural historical method among many physicians. Only a well-prepared and knowing mind could discern the true from the untrue "fact," the natural historians argued; only they could derive useful rules for treatment.

3. CONCLUSION

By the end of the seventeenth century, then, the attack on learned physic had succeeded almost entirely. The foundational principles of natural philosophy ("physical" *theoria*) from which scholastics could derive rules for understanding individual cases (*practica*) had been dropped. Even the last vestiges of academic learning in physic, the rules of individualistic hygiene, were under attack from those who privileged experience-derived medical therapy. As physic declined, the clinic, where medicine could be learned through experience, became essential to the training of medical practitioners. Even studying books could help the educated physician, not by supplying philosophical certainty, but by supplying examples of previous cases, by extending the learned man's clinical experience.[61] Physic had become something more like our medicine.

Viewed through the eyes of the university-educated physicians, then, there was indeed something that approaches a "scientific revolution" in the early modern period—although it occurred over a long period. That fundamental change was connected to the rising importance placed by physicians on the experience of nature rather than on its universal principles: on natural history rather than natural philosophy. The scientific revolution occurred not so much in the details of natural history or natural philosophy (however important these precise changes were) but in a reordering of intellectual values. The revolution came in giving primary intellectual value to those things that had formerly been valued less. Whereas in the scholastic tradition certainty had been found in the principles of philosophy, in the natural historical

---

61. For example, see John Freind, *The History of Physic* 1 (London, 1725), pp. 309–10.

tradition certainty was found in the investigation of the "facts" of nature. In physic, certainty no longer stemmed from the study of natural philosophy but from the study of what had been "practice"—even, for some people, in the study of therapy alone.

The significance of this transformation in the categories of medical knowledge is fundamental to understanding the transformations wrought in physic by the "new" philosophy. The new philosophy was new because it tended to place a knowledge of natural history (or "practice") close to the top of the hierarchy of knowledge, and in so doing, shifted the content of the established meanings of "theory" and "practice" to those that we are more comfortable with today. Thus, the new philosophy did not emphasize mere empiricism; it rather emphasized the ability to engage in what the scholastics had called "practice," that is, the ability to connect universal and particular; but when universals were in doubt (as they were), they had to be derived from the particulars, not vice versa. The point is that, by emphasizing the "natural historical," "practical," or "art" at the expense of the "natural philosophical," "theoretical," or "science," the new philosopher-physicians helped cascade the knowledge of physic down the ladder of certainty, with things of formerly less intellectual value becoming more important. Physic was becoming medicine.

The natural historical endeavors of the physicians, then, do not arise simply out of the botanical interests of people who still found medicaments in plants; they reflect a larger shift in intellectual values, a shift that had implications of great magnitude. Much of the new philosophy, we might say, had to do with paying closer attention to natural historical details, for the sake of edification as well as for utility. While some physicians encouraged these endeavors, they did so, not so much because there was a necessary connection between curing and collecting, but because they responded to the changing intellectual climate: they responded to the growing search for certainty in natural events rather than in philosophical principles. Perhaps we should not separate their natural historical efforts from the new philosophy as a whole.

4   A New Science of Geology in the Seventeenth Century?

ROGER ARIEW

A brief outline of the history of twentieth-century views concerning the continuity or discontinuity of medieval and early modern science would look roughly as follows: Before Pierre Duhem—that is, before a *mature* Duhem, or before 1906, the year of publication for vol. 2 of *Les origines de la statique* and vol. 1 of *Etudes sur Léonard de Vinci*—most historians of science accepted the doctrine of a gap between ancient science and modern science, or a thesis of noncontinuity between the theories of medieval and modern science. Duhem and others showed the thesis false by documenting medieval science in great detail, demonstrating that much of it was subsumed into modern science, often without the medieval achievements being credited, and sometimes while they were being denied altogether. So the historiographical thesis of noncontinuity was transformed; most historians accepted the existence of a medieval science, but some rejected its influence: when confronted with similar or near-identical theories, they held steadfast to their thesis, changing it into a thesis about methodology. What was different about modern science, they argued, was its espousal of an empirical methodology, an emphasis on observational evidence (or experiments) and hypothetico-deductive form, or its rehabilitation of mathematics—as the language of nature or as the model for certain sciences.[1] In its most radical form, the thesis could hold that there is a discontinuity between medieval and modern science even when their theories are identical, because there is a discontinuity between the methodologies of medieval and modern science. Naturally, the thesis of methodological discontinuity has also been denied by many able his-

---

1. Spokesmen for modern science such as René Descartes can provide evidence that this is how many early modern scientists viewed themselves and their relations with their predecessors—cf. the *Letter to Mersenne,* 28 October 1638, in Descartes, *Oeuvres* 2 (Paris: Vrin, 1964–74), pp. 38off.

torians—Wallace and Crombie, for example.[2] So the thesis was transformed again; in the fashionable language of revolution and incommensurable paradigms, seventeenth-century science is said to be discontinuous from earlier science because of changes in some of the epistemic values of science, or even because of changes in its social milieu, that is, because of the institutionalization of science in the seventeenth century.

Naturally, the same would hold true for geology. Most commentators depict geology as a science that begins in the seventeenth century. In the older secondary literature, writers stressed the difference in methodology between the seventeenth-century geologists and earlier naturalists. Such thinkers as Nicolaus Steno, proclaimed by most as the father of modern geology, were said to be more empirical, to have relied more on observations.[3] But this thesis could not withstand much scrutiny. It quickly became evident that even the opponents of the new geology—writers such as Athanasius Kircher and Niccolo Cabeo—cloaked themselves with the mantle of empiricism.[4] More recent works have stressed the emerging consensus of the seventeenth-century scientific community in the birth of geology. That is the theme of Paolo Rossi's latest book, *The Dark Abyss of Time, the History of the Earth and the History of Nations from Hooke to Vico*. In a series of questions laden with presuppositions Rossi asks: "How is it that the same *thing* could seem to some people a remnant, or a document in a history of nature, and to others, during the same years, as simply one of the many objects or the many forms that abound in nature? What were the effects on the image of man and the investigation of nature of the adoption of a chronological scale enormously vaster than the traditional one? . . . Why is it that only when the metaphysical hypotheses underlying Burnet's 'romance' were accepted could the books on modern 'geology' be written?"[5] Using the

2. See, for example, William Wallace, *Galileo and His Sources* (Princeton: Princeton University Press, 1984), and A. C. Crombie, *Robert Grosseteste and the Origins of Experimental Science 1100–1700* (Oxford: Oxford University Press, 1953).

3. See, for example, the foreword to the 1916 English translation of the *Prodromus*. Given that we are dealing with geology, I exclude discussion of the mathematization of nature as irrelevant here.

4. Cf. Thorndike, *History of Magic and Experimental Science* 8, chap. 29: Other Exponents of Experimentation (New York: Columbia University Press, 1923–58); or Cabeo, *Meteorologicorum Aristotelis commentarii* 1 [Rome, 1646], p. 254).

5. Rossi, *The Dark Abyss of Time* (Chicago: The University of Chicago Press, 1984), p. vii. The allusion to Kuhnian revolutions is unmistakable: to claim that the same fossil is seen by some as a remnant and by others as a natural object is to claim that the two groups hold incommensurable paradigms. For such extended perceptual talk in Kuhn, or for gestalt switches, cf. Kuhn, *Structure of Scientific Revolutions*, chaps. 8 and 10 (Chicago, 1962). Naturally, there are plenty of references to Kuhn and scientific revolutions throughout Rossi's preface.

language of "seeing as," of scientific revolution and incommensurable paradigm, Rossi represents the view that there is a new science of geology emerging in the second half of the seventeenth century, a science that can be read in the works of Thomas Burnet, Nicolaus Steno, Agostino Scilla, and G. W. Leibniz.[6]

I propose to investigate some related topics in the works of the contemporaries and immediate predecessors to late seventeenth-century geology: the earlier geology and its seventeenth-century successors referred to directly or indirectly by the later geologists. I shall use Leibniz's natural history as the point of reference. Leibniz is ideal for this kind of study because, unlike some of his contemporaries, he does not choose to conceal the influences on his thought; on the contrary, he seems to enjoy letting his reader know how widely he reads. Moreover, since Leibniz's main work on natural history, the *Protogaea*, was composed in the 1680s, it was written sufficiently late to reflect the work of the other early modern naturalists. I shall work backward through the authors Leibniz mentions, from the contemporaries he cites with approbation and uses as authorities to the contemporaries and older naturalists with whom he chooses to dispute, namely, the Jesuit natural philosophers—Athanasius Kircher, Niccolo Cabeo, and Paul Guldin, among others.[7] But I shall restrict my discussion to the following related (and representative) questions: (1) what are fossils and how are they produced? and (2) how is it that one can find fossils (or shells) on the tops of mountains? Ultimately, this question asks about the mechanism for the flood, or floods.[8] Moreover, I also propose to compare the Jesuit theories with those of the thirteenth- and fourteenth-century natural philosophers—of Albertus Magnus and John Buridan, for example—and with those of the late seventeenth century.[9] These comparisons should enable me to answer whether, and in what way, the geology of Steno, Leibniz, et alii, is a new science of geology.

6. See Burnet, *The Sacred Theory of the Earth* (London, 1684), Steno, *Prodromus* (1916), Scilla, *La vana speculazione* (Naples, 1670), and Leibniz, *Protogaea* (Göttingen, 1749); also Woodward, *An Essay toward a Natural History of the Earth* (London, 1695).

7. See Kircher, *Mundus subterraneus* (Amsterdam, 1644); Cabeo, *Philosophica magnetica* (Ferrara, 1629) and *Meteorologicorum Aristotelis commentarii* (Rome, 1646); Guldin, *Dissertatio physico-mathematica de motu Terrae* (Vienna, 1622) and *Centrobaryca* (Vienna, 1635); Becher, *Physica subterranea* (1669); Vasquez, *Commentariorum ac disputationum in primam partem Sancti Thomae* (Antwerp, 1590).

8. But it does not do so universally, as Rudwick points out in *The Meaning of Fossils*, chaps. 1 and 2 (Chicago, 1985); some thinkers, like Robert Hooke, argued that the flood was too brief an event to account for marine fossils in rocks far from the sea—cf. pp. 74–75.

9. Albertus Magnus, *De mineralibus* (Oxford, 1967); Buridan, *Questiones super tres primos libros metheororum* (1350) and *Quaestiones super libri quattuor De Caelo et Mundo*

1. FOSSILS

The collection of Leibniz's writings published in the eighteenth century by Louis Dutens, from the most readily available sources, gives three items dealing with fossils: a letter on fossils,[10] a report to the Académie des Sciences de Paris about fossils[11] (both items having been published by Leibniz during his lifetime), and the unpublished book-length manuscript called the *Protogaea*. All three pieces propound the same theory about the formation of fossils. I quote the *Protogaea:*

I do not pretend to demonstrate for now what cause could have thus transformed some earth into slate and introduced some metal into it. However, in the same way that, in our furnaces, our craft imparts the consistency of stone to clay, is it not possible that nature, by means of a more powerful fire, transformed various sorts of earths and mixtures into slate, or alabaster, or some other kind of stone, and further, at the same time, the metallic matter that was scattered in the silt, having been fused, came to occupy the voids left behind by the decayed flesh of the fish, a flesh that was easily destroyed by time and heat. The goldsmith's craft gives us something analogous, and it is with pleasure that I compare the hidden operations of nature with the overt labors of men. After having covered a spider, or some other insect, with some matter appropriate to this end, though leaving a narrow outlet, goldsmiths harden this matter with fire, then with the help of some mercury that they introduce in it, they drive away the ashes of the animal through the small opening, and in their place, they pour some molten silver through the same opening, and finally, breaking the shell, they discover a silver animal with all its assemblage of feet, antennas, and filaments with an astonishing similarity [to the original animal].[12]

Leibniz's cleverness is evident in this analysis of fossils. For Leibniz, fossils are, in some sense, the remains of animals; they are the real products of a natural furnace, the earth, products to be explained. As Leibniz himself indicates, his thesis is a conscious attempt to oppose the then fashionable views of Athanasius Kircher, Joachim Becher, and others, that fossils are mere games of nature (*lusus naturae*) produced by nature's power of making stones (perhaps simply by mimicking the forms about it)—the *vis lapidifica*—and requiring no further explanation: "Those who hold a different opinion from ours have let themselves be seduced by the frivolous accounts, set out in a somber fashion, in the

---

(Cambridge, 1942); also Oresme, *Livre du ciel et du monde* (Madison, 1968); and Albertus de Saxonia, *Quaestiones in libros de Caelo et Mundo* (Venice, 1492).

10. *Epistola ad autorem dissertationes de figuris animalium quae in lapidibus observantur, & lithozoorum nomine venire possunt*, in *Leibnitii opera omnia* 2, part 2 (Geneva, 1768), pp. 176–77.

11. *Mémoire sur les pierres qui renferment des plantes & des poissons dessechés*, in *Leibnitii opera omnia* 2, part 2 (Geneva, 1768), pp. 178–79.

12. *Protogaea*, sec. 18, p. 31.

writings of Kircher, Becher, and others, who speak of *marvelous games of nature* and of *nature's formative power [de miris naturae lusibus & vi formatrice].*"[13] The thesis is not original with Leibniz; he probably owes it to Agostino Scilla, to whom he refers.[14] Leibniz also shares his thesis with Nicolaus Steno who, for a short time, was his colleague at Hanover and whose authority he is pleased to invoke. Steno, Scilla, and Leibniz all hold the same general thesis that fossils are remains of animals (or, to put it negatively, that fossils are not games of nature). Yet they differ in how they account for the petrification of those remains: Steno thinks he can account for fossils by stratification and the effects of water; Scilla does not really propose any account; and Leibniz believes he needs the effects of fire.

Although the immediate context for the theories of Scilla and Leibniz on the formation of fossils is the works of Kircher and Becher, which treat fossils as the artifacts of a playful nature, those works do not make up the broader context. Leibniz limited his attack to denying that fossils are games of nature, going so far as to suggest that he could accept fossils as remains of creatures *transformed* by some petrifying virtue:

If someone refuses to admit that nature formed these stones by cooking them, and prefers to assume that after some silt has covered the fish, the silt was changed into stone, either by the effect of its own constitution, or by a kind of petrifying virtue, or by some other cause, and that afterwards, the compressed metallic matter came to occupy the mold formed by the still soft and penetrable body of the fish—although that is hard to understand—I would not deny this. I do not claim to establish anything in this respect, other than that *these impressions come from real fish.*[15]

Steno does not so limit his criticism; he tries to attack the thesis that fossils are produced by some kind of petrifying force that place possesses. He first examines the thesis that *glossopetrae* are produced by the earth and argues that "if we grant the earth the power of producing these bodies, we cannot deny to it the possibility of bringing forth the rest."[16] Similarly, with other bodies dug up from the earth, "if one should say that these bodies were produced by the force of the place,

13. *Protogaea*, sec. 29, pp. 44–45; for Kircher, see *Mundus subterraneus* 2, chap. 6 (Amsterdam, 1644). See also Leibniz, *Protogaea*, sec. 20: In which we show that the figures of fish imprinted into clay come from real fish and are not mere games of nature; and sec. 27: *Glossopetrae* [and other such things] . . . are teeth, shells, the remains and small bones of marine animals, and not games of nature.

14. 14.Cf. Leibniz, *Protogaea*, sec. 29, p. 45: "I invoke the testimony of a learned painter against these assertions." Scilla was an Italian painter; he had rejected Kircher's thesis by affirming that fossils are the remains of real creatures—Scilla, *La vana speculazione*, pp. 74–75 and elsewhere.

15. *Protogaea*, p. 33.   16. *Prodromus*, p. 9.

one must confess that all the rest were produced by the same force."[17] And if that is so, we should be able to ascertain "whether a fossil was produced in the same place in which it is found; that is, one must investigate not only the character of the place where it is found, but also the character of the place where it was produced."[18] Ultimately, Steno holds that "he who attributes the production of anything to the earth names the place indeed, but since the earth affords place, at least in part, to all the things of earth, place alone does not account for the production of the body."[19] The doctrine discussed by Steno is part of what is normally invoked by those who conceive of fossils as games of nature, but it is also invoked by others who hold that fossils are remains of creatures. In fact, most scholastics held that fossils are the remains of animals transformed by the petrifying force of the place.

The standard scholastic doctrine, as represented by the tradition of commentaries on Aristotle's *Meteorology* or such scholastic treatises as the *De mineralibus*, is that fossils are the remains of animals.[20] The scholastic tradition then has the difficulty of explaining how it is possible that these remains are petrified, or are constituted by some matter different from that of the original animals. The responses to such questions were seldom satisfactory: it is not obvious how (or even whether) the original animals could have left their form but not their matter or could have transferred their form from one matter to another. The view that had great currency is that the remains of the animals were turned into stone by the petrifying power of the place, the *vis lapidifactiva* or *virtute quadam minerali lapidifactiva* (as opposed to Kircher's *vis lapidifica*). That is the view shared by Avicenna, Albertus Magnus, and others.[21] Thus, the initial doctrine was that fossils are the remains of

---

17. Ibid.   18. Ibid., pp. 8–9.
19. Ibid., p. 15.

20. See, for example, Avicenna, *De congelatione et conglutinatione lapidum* (Paris, 1927), pp. 46–47; Albertus Magnus, *Metheororum* 4, 1, 379 b 7, or *De mineralibus* 1, i, chap. 2 and ii, chap. 8, pp. 13–14, 52–53. See also Duhem, *Le Système du Monde* 9, chap. 18 (Paris, 1914–59).

21. Albertus Magnus, *Book of Minerals* (Oxford, 1967), p. 52 is a typical passage: "It seems wonderful to everyone that sometimes stones are found that have figures of animals inside and outside. . . . And Avicenna says that the cause of this is that animals, just as they are, are sometimes changed into stones, and especially [salty] stones. For he says that just as Earth and Water are material for stones, so animals, too, are material for stones. And in places where a petrifying force is exhaling, they change into their elements and are attacked by the properties of the qualities which are present in those places, and the elements in the bodies of such animals are changed into the dominant element, namely Earth mixed with Water; and then the mineralizing power converts [the mixture] into stone, and the parts of the body retain their shape, inside and outside, just as they were before." See also Duhem, *Le Système du Monde* 9, chap. 18, sec. 6 and 9.

animals petrified by the power of the place. It is easy to see how this doctrine could evolve to become the doctrines of Kircher and Becher, that fossils are the creation of the power of the place mimicking animals, without there being any actual remains of animals. The views of Kircher and Becher, therefore, should be understood as attempts to improve upon the standard scholastic doctrines. They provide a ready answer for the obvious differences between fossils and living creatures, including the problem of the stony matter of the fossils. As a result, these views are not saddled with the same difficulties as the scholastic doctrines, but they achieve this status at the cost of severing links between creatures and fossils and rejecting any historical account for the genesis of fossils.[22]

In this broader context, the seventeenth-century doctrines of Steno, Scilla, and Leibniz should be considered, in part, as a return toward the older theories of Avicenna and Albertus Magnus, that fossils are the remains of animals, but with a different, mechanistic account (as opposed to an account based upon some kind of virtue, force, or power) for the process of petrification. The three seventeenth-century theories, though in disagreement about the process of petrification, are also developments within the mechanistic (or Cartesian) picture of the world. The evidence for Steno, Scilla, and Leibniz being indebted to Cartesianism or being counted as Cartesians is very strong. In the *Prodromus* Steno adopts the outlines of the corpuscular theory of matter: a body is an aggregate of particles; a fluid differs from a solid in having its particles in constant motion.[23] Steno even cites Descartes as an authority: "And Descartes also accounts for the origin of the earth's strata in this way."[24] Scilla is an eclectic, but there are clear signs of his having read Descartes, including the claim that it is important to question everything.[25] Leibniz is more problematic. Leibniz, of course, denies being a Cartesian, and does reject some elements of Cartesian mechanism; he even introduces some souls, forms, or forces into the account of body. However, those souls cannot transmigrate, so they cannot be used in the account of fossils. Moreover, in many other respects, Leibniz is clearly indebted to Descartes. He accepts much of Cartesian cosmology, including Descartes' vortex theory and account of the formation of the earth and sun.[26]

22. There is an even more sympathetic account of the rejection of the organic theory of the origin of fossils, in Rudwick *The Meaning of Fossils*, chaps. 1 and 2 (Chicago, 1985), esp. pp. 60–68.

23. *Prodromus*, pp. 10–11.   24. Ibid., p. 28.
25. See Rossi, *The Dark Abyss of Time*, chap. 4 (Chicago, 1984).
26. See Leibniz, *Protogaea*, sec. 3 and elsewhere.

2. THE FLOOD

Burnet, Steno, Scilla, and Leibniz all expound upon the much discussed question about the origin of shells upon the tops of the mountains; here is Leibniz's account:

In the same way that, at the beginning, everything was prey to fire before light was separated from darkness, we believe also that everything was submerged by the waters after the extinction of this conflagration. The fact is conserved by the monuments of our holy religion, and the most ancient traditions of various peoples are unanimous on this point; and even when this is not in our minds, the *traces left by the sea in the midst of the earth* would settle our uncertainty, for *there are shells scattered upon the mountains.*[27]

Leibniz's answer to the problem of the origin of the shells is the flood—that is, the Flood—the tops of the mountains were flooded at one time.[28] Leibniz does not even mention the once-popular answer that the shells are the garbage deposited by travelers during their meals.[29] Obviously, Leibniz is pleased to be able to give an answer in harmony with biblical accounts. But it is interesting to note that the question about the origin of shells on the tops of mountains becomes pressing only after one gives an account of those shells as remnants of sea creatures. (The question is not as pressing if one thinks of those shells as stones having only accidental resemblance to sea creatures.) Hence, the fact that the question is discussed equally by Burnet, Scilla, Steno, and Leibniz should not be a historical accident. Similarly, one can predict that the question has had numerous precedents in the scholastic literature (and that Burnet et alii are not terribly original in their metaphysical interests). In any case, the shells were deposited during the flood, so the question arises: *"Where could such a quantity of water capable of submerging the mountains*

27. *Protogaea,* sec. 6, p. 9.
28. But Leibniz does provide for the possibility of multiple floods, with the ungainly hypothesis of multiple caverns, some filled with air (*Protogaea,* sec. 6, p. 12).
29. On the other hand, Leibniz realizes that the answer must contain a fair amount of speculation, so he attempts to place limits upon the range of such speculation: "I am aware that there are some who push their conjectures to such an extent as to think that, when the ocean covered everything, the *animals* inhabiting the world today *were aquatic,* that they became *amphibious* in proportion to the receding of the waters, and that their posterity finally abandoned their primitive habitations" (*Protogaea,* pp. 9–10; Leibniz calls the hypothesis "Lucretian," in the 1710 letter, *Epistola ad autorem dissertationes de figuris animalium,* in *Leibnitii opera omnia* 2, 2, pp. 176–77). The limit of such speculation, according to Leibniz, is its close fit with the sacred Scriptures: "But, other than the fact that these opinions are in opposition to the sacred Scriptures, from which we should not deviate, the hypothesis, considered in itself, offers inextricable difficulties" (*Protogaea,* p. 10; the hypotheses that the globe had been covered by the sea and the *possibility* that terrestrial animals had descended from amphibious animals seems to have been accepted by Leibniz in the letter of 1710).

have come from, and how did this water recede in such a way as to leave part of
the earth bare[?]"[30] Leibniz gives the following tentative answer:

Some people, accounting for this phenomenon with more cleverness than co-
gency, derive this effect from the *displacement of the center of the earth* alone. In
this way, the inclination of heavy bodies changes direction; the surface of the
earth remains the same, but the height or depth of each place would be modified, for heights or depths are not determined in themselves but according to
their relation with the center.[31]

The hypothesis of the ebb and flow of waters depending upon the
displacement of the center of the earth is John Buridan's hypothesis or
its successor, a hypothesis that had great currency during the sixteenth
and seventeenth centuries in the debates of the Jesuit natural philosophers—Gabriel Vasquez, Niccolo Cabeo, and Paul Guldin, among others.[32] Leibniz seems extremely impressed by the hypothesis; his reasons
for rejecting it are interesting: "This hypothesis could hold if the seas
and mountains occupied separate regions of the globe, and were not
intermixed on the same hemisphere. Although, even in this case, one
could allow an *oscillation of the center,* directed successively in various
directions; this would allow elevations and depressions alternatively ev-
erywhere."[33] In other words, Leibniz could have accepted the hypothesis if it were the case that all lands are massed on one hemisphere; he
argues against the hypothesis on account of the discovery of the American continent (on the other hemisphere). According to the hypothesis,
the flood is produced by the displacement of the center of the earth
from the natural place of earth without a similar displacement of the
natural place of water. This displacement would cause a flood on one

30. *Protogaea*, p. 10.   31. Ibid.
   32. In vol. 2, chap. 12 (especially sec. 10–11) of the *Etudes sur Léonard de Vinci* (Paris,
1916), Duhem sketched Buridan's theory back from Leonardo da Vinci to Albert of
Saxony. In the *Le Système du Monde* 9, chap. 18 (Paris, 1914–59) (esp. sec. 13–17), Duhem
traces the theory back to John Buridan, Albert of Saxony's teacher at the University of
Paris, and forward to the sixteenth and seventeenth centuries. There are minor differences among the hypotheses of Buridan, Guldin, and Leibniz; however, all of them
share the thought that some motion of the earth can be brought about through the
displacement of the center of the earth, without necessarily bringing about any motion
of the sphere of water. Buridan's theory, adopted by Guldin, requires the motion of the
earth as a whole whenever any part of the earth moves. If a stone is moved, the center
of the earth would be displaced and would no longer coincide with the center of the
universe, so the whole earth would have to shift to remedy the defect. Cabeo thinks this
absurd and ridicules it; he calls the opinion "monstrous," and asserts "they seem to have
demonstrated through this argument a strange motion of the earth and a perpetual
fluctuation that even the flight of a bird would cause" (Cabeo, *Philosophica magnetica*, lib.
I, cap. XVIII [Ferrara, 1629], pp. 66–71). But Cabeo thinks the absurdity is due to
taking the center of the universe mathematically, instead of thinking of it physically,
with respect to its physical distance from heaven.
   33. *Protogaea*, p. 10.

hemisphere and a land to arise on the other hemisphere—that is, given the hypothesis, there cannot be a flood on both hemispheres at the same time. So the flood would not have been a universal phenomenon. But Leibniz also provides a minor emendation for the hypothesis, an oscillation of the center, as opposed to a simple displacement of the center, that would allow the hypothesis to account for the counterexample. Further, Leibniz even works out the mechanism for such an oscillation, and attempts to place it in the context of more modern theories; discarding the connections to the doctrines of natural motion and natural place, Leibniz asserts:

There are others who come to believe, *given the experiments with magnetic variation,* that another large body is enclosed within the earth, like an almond in its shell, a body having its own proper motion. As a result, in our research, we should consider the center for the *attraction of heavy bodies* in the same way that we consider the poles for *magnetic attraction;* [that is, we should consider that] this center is prone to changes together with the [enclosed] body that has not yet been completely fixed.[34]

Ultimately, Leibniz rejects the hypothesis because he does not find it sufficiently plausible; he thinks there is a more admissible or more plausible hypothesis that can account for the same phenomenon:

There is an easier explanation for where the extra waters shed by the earth have flowed. In fact, it is very possible that, in the beginning, *hidden outlets* were formed, and the water was received *in deep caverns and had penetrated inside the globe,* since the water necessary to overflow the present mountains, even assuming that their height is four-thousand German miles above sea-level, would still be only one-seventieth of the remainder of the globe. . . . I would not dare account for this through external causes, such as *a comet passing through* the neighborhood of the earth, or *the moon getting nearer* [the earth] and raising the waters by an attractive force. The change of direction of heavy bodies and the displacement of the center do not satisfy me either; and if we must insist on this inquiry, what appears most admissible to me is that the vault of the earth was broken where it offered the least resistance, and enormous masses were thrown into the depths where the water had been received, and these waters, chased violently from their retreats, were elevated above the mountains.[35]

The more admissible hypothesis accepted by Leibniz is Steno's, a return to what was a standard doctrine before Buridan, of outlets or caverns in which the waters covering the mountains had receded.[36] The account is also shared by Scilla and Burnet. The doctrine was used generally to

34. Ibid. Cabeo would not have accepted this emendation, since he denies that the earth is a magnet—see Cabeo, *Philosophica magnetica.*
35. *Protogaea,* pp. 11–12.
36. Steno, *Prodromus,* pp. 73–74.

explain the formation of mountains and the mechanism for a natural deluge, at least as early as Avicenna.[37]

Here again there is a return toward older theories. This time, however, the thesis of continuity is almost complete. The hypothesis being rejected could just as easily have been accepted. The hypothesis of a flood being caused by a displacement of the center of the earth could have fit into a Cartesian framework as easily as it had fit into a scholastic one. In fact, Leibniz's attitude toward the hypothesis is reminiscent of Nicole Oresme's reaction to it in the fourteenth century. Oresme called the hypothesis "une belle ymagination," and rejected it because he thought that the air would resist the motion of the earth.[38]

There is little that would warrant the talk of discontinuity, of a new paradigm, in our examination of related topics in medieval and early modern geology, unless, of course, it is the shift from a scholastic framework to a Cartesian one. In both of our cases, there was a return toward older theories. In the case about fossils, scholasticism could accommodate two different hypotheses in succession, or at the same time, while Cartesianism could countenance only one of them. This case does not warrant the kind of talk about "seeing" the same *thing* as a natural form or as a document in a history of nature. In the case of the flood, both scholasticism and Cartesianism could accommodate the hypothesis. The hypothesis was deemed plausible by both, but was not deemed as plausible as another hypothesis, which was also able to be accommodated by both.[39] Moreover, the kind of evidence normally given as grounds for a change in paradigm is missing. For example, it is clear that Leibniz considers his various hypotheses, whether of medieval or Cartesian origin, to be roughly similar. He does not suggest that his contemporaries use different methods, are better observers or more

37. Cf. Avicenna, *De congelatione et conglutinatione lapidum* (Paris, 1927), pp. 17–32; see also Duhem, *Le Système du Monde* 9, chap. 18, sec. 6 and 9 (Paris, 1914–59); for a seventeenth-century scholastic representation of such views, see Eustachius of Sancto Paulo, *Summa Philosophiae Quadripartita* 2, tract. ii, cap. 7 (Cologne, 1629), p. 145.

38. Oresme, *Livre du ciel et du monde*, book I, chap. 36 (Madison, 1968).

39. Similarly, I find much to challenge in the presuppositions of Rossi's other questions. Briefly, there were many differences of opinion about chronological scales (from Aristotelian doctrines requiring a potential infinity of time past, to Stoic doctrines of eternally repeating temporal cycles, to their medieval successors and critics), so that it is difficult to contrast the modern and medieval views as simply two chronological scales, one being "vaster than the traditional one." Obviously, I would begin my story earlier than Stephen Jay Gould begins his (in *Time's Arrow, Time's Cycle* [Cambridge, 1987]). Moreover, I think that Rossi's emphasis on Burnet's *Sacred History of the Earth* (London, 1684) is misplaced, unless, of course, it is an emphasis on Burnet's Cartesianism (which Rossi acknowledges). The enterprise of constructing naturalistic reconstructions of biblical events such as the flood is at least as old as some of the Church Fathers—cf. Duhem, *Le Système du Monde* 9, chap. 18, sec. 8 (Paris, 1914–59).

empirical (except, perhaps, with respect to microscopic observations). He is clearly impressed with some remnants of the older geology. That is evident in the respect he shows Buridan's theory of the cause of the flood (or floods) or its successor; he ultimately abandons this theory only after attempting to improve upon it, and only because he does not find it plausible.[40]

On the other hand, we should not go too far in denying noncontinuity. The denial of a strong thesis of noncontinuity does not entail an affirmation of continuity; moreover, continuity does not entail identity. A study of early modern geology and its antecedent theories is not likely to confirm Duhem's uncharitable comment that there is nothing novel in modern science, that modern thinkers are often mere continuators, and sometimes plagiarists.[41] At the very least, the older theories are set into more modern frameworks. Although one can show some continuity of theories at the micro-theory level, one might be able to show discontinuity of theories at the macro-theory level. Early modern geology is discontinuous from the older natural history to the extent that Cartesianism is discontinuous from scholasticism. Now, there may not be great novelty in Cartesianism (though that is another story), but at the very least one can claim that Cartesianism embraces novelty. Given that Cartesianism sees itself as novel, it might be more accurate to say that what is novel in modern science is the claim that novelty is an epistemic value, or what is novel is novelty itself.

40. It could be thought that I have been unfair to Rossi. One might argue that he is only talking about similarities and differences among contemporaries and that it is not his purpose to investigate the medieval origins of their debates—a fair criticism if all I had shown is that the seventeenth-century debates are continuous with older debates. The situation would be similar to two twigs on the same branch. One could talk about how the twigs are discontinuous at a given area, and I could point to their continuity, tracing them down the trunk. But that would be to miss the real significance of scholasticism for the seventeenth century. As Leibniz clearly indicates, the scholastic explanation is still a live option for him. Hence, one cannot be so cavalier about not being interested in the medieval origins of the debates. Whether one is interested in them or not, they should be accounted for (or at least their numerous successors in the seventeenth century should). They provide an intermediary between the alleged discontinuities, the artificial dualism of "document in a history of nature" or "form that abounds in nature." The seventeenth-century scholastic accounts are continuous with both positions, in some respects, *during the same years.*

41. Duhem, *Le Système du Monde* 7 (Paris, 1914–59), p. 3.

PART III

ASTRONOMY AND PHYSICS

5   Innovation and Continuity in the History of Astronomy: The Case of the Rotating Moon

ALAN GABBEY*

1. PROLOGUE

The cover is all that remains of a letter Isaac Newton received some time between 1719 and 1725 from Newton Chapman, a young relative on the Ayscough side of the family. Yet this cover is a document of considerable significance, for written upside down *recto,* in Newton's hand, is the following chauvinistic memorial:

> The Variation of ye Variation to ye English
> The circulation of the blood to ye English
> The Libration of the Moon
> Telescopical sights
> The Micrometer.[1]

*The "pre-Newton" parts of this study were first presented in a skeletal form at the XVI International Congress of the History of Science, Bucharest, 1981. Enlarged versions of the whole study were aired at the conference on "Newton and Halley, 1686–1986," Clark Memorial Library, Los Angeles, August 11–14, 1985, and at the conference on "Pierre Duhem: historian and philosopher of science," Virginia Polytechnic Institute and State University, Blacksburg, March 16–18, 1989.

I wish to record my gratitude to the Council of the Royal Society for a History of Science Research Grant, which enabled me to carry out the research involved in preparing the study; again to the Council of the Royal Society, to the Academic Council of Queen's University Belfast, and to the organizers and sponsors of the "Newton and Halley" conference, for financing my 1985 visit to Los Angeles; to the organizers of the Duhem conference for financing the 1989 visit to Blacksburg. Thanks are due also to those librarians and archivists who facilitated my researches, and to colleagues who advised and otherwise assisted: Dr. Godfrey Waller, Manuscripts Room, Cambridge University Library; Dr. Timothy Hobbs, Trinity College Library, Cambridge; Mr. P. Woudhuysen and Mr. A. Bennett, Fitzwilliam Museum, Cambridge; Ms. Janet Dudley, Royal Greenwich Observatory Archives; Mr. N.H. Robinson and Ms. Sally Grover, Royal Society Archives, London; Eric Aiton, Roger Ariew, Sir David Bates, Jim Bennett, Judith Field, Bernard Goldstein, George Huxley, Richard Jarrell, Nicholas Jardine, John North, Simon Schaffer, Victor Thoren, Sam Westfall, Robert Westman, and Tom Whiteside.

1. Cambridge University Library, Add. MS 3965 (13), f. 479r. The cover bears no

The last three items are also clearly "to the English." Here was Newton in his last years surveying with pride some of the scientific achievements of his nation: Henry Gellibrand's discovery of the secular variation in magnetic declination (for whose explanation Edmond Halley was later to offer his magnetic quadripole shell-and-nucleus hypothesis); Harvey's great contribution to physiology; William Gascoigne's invention of both the micrometer and telescopic sights (an important share in the development of both inventions being due also to Englishmen such as Richard Towneley, Robert Hooke, and John Flamsteed).

What of the English claimant to "the Libration of the Moon"? The libration in question is of the moon itself, so Newton would not have had in mind (say) the apsidal libration introduced by Jeremiah Horrocks in his lunar theory. Neither can he have been referring to the *phenomena* of lunar libration, since these were all known to have been discovered by foreigners. Galileo gave his account of parallactic lunar libration in the *Dialogo* of 1632, and in correspondence of 1637–38 (albeit probably unknown to Newton) announced further parallactic libratory effects. Libration in latitude and longitude was discovered by Hevelius, who produced the first maps of libration in *Selenographia* (1647). In 1648 Hevelius hit on the idea of explaining longitudinal libration by having the Moon face always its eccentric point, not the Earth. This idea occurred independently to Riccioli, who published it in *Almagestum novum* (1651), together with libration maps and a general survey of the subject. However, Riccioli was unhappy with the eccentric hypothesis, since the observed librations were not quite those predicted from the accepted value for lunar eccentricity, a weakness that Hevelius recognized, though he favored retaining the hypothesis until something better came along.[2] Newton was thinking therefore of a theory

date, but the address is to "The Right Worshipfull s^r Isaac Newton Neare Lesterfields," and Newton lived in St. Martin's Street (south of Leicester Fields) from September 1710 until January 1725. Also, Chapman "had grown into a servile young man" (Westfall) by 1719, and two other known letters from him to Newton are dated 16 April 1719 and 23 October 1725. Accordingly, this third letter probably dates from between 1719 and January 1725. These biographical details on Newton and Chapman are taken from Richard S. Westfall, *Never at Rest: A Biography of Isaac Newton* (Cambridge: Cambridge University Press, 1980), pp. 670, 857, 866. For the letters of 16 April 1719 and 23 October 1725, see *The Correspondence of Isaac Newton*, H. W. Turnbull, J. F. Scott, A. R. Hall, and Laura Tilling (eds.), 7 vols. (Cambridge: Cambridge University Press, 1959–77), vol. 7, pp. 33–34, 335. The editors of the *Correspondence* do not mention the textless Chapman fragment in Add. MS 3965 (13). I thank the Syndics of Cambridge University for permission to quote the fragment.

2. *Jeremiae Horroccii Liverpoliensis . . . Opera posthuma*, John Wallis (ed.) (London, 1672), pp. 467–68; 2d ed. (London, 1673), pp. 467–72. On Flamsteed's reinterpretation of Horrocks's lunar model see *The Gresham Lectures of John Flamsteed*, Eric G. Forbes (ed.) (London: Mansell, 1975), pp. 50–54; cf. D. T. Whiteside, "Newton's Lunar Theory:

that in his view did a better job of explaining the phenomena, and of the Englishman who first gave such a theory to the astronomical world. In short, he was thinking of himself.

2. INNOVATION: NEWTON AND LUNAR LIBRATION

The earliest version of Newton's libration theory appeared in the Appendix to Nicholas Mercator's *Institutionum astronomicarum libri duo,* which was published in 1676.[3] There Mercator presented the explanatory model for lunar libration that Newton explained to him in a letter and perhaps also verbally either in Cambridge or during one of Newton's visits to London for Royal Society meetings. I have been unable to discover exactly when Newton devised his libration model, or when he told Mercator about it. According to Edward Sherburne, in 1673 Newton had ready for the press *"divers Astronomical Exercises, which are to be sujoyned to Mr. Nicholas Mercator's Epitome of Astronomy,* and to be printed at *Cambridge."*[4] There are no "Exercises" by Newton at the end

From High Hope to Disenchantment," *Vistas in Astronomy* 19 (1976): 317–28; *Isaac Newton's Theory of the Moon's Motion (1702),* bibl. and hist. introd. by I. Bernard Cohen (London: Dawson, 1975), esp. pp. 62–80. Galileo Galilei, *Dialogue concerning the Two Chief World Systems—Ptolemaic and Copernican,* trans. Stillman Drake, foreward by Albert Einstein, 2d ed. (Berkeley and Los Angeles: University of California Press, 1967), pp. 65–67; *Le Opere di Galileo Galilei,* Ed. Naz., ed. Antonio Favaro, 20 vols. (Florence, 1890–1909); vol. 7, pp. 89–91; correspondence with Fulgenzio Micanzio and Alfonso Antonini, November 1637–April 1638, *Le Opere* 17, pp. 211–15, 230–31, 259–60, 269–71, 291–97, 305–06, 329–30. Johannes Hevelius, *Selenographia: sive, lunae descriptio* (Danzig, 1647), Cap. 8, pp. 204–72: pp. 235 et seq., maps at pp. 222, 226, 262. Giovanni Battista Riccioli, *Almagestum novum* (Bologna, 1651), vol. 1 (announced vols. 2, 3 not published), Pars 1, Lib. 4, Cap. 9, pp. 208–15, map at p. 204; Cap. 18, p. 239. Hevelius, *Epistolae II. prior: de motu lunae libratio, in certas tabulas redacto. Ad . . . D. Johannem Bapt. Ricciolum Soc. Jes . . .* (Danzig, 1654), pp. 8, 46–48 (letter dated Autumn Equinox 1654).

3. *Nicolai Mercatoris Holsati, è Soc. Reg. Institutionum astronomicarum libri duo, De motu astrorum communi & proprio, secundum hypotheses veterum & recentiorum praecipuas; deque hypotheseon ex observatis constructione: cum tabulis Tychonianis solaribus, lunaribus, lunae-solaribus, et Rodolphinis solis, fixarum, et quinque errantium; earumque usu praeceptis & exemplis commonstrato. Quibus accedit Appendix de iis, quae novissimis temporibus coelitus innotuerunt* (London: William Godbid, for Samuel Simpson in Cambridge, 1676), pp. 285–88. The Preface to the reader is dated 13/23 March 1675–76 from Westminster. Not surprisingly, Newton possessed a copy of Mercator's treatise: Trinity College Library, Cambridge, Newton's Library, NQ.10.52. There are no marks, no notes, no dog-ears, relating to the libration hypothesis on pp. 285–88.

4. *The Sphere of Marcus Manilius made an English Poem: with Annotations and an Astronomical Appendix, by Edward Sherburne, Esquire* (London, 1675), p. 116 (quoted in Westfall, *Never at Rest,* p. 258). The news about Newton's astronomical projects is dated 1673 and appears in "A Catalogue of Astronomers Ancient and Modern," which forms part (pp. 6–126) of the valuable "Astronomical Appendix." It is worth noting that if Newton's libration theory was indeed among the "divers Astronomical Exercises" of 1673, the evidence of Sherburne's *Sphere of Manilius* indicates that he had no knowledge of what the theory contained. Newton's theory depends centrally on the idea of an axially

of Mercator's *Institutiones,* but the libration model at least is an exercise in astronomical theory, and whatever other promised exercises may have fallen by the wayside between 1673 and 1676, it is possible that Newton explained the theory to Mercator in 1673 (or earlier), its publication in the *Institutiones* being perhaps the result of an agreement or understanding between them. Newton's letter to Oldenburg for Huygens of 23 June 1673, which is quoted and discussed below in a different context, strongly suggests that the theory itself was in existence by that date at the latest.

On the other hand, some unpublished manuscript evidence raises the possibility that Newton furnished Mercator with the details of the theory in 1675. Among the draft additions for the third edition (1726) of *Principia Mathematica* is the following passage from Prop. 17 of Book 3 (see also below):

... Hanc Librationis Lunaris Theoriam [a me acceptam *del.*] [anno 1675 a me acceptam *interl.*: anno 1675 *del.*] a me acceptam, D. Mercator in Astronomia sua, [initio anni 1676 edita *del.*] [ante annum 1676 compositam [?] [*one or two illegible words*] *interl. and del.*] initio anni 1676 edita [initio anni 1676 edita *interl.*] plenius exposuit [mihique attribuit *interl. and del.*]. ...⁵

Whether Newton's deletion of "anno 1675" signifies an uncertain recollection of the date of something that took place at least forty years pre-

---

rotating moon (see below in text), yet Sherburne makes no mention whatever of lunar rotation, even though he provides fold-out plates of (Hevelian) lunar libration (pp. 169–80), and surveys previous astronomers' estimations of the rotation periods of the sun, Mercury, Venus, Mars, Jupiter, Saturn, and even of some variable stars, their variability being attributed by Riccioli to "a circumrotation about their own proper axes" (pp. 165–67, 180–88).

5. Cambridge University Library, Add. MS 3965 (13), f. 446r. The form of words "a me acceptam" suggests a meeting or meetings between Newton and Mercator, yet we have the explicit statement in the *Principia* (Book 3, Prop. 17, Theor. 15) that Mercator's explanation of the libration theory was taken "ex literis meis": *Isaac Newton's Philosophiae Naturalis Principia Mathematica,* 3d ed. (1726) with variants, ed. Alexandre Koyré and I. Bernard Cohen, assist. Anne Whitman, 2 vols. (Cambridge, Mass.: Harvard University Press, 1972), vol. 2, p. 592 (hereafter cited as *Principia* (KC). *Sir Isaac Newton's Mathematical Principles of Natural Philosophy and his System of the World,* trans. Andrew Motte, ed. Florian Cajori (Berkeley: University of California Press, 1934), p. 423 (hereafter cited as *Principles* (MC)). Of course, communication by letter and by verbal discussion are by no means mutually exclusive. This draft addition for the third edition may be read in conjunction with the following remark Newton scribbled upside down on the first page of a letter, dated 26 March 1717, he received from a William Newton: "Sententiam meam de libratione utraque D. Mercator in Astronomia sua Anno 1676 [1676 edit] *apparently changed to* 1675 edita *changed to* 1675 scripta *changed to* 1676 edita]" (C.U.L., Add. MS 3965 (17), f. 638r; cf. *Correspondence* 6, pp. 381–83, where there is no mention of this scribble). Since Newton did not mention Mercator (in the context of lunar libration) until the third edition of the *Principia* (*Principia* (KC), vol. 2, p. 592), this note on a letter of March 1717 could constitute a modest contribution to the difficult question of when exactly Newton began to prepare the last edition of his masterpiece.

viously, or a decision to mention only the date of publication, it seems safe to link Newton's lunar libration theory, his first published contribution to astronomy,[6] with the years 1673–75.

The occasion for Mercator's presentation of the theory was his inclusion of the phenomena of lunar libration among those things "quae novissimis temporibus coelitus innotuerunt" (see note 3), which included also new stars, sunspots and solar rotation, the rotations of Mars and Jupiter, the satellites of Jupiter and Saturn, the phases of Mercury and Venus, and comets. Beginning therefore with an account of the libration phenomena, Mercator invites the reader (I paraphrase) to take a terrestrial globe and tilt it so that its north pole coincides with the zenith (its equator correspondingly coinciding with the horizon). Select some island lying on the equator, say Sao Tomé,[7] turn the globe so that the island lies at the south cardinal point, and look down at the island from a position vertically above. Now imagine Sao Tomé to be a spot at the center of the *lunar* face, and oscillate the globe about its axis so that this spot shifts along the horizon about 7° first to the left then to the right of the south cardinal point (figure 1). All visible parts of the moon will therefore be seen to move eastward or westward, in step with the motion of the spot. "Such is the libration of the moon eastward and westward [i.e., in longitude]," explains Mercator, "which is effected while the moon goes from apogee back to apogee, an eighth part nearly of the moon's diameter being traversed near its center."[8] Now hold Sao Tomé on the meridian, he continues, and block the globe's rotation, observing the island from the same position as before and still imagining it to be a spot on the lunar face. Oscillate the (north) pole along the meridian ring so that it moves 7° or 8° on either side of the zenith first away from then toward you. You will see the spot and the other parts of the globe withdrawing north or returning south. This nodding of the moon's face is the libration in latitude, which is (so Mercator claims) a little larger than the libration in longitude, and is observed as the real moon moves from ascending node back to ascending node. Mercator notes that although these two librations per se are not bound into an

---

6. This interesting fact about the theory has been noted by I. Bernard Cohen, "Newtonian astronomy: the steps toward universal gravitation," *Vistas in Astronomy* 20 (1976): 85–98.

7. In the Gulf of Guinea, 0° 10'N, 7° 0'E.

8. *Institutiones*, p. 285. Throughout this paper all translations are my own, unless otherwise stated. In fig. 1, EW represents the total observed libration of the spot S on the lunar equator ESWQ. The axis of rotation passes through the center O perpendicular to the paper, and the moon oscillates 7° on either side of the meridian plane OS. Hence EW = 2.OE sin 7° ≈ 1/8 × diameter of the moon.

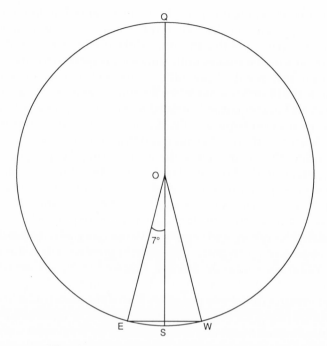

FIGURE 1. Libration ESW of spot S on lunar equator (see note 8).

association with the sun, the libration in longitude increases slightly when the moon is in quadrature.

There is finally what Mercator sees as a third libratory effect that does depend directly on the relative positions of the sun, moon, and earth. Employing now the terrestrial globe tilted so that the poles lie on the horizon ring [*horizon ligneus*], he asks the reader to illuminate one half of the globe so that the meridian ring [*meridianus aeneus*] becomes the terminator (that is, at half-moon). If you oscillate the globe a little on either side of the meridian ring so that its own meridian moves alternately into light and darkness, you will then have a representation of periodic changes in visibility, depending on phase and orbital position, of those parts of the lunar surface that lie within a maximum of 5° (Hevelius's approximate figure)[9] on either side of mean terminator positions.

To explain the causes of these librations ("as various as they are complicated"), Mercator then presents Newton's "*hypothesis elegantissima.*"

9. As one would expect, this figure equals the inclination of the lunar orbit to the plane of the ecliptic.

He invites the reader (again I paraphrase) to imagine that the terrestrial globe now represents a sphere centered on the earth and bearing the moon's orbital path, and that the lunar globe itself possesses poles and an axis about which it rotates uniformly with respect to the fixed stars once per sidereal month. Again tilt the north pole to coincide with the zenith (as in the illustrative model for libration in longitude), and let the newly created lunar equator, extended to the celestial sphere, lie in the plane of the horizon (equivalently, the plane of the celestial equator), with the lunar axis thereby pointing zenithward (northward).

Now incline the lunar orbit to the horizon ring (keeping the lunar equator parallel to the horizon plane), just as the ecliptic is so inclined (Mercator notes parenthetically that the inclination of the lunar equator to its orbit is perhaps not as large as the model implies)[10] (figure 2).[11] Imagine two small equal globes, each of them equipped with poles, axis, and first meridian, and suspended from (parallel) cords attached to their respective north poles. One globe represents a fictitious moon carried uniformly round the horizon ring (that is, in its own and the earth's equatorial plane) once per sidereal month, and which in the same time rotates uniformly just once, so that the plane of its first meridian always passes through the earth's center. The other small globe represents the true moon and is carried in its inclined orbit with non-uniform motion in such a way that its equatorial plane remains parallel to the horizon ring, that is, parallel to the celestial equator,[12] with the

10. In the model as described the lunar equator is always parallel to the celestial equator (equivalently, to the horizon ring) (fig. 2). Hence: inclination of lunar equator to lunar orbit = inclination of ecliptic to celestial equator + inclination of lunar orbit to ecliptic = 23.5° + 5° = 28.5°. It was not until c. 1685 that Cassini first produced an observationally grounded figure (7.5°: modern figure 6° 40′44″) for the inclination of lunar equator to orbit, so Mercator and Newton were clearly aware that the latter's 1673–75 libration model could not claim to be more than a qualitative account of the phenomena. Mercator's reference to the ecliptic suggests he is now thinking of an armillary sphere, which in fact serves his explanatory purpose better than (and is in fact equivalent to) his conceptually extended terrestrial sphere. Presumably the tilt of the North Pole to coincide with the zenith means that the ecliptic is also understood to be tilted so that it makes an angle of 23.5° with the coincident horizon and celestial equator. For Cassini's figure of 7.5° (minus the observations) see his "De l'origine et du progrès de l'astronomie, et de son usage dans la géographie et dans la navigation," *Mémoires de l'Académie royale des sciences 1666–99*, 9 vols. (Paris, 1729–33), vol. 8 ("Oeuvres diverses de M. I. D. Cassini, de l'Académie royale des sciences"), pp. 41–43.

11. Mercator provides no diagrams to illustrate Newton's libration model, on the curious grounds that "plane figures are scarcely sufficient for this matter, and in any event this handbook abounds in them already" (*Institutiones*, p. 286). I have supplied figures that I hope will be sufficient for the business of clarifying Newton's intentions.

12. Note that Mercator's posited congruence of the planes of the horizon and the celestial equator is (I think) introduced only to simplify the explanation. The substantive content in this component of the theory is that the equatorial planes of moon (true and fictitious) and earth are parallel. (If the lunar equator were taken to be in fact parallel

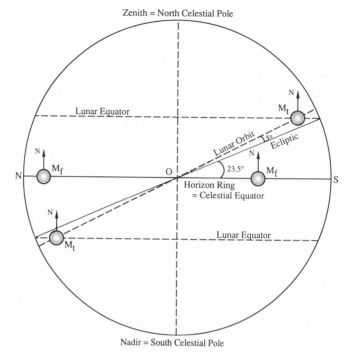

FIGURE 2. Newton's early libration model.

plane of its first meridian always parallel to the first meridian plane of the fictitious moon.

Mercator then continues:

Thus it happens that the fictitious moon [M$_f$], always turning the same face to us [E], is subject to no libration whatever [figure 3.]. But the true moon [M$_t$] precedes the fictitious Moon as it goes from perigee [P] to apogee [A], and so its first meridian [TZ] extends into the left-hand half of its disc by as many degrees [EŤZ] from the middle as there are between the longitudes of the true and fictitious moons [♈ÊT−♈ÊF=FÊT]. On the other hand, as the true moon descends from apogee to perigee it follows the fictitious moon, and the first meridian of the true Moon then withdraws from its midpoint to the right [ÊT$_1$Z$_1$], that is, all the spots incline toward the west. And since the difference between the mean and true longitudes of the moon turns out to be greater in the quadratures, on account of the ejection of the lunar system from the center of the earth, the librations in longitude are observed to be greater in the quadratures.[13]

---

to the real horizon, one would have the absurd consequence that the orientation of the axis of lunar rotation depends on the observer's latitude.)

13. *Institutiones*, p. 287. The lunar evection (discovered by Ptolemy) is the amount by which the difference between mean and true positions at quadrature exceeds the difference between mean and true positions at syzygy.

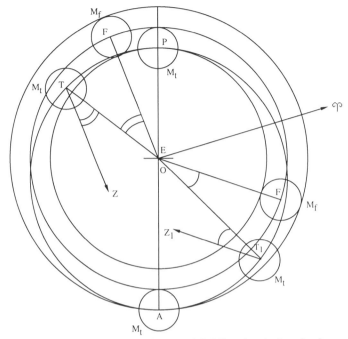

FIGURE 3. Newton's early libration model. Libration in longitude, employing true moon M_t and fictitious moon M_f.

As for the libration in latitude, this occurs when visible parts of the lunar surface seem to move toward the lunar north as the moon ascends into the northern half of its orbit (figure 4). The opposite happens as it descends into the southern half of its orbit. The third libratory effect is superimposed on the other two, and consists in variations in illumination of the lunar face as the sun and moon change position with respect to each other on their mutually inclined respective orbits.

There are no extant letters between Newton and Mercator, nor does there seem anything relevant in Newton's surviving manuscripts (at least those that are not hoarded within private walls, safe from the grubby paws of historians) that might tell us how closely this account of the libration theory of 1673–75 matches the theory as Newton told it to Mercator. Is this a verbatim report or an interpretation designed for insertion in an astronomical textbook? One wonders if the illustrative device of the two small globes, for example, is due to Mercator and intended as an explanatory aid for his readers or if it is Newton's own

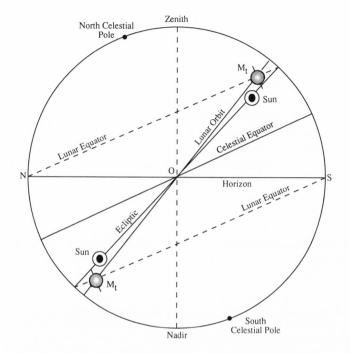

FIGURE 4. Newton's early libration model. Libration in latitude.

invention. It is a pleasing model, linking the three different key components of the new theory: the equation of the center,[14] the sidereal rotation of a moon in orbital motion subject to this traditional longitudinal correction, and the inclination of the lunar equator to both ecliptic and lunar orbit. Hence the employment of *two* small globes: readers of Mercator's *Institutiones* might have had difficulty seeing clearly what was involved had they been presented with a single globe illustrating the librating real moon. However, it is hard to believe that Newton himself needed such aids in visualizing the composition of motions entailed by his theory, which reappears without benefit of instrumental representation in its later improved versions in his own writings.

The first of these writings, in chronological order of composition,

14. Ptolemy's *prosthaphairesis* (of anomaly), the difference between mean and true longitude (TEF in fig. 3). Its maximum value for the moon (in the quadratures) was 7° 40′ for Ptolemy and Copernicus, 7° 28′ for Tycho Brahe. This explains the choice of "about 7°" for the angular swing on either side of the South cardinal point in the illustration of libration in longitude (see above).

was the early version of Book 3 of the *Principia,* which was written some time before the autumn of 1685 but first published in 1728 under the title *De mundi systemate liber Isaaci Newtoni* (an English translation also appeared the same year).[15] Here Newton coupled his libration theory, in abbreviated form, with affirmations that the earth, sun, Venus, Mars, and Jupiter also rotate uniformly with respect to the fixed stars, and he included figures for their sidereal rotational periods culled from reports of the observational work of Giovanni Domenico Cassini and others. Unlike the situation in the three editions of the *Principia,* however, he devoted a separate article to lunar libration (number 36 in the Motte-Cajori edition) in *De mundi systemate,* an indication perhaps of the relative prominence the problem had in his mind during the composition of the *Principia.*

Beginning with the sidereal rotation of the moon in one sidereal month, Newton notes that this rotation is therefore equal to the moon's mean motion, so:

Upon this account the same face of the moon always turns towards the center about which this mean motion is performed, that is, the exterior focus of the moon's orbit, nearly; and hence arises a deflection of the moon's face from the earth, sometimes towards the east, and other times towards the west, according to the position of the focus toward which it is turned; and this deflection is equal to the equation of the moon's orbit, or to the difference between its mean and true motions; and this is the moon's libration in longitude: but it is likewise affected with a libration in latitude arising from the inclination of the moon's axis to the plane of the orbit in which the moon is revolved about the earth; for that axis retains the same position to the fixed stars, nearly, and hence the poles present themselves to our view by turns, as we may understand from the example of the motion of the earth, whose poles, by reason of the inclination of its axis to the plane of the ecliptic, are by turns illuminated by the sun. To determine exactly the position of the moon's axis to the fixed stars, and the variation of this position, is a problem worthy of an astronomer.[16]

The fictitious moon of 1676 has been discarded, and the center of uniform angular motion is now located at the upper focus of an (elliptical) orbit in which the real moon moves in accordance with (one infers) the prescriptions of Horrocksian-Flamsteedian lunar theory.[17] The true moon of 1676 rotated uniformly via a purely kinematic in-parallel coupling (a standard dodge in traditional planetary theory) with the rotation of the fictitious moon. Now the real moon is part of a dynamical

15. See I. Bernard Cohen, *Introduction to Newton's Principia* (Cambridge: Cambridge University Press, 1971), pp. 109–15, 327–35.

16. *Principles* (MC), p. 580. The article dealing with the rotation of the other celestial bodies is on pp. 579–80.

17. *Principles* (MC), p. 577; D. T. Whiteside, op. cit. (note 2), pp. 318–20.

system, rotating "inertially" (see below) and moving under a gravitational force centered at one focus of the ellipse (the earth), from which it follows that the angular motion about the other focus is approximately uniform.[18] Newton does not present the improved theory in those terms, but given that the above passage was written during the composition of the *Principia*, it is clear that the emergence of dynamical considerations marks the difference between what are two versions of essentially the same kinematic description. However, as I shall argue later, the shift from a kinematic to a dynamical theory of libration bears a significance that goes well beyond its role in the development of lunar theory.

As for the libration in latitude, the assumption in 1676, no doubt pro tempore, that the lunar and terrestrial equatorial planes are parallel, was a natural one to make. Why not accept, as Copernicus and others did, the close kinship between moon and earth, the similarity in substance and behavior, and why not assume therefore a parallelism in their rotations, the moon's mirroring that of the earth by sharing the same direction toward the fixed stars? In 1685 that assumption has gone, the lunar equator is now inclined to the lunar orbit, and the axis of rotation maintains the same sidereal direction (or nearly so, presumably because of the precession of the lunar nodes). But Newton has no angle of inclination to offer, only the challenge of the last sentence in the passage.[19] Yet that challenge might already have been successfully met by the time he had written these lines, since it was in fact about 1685 that Cassini deduced a figure of 7.5° (modern value 6°40'44") for the inclination between lunar equator and orbit (see note 10).

In contrast to Mercator's account, and even to *De mundi systemate*,

18. For a body moving in an ellipse under an attractive force centered at one focus, at any point P in the orbit the component of speed perpendicular to the *other* focal radius varies directly as the length of that radius times $\cos^2 \theta$, where θ is the angle between the radius and the normal at P. Equivalently, the angular speed about the other focus varies inversely as the square of the normal at P. Since $\cos^2 \theta$ is close to unity for orbits of small eccentricity (such as that of the moon), and since the normal meets the major axis at a point lying between the foci, it follows that in these conditions the angular speed about the other (upper or empty) focus is nearly constant. It is odd that Newton does not demonstrate any of these propositions in any edition of the *Principia*, or in *De mundi systemate*; yet they are entailed by what I take to be the new dynamical context of post-1685 formulations of his libration theory, and indeed follow directly from Prop. 16, Theor. 8, of Book 1.

19. One senses the same challenge in the following remark from a version of Prop. 17, Book 3 of the *Principia* in an interleaf of Newton's own interleaved and annotated copy of the first edition: "Hanc autem esse causam librationis hujus significavi olim D. N. Mercatori, ut Astronomi [*changed from* ejus Lectores Astronomici] positionem axis Lunae ad planum Eclipticae et positionis variationem per observationes suas determinarent." *Principia* (KC), vol. 2, p. 592, variant for lines 6–17.

Book 3 of the first and second editions of *Principia* (1687, 1713) offered only the briefest statement of lunar libration theory, coupling it with the uniformity of planetary rotations in the single Proposition 17.[20] The theory per se was the same as that in *De mundi systemate*, except for the puzzling omission of the "nearly" at the beginning of the above passage qualifying the invariability of the lunar face presented to the upper focus.[21] But no relation between libration in longitude and the equation of the center was mentioned, and no rotation periods of the planets were quoted. As in *De mundi systemate*, there was no mention of the earlier version of the theory given in Mercator's *Institutiones*.

These omissions of a reference to the earlier version were repaired in the third edition of *Principia* (1726), in the same Proposition 17 (Theorem 15) of Book 3, where Newton also restored the planetary rotations of the outermost satellites of Saturn and Jupiter that Cassini had ingeniously deduced from his observations. There was still no mention of a relation between libration in longitude and the equation of the center, however, and there was an unexpected lapse from precision in the account of libration in latitude: "The libration in latitude arises from the latitude of the moon and the inclination of its axis to the plane of the ecliptic."[22] Since it is the inclination of axis to orbit that is the prime cause of the libration, the formulation in *De mundi systemate*, repeated in the first and second editions of *Principia* (above and note 20), was preferable. Apparently wishing to include in the third edition the libratory effect arising from the moon's latitude and its position relative to the sun, Newton gave a contracted version of a better form of words it seems he thought of using at one stage in the preparation of the third edition: "the inclination of the lunar axis to the plane of the orbit, and the plane of the orbit to the plane of the ecliptic."[23]

In no edition of the *Principia*, therefore, did Newton discuss or even mention a relation between libration in longitude and the equation of the center. Give the unending difficulties he had with the intractably complicated theory of the moon's motion, the only study that ever

20. "Propositio XVII. Theorema XV [XVI *in 1st ed.*]. *Planetarum motus diurnos uniformes esse, & librationem lunae ex ipsius motu diurno oriri.* Patet per motus legem I, & corol.
22. prop. LXVI, lib. I. Quoniam vero lunae circa axem suum uniformiter revolventis dies menstruus est: hujus facies eadem ulteriorem umbilicum orbis ipsius semper respiciet, & propterea pro situ umbilici illius deviabit hinc inde a terra. Haec est libratio in longitudinem: Nam libratio in latitudinem orta est ex inclinatione axis Lunaris ad planum orbis. Porro haec ita se habere, ex Phaenomenis manifestum est." *Principia* (KC), vol. 2, pp. 591–92, text + variants.
21. Due restoration was made in the third edition: ibid., p. 592, variant at lines 2–3.
22. Loc. cit., lines 4–6.   23. Loc. cit., variants for line 6.

made his head ache,[24] we can perhaps understand a wish to give a qualitative description of lunar libration to accompany published attempts to hammer out a mathematically accurate account of lunar position. As far as I am aware, Newton never attempted to integrate his theory of libration into his "lunar theory" as conventionally understood,[25] nor is there any sign in the available extant writings, including of course the original theory in *Institutiones,* that he addressed himself to the task of confronting the predictions of his libration theory with the observational data.

Yet this need not be taken to mean that he ever relinquished the relation explicitly stated in Mercator's *Institutiones* and in *De mundi systemate.* We have reason to believe otherwise from the evidence of the following draft (subsequently deleted, but see note 27) of Proposition 17, written with the third edition in mind, probably not long after the publication of the edition of 1713:

The diurnal motions of the planets are uniform and the libration of the moon arises from its diurnal[26] motion.

This we have numbered among the phaenomena. That these motions themselves must be uniform, however, appears more fully from Lex I and Coroll. 22 of Prop. LXVI of Bk. I. If the moon in its orb were borne around the earth with uniform motion without any latitude, it would look toward the earth always with the same face. [However,] its libration in longitude and in latitude arises from the inequality of this motion. And the libration in longitude is in fact equal to the error of the moon in longitude, which they call prosthaphaeresis or the equation of the center. The libration in latitude, on the other hand, is equal to the latitude of the moon.

Mr. Mercator expounded more fully this theory of the moon's libration, which he received from me, in his astronomy, which was published at the beginning of 1676.[27]

24. Whiteside, op. cit. (note 2), p. 324.

25. Indeed, in Newton's *A New and most Accurate Theory of the Moon's Motion* (1702) there is no mention of lunar libration at all, nor is there even the slightest hint that the theory of the moon's motion might be anything other than an exercise in positional astronomy. See Cohen's edition of this work (note 2).

26. How should *diurnus* be translated in this context? Clearly it does not mean "daily," or "lasting a day," but carries the transferred meaning of "every planetary or lunar day" or possibly "lasting a planetary or lunar day." However, since the central idea of the proposition is axial rotation, the presence of the sun is irrelevant as far as the rotation is concerned, and its relevance is strictly post hoc in the explanation of libration. This transferred use of *diurnus* (and "diurnal") to describe these new phenomena requires further examination. Suppose there were a planet circling the sun with one hemisphere always pointing towards it. Would such a planet possess uniform "diurnal" motion?

27. "Propositio XVII Theor. XV. Planetarum motus diurnos uniformes esse & librationem Lunae ex ipsius motu diurno oriri."

Haec inter Phaenomena numeravimus. Motus autem hosce uniformes esse debere [debere *interl.*] plenius patet per motus Legem I & Coroll 22 Prop LXVI Lib. I [Si Luna in Orbe suo [uniformi cum motu *del.*] circam terram [ferretur *del.*] uniformi cum

I have quoted this manuscript draft in its entirety because it raises two other important points, the second of which will lead us into the second part of this paper. First, the draft contains the third trial we have seen so far (compare note 5 and the corresponding quotation in the text) that Newton made to find a suitable form of words in which to remind readers of the third edition of the *Principia* of the date of the first appearance of his libration theory. There were (to my knowledge) two other such trials, not including the draft for Proposition 17 as it appeared in 1726.[28] What prompted these persistent efforts in Newton's old age to get right for first publication the wording of what we now realize is a priority claim? It seems to me that Newton had read or heard about (I cannot say when) the theory of lunar libration in longitude that Cassini had published in 1693. Cassini's theory seems to date also from circa 1675 and also employed, though none too clearly, the model of a moon rotating uniformly about an axis inclined at 7.5° to the orbital axis (and at 2.5° to the ecliptic axis), and moving nonuniformly in its orbit.[29] No doubt it was G. D. Cassini (whose striking contri-

---

motu sine Latitudine [Latitudine *interl.*] ferretur haec Terram eadem sui facie semper respiceret. Ab inequalitate hujus motus oritur Libratio ejus in Longitudinem & Latitudinem. Et Libratio quidem in Longitudinem aequalis est errori Lunae in Longitudinem quem Prostaphaeresin seu aequationem centri vocant, Libratio vero in Latitudinem aequalis est Latitudini Lunae].

Hanc Librationis Lunaris Theoriam a me acceptam D. Mercator in Astronima sua initio anni 1676 plenius exposuit." C.U.L., Add. MS 3965 (13), f. 446v.

Note that this draft appears on the verso of the draft for the same proposition quoted in part near the beginning of this study. The square brackets at "Si Luna" and "Latitudini Lunae" are Newton's and the deletion, which strictly speaking does not include the last sentence, is effected by a single stroke of the pen. It scarcely needs remarking that the deletion of a passage does not necessarily imply that the author no longer agrees with its contents. Whatever Newton's reasons for deleting this draft for Prop. 17, there is nothing in the passage or elsewhere in his writings to suggest that it was because he had doubts about the equality between longitudinal libration and the equation of the center.

28. *Principia* (KC), vol. 2, p. 592, variants for lines 6–17. C.U.L., Add. MS 3965 (13) f. 520r (draft for Prop. 17 of third edition).

29. Cassini's libration theory appeared in the work cited in note 10 (same page references). The first edition of this work appeared in *Recueil d'observations faites en plusieurs voyages par ordre de Sa Majesté pour perfectionner l'astronomie et la géographie, avec divers traités astronomiques par Messrs de l'Académie royale des sciences* (Paris, 1693), pp. 1–43. The text on the libration of the moon (pp. 34–36) does not differ materially from that given in the *Mémoires*. According to Jean Dominique Cassini, Giovanni Domenico's great-grandson, the latter's "Hypotheses circa motus librationes lunae" date from 1675, but he provides no further details: *Mémoires pour servir à l'historie des sciences et à celle de l'Observatoire royal de Paris, suivis de la vie de J.-D. Cassini, écrite par lui-même . . .* (Paris, 1810), p. 332. There is no evidence that Cassini published anything on lunar libration in 1675. He did however read a treatise on the subject to the Académie in 1685: *Histoire de l'Académie royale des sciences* (Paris, 1729–33), vol. 1, p. 441 (no text given). See also Jacques Cassini (Giovanni Domenico's son), *Élémens d'Astronomie* (Paris, 1740), Book 3, Chap. 3 ("De la libration apparente de la Lune, ou de la révolution de la Lune autour de son axe"), pp. 255–56. Giovanni Domenico Cassini's studies of lunar rotation and

butions to the observational study of ultra-lunar rotations Newton recognized and used) who was the implied foreign rival claimant to the libration of the moon to whom Newton wished to deny priority in the third edition of the *Principia* and, more pointedly, in the list of English scientific discoveries he jotted on the cover of the letter from Newton Chapman.

The second point arising from the above passage is that the first two sentences convey a train of thought that is not evident from a reading of the printed versions of the proposition. The opening sentence is tantalizingly ambiguous. Does it mean that at the moment of writing Newton had decided on or at least was contemplating for the third edition the inclusion of the rotation of celestial bodies among the *"Phaenomena"* of Book 3? Or is the sentence simply another way of saying these rotations "ex phaenomenis manifestum est," as Newton does in the proposition as given in all three editions?[30] Whichever reading one favors, the second sentence of the passage is a significant complement to it. Observation has shown (more correctly, it has been inferred from observation) that these celestial rotations are uniform: the laws of mechanics now show that they *must* be uniform. The 1673–75 libration model offered a kinematic account of the phenomena, without any appeal to mechanical law. In *De mundi systemate* Law I was at most only implied, via a reference to Corollary 22 of Proposition 66 of Book 1,[31] in the article on the uniform rotations of the planets, and accordingly in the article on lunar libration.[32]

---

libration evidently require serious investigation, especially the three laws (uniform lunar rotation, constant inclination of lunar equator to ecliptic, and coplanarity of axes of rotation, orbit, and ecliptic) attributed to Cassini in modern studies on the Moon. See *The Moon, Meteorites and Comets,* B. M. Middlehurst and G. P. Kuiper (eds.) (Chicago: Chicago University Press, 1963), p. 58; Zdenek Kopal and Robert W. Carder, *Mapping of the Moon, Past and Present* (Dordrecht: Reidel, 1974), pp. 50–51.

30. *Principia* (KC), vol. 2, pp. 550–51, 556–63, 591–92.

31. Law 1 is of course Newton's "principle of inertia." The relevant part of the corollary is the claim that if a homogeneous globe is impelled into both rotatory and translational motion in free space, then "since this globe is perfectly indifferent to all the axes that pass through its center, nor has a greater propensity to one axis or to one situation of the axis than to any other, it is manifest that by its own force it will never change its axis, or the inclination of its axis. . . ." Also, even if a second impulse acts on such a globe in a new direction, a "homogeneous and perfect globe will not retain several motions distinct, but will unite all those that are impressed on it, and reduce them into one; revolving, as far as in it lies, always with a simple and uniform motion about one single given axis, with an inclination always invariable. And the inclination of the axis, or the velocity of the rotation, will not be changed by centripetal force [arising from the rotation] . . . " (*Principles* [MC], p. 188). Law 1 is not mentioned anywhere in this corollary, but it is clearly present, since the spinning top is cited as empirical support in the statement of the law itself (ibid., p. 13).

32. "[35]. *The planets rotate around their own axes uniformly with respect to the stars; these motions are well adapted for the measurement of time* . . . and those motions are neither

In all three editions of the *Principia,* however, we have the explicit invocation, in the first line of the "demonstration" of Proposition 17, of Law I and the corollary as nomological grounds for the observed uniformity of lunar and planetary rotations (compare note 20). This recourse to Law I (directly and via the corollary) is as unavailing for Newton's purpose as the unexplained appearance of the spinning top as an empirical exemplification of the law itself,[33] but the point here is Newton's concern, in and after 1685, to present and justify his libration theory within the mechanical framework provided by his laws of motion, and thereby to contribute to his general program of elaborating a thoroughly mathematical and mechanical treatment of celestial motions.

However, those who are familiar with the development of Newton's mechanics will have realized that the story so far is not as simple as I have made it. Leaving aside the problem of lunar *libration,* Newton had a purely mechanical explanation of the everyday *invariability* of the moon's face some years before he devised the libration model of 1673–75 and gave a different mechanical explanation of the same phenomenon which appeared in all three editions of the *Principia* (though not in *De mundi systemate*). To take up this part of the story, and to evaluate its relation to the libration theory, we must look closely at what to me is the latter's most significant component, the one signal feature whose implications are of wider significance than simply its explanatory role in Newton's theory. I refer to the fact that Newton's libration theory depends centrally on the recognition that, in his own (translated) words, "the moon *rotates* around its axis . . . by a motion most uniform in respect of the fixed stars, viz., in 27d. 7h. 43 m., that is, in the space of a sidereal month" (reference in note 32; my italics).

accelerated nor retarded by the actions of the centripetal forces, as appears by Cor. 22, Prop. 66, Book 1; and therefore of all others they are the most uniform and most fit for the measurement of time; but those revolutions are to be reckoned uniform not from their return to the sun, but to some fixed star. . . . [36] *In like manner the moon rotates around its axis by diurnal motion; hence arises its libration.* In like manner is the moon revolved about its axis by a motion most uniform in respect of the fixed stars, viz., in 27d. 7h. 43m . . ."*Principles* (MC), 579–80.

33. See note 31. The required law in these cases is the principle of conservation of angular momentum, which Newton did not state in the *Principia.* As Herivel has shown, a fine start on the problem of rotating bodies is contained in the early (pre-1669) "Laws of Motion" paper, in which Newton introduces "the angular quantity of body's circular motion," its "radius of circular motion," and the persistence (if unimpeded) of a body's "real quantity of circular motion," and of its axis of rotation. Yet given the evidence of Law 1 and its spinning top, one cannot infer from this paper that Newton has in mind a new principle to set alongside Law 1; rather, he sees the persistence of free rotation as a simple consequence of the law. See John Herivel, *The Background to Newton's Principia: A Study of Newton's Dynamical Researches in the Years 1664–84* (Oxford: Clarendon Press, 1965), pp. 81–84, 208–18.

Some readers will wonder why I am fussing about Newton's recognition of something so obvious. The reason is that the sidereal rotation of the moon, that is the real per se motion with respect to the fixed stars, was anything but obvious prior to the mid-seventeenth century. To see why this was so, let us explore "pre-Newtonian" interpretations of the commonplace of everyone's experience: the fact that we always see the same side of the moon.

3. ARISTOTELIAN PORTRAITS OF THE MOON: ARISTOTLE TO MERCATOR

Suppose we forget for the time being what we have learned from the preceding section, or forget whatever we might have already known about the moon's behavior. Suppose further that we then look at the moon in an unreflectively empirical mood. We will see that the moon does not rotate on an axis. All sides of a rotating object are successively visible, but no one has ever seen the other side of the moon, so every human being in the history of our species has been able to *see* that it does not rotate. At the same time, every human has been able to see that the moon moves across the sky while looking at us with the same face. Accordingly, the natural and simplest solution to the puzzle of what carries the moon in its path was to imagine a solid spherical transporter to which the moon is rigidly attached in some way. Hence the (psychological?) origin of the celestial orbs.

We should try to imagine how the histories of cosmology, astronomy, and celestial mechanics might have turned out had the moon been always *seen* to rotate, just as the earth is in fact seen to rotate for an observer on the near side of the moon, had (for example) the moon's present state of captured equilibrium in the earth-moon gravitational field not come about. In such peculiar circumstances, it is most improbable that Aristotle would have championed the hypothesis of solid orbs carrying not only the moon but also consequently all celestial bodies in their periodic motions across the celestial vault. It taxes the mind to work out what Aristotle's cosmological doctrines might then have been, but it is safe to surmise that the solid Aristotelian orbs we are familiar with would not have been explanatory counters in a strangely and unpredictably different *De caelo*. Long ago Heath showed the way by suggesting "that the fact of the Moon always showing us one side was one of the considerations, if not the main consideration, which suggested to Aristotle that the stars were really fixed in material spheres concentric with the Earth."[34]

34. T. Heath, *Aristarchus of Samos, the Ancient Copernicus: A History of Greek Astronomy*

So is not the invariable face of the moon to the histories of the celestial sciences what the pensive Pascal surmised the length of Cleopatra's nose was to the general history of the world? Was this the face that launched humanity on over a thousand years of cosmologies wedded to the concept of deferent celestial spheres, all because it was thought that the moon moving in a simple geocentric orb does not turn on an axis, a belief that did not begin to be questioned until the mid-seventeenth century, after the dust of the Copernican Revolution had begun to settle?

Some will want to remind me that I am overlooking the strikingly different views in the *Timaeus* (34a, 39a–40c), where Plato argues that the fixed stars, as "divine and eternal animals" possessing intelligent souls, move with rotational as well as progressive motion. And Proclus assumed that Plato intended the same to hold for the planets, also possessed of intelligent souls. Yet Cornford implies that according to Proclus the moon in particular must rotate on its axis; otherwise the same face would not continually turn towards earth.[35] I can find no indication of this idea in Proclus's *Commentary*, nor in the *Timaeus* itself, nor anywhere else in Plato. Besides, if we recall that Aristotle's moon argument was itself advanced partly as an empirical refutation of the rotating stars of *Timaeus*, via an application of "the cosmological postulate," a telling idea suggests itself. We cannot freely speak of a rotating moon, since the immediate empirical evidence advises us otherwise: we can speak freely and imaginatively of rotating stars and sun, since the immediate empirical evidence does not advise us one way or the other.

So let us return to Aristotle's moon argument. In *De caelo*, Book 2, Chapter 8, he invokes the invariability of the moon's face to argue that the stars have no motion of their own but participate in the rotations of the spheres to which they are attached. The argument depends in part on the distinction between *whirling* or *rotating* (δίνησις) and *rolling* (κύλισις). The stars are spherical, and the two motions proper to the spherical as such are rolling and axial rotation, either of which would thereby be the motion of a star moving autonomously. But if the stars rotated, they would, Aristotle imagines, not change place, which is obviously counter to fact. "On the other hand it is equally clear that the stars do not roll. Whatever rolls must turn about, but the moon always shows

---

to *Aristarchus Together with Aristarchus's Treatise on the Sizes and Distances of the Sun and the Moon* (Oxford, 1913), pp. 234–35.

35. *Procli Diadochi in Platonis Timaeum Commentaria*, ed. E. Diehl, 3 vols. (Leipzig, 1903–06), vol. 3, p. 128; and *Proclus, Commentaire sur le Timée*, trans. and notes by A. J. Festugière, 5 vols. (Paris, 1966–68), vol. 4, p. 164. (See further A. E. Taylor, *A Commentary on Plato's Timaeus* (Oxford, 1928), pp. 225–26.) F. M. Cornford, *Plato's Cosmology: The Timaeus of Plato*, translated with a running commentary (London, 1937), p. 119.

us its face (as men call it). Thus: if the stars moved of themselves they would naturally perform their own proper motions; but we see that they do not perform these motions; therefore they cannot move of themselves" (290a 25–30).[36]

Contrary to Heath's accommodating interpretation of this argument (note 34), there is nothing in it to justify his assumption that when Aristotle says the moon does not perform "its own proper motions," which would be rotational, he also believes that the moon's participation in the motion of its sphere incidentally entails a real sidereal rotation about an axis in its own body. Aristotle does not mention any such rotation, and there are several reasons why he would not have done so. First, recall how the argument goes. Whatever rolls must rotate while moving from place to place, but the moon always presents the same face, therefore, it cannot be rotating, and neither can it therefore be rolling while changing position in the sky. Hence Aristotle can argue, via "the cosmological postulate," that *no* star rolls while being carried in its sphere.

Second, it follows from Aristotle's conception of local motion as change of place, defined as "the innermost motionless boundary of that which contains" (*Physica* 212a 20), that the Moon does not move locally per se, but only per accidens, since it is carried by its sphere, which does move per se (243b 15–20). Third, if the moon were rotating on its axis, what would cause the rotation? The moon has no per se motion, and "twirling [rotation] is a compound of pulling and pushing, for that which is twirling a thing must be pulling one part of the thing and pushing another part, since it impels one part away from itself and another part towards itself."[37]

I have lingered on Aristotle's account of the invariable lunar face because it provided the principal framework for subsequent discussions on the topic up to the time of Newton. The Platonic ideas mentioned earlier were an influential factor in sixteenth-century discussions on the rotating sun,[38] and perhaps also in the emerging awareness of the possibility of planetary rotation, yet the moon remained solidly Aristotelian in this respect until the mid-seventeenth century.

Let us examine some Medieval and Renaissance discussions centering on the Aristotelian *locus classicus*. In his commentary on *De caelo*, Averroes presents with approval a more elaborate account of Aristotle's

36. *On the Heavens*, ed. and trans. by W. K. C. Guthrie (Loeb Classical Library, London and Cambridge, Mass., 1939), pp. 188–89.

37. *Physica*, trans. R. P. Hardie & R. K. Gaye, in *The Works of Aristotle*, W. D. Ross (ed.), vol. 2 (Oxford, 1930): 244a 1–5.

38. See Michel-Pierre Lerner, " 'Sicut nodus in tabula': de la rotation propre du soleil au seizième siecle," *Journal for the History of Astronomy* 11 (1980): 114–29.

thesis of the stars' immobility *in loco,* and explores that point in the argument that assumes that what is true per se of one star is true of them all (the cosmological postulate), an assumption deriving from the more fundamental tenet that all stars are of the same species, with which Averroes counters Avicenna's opinion that the stars are merely of the same genus.[39] Albertus Magnus, in his *De caelo et mundo* (early 1250s),[40] also retails with approval Aristotle's argument, and presents at some length the same point of difference between Averroes and Avicenna. In Aquinas's commentary on *De caelo* (1271–74) the Aristotelian argument is given briefly and defended.[41]

However, the thirteenth century also saw new developments that complicated the issue in important ways. These developments arose from the debate between those who, following Ptolemy, advocated the use of epicycles in planetary theory and those who, following Ibn Bajja and al-Bitruji, claimed that astronomy could do without them, arguing their incongruity with the simple orbs of Aristotelian cosmology. According to Duhem, it was Roger Bacon, in his *Opus tertium,* who first thought of "une très forte objection" against the hypothesis of solid planetary epicycles entertained by the Ptolemaists, especially Ibn al-Haytham. If the moon were fixed in an epicyclical orb, then we ought to see all its sides. But we see only one side, so, concludes Bacon, "this difficulty cannot be avoided except by attributing to the [Moon] a proper motion about its center, which is contrary to Aristotle's teaching in *De caelo.*"[42] Bacon was therefore unsure as to which of the opposing camps had the better answer. Later, Levi ben Gerson was to use the same argument to reject epicycles *in toto* from astronomy.[43]

Two of Bacon's fellow Franciscans were in doubt as to the validity of the Ptolemaic approach. Bernard of Verdun, in his *Tractatus super totam astrologiam* (end of thirteenth/beginning of fourteenth century), argued that only Ptolemaic methods can save astronomical phenomena, and in the case of the invariable lunar face, he suggested that this partic-

39. *Aristotelis Opera cum Averrois Commentariis,* 9 vols. + 3 supp. vols. (Venice, 1562–74; Minerva reprint, Frankfurt-am-Main, 1962), vol. 5, ff. 131 recto–132 recto.

40. Lib. 2, Tract 3, Cap. 7–8: *Alberti Magni . . . Opera omnia,* gen. ed. B. Geyer, Albertus Magnus Institute of Cologne (Aschendorff, 1951-), vol. 5, Pt. 1 (ed. P. Hossfeld), pp. 155–61.

41. Lib. 2, Cap. 8, Lect. 12: *Sancti Thomae Aquinatis . . . Opera omnia,* Leonine ed., 16 vols. to date (Rome and Turin, 1882-), vol. 3, pp. 167–68.

42. Pierre Duhem, *Le système du monde: histoire des doctrines cosmologiques de Platon à Copernic,* 10 vols. (Paris, 1913–59), vol. 3 (2), p. 437.

43. Bernard R. Goldstein, "Theory and observation in medieval astronomy," *Isis* 63 (1972): 39–47. See also *The Astronomy of Levi ben Gerson (1288–1344): A Critical Edition of Chapters 1–20, with Translation and Commentary by Bernard R. Goldstein* (New York, Springer-Verlag, 1985), p. 193.

ular phenomenon would require the moon to be given an extra motion "by means of the motion of certain orbs containing the epicycle within them." However, Duhem notes that Bernard provides no further details on these orbs.[44] Richard of Middleton (died c. 1300) shared Bernard's enthusiasm for Ptolemaic astronomy and was more forthcoming, though unaccountably confused, about the additional motion required by the epicyclical moon. In his *Quaestiones* (c. 1281) on the *Sentences* of Peter Lombard, Richard seems to be happy with the idea of a rotating sun, is uncertain about the rotation of the five planets, and as for the moon:

Many find a proof of the motion of the moon about its own center, between the external and internal surface of the epicycle, in the fact that the spot on the moon never appears to us upside down [*sic*], that the part of the spot near the bottom at a given moment is not found near the top at another moment, and inversely. . . . In fact, if the moon did not move with its own proper motion, this spot would become transposed through the motion of the epicycle; the lower part would become the upper part, and inversely. The moon must therefore move about its own center in a direction contrary to the motion of the epicycle, and in such a proportion that the transposition the spot would undergo as a result of the epicyclical motion is exactly compensated by the moon's proper motion in the opposite direction.[45]

For a more closely considered account of the matter I turn to Jean Buridan. In *Quaestiones super libris quattuor de caelo et mundo* (c. 1340), Lib. 2, Q. 18, Buridan asks "whether the stars move per se or following the motion of their spheres." He notes that any presumed "circumgyratory" motion of a star about its own center would not be the motion by which the star revolves diurnally, because of the sense in which a rotating body does not change place, and he repeats Aristotle's moon argument against such circumgyration. Then he continues with the objection that on the contrary the moon *in an epicycle* must move with a circumgyratory motion; otherwise we would see a succession of different faces. However, note how this counterargument works:

The motions with which the moon is observed to move cannot be saved except by assuming an epicyle bearing the moon, as postulated by Ptolemy and other modern astronomers. But due to the motion of the epicycle, the moon must turn around [verti], unless the moon itself turns per se about its own center, so that it rotates [*vertatur*] [to face in] one direction just as much as it is turned round by the epicyle to [face in] another [figure 5].
From this controversy I draw kindred conclusions [*conclusiones copulativas*]. One is that either the moon does not have an epicycle as postulated by Ptolemy, or that the moon itself moves per se with such a circumgyratory motion. The

44. Duhem, op. cit., vol. 3 (2), p. 455.   45. Ibid., pp. 487–88.

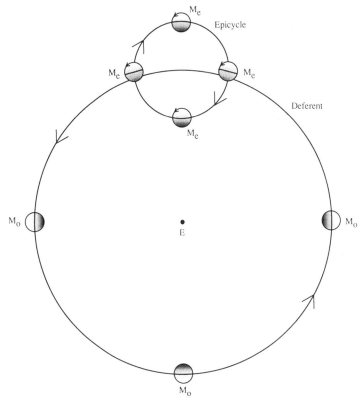

FIGURE 5. Epicyclical lunar model. To maintain same side facing earth, the moon (M_e) rotates in epicyclical orb in direction opposite to and with the same period as epicyclical revolution. Without epicycle, the same side of Moon (M_o) faces earth without rotation in deferent orb. (Not to scale.)

other conclusion is, equivalently, that if the moon has an epicycle, it does move per se in that way, and if it does not have an epicycle, it does not move per se in that way.[46]

Later Buridan argues that the moon cannot *roll* if an epicyclical lunar model be assumed: there would be a thousand rotations,[47] he claims,

46. *Iohannis Buridani Quaestiones super libris quattuor de caelo et mundo*, E. A. Moody (ed.) (The Medieval Academy of America, Pub. No. 40, Studies and Documents, No. 6: Cambridge, Mass., 1942), pp. 210–11.

47. A more modest estimate might be 17.6, the ratio of epicyclical to lunar circumference that follows from the figures given by Campanus of Novara in his *Theorica planetarum* (1260s): circumference of epicycle 105,400 miles (1 mile = 4000 cubits), circum-

entailed in the moon's rolling once round its epicycle. However, whether or not the moon rotates depends on whether or not the epicyclical model is valid, and this is a question on which Buridan suspends judgment, because of the geometrical equivalence between the epicyclical model and one based on an appropriate moving deferent. Although he does not elaborate on the physical differences hiding behind this geometrical equivalence, Buridan presumably has in mind the fact that the moving deferent model would not require the moon to rotate per se to maintain the same face toward the earth. This is a striking case where geometrical equivalence between models carries quite different physical implications.

Now Buridan and Richard of Middleton are quite unequivocal: if the epicyclical lunar model be granted, then the moon must rotate on its axis. Yet these replies to Aristotle are not as radical as they might seem, since their conceptual framework is that of Aristotle's original argument. The moon is carried round by its epicyclical sphere, in which it is embedded, and therefore if undisturbed it would turn the same face to the center of the epicycle. But that means that we, outside the epicycle, should see all parts of the moon's surface during a given anomalistic month (= the epicyclical period). In fact we always see the same side; hence the moon must rotate axially in the direction opposite to its epicyclical revolution, and with a rotation period equal to the epicyclical period. But note how this counterargument works. For a body fixed on the equator of a rotating spherical orb, whether epicycle or deferent, any proper rotation it might have is taken to be with respect to the body's place in the orb, which was precisely the principle on which Aristotle based his argument that the moon, moving in a simple geocentric orb, does *not* rotate. Here an Aristotelian physical principle has been used in one context to negate an implication of the same principle used in another, the same phenomenon being the *explanandum* in each case.

The rotating epicyclical moon appeared in various texts from Buridan's time right up to the second half of the seventeenth century. It turned up in Oresme's *Le livre du ciel et du monde*, the substance of the argument being that of Buridan's, written some thirty to thirty-five years previously, except that Oresme, unlike Buridan, was convinced of the moon's epicyclical per se rotation, since he claimed that it is impossible to account for the moon's motions without assuming an epicy-

ference of moon 5,961.3 miles. *Campanus of Novara and Medieval Planetary Theory: Theorica planetarum*, F. S. Benjamin and G. J. Toomer (eds. & trans.) (Univ. of Wisconsin Publications in Medieval Science, Madison, 1971), pp. 190–91, and editors' note 67 on p. 399.

cle.[48] Not unexpectedly, the rotating moon reappeared in one of the most influential writings of one of Buridan's pupils, the *Questiones subtilissimae in libros de caelo et mundo* of Albert of Saxony. Here Albert repeats the substance of Buridan's argument, but he brings out the additional point, apparently missing from Buridan's text, that there were those who rejected Ptolemy's epicycles on the grounds that the entailed rotation of the epicyclical moon is an impossibility, since the moon is of the same nature as the nonrotating stars. Albert's own view was that the moon is not entirely of the same nature as the stars, so he accepts epicycles and the consequent "special motion the moon has about its own center contrary to the motion of the epicycle."[49]

Among the many students of Albert's *Questiones* was Galileo, so the rotating epicyclical moon duly makes an appearance, not in the early commentary (c. 1590) on Aristotle's *De caelo*,[50] as we might have expected, but in the First Day of the *Dialogo* (1632). Taking their cue from Kepler,[51] Salviati, and Sagredo, discussing the similarities between earth and moon vis-à-vis their reciprocal relation to the sun, agree that the inhabitants of the moon observe phases of the earth, just as we observe phases of the moon, though in the reverse order. Yet earth and moon differ in that a moon-dweller observing the earth would see its whole surface within a period of twenty-four hours, "but we shall never see more than half the moon," notes Sagredo, "since it makes no revolution of its own, as it would have to do for all of it to show itself." Salviati replies: "Provided that the very opposite is not implied; namely, that its own rotation is the reason that we do not see the other side—for such would have to be the case if the moon should have an epicycle."[52]

One appearance of the rotating epicyclical moon in a seventeenth-

48. Book 2, Chap. 16: Nicole Oresme, *Le livre du ciel et du monde*, A. D. Menut and A. J. Denomy (eds. & trans.) (Univ. of Wisconsin Publications in Medieval Science, Madison, 1968), pp. 450–51, 452–61.

49. *Questiones subtillissimae Alberti de Saxonia in libros de caelo et mundo* (Venice, 1492): Lib. 2, Q. 7 (unpaginated). In Chap. 11 of his *De motibus corporum coelestium juxta principia peripatetica sine eccentricis & epicyclis* (Paris, 1540) (unpaginated), J. B. Amicus, after rejecting lunar epicycles and lunar rotation, goes on to reject planetary epicycles too on the grounds that the planets and the moon are of the same or an analogous nature (*natura analoga*).

50. See William A. Wallace's English translation: *Galileo's Early Notebooks: The Physical Questions: A Translation from the Latin, with Historical and Paleographical Commentary* (University of Notre Dame Press, 1977), pp. 25–58.

51. Galileo had learned of Kepler's meditations on Levanian astronomy twenty years earlier, on reading the *Dissertatio cum Nuncio Sidereo . . .* (Prague, 1610): *Le Opere* 3 (1), pp. 114–16. See also E. Rosen's translation, *Kepler's Conversation with Galileo's Sidereal Messenger* (The Sources of Science, no. 5: Johnson Reprint Corp., New York 1965), pp. 26–30.

52. *Dialogue concerning the Two Chief World Systems* (note 2), p. 65. *Le Opere* 7, pp. 89–90.

century text contains a nice irony, and in its way symbolizes the general argument I wish to advance. In the chapter on lunar phases in his *Institutiones,* Nicholas Mercator raises the traditional objection against epicycles on the grounds of their imcompatibility with the invariability of the lunar face, and dispatches the objection with the same reply made three centuries earlier by Buridan, Oresme, and Albert of Saxony.[53] He makes no forward reference to Newton's libration theory in the Appendix.

4. CONTINUITY IN THE MIDST OF INNOVATIVE CHANGE: KEPLER, DESCARTES, AND BACK TO NEWTON

This Aristotelian rondo reminds us that the pale face of the moon peers into the house of positional astronomy as an outsider. For those working within the Ptolemeo-Copernican tradition of positional astronomy, "lunar theory" was not to be equated with "the theory of all things lunar." Even in writings outside that tradition, whenever the rotating moon did put in an appearance, it was usually as a stick to beat anti-epicyclists with, an exercise in ad hoc kinematic invention, intended to legitimize the use of epicyclical models. Thus the problem of explaining the invariability of the moon's face, and the associated question of the moon's possible per se rotation, are not even mentioned, let alone discussed, in (for example) Ptolemy's *Almagest,* in Copernicus's *De revolutionibus,* or in the works of Tycho Brahe. As a natural consequence, we do not find discussions of these lunar questions in modern histories of astronomy, however supposedly comprehensive or allegedly "general," or in the work of most modern historians of pre-eighteenth century lunar theory.[54]

---

53. Op. cit. (note 3), Lib. 2, Sec. 2, Cap. 5, pp. 85–89.

54. The only serious accounts of the topic are to be found in the older histories of astronomy. Jean-Sylvain Bailly, *Histoire de l'astronomie moderne depuis la fondation de l'Ecole d'Alexandrie, jusqu'à l'epoque de MDCCXXX,* 3 vols. (Paris, 1779–1782): vol. 2, pp. 132–34 (Galileo), 396–98 (Jean-Dominique and Jacques Cassini, Hevelius); vol. 3, pp. 165–67 (Newton, d'Alembert, Lagrange). Jean-Baptiste Joseph Delambre, *Histoire de l'astronomie moderne,* 2 vols. (Paris, 1821; Johnson Reprint, New York and London, 1969): vol. 1, pp. 623–26 (Galilean libration); vol. 2, pp. 210 (Descartes), 544–46 (translation of Newton's theory in Mercator, Cassini, mistaken attribution of Newton's and Cassini's hypothesis to Kepler). A short but informative summary of the history of libration theory from Galileo to Lagrange is given in Robert Grant, *History of Physical Astronomy, from the Earliest Ages to the Middle of the Nineteenth Century* (London, 1852), pp. 72–76. The earliest (and perhaps the only) "monograph" on the subject of the moon's rotation seems to be D'Ortous de Mairan's "Recherches sur l'équilibre de la lune dans son orbite. Premier Mémoire: De la rotation de la lune," *Histoire et Mémoires de l'Academie royale des sciences: Mémoires de mathématique et de physique,* 1747 (Paris, 1752), pp. 1–22. De Mairan promised a second *mémoire* on lunar libration, but this seems not to have been pub-

There remains however the case of those astronomers and writers on astronomical themes who saw astronomy as a discipline combining the traditional ideals of positional astronomy with new and more effective physical and mechanical theorizing, or who had new cosmologies to offer.[55] So we turn to Kepler, the creator of *physica coelestis*, in whose *Epitome astronomiae Copernicanae* (1618–21) we find the promising question (IV, 6): "By what arguments is it made probable that the primary planets share their own movements around themselves with the secondary planets, and especially the earth with the moon?" Kepler argues that the sun's rotation (mooted by him before the telescopic discoveries of Galileo and Scheiner) is the cause of the orbital revolution of the planets, that the rotation of Jupiter is to be inferred from the orbital revolutions of its satellites, and that the rotation of the earth is the cause of the orbital revolution of the moon. Generally, the rotation of primary bodies is the cause of the revolutions about them of secondary bodies. As for the moon itself, however, "that the moon in turn does not wheel around the axis of its own body is argued by the spots. But why is this so? If not because no further planet is seen to go around the moon. Accordingly the moon has no planet to which it gives movement by the rotation of its body. Accordingly, in the moon, the rotation was left out, as being superfluous."[56] Note that for Kepler the divinely decreed purpose of such a lunar rotation would have been to provide the magnetic source of a lunar satellite's motion, and since the satellite would have been real, so also would have been the lunar rotation, which Kepler here denies.

He denied lunar rotation also in the 1609 drafting of his *Somnium seu Opus posthumum de astronomia lunari*, in which he describes how the heavens and the earth would appear to the Levanians, the inhabitants of the moon. Levanians living on our side of the moon (the Subvolvans) would see the earth rotating once every earth day, so the very name

---

lished. Though he misreads some of the historical evidence, de Mairan provides a fascinating account of the two views concerning the moon's rotation: nonrotation (with respect to the orbit) v. physical rotation (with respect to "l'espace infini & immobile" and evinced by centrifugal forces). See also the useful (but historically inaccurate) "History of lunar mapping: 1600–1960" in Zdenek Kopal and Robert W. Carder, op. cit. (note 29), pp. 1–49. For bibliographical orientation, see the relevant sections of J. C. Houzeau and A. Lancaster, *Bibliographie générale de l'astronomie*, 2 vols. in 3 (Brussels, 1882–1889; reprint London: Holland Press, 1964).

55. The operative words here are "new and effective": I am not at all implying that astronomy as a "mixed" discipline did not exist long before Kepler. The examples of Sosigenes and Ptolemy (*Planetary Hypotheses*) show otherwise: see Eric J. Aiton, "Celestial spheres and circles," *History of Science* 19 (1981): 75–114.

56. *Epitome of Copernican Astronomy, IV and V*, trans. Charles Glenn Wallis, in *Great Books of the Western World*, vol. 16 (Ptolemy, Copernicus, Kepler) (Encyclopaedia Britannica and University of Chicago, 1952), pp. 919–20.

they have for the earth—Volva—implies that Kepler thinks the moon does not rotate, and he attributes to the Subvolvans the observation that though their Volva "does not seem to have any motion in space, nevertheless, unlike our Moon, it rotates in its place." In the "Notes" Kepler added to the original *Somnium* over the period 1620–30 there is the following significant physical analogy: "The Moon always turns the same spots towards us earth-dwellers. Hence we know that it revolves around the Earth just as though it were attached to the Earth by a cord, and that whereas its upper part never faces the earth, its lower part or hemisphere always does so."[57]

Now Kepler had abandoned solid celestial orbs by October 1595, and his *physica coelestis* is vastly different from the cosmology of *De caelo;* yet these inferences drawn from the invariability of the lunar face, supported by teleogical considerations, are *structurally* similar and kinematically identical to those made by Aristotle. This is not surprising, given Kepler's Aristotelian conception of motion as "the separation of the mobile, in so far as it is movable, from its place and its translation into another place."[58]

In his *Principia Philosophiae* (1644), Descartes asked why approximately the same side of the moon always faces the earth. "We shall easily judge that this occurs," he explains, "because the far side of the moon is somewhat more solid [than the near side], and therefore must complete a larger orbit as it revolves around the earth. . . . And certainly those innumerable inequalities, resembling mountains and valleys, which are observed on the near side with the aid of a telescope, seem to prove that that side is less solid." There is no hint of lunar rotation here or elsewhere in Descartes' writings, and again that is something that need not surprise us, given Descartes' vortex theory of planetary and lunar motions and (as with Kepler) his Aristotelian definition of "true" local motion as "the transference of one part of matter or of one body from the vicinity of those bodies immediately contiguous to it and considered as at rest, into the vicinity of others."[59] We have now reached 1644, and the moon's kinematic behavior is still in line with the teachings in *De caelo* and the *Physica.*

Descartes does not say so in the article on the moon's face, but the

57. *Kepler's Somnium: The Dream, or Posthumous Work on Lunar Astronomy,* trans. and commentary by Edward Rosen (Madison: University of Wisconsin Press, 1967), pp. 23, 77–78.

58. *Epitome astronomiae Copernicanae,* Book 1, Pt. 5 ("De motu terrae diurno"): *Opera Omnia,* ed. by C. Frisch, 8 vols. (Frankfurt-Erlangen, 1858–1871), vol. 6, p. 169.

59. *Principles of Philosophy,* trans. with explanatory notes by Valentine Rodger Miller and Reese P. Miller (Collection des Travaux de l'Académie internationale d'histoire des sciences, No. 30: Dordrecht, Reidel, 1983), Pt. 3, Art. 151, p. 174; Pt. 2, Art. 25, p. 51.

more solid side traces out a larger path because of its greater centrifugal *conatus recendendi* from the center of its orbit.[60] Thus we find ourselves back with Newton, a surprisingly different Newton this time, given the earlier part of our investigation, a Newton who presents the same Cartesian argument in his letter to Oldenburg (for Huygens) of 23 June 1673:

> I am glad that we are to expect another discours of the *vis centrifuga,* which speculation may prove of good use in naturall Philosophy and Astronomy as well as mechanicks. Thus for instance if the reason why the same side of the Moon is ever towards the earth be the greater conatus of the other side to recede from it; it will follow (upon supposition of the Earths motion about the Sun) that the greatest distance of the sun from the earth is to the greatest distance of the Moon from the earth, not greater than 10,000 to 56. . . . Because were the sun's distance less in proportion to that of the Moon, she would have a greater conatus from the sun than from the earth. I thought also sometime that the moons libration might depend upon her conatus from the Sun and Earth compared together, till I apprehended a better cause.[61]

—the "better cause" being that explained in the libration model published three years later by Mercator. In an earlier manuscript (1667–69) setting out a quantitative assessment of the ratio between the centrifugal force at the earth's surface and that of the moon away from the earth's center, Newton produces the same argument, and rounds out for us the implication of the penultimate sentence in the above passage: "And if the Moon's endeavour from the Earth is the cause of her always presenting the same face to the Earth, the endeavour of the lunar and terrestrial system to recede from the Sun ought to be less than the endeavour of the Moon to recede from the Earth, otherwise the Moon would look to the Sun rather than the Earth."[62] Note that there is no suggestion in either of these passages of a lunar rotation with respect to the fixed stars or to any sort of immutable space.

This affinity between Newton and Descartes on the explanation of the invariability of the lunar face reflects the well-established influence of the latter on the early development of Newton's thought,[63] and in particular attests Whiteside's reading of Newton's celestial mechanics up to the moment of the preparation of the *Principia.* Whiteside has

60. For Descartes' presentation of his conception of centrifugal force, see ibid., Pt. 2, Art. 39, pp. 60–61, and Pt. 3, Arts. 55–64, pp. 111–18.

61. *Correspondence* 1, p. 290; Herivel, op. cit. (note 33), pp. 236–37. The reference to *vis centrifuga* is explained by the fact that Newton had just received a presentation copy of Huygens' *Horologium oscillatorium* (1673). For some reason Oldenburg omitted the quoted passage from the copy he forwarded to Huygens.

62. C.U.L., Add. MS 3958(5), ff. 87, 89. Latin text and translation in Herivel, op. cit., pp. 192–98, quotation on p. 196.

63. See Herivel, op. cit., pp. 42–53.

shown that quasi-Cartesian deferent vortices formed part of Newton's thinking up to a relatively late stage, probably about 1682, in the development of his celestial mechanics, with 1684 as "the true *annus mirabilis* in which the basic tenets of the *Principia* . . . were clearly conceived." Also, continually present in Newton's writings up to 1684 (and indeed beyond, though then infrequently) was the notion of *centrifugal* force, an integral component of the Cartesian account of the motion of bodies carried around in a vortex, as opposed to the *centripetal* attractive force, first employed by Newton in 1684, which diverts moving bodies from their natural rectilinear paths toward the center of force.[64] Leaving aside the manifold differences between Newton and Descartes in their accounts of planetary and lunar motion, whether in their mechanical principles or their application of them, or in the details of their celestial mechanisms (for a long time Newton entertained both vortices and an attractive force), we may suspect that until the eve of the *Principia* Newton shared with Descartes the basic idea of the moon being embedded in and borne round the earth (at least partly) by a vortex, the action of a centrifugal force accounting for our seeing always the same face.

In the *Principia* itself (all three editions), however, and in *De mundi systemate,* the centrifugal force hypothesis had disappeared, to be replaced (though only in *Principia*) by quite a different mechanical account of the invariability of the moon's face. The new hypothesis appears in Proposition 38 of Book 3, immediately following the two propositions and their corollaries in which Newton finds the respective forces of the sun and moon "to move the sea." In this proposition, "to find the shape of the moon's body," Newton argues that if it were a fluid, "like our sea," then the ratio of the forces of the earth to raise the lunar fluid and of the moon to raise the sea would equal that of the moon's diameter times its accelerative gravity toward the earth to the earth's diameter times its accelerative gravity toward the moon. Since this ratio is 1081/100, and since the moon (he imagines) raises the sea 8 3/5 feet, the lunar fluid would rise 93 feet under the earth's gravitational force, and the moon would become spheroidal in shape, with its maximum diameter produced passing through the earth's center. Then out of the blue comes the conclusion: "The moon therefore strives to acquire such a shape, and it must have assumed this shape

64. D. T. Whiteside, "Before the *Principia:* the maturing of Newton's thoughts on dynamical astronomy, 1664–1684," *Journal of the History of Astronomy* 1 (1970): 5–19; cf. Whiteside, op. cit. (note 2), p. 318; Herivel, op. cit., pp. 54–64. See also I. Bernard Cohen, *The Newtonian Revolution, with Ilustrations of the Transformation of Scientific Ideas* (Cambridge: Cambridge University Press, 1980), pp. 93, 115–116.

from the beginning. Q.E.I." Less out of the blue, perhaps, is the corollary:

Hence it happens that the same face of the moon is always turned toward the earth. For in another position the lunar body cannot be at rest, but while always oscillating will return to this position. But then the oscillations would be exceedingly slow, on account of the smallness of the activating forces, so that the face, which should look always toward the earth, may look toward the other focus of the lunar orbit (for the reason given in Proposition 17), and not be immediately drawn back therefrom and turned back earthward.[65]

That the spheroidal moon just happens to point (in our era) to the upper focus is a piece of fancy, quite untempered by the involvement of Proposition 17, which offers no sort of reason of the kind required; and one would have liked a mite more information on Newton's prophetic hypothesis of the originally liquid moon, and on whether this is the heterogeneous moon of 1667–69 and 1673. Yet for all its loose ends this corollary marks a new departure, not only because it constitutes a modest qualitative prelude to the serious eighteenth-century researches on libration theory and the invariability of the lunar face,[66] but (of closer concern in the present context) because it signals a *terminus ad quem* in "the dissolution of the celestial spheres," and a *terminus a quo* in the emergence of modern celestial mechanics.

I return for a moment to the vortex-embedded moon that I have attributed to Newton up to the early 1680s. As we can see from the preceding pages, it resembled Kepler's moon on a cord, and the moon of Aristotle and his successors (whether astronomical or cosmographical), including of course Descartes, in the sense that they too all shared the notion of a moon lodged or held in some way in a deferent (or epicyclical) *orbis—sicut nodus in tabula*—whatever the orb's physical makeup and geometrical form, and whatever causal scheme associated with the moon's orbital transportation. This part of the story does not square with the kinematic libration theory of 1673–75, but suggests rather a mismatch between Newton's theory of *libration* and his pre-

---

65. *Principia* (KC), vol. 2, pp. 673–74. My translation. The Motte-Cajori translation of this Corollary, and of the last line of the Proposition, is unsatisfactory (the original Motte translation being preferable): *Principles* (MC), pp. 484–85.

66. Here the principal figures are Euler, d'Alembert, Clairaut, Mayer, Gottfried Heinsius, and Lagrange, whose "Recherches sur la libration de la lune" (1763) was written in response to the prize question, set by the Académie des Sciences in 1762 for 1764, asking whether a physical explanation can be given of why the moon always presents nearly the same face to us, and whether the motion of its axis can be determined. The more complete "Théorie de la libration de la lune et des autres phénomènes qui dépendent de la figure non sphérique de cette planète" appeared in 1782. See *Oeuvres de Lagrange*, ed. by J. A. Serret, 14 vols. (Paris, 1867–1892), vol. 5, pp. 5–123; vol. 6, pp. 5–61.

Principia account of the *invariability* of the lunar face. In the 1673 letter
to Oldenburg, he has abandoned what we may call the "centrifugal hypothesis" of libration in favor of (one assumes) the theory in Mercator's
Institutiones, but there is no suggestion that at that time he has abandoned also the centrifugal hypothesis as an explanation of why the
moon always presents us with the same face.

Assuredly, there is no outright contradiction between the two parallel strands in the story. One can say that there are simply two levels of
explanation at work, the Mercator text offering a kinematic account of
the libration phenomena (whatever the underlying mechanical causes),
the Oldenburg letter and the above manuscript of 1667–69, on the
other hand, offering a possible mechanical explanation of the observed
effects of the moon's consequent monthly sidereal rotation. And in the
letter to Oldenburg, Newton cannot have forgotten the 1673–75 lunar
rotation with respect to the fixed stars.

Yet the available texts do not collectively present matters that way.
They do not suggest a unified physical theory of the moon and its
librations in which kinematic and mechanical considerations play their
interrelated respective roles. Despite the promise of the "Laws of Motion" paper (see note 33), Newton does not tell us how the rotating
moon of 1673–75, behaves mechanically as a (presumably) heterogeneous body within Cartesian deferent vortices. He does not indicate
how the centrifugal hypothesis of the late 1660s and 1673 relates to
the non-Cartesian sidereal "inertial" rotation we assume he also then
understood the moon to undergo. I see no reason for thinking that
Newton's reticence on these questions is explained by saying (as some
will) that he had no need to tell us any of these things, since of course
he saw all the ins and outs of the business, but simply did not think
it sufficiently important to commit to paper. (In the same way, some
historians of astronomy will claim that of course everyone since the beginning of astronomical thought always knew that the (nonepicyclical)
moon rotates sidereally, but thought it too trivial to mention anywhere
in any text.) Taking the Newton texts we do have, and lending them
some degree of importance, we should see in them rather the tensions,
the partial explanations and partially articulated theorizing, whose outcome was the "confluence" of theory and concept in the corollary to
Proposition 38. Despite the makeshift effect qua consequence of the
proposition, it allows us to reconstruct Newton's final position on the
theory of the physical moon (as opposed to his "lunar theory"). No
doubt for the reasons touched on earlier in this study, and no doubt
also because of the speculative nature of Proposition 38, Newton did
not attempt a formal presentation of such a theory. We are left wonder-

ing if he saw in his libration theory, and in his awareness of the moon's rotation, something of greater historical significance than simply the subject of a priority claim.

The overall background to the corollary is Newton's emancipation from the pressing difficulties of Descartes' vortex theory, which he dispatches at the end of Book 2 of the *Principia*. In Section 9 ("The circular motion of fluids"), Proposition 52, Newton shows that the periodic times of the parts of a vortex vary as the squares of their distances from the center of motion, a result quite out of keeping with Kepler's Third Law. Then in Proposition 53 he shows that bodies carried in a vortex in the same orbit must be of the same density as the vortex and must move with the same motion, from which it follows that denser bodies cannot move in ellipses, but must recede from the center of the vortex in a spiral path. The conclusion is that "the hypothesis of vortices is utterly irreconcilable with astronomical phenomena, and rather serves to perplex than to explain the heavenly motions."[67]

Given therefore the exorcism of vortices finally from Newton's celestial mechanics, we find that the solid moon has become a dynamical entity independent of orb or vortex, emulating thereby the post-Copernican earth. That is, it has become a physical body revolving and rotating under inertial and gravitational forces, and governed by laws of motion operating within a framework of absolute space and time. The moon's monthly sidereal rotation, now a natural inference from the invariability of the face, would be uniform were the moon a perfect sphere, and would therefore be uninfluenced by external gravitational attraction, a result that follows (though Newton, rather oddly, does not demonstrate it) from the arguments in Section 12 of Book 1 of the *Principia*, especially Proposition 75, according to which, in effect, bodies attracting each other under the inverse square law do so as though their masses were concentrated at their respective centers of gravity. However, by Proposition 38 (above) the moon is not a perfectly spherical body; and only through pure coincidence could such a body rotate in exactly one sidereal month. So Newton offers a hypothesis of the moon's original and present shape that yields a physical explanation of the invariability of the lunar face[68] and therefore (albeit with its own

---

67. *Principia* (KC), vol. 1, pp. 537–47; *Principles* (MC), pp. 387–96.
68. Yet we cannot exclude a higher teleological element in Newton's thinking. Note the following revealing piece of information from David Gregory, writing in June 1698: "Mr. Newton thinks that the same side of the secondarys [satellites] are [*sic*] still towards their primarys [planets], because if the secondary should roll about its axis [i.e. oscillate] the tide of its fluide would by such rotation cover all the arida to a great height & disturbe all its economy . . . " Royal Society Archives, MS 247 (The Gregory Volume), f. 61. It is worth noting that lunar rotation in *Principia* caught Gregory's attention when

substituted coincidence) of the phenomena of libration. Superimposed therefore on the uniform rotation is effectively a secular variation in rotational speed.

Neither Newton nor Cassini was the first to attribute per se rotation to the nonepicyclical moon. Earlier in the century the Neapolitan astronomer Francisco Fontana and (perhaps surprisingly) Thomas Hobbes had the same insight. But it is not primarily a question of who got the idea first but of who did something interesting with it when he got it. Neither Fontana nor Hobbes applied their insight to the problem of lunar libration; nor did either of them create a celestial mechanics (neither did Cassini) or a cosmology within which their insight might have found a significant role. They did not follow through the implications of their discovery, although Fontana was well aware that at least in the case of Jupiter, whose rotation he inferred from his telescopic observations, it followed that "the Jovian globe is not part of a deferent celestial orb [deferens coelum], but exists wholly in its own right."[69] At all events, Newton was one of only a handful of people whom the general bewitchment with celestial orbs, and with their ancient psychological source, the ever-constant lunar face itself, did not prevent attributing to the moon the rotation that everyone else, after the Copernican earth and the rotating sun, had begun to attribute (supported by observation) to all sorts of other heavenly bodies: planets, their satellites, even variable stars.[70]

5. CONCLUSION

Newton's dismissal of Cartesian vortices was a late episode (and not the last, in the eyes of some early eighteenth-century Cartesians) in the

he was drafting his "Notae" on the work between 1687 and 1694. In the nota on Prop. 66, Cor. 22, of Book 1, he asks what the motion would be of a globe whose center is carried along the circumference of a circle with no other motion: "Would the same point of the globe face towards the center? This cannot happen, since for that it would require another motion to be impressed on the globe, for the Moon needs an [additional] motion so that the same face turns towards the Earth. . . ." The nota on Prop. 17 of Book 3 consists of an exposition in Gregory's own words, complete with diagram, of the brief first-edition statement of the proposition, and unlike Newton in the first edition, Gregory links the proposition with the theory in Mercator's 1676 treatise. He also discusses Riccioli's eccentric hypotheses (see above and note 2), Royal Society Archives, MS 210, pp. 46, 128.

69. *Novae coelestium terrestriumque rerum observationes, et fortasse hactenus non vulgate, a Francisco Fontana, specillis a se inventis, et ad summam perfectionem perductis, editae* (Naples, 1646), pp. 26–27, quotation at p. 108; Hobbes, *De corpore* (1655), Pars 4, Cap. 26, Art. 9.

70. Twentieth-century accounts of these matters are useless (where they exist), but for a synoptic survey of the celestial rotations (including the hypothesized rotation of variable stars, attributed to Riccioli), see Edward Sherburne, op. cit. (note 4).

long, complicated story of the disintegration of the Ptolemeo-Aristotelian cosmos of celestial spheres and material deferents (one version of which in this context was the Cartesian cosmology).[71] Yet with the emergence of Newton's new, autonomous "dynamical" moon disappears what we now see was the last vestige of the old doctrine of celestial spheres: the lunar orb itself, in whatever material form. The sidereally rotating moon in the *Principia* marks the final stage in the effective demise of the Ptolemeo-Aristotelian cosmos, as well as the inauguration, for the great theorists of the eighteenth century, of the problem of the lunar body and its motions. That is why the moon's per se sidereal rotation turns out to be the most significant feature of Newton's libration theory, and why this curious subplot in the histories of astronomy and cosmology signals a caveat for those who might undervalue or overlook the significance of continuities in the midst of change and revolution. Our story has also shown that innovative discontinuities are sometimes not to be found where they might have been expected.

71. See William H. Donahue, *The Dissolution of the Celestial Spheres 1595–1650* (New York, Arno Press, 1981); Nicholas Jardine, "The Significance of the Copernican Orbs," *Journal for the History of Astronomy* 13 (1982): 168–94; Edward Rosen, "Francesco Patrizi and the Celestial Spheres," *Physis* 26 (1984): 305–24; "The Dissolution of the Solid Celestial Spheres," *Journal of the History of Ideas* 46 (1985): 13–21. On the fortunes of the vortex theory in the eighteenth century see Eric J. Aiton, *The Vortex Theory of Planetary Motions* (London and New York, 1972).

6   The Heavens and Earth: Bellarmine and Galileo

JOSEPH PITT

One way to characterize the transformation we refer to as the Scientific Revolution is as that period of time during which Aristotle's fragmented view of the separation of the heavens and the earth was replaced by Newton's mathematically unified one. In order for this to have occurred, several different things had to happen: mathematics had to be accepted as the appropriate medium for the expression of scientific truths; the Aristotelian framework had to be convincingly shown to be inadequate to the job at hand; an alternative framework denying the fundamental metaphysical distinction between the heavens and the earth had to be established. It is no secret that Galileo's *Dialogue on the Two Chief World Systems* and others of his works contributed significantly to achieving all three of these objectives. This is not to say that Galileo had these objectives in mind, but rather that his work was instrumental in achieving the intellectual reorientation necessary for the final set of moves culminating in Newton's work. Along with recognizing the role Galileo played in the scientific revolution, it is also widely believed that Galileo's strong opposition to Aristotelianism was the major reason for his downfall at the hands of the Jesuits. These last two claims, that Galileo was a major contributor to the overthrow of the Aristotelian framework and that his opposition to it was the cause of his troubles with the Church, now seem less significant than previously thought.

The source of these new doubts is the publication of Cardinal Bellarmine's *Louvain Lectures* and some comments in the Introduction to that volume by Ugo Baldini and George V. Coyne.[1] While it has been known for some time that there was a non-Aristotelian tradition in cosmology prior to Galileo, it now appears that it flourished among, of all groups,

---

1. R. Bellarmine, *The Louvain Lectures*, U. Baldini and G. V. Coyne (eds.) (Vatican City: Vatican Observatory Publications, 1984). Studi Galileiani, vol. 7, no. 2.

the Jesuits and, furthermore, one of its major proponents was Roberto Bellarmino, cardinal inquisitor and official theologian to Pope Paul V during the Vatican's dispute with Venice in 1606. This was the same Bellarmine who wrote the famous letter to Foscarini in 1615 complimenting him and Galileo for speaking only hypothetically of the Copernican system and who later in 1616 also informed Galileo of the pope's decision that Copernicus's views were contrary to the teachings of the Church and that he was not to defend Copernicanism. The fact that Bellarmine was opposed to the Aristotelian distinction between the heavens and the earth and yet acquiesced in the Edict of 1616 presents a number of difficulties for the traditional account of Galileo's conversion to Copernicanism and in turn raises some interesting questions about Pietro Redondi's new account of the Jesuit conspiracy to get Galileo. In particular, the view of Galileo as breaker of tradition and destroyer of the Aristotelian world view is in doubt. In fact, if, as it turns out, anti-Aristotelianism was strongly evident in the sixteenth century, then the extent to which Galileo's anti-Aristotelianism is significant needs to be rethought. Although the complete case cannot be made here, I will suggest that the importance of Galileo's anti-Aristotelian work lies not in his originality but in its having brought together in a systematic manner a number of considerations under examination in a variety of forms at the time. The primary foci of this paper are two claims: that the ideas found in Galileo's defense of Copernicanism were not without precedent, and that Galileo's conversion to Copernicanism was not so much a matter of his being convinced by the sheer intellectual beauty of Copernicus's scheme as it was a form of scientific opportunism.

Before going any further let me clarify in a preliminary way certain aspects of my thesis. As far as I have been able to determine up to this point, there is no concrete evidence in the form of reports of meetings or exchanges of letters to show that Bellarmine's views on the relation between the heavens and the earth had any direct impact on Galileo. The case presented here rests on circumstantial evidence. On the other hand, previous answers to the question of the circumstances under which Galileo adopted Copernicanism have also relied rather heavily on circumstantial evidence. So, the story I am telling should not be dismissed out of hand any more than some others currently on the books. Consider the following example as support for this plea for tolerance.

1. GALILEO'S CONVERSION TO COPERNICANISM.

Stillman Drake gives an account of the circumstances under which Galileo became a Copernican.[2] Drake finds his "clue" in an anecdote taken from the *Dialogue*. The speaker is Sagredo:

> This is the time for me to tell you a few of the things that happened to me when I first began to hear these opinions spoken of. I was then a youth who had scarcely finished the course in philosophy, giving this up in order to apply myself to other activities. It happened that a certain foreigner from Rostock, whose name I believe was Christian Wursteisen, a supporter of the Copernican opinion, arrived in these parts and gave two or three lectures in an academy on this subject. He had a throng of hearers, more from the novelty of the subject than for any other reason, I think. I did not attend them, having formed a definite impression that this opinion could be nothing but solemn definite foolery. Later, asking about it from some who had gone, I heard them all making a joke of it except one, who told me that the matter was not entirely ridiculous. Since I considered this person an intelligent man and rather conservative, I was sorry that I had not gone; and from then on, as I happened from time to time to meet anyone who held the Copernican opinion, I asked him whether he had always believed in it. Among all the many whom I questioned, I found not a single one who did not tell me that he had long been of the contrary opinion, but had come over to this one, moved and persuaded by the force of its arguments. . . .

Now this sounds harmless enough, but Drake then proceeds to go to work on the accuracy of Sagredo's claims. Drake wants to prove that the conservative stranger is Galileo and that Galileo was introduced to Copernicanism when he attended these lectures given by this Christian Wursteisen. But Drake notes a problem. It seems that while Christian Wursteisen from Rostock did exist, he never visited Padua and he died in Switzerland in 1588. On the other hand, it seems there was a *Christopher* Wursteisen enrolled at Padua in November 1595. Drake then speculates that Galileo, writing the *Dialogue* some twenty-five years later, probably confused the names and, hence, we have Galileo hearing about Copernicus from a man from Rostock, who "quite possibly" went there from not just Rostock now, but from the *University* of Rostock and who "quite probably" went to Padua to complete a course in law and gave a series of lectures on Copernicus, since the University of Rostock was the first place where Copernicanism was taught.

This is, no doubt, an interesting thesis. But it does seem to contain some adventuresome logical moves. To begin with, we have the wrong man in Christian Wursteisen. Second, we do not have Galileo clearly identified. Third, we have no real reason to believe that someone

2. S. Drake, *Galileo at Work* (Chicago: University of Chicago Press, 1978). The following quotation is from p. 128.

named Wursteisen from Rostock gave the lectures. Now, if Sagredo had, in mentioning this conservative stranger, gone on to mention that this person had since become a close friend from whom he continued to learn many things of novelty, then we might want to give Drake his clue. But even then, when the entire text is considered, that would still be a fairly weak case. For the conclusion of Sagredo's story is not that in hearing the evidence for Copernicanism he immediately adopted the view. Rather, he tells us that in speaking with those who had converted to Copernicanism, he found they all did so on the strength of the arguments and that they knew those arguments, whereas few if any of the Peripatetics and Ptolemaics had even seen Copernicus's book and none had understood it, and, furthermore, none had converted to the Ptolemaic view from Copernicanism. From these investigations Sagredo concludes: "[O]ne who forsakes an opinion which he imbibed with his milk and which is supported by multitudes, to take up another that has few followers and is rejected by all the schools and that truly seems to be a gigantic paradox, must of necessity be moved, not to say compelled, by the most effective arguments."[3] One might assume that Sagredo is admitting to his having been converted by the strength of the arguments also, but this is *not* so. He goes on to tell Simplicio and Salviati how grateful he is to have met them so they can clarify the subject and put him "into a position of certainty."[4] So Sagredo did not convert to Copernicanism after hearing about it; he is still waiting to be set straight. Thus, it appears that Drake's account fails to establish the legitimacy of his clue. But Drake should not be singled out here. He at least attempts to give an account of the circumstances under which Galileo might have first encountered Copernicanism. Clavelin doesn't even engage the issue. He merely notes that in 1597 Galileo wrote first to his friend Mazzoni and then to Kepler announcing he was a Copernican.[5]

I will call the standard view the "presto-converto" thesis. On this view, Galileo simply heard about Copernicus, or attended some lectures, or maybe even read *De revolutionibus*. Then by virtue of the clarity of the arguments (!), which Copernicus himself admitted would only be fully apparent to accomplished mathematicians, presto! Galileo was convinced. By contrast, I want to argue for an account in which sociological factors have an important role to play. There are also several additional troublesome points about the presto-converto thesis. Let me briefly enumerate these and then continue with some socio-history.

3. Ibid., p. 128.   4. Ibid., p. 129.
5. Ibid., p. 120n. M. Clavelin, *The Natural Philosophy of Galileo* (Boston: MIT Press, 1974).

The first problem is this: Galileo had no initial need to adopt Copernicanism. In 1597 he was not an astronomer; he held the chair of mathematics at Padua. Second, and perhaps more important, we have no other evidence than his claims in these two letters that he follows Copernicus—no notes, no treatises drafted, no public defenses. That leaves the question as to why he would write Kepler, of all people, and make this strange announcement. I will return to these worries later, after having discussed Bellarmine and thereby, I hope, developed a plausible basis for addressing these problems.

2. BELLARMINE ON THE HEAVENS

That there were a number of anti-Aristotelian traditions flourishing in the sixteenth century is no news to historians of the period. The publication of Bellarmine's *Louvain Lectures,* however, is news, in that it reveals a leading Jesuit theologian and prominent member of the Inquisition as an anti-Aristotelian; on traditional versions of the story, being an anti-Aristotelian is precisely what got Galileo into trouble with the Inquisition. The material, translated by Ugo Baldini and George V. Coyne and published by Vatican Observatory Publications, is the first section of the course Bellarmine gave at Louvain; it deals with the question of the incorruptibility of the heavens.[6] It follows the standard procedure of taking Aquinas's work as its starting point and worries the following four questions:

1. Whether by its nature the sky is corruptible
2. Whether the sky may in fact be corrupted.
3. On the work of the second day: what was made on that day.
4. On the work of the third day.

In the lectures Bellarmine appeals to the work of established Patristic writers who have questioned Aristotle's and Aquinas's views and to the text of the Bible for evidence for the claims that the heavens and the earth are continuous and that the heavens are corruptible. It is a detailed work requiring intense analysis. Baldini and Coyne supply eighteen pages of notes for eight pages of text and it would be foolish of me to try even to summarize the arguments. But its thrust is clear: according to the Holy Fathers and the Bible itself, Aristotle's distinction between the perfection of the heavenly sphere and the imperfection of the terrestrial domain is untenable. This is not asserted hypothetically, although at one point Bellarmine does issue a caution:

6. Bellarmine, *Louvain Lectures,* note 1 above.

It is not our task to decide which is the truer of the above opinions. As St. Basil in his third homily on the six days and Chrysostom in his homily on Genesis say: with respect to the divine works, we should not conjecture except as allowed; therefore, we know that the firmament exists; what it is and how it is we will know when we have gone up there.[7]

This is an appropriately modest and obviously protoscientific stance. Nevertheless, despite this single caveat, the tone of the arguments is positive and argumentative.

The importance of this material is just beginning to be understood. Consider some of the issues and topics on which it bears. Bellarmine was an extremely important figure in Galileo's affairs. Not only was he responsible for informing Galileo of the Edict of 1616, which forbade the teaching of Copernicus as true and any attempt to reconcile Copernicus's views with the Bible (but which, contrary to popular opinion, did *not* ban Copernicanism); he was also responsible for providing Galileo with a handwritten letter affirming that although Galileo was informed of the Edict he was not asked to abjure. This document proved extremely important in the trial of 1632. One of the other few documents of Bellarmine's we have is his wellknown letter of 1615 to Foscarini, in which he commends both Foscarini and Galileo for discussing the Copernican theory only hypothetically. Thus, from such previously available material we have the idea of a cautious supporter of Galileo, who nevertheless is careful to keep intact his role as defender of the faith. This does not square with the picture of Bellarmine, the radical of Louvain.

3. THE IMPORTANCE OF SCHEINER

But there is more, and unfortunately this additional information only confuses matters further. One of Galileo's long-standing enemies was the Jesuit Christopher Scheiner, whose arguments with Galileo over the sunspots brought Galileo fame and Scheiner's unending enmity. Drake has suggested that the depth of Scheiner's antagonism toward Galileo was such that he can be held directly responsible for Galileo's 1632 trial. Furthermore, Scheiner's attack on Galileo continued after the trial; the invective of Scheiner's final work was such that the Church suppressed its publication until after his death. But, prior to that final effort, between 1626 and 1630 Scheiner published a long work entitled *Rosa Ursina* (Bracciani, 1626–30). In this book he continued his attack on Galileo, begun during the debate over the sunspots,

7. Ibid., p. 14.

and also developed a theory of the universe that is anti-Aristotelian; in support of his theory he cited Bellarmine's Louvain Lectures. In so doing, as Baldini and Coyne note, "He shows there exists an alternative theory, acceptable theologically and compatible with the strange phenomena which had been observed during the previous decades, beginning with the appearance of the Nova of 1572."[8] This last bit is important because Bellarmine's Louvain lectures were given in 1570–72— prior to the appearance of the nova of 1572 and the later comets. In other words, Bellarmine's discussion was not motivated by external events—it was not a response to the empirical inadequacy of the Aristotelian position. Bellarmine's attack was apparently motivated by the failure of the Aristotelian-Thomistic view to accord with the teachings of the Church, pure and simple.

The significance of Scheiner's *Rosa Ursina* reference to the position Bellarmine's detailed in the *Louvain Lectures* goes beyond the demonstration of an available alternative to Copernicus and Aristotle. It suggests that Bellarmine's views were still acceptable in 1630. This suggests in turn that, contrary to Pietro Redondi's claims in his flashy *Galileo Heretic*, the Jesuits were not obviously retrenching theologically in the 1620s and 1630 as part of the Counter-Reformation.[9] Let us look first at the implications of Scheiner's references to Bellarmine and then at the Redondi thesis.

There is a certain irony in the fact that Scheiner and Galileo were both aiming at the same result: that is, the overturning of what Kuhn calls the two sphere universe.[10] And yet, perhaps it is not so strange. Their dispute begins, after all, over a priority claim about the discovery of the sunspots. So, irrespective of who discovered what, the fact is that both Galileo and Scheiner were interested in the sunspots and for the same reason: their existence cast empirically based doubts on the Aristotelian thesis concerning the perfection of the heavens.

Galileo and Scheiner also share another feature: their motivation for attempting to replace the Aristotelian view is not clear. I think an explanation for Galileo's behavior can be made. The case for Scheiner's efforts is less obvious.

8. Ibid., p. 3.
9. P. Redondi, *Galileo Heretic*, R. Rosenthal (tr.). (Princeton: Princeton University Press, 1987).
10. T. S. Kuhn, *The Copernican Revolution* (Cambridge: Harvard University Press, 1957).

4. GALILEO'S CONVERSION TO COPERNICANISM
RECONSIDERED

Galileo's reasons for attacking the Aristotelian distinction between the domain of the heavens and the domain of the terrestrial can be broken down into two different categories, both of which are self-serving. In the first case, it was important to Galileo to have this distinction abandoned in order to establish the credibility of his telescopic observations. In this case it is important to note that these observations of Galileo were not in doubt in the astronomical community. It was the philosophers who constituted his opposition here. It is, furthermore, interesting to note that his telescopic discoveries were confirmed by the Jesuit astronomers and were not viewed as problematic once they gained access to telescopes themselves.

The main opposition to Galileo's telescopic revelations was in a philosophical community whose methodology can be described as a form of blind adherence to the truth of Aristotle's writings; there was no tolerance for any appeal to empirical evidence as a basis for doubting those works. The Jesuit scientific community, on the other hand, appeared to be rather openminded about new empirical data. Thus, we find Galileo forced to engage the philosophers on their own ground, using logic and appeals to Aristotle's original method as a basis for undermining, not just Aristotle's false cosmological claims, but the methodology of these degenerate latter-day Aristotelians. Galileo's major efforts along these lines are to be found in the *Dialogue*, specifically in the first two days.[11]

Now it might be asked why Galileo felt compelled to take on the philosophical community when the astronomers were not objecting to his new discoveries about the heavens. In part his efforts to refute the philosophers were probably a function of his generally combative nature. He enjoyed a fight, especially if thought he could win it. And in this case, the methodology of the philosophers was particularly galling to someone who delighted as much as he did in the degree of certainty his own developing methodology provided. So, in the first place, Galileo wanted his telescopic discoveries accepted universally and this meant defusing the objections of the philosophers.

The second reason Galileo needed to overturn the Aristotelian commitment to a fundamental distinction between the heavens and the earth had to do with his theory of the tides. We are most familiar with

11. J. C. Pitt, "Galileo, Rationality and Explanation," *Philosophy of Science* 55 (1988): 87–103.

Galileo's theory of the tides as it appears in the fourth and final day of the *Dialogue*.[12] The usual account of its role in the *Dialogue* is that it is Galileo's *piéce de resistance* in his defense of Copernicanism. In the *Dialogue* he spends Day One establishing the priority of geometric proof over Aristotelian appeals to first principles. Day Two is devoted to showing that empirical arguments employing earth-based observations cannot prove that either the Aristotelian or the Copernican view is correct. In Day Three he lays out the Copernican view and in Day Four he argues that, on the assumption of the Copernican motions of the earth, he can explain a phenomenon that has hitherto resisted explanation, namely, the motions of the tides. Hence, we are left to conclude that, given its superior explanatory power, Copernicanism must be correct.

Thus, once Galileo has his telescopic observations to defend, a fairly strong case can be made for his need for Copernicus's theory. But what is sometimes overlooked is that Galileo had developed his theory of the tides *before* he discovered the telescope and, I want to argue, before he had converted to Copernicanism. That is to say, Galileo developed a theory of the tides that required the motion of the earth before he adopted the Copernican account of how the earth moved.

The first overt mention of his theory of the tides occurs in Galileo's 1610 application for a position at the University of Pisa, where he lists a treatise on the tides as one of his many works. But, although that constitutes Galileo's first public reference to his theory of the tides, there is evidence of his interest in the tides dating back at least to 1597, when he writes to Kepler, expressing delight at Kepler's support of Copernicus. In that letter he says that he has for many years been a supporter of Copernicus since (among other reasons) "from that position I have discovered the causes of many physical effects which are perhaps inexplicable on the common hypothesis."[13] Kepler apparently guessed that Galileo was here referring to the tides, and that is all we can do as well. But Stillman Drake offers one further bit of speculation, which this time seems reasonably well founded and which lends support to the idea that Galileo had his basic theory of the tides in hand before he adopted Copernicanism. It seems that in 1595, Galileo's good friend Fra Paolo Sarpi entered in his notebooks a description of a theory of the tides identical to Galileo's. Furthermore, Sarpi's biographer made observations of the tides in Venice on Galileo's behalf in 1595.[14] Thus, the evidence points to the possibility that Galileo had developed

---

12. G. Galilei, *Dialogue on the Two Chief World Systems* (Berkeley: University of California Press, 1967).

13. Drake, *Galileo at Work*, p. 41.

14. As reported by Drake in *Galileo at Work*, p. 37.

his theory of the tides as early as 1595, prior to his confession to Kepler of his conversion to Copernicanism.

But if, before becoming a Copernician, Galileo adopted a theory of the tides that required that the earth moved, and if the standard cosmological position had the earth stationary at the center of the universe, are we suggesting that Galileo came on his own to the conclusion that the earth moved? No, that is not what is being claimed, although it is clear that the question now before us concerns the origin of Galileo's theory. It seems fairly certain that minimally Galileo and Sarpi worked on the tidal theory together. Honesty may even compel us to suggest that Galileo adopted Sarpi's theory. Nevertheless, even if Galileo borrowed Sarpi's theory, we are still left with an important problem: What led Sarpi to postulate the movement of the earth? At this point the trail gets fairly cold. But it seems to me that in the absence of clear-cut causal relations, we can still arrive at an understanding of the development of ideas and theories by getting a feel for the general intellectual climate. In this case, if it could be shown that heterodox cosmological views were in the air, then my contention that Galileo developed his theory of the tides before he became a Copernican and, furthermore, that he adopted Copernicanism because it supported his theory of the tides (and not the other way around, i.e., that he developed his theory of the tides in order to have a conclusive proof for Copernicus's views), would be reinforced, if not vindicated.

Interestingly enough, Pierre Duhem, in a piece on the history of physics, notes that Celio Caleagnini attributed a daily rotation to the earth in a work written around 1530 but not published until 1544. This view of Caleagnini also postulated an oscillation that was invoked to explain the precession of the equinoxes. In addition, he postulated an additional oscillation that accounted for the tides. According to Duhem, the postulation of this second oscillation was picked up by Andrea Cesalpino and is supposed to have "inspired" Galileo.[15]

Even though Duhem did not lead us to Sarpi, that he identifies Caleagnini and Cesalpino as holding unorthodox views about the motion of the earth is enough to suggest that these sorts of ideas were being discussed and were available to Galileo and Sarpi. Thus, given that Sarpi's theory and Galileo's interest in the tides date back to 1595 and the first expression of Galileo's interest in Copernicus's views comes in 1597, it seems reasonable to claim that it is certainly possible and appears increasingly likely that Galileo had his theory of the tides before he was a Copernican.

15. P. Duhem, "History of Physics," in *The Catholic Encyclopaedia* 12 (New York: Robert Appleton Co., 1907–22), pp. 47–67, esp. p. 53.

This also means that Galileo had developed an explanation for the motion of the tides as early as 1595, before he converted to Copernicanism; but he had no theoretical justification for his explanation. Copernicanism, however, with its motions of the earth, provided just such a grounding. Hence, it seems that Galileo was attracted to Copernicanism because it provided a theoretical justification for his theory of the tides, to which he had a long-term attachment, which continued up to his writing of the *Dialogue*. In fact, Galileo's original title for the *Dialogue*, which the papal censors forced him to change, was *Dialogue on the Tides*. . . . Thus, Galileo had two reasons to oppose the Aristotelian distinction between the domains of the heavens and the earth: his concern over his telescopic discoveries and his belief in the truth of his theory of the tides.

Now when we turn to Scheiner, the motivation for challenging the Aristotelian cosmology is less clear. We can attribute his efforts either to his enmity toward Galileo or to his zeal for biblical accuracy. If he is really just after Galileo, then this seems like a strange way to achieve that goal, since it means rejecting Aquinas's position, a view that has in general been adopted by the Jesuit theologians. That he is concerned with biblical accuracy seems more likely, especially since he cites Bellarmine and Bellarmine is operating in a context, i.e., the giving of lectures at a major Catholic university, in which that can be his only motivation for challenging Aristotelian orthodoxy. And this leads us back to Bellarmine.

5. GALILEO AND BELLARMINE

Did Bellarmine influence Galileo on this question? The answer is probably no if we are trying to establish direct influence. But Galileo did meet Bellarmine in Padua in the early 1590s. Furthermore, we know that although no direct exchange of letters took place, Bellarmine carried on a correspondence with Galileo through intermediaries (witness the letter to Foscarini and Galileo's reply, as well as Bellarmine's willingness to write out a letter in which he denies that Galileo had to abjure in his hands any adherence to the Copernican system. Finally there is the fact that significant Jesuits continued to develop anti-Aristotelian views, even after Bellarmine's death—a point that counts against Redondi's view that the Jesuits in the 1620s and 1630s were retrenching theologically to a narrow Aristotelian position in order to bolster the Church's Counter-reformation.

Despite the high probability that the specific early views of Bellarmine did not directly influence the young Galileo, that Bellarmine was publicly lecturing on such an anti-Aristotelian theme itself raises some further questions. If Bellarmine was arguing this way, were others, and if so, how widespread was the anti-Aristotelian movement? I cannot answer these questions completely here, but there is enough evidence to suggest that anti-Aristotelian cosmologies were in the air at the time Galileo was first introduced to Copernicus's views at Padua. Therefore, it seems that it was neither irrational nor a sign of extraordinary perspicacity on the part of Galileo to consider adopting one of these views, especially if he could use it to help support a pet theory of his own.

In conclusion, I have been arguing that a number of the ideas that Galileo used in his attack on Aristotelian cosmology and methodology were already available in the sixteenth century and that their appearance in Galileo's work does not occur *ex nihilo*. More importantly, that such anti-Aristotelian sentiment existed and that some of it could even be found in the Church suggests that excessive claims for Galileo the dragon slayer have to be muted. This is not to diminish the significance of Galileo's achievement in putting these various anti-Aristotelian viewpoints together to form a coherent overall attack on the Aristotelian point of view. But let us be clear concerning what can be claimed by way of originality.

Finally there is the question of the motivation for Galileo's Copernicanism. I have suggested that Galileo's defense of Copernicus's theory in the *Dialogue* is a form of opportunism. Galileo had a theory of the tides—perhaps not his—and his telescopic observations; he needed a justificatory framework within which to embed them. Copernicus's theory provided such a framework. This is not to condemn Galileo for seizing the opportunity to systematize his views in this fashion. But when we reflect on the manner in which Galileo's overall account developed, we should note the variety of sources, reasons, and causes at work. No simple argument from one feature of Galileo's work to his final fate at the hands of the Church can do justice to the wealth of the historical context. In this there is a lesson for philosophical theories of scientific change: paraphrasing Hume, reason is and ought to be the slave of people and events.[16]

16. I wish to thank my colleagues Roger Ariew and Peter Barker for their assistance in helping me attempt to make a philosophical thesis historically respectable. They are not to be blamed if they failed. In addition, I would like to express my appreciation to the participants in the Conference on Continuities and Discontinuities in Science for their input and assistance.

7   The Blasphemy of Alfonso X: History or Myth?

BERNARD R. GOLDSTEIN

> [Dupleix] est le premier historien qui ait cité en marge
> ses autorités; précaution absolument nécessaire quand
> on n'écrit pas l'histoire de son temps, à moins qu'on
> ne s'en tienne aux faits connus.
>
> VOLTAIRE, *Catalogue des écrivains*

1. INTRODUCTION

It is often said that many adherents of the Ptolemaic System took it to be merely a calculational device and not a true representation of the heavens. Among the small number of medieval scholars who worked in astronomy and for whom there is some evidence to suggest that they held this position is King Alfonso X of Spain (died 1284), best known in the history of astronomy for his sponsorship of the Alfonsine Tables whose popularity continued even as late as the seventeenth century.[1] For example, in Dreyer's still standard history of astronomy we read: "Apparently the King [Alfonso X of Castile] must have had his doubts about the physical truth of the [Ptolemaic] system, judging from his well-known saying that if God had consulted him when creating the world,

This study was stimulated by a question concerning the ultimate source of Kuhn's reference to Alfonso which J. Haugeland asked me in 1984. (As a result of this research I won a wager of five dollars from him.) In addition, I am grateful to Alan C. Bowen, Jerry R. Craddock, and Keith McDuffie for their translations of the Latin and Spanish texts, and to Stephen C. Wagner for locating the passage in Montucla 1799. Moreover, Michael Crowe kindly called to my attention some passages in H. Blumenberg's recent book in which references to Bayle and Fontenelle are found (*The Genesis of the Copernican World* [Cambridge, MA, 1987], pp. 259–60).

1. For a recent claim that the Alfonsine Tables first appeared in Paris ca. 1320 and that there was no antecedent Castilian text composed at the court of Alfonso, see E. Poulle, "The Alfonsine Tables and Alfonso X of Castille," *Journal for the History of Astronomy* 19 (1988): 97–113 (where an earlier version of my paper is cited on p. 113 as: "Voltaire and the Blasphemy of Alfonso X").

he would have given Him good advice."[2] There are two components of this story: (1) that Alfonso uttered a phrase something like, if God had consulted me concerning the creation, I would have given Him good advice, and (2) that this statement was intended to reflect his skepticism concerning the reality of the Ptolemaic description of the heavens. Since this passage continues to be cited despite, as we shall see, its lack of historical foundation, it seems worthwhile to trace the elements of the story. It is noteworthy that almost all the citations fail to include their sources, and this has impeded the identification of the lines of transmission. Despite such difficulties, it will be shown that the statement attributed to Alfonso first appears in a polemic against him some sixty years after his death and lacks support in any thirteenth-century document (see section 2, below), and that its astronomical interpretation indicating skepticism about the Ptolemaic System goes back no earlier than the latter part of the seventeenth century (see section 4, below).

2. THE CHRONICLE OF PEDRO AFONSO

There is no contemporary source that ascribes this saying to Alfonso; its earliest occurrence, according to Craddock, who only recently identified the relevant passages, is in the *Crónica geral de 1344*, a polemical tract against the dynasty of Alfonso X, which was composed by Pedro Afonso (died 1354), conde de Barcelos and bastard son of King Dinis of Portugal (reigned: 1279–1325).[3] The story here is filled with legendary features involving the parents of Alfonso, predictions about Alfonso by a Greek woman fortune teller, a friar who demands that Alfonso admit his sin and do penance, and a bolt of lightning that leads the king to confess the "sin of blasphemy he had uttered against God." The blasphemous phrase is reported as: "If he were with God when He made the world, He would have corrected (*emendara*) many things to make them better than what was done." For us it is important to note that no connection with astronomy or astrology is present in this account.

Because Pedro Afonso's version of this story is the earliest source, we offer here an extended set of excerpts from it (page numbers refer to Craddock 1986; some passages have been summarized).

The narration (214–16) tells how the queen explains her tears to Don Fernando, the king. In effect, these are her words (she has explained

2. J. L. E. Dreyer, *A History of Astronomy from Thales to Kepler* (Cambridge 1906, 2d ed., New York 1953), p. 273.

3. J. R. Craddock, "Dynasty in Dispute: Alfonso X el Sabio and the Succession to the Throne of Castile and Leon in History and Legend," *Viator* 17 (1986): 197–219.

that a Greek woman foretold that she would marry a great king, and that their son [Alfonso X] would be even greater).

And that he would be even more powerful and honored than his father and would reign a long time. And that because of a word of arrogance he would speak against God, he would be disinherited of his land except for a city in which he would die.[4]

King Fernando learns of a misappropriation of war funds by his son Alfonso (216).

And [the king told them how] because of that word that [Alfonso] was to utter against God he was to be disinherited. That [word] was to cause Him [God] the greatest grief ever caused Him by men up to that time since the death of Jesus Christ. And this certainly appeared [to the king] to be the truth, in view of the deeds he [Alfonso] was committing against him and against those who were in the service of God.[5]

The knight Pero Martínez learns of the incident in a vision.

And the knight [Pero Martínez] asked how it came about that God was so angry. And that man who appeared to him said: Don Alfonso, while in Seville, said in public [*lit.* in the plaza] that if he were with God when He made the world that He would have corrected many things to make them better than what was done and that for this reason God was angry with him.[6]

The friar who saw the same vision as the knight comes and tells King Alfonso that he must confess his sin and do penance (218).

And he came to the king and told him to do penance for the sins he had committed and that it would be to his advantage [to do so], and especially those damned and perverse words full of great pride and said with great arrogance and vanity, which he had said many times in public saying that if he had been God's advisor when He made the world, and God had taken his advice, He would have done it better than He did. If he did not [repent] he should not doubt that God would reveal his power over him. And he responded with great anger and words of fury and said: "I tell you the truth in what I say. And for what you say I consider

4. "& que seria avn mas poderosa & ho⟨n⟩rrado que su padre & asi duraria gra⟨n⟩ tienpo. Et que por vna palabra de soberuja que diria contra dios auja de ser deseheredado de su tierra saluo de vna c'ibdat enque auja de morir" (Craddock, 1986), p. 215.
5. "Et co⟨m⟩mo por aquella palabra q⟨ue⟩ auja de dez'ir contra dios auja de ser deseredado. Ca le faria el mayor pesar que nunca le om⟨n⟩e fiz'iera desde la muerte de ih⟨es⟩u xp⟨ist⟩o fasta entonc'es. Et que esto paresc'ia muy bien ser verdat por las obras quele el faz'ia contra el & contra aquellos que estaua⟨n⟩ enel serujc'io de dios" (Craddock, 1986), p. 216.
6. "Et el cauallero [Pero Martínez] pregu⟨n⟩tole por que era esto que ios del tal sa⟨n⟩na auja. Et aquel om⟨n⟩e quele aparesc'io le dixo. don alfonso estando en seujlla dixo en plac'a que si el fuera con dios quando faz'ia el mundo q⟨ue⟩ muchas cosas emendara enque se fiz'iera mejor quelo que se fiz'o & que por esto era dios yrado contra el" (Craddock, 1986), p. 217.

you a fool and a dullard." And the friar quit his presence and left. And the following night God sent such a great storm of lightning and thunder that it was a great marvel. And in the bedchamber where the king lay with the queen a lightning bolt fell which burned the queen's headdress and most of the other things that were in the room.[7]

Alfonso confesses his erring ways right after the fire.

And the next day the king prayed and confessed publicly that sin of blasphemy that he had uttered against God.[8]

3. THE ANNALS OF GERÓNIMO DE ZURITA

The next version of this story cited by Craddock occurs in a sixteenth-century printed text by Gerónimo de Zurita (1512–80), a noted Spanish historian of the period, which depends on the *Crónica geral de 1344*. The passage that interests us appears in his *Los Anales de la Corona de Aragón* and serves to explain the disastrous character of Alfonso's reign:

For this reason [explaining why Alfonso's reign was a disaster], some authors, among them King Pedro IV of Aragon, write that he was so insolent and arrogant, because of his great knowledge of human sciences, and for his knowledge of the secrets of nature, that he arrived at the point of scorning the providence and supreme wisdom of the universal Creator, and [said] that if he had been His advisor at the time of the general creation of the world and of all the things in it, and had found himself with God, some things would have been produced and formed better than they were made; and others would not even have been made, or they would have been improved and corrected (*emendaran y corrigieran*); and so it seemed clearly that through this very great blasphemy, Our Lord let it be known how perverse his judgment and understanding were, and he was disinherited of his kingdoms, and abandoned by all the Christian princes, and his line of successors was broken in the fourth generation, and this is what an old author of the things of Portugal recounts, [all of] which was revealed to

7. "& vino al rrey & dixole q⟨ue⟩ fiz'iesse penjtenc'ia delos pecados que auja fechos & que faria su pro & majormente de aquellas malditas & descomulgadas palabras conplidas de mucha soberuja & dichas con grant presunc'ion & vanjdat (^de)las quales dixera muchas vez'es en plac'a diz'iendo quesi fuera consejero de dios quando fiz'iera el mundo & lo q⟨u⟩i'siera creer quelo fiz'iera mejor que lo fiz'o. si non que non dubdase que dios sobre el non mostrase el su poder. Et el rrespondio con vulto yrado & palabras de sa⟨n⟩na & dixo yo digo verdat enlo que digo. Et por lo que vos dez'ides tengo uos por nesc'io & por sinsabor. Et el freyre partiose delante del & fuese luego. Et esa noche sigujente enbio dios tan grant tenpestad de rrelanpagos & truenos que esto era vna grant maraujlla. & enla camara donde el rrey yaz'ia conla rreyna cayo vn rrayo que q⟨ue⟩mo las tocas dela rreyna & grant parte delas otras cosas que ay estaua⟨n⟩ ensu camara" (Craddock, 1986), p. 218.

8. "Et otro dia pedrico el rrey & confeso publica mente aquel pecado de blasfemja que dixera contra dios" (Craddock, 1986), p. 219.

the Queen Doña Beatriz his mother, by a Greek woman, a sorceress, and by different visions, that he would die disinherited.[9]

The dependence on the *Crónica geral de 1344* is seen most clearly from Zurita's final remark, "This is what an old author of the things of Portugal recounts." Although Alfonso's words are specifically called blasphemous, again there is no mention of astronomy or astrology.

4. THE HISTORY OF SPAIN BY RODRIGO SANCHEZ DE ARÉVALO

Another early printed source gives a slightly different version of this story. In 1579 Robert Bell published a collection of Spanish histories that included a work by Rodrigo Sanchez de Arévalo (1414–70), who was successively bishop of Oviedo, Zamora, Calahorra, and Palencia. Rodrigo tells us that Alfonso was called "astrologus," a term that signifies either an astronomer or an astrologer or both. Moreover, in the same chapter, to account for the misfortunes of Alfonso's reign, he describes Alfonso's blasphemy, leaving out most of the details and referring only to the "annals of the Spaniards" as his source. Here the blasphemy is given as, "For he used to state openly in blasphemous speech that if he had taken part in a council of God, Most High, at the beginning of the creation of Man, some things would have been arranged better and in a more orderly fashion." It is important to see that this statement is not linked to astronomy or astrology, and that there is certainly no hint of criticism of Ptolemaic astronomical principles.

Rodrigo's version of the story is the following:

Chapter 5: Why this Alfonso X was called an *astrologus*, how he declared arrogantly that the works of God could have been made better, the manner in which he was ruined from on High, concerning the misfortunes and disasters he suffered for this reason, and concerning the other events in his time.

9. "Por esta causa escriven algunos autores, y entre ellos el Rey don Pedro el quarto de Aragon, que fue tan insolente y arrogante, por la grande noticia que tuuo de las sciencias humanas, y por los secretos que supo de naturaleza, que llego a dezir en menosprecio de la prouidencia y suma sabiduria del vniuersal Criador, que si el fuera de su consejo al tiempo de la general creacion del mundo, y de lo que en el se encierra, y se hallara con el, se vuieran produzido y formado algunas cosas mejor que fueron hechas; y otras, ni se hizieran, o se emendaran, y corrigieran; en que parecio manifiestamente, que por tan grande blasfemia como esta permitio nuestro Señor, que se conociesse, quan peruerso juyzio y entendimiento fue el suyo, y fue desheredado de sus Reynos, y desamparado de todos los Príncipes Cristianos, y que faltasse en la quarta generacion la linea de sus sucessores, y assi cuenta vn Autor antiguo de las cosas de Portogal, que fue revelado a la Reyna doña Beatriz su madre, por vna Griega gran hechizera, y por diuersas visiones, que auia de morir desheredado" (Gerónimo de Zurita, *Anales de la Corona de Aragón* [Zaragoza, 1585]), book 4, chap. 47, pp. 274v–275r.

Not only did this Alfonso X strive to enhance the glory of his name in establishing laws, in pouring out his wealth, and in other splendid deeds, but either out of arrogance or because his nature drove him to it, he used to delight in astronomy. Wherefore, he was called an *astrologus*. In his name, perhaps through his knowledge, the Alfonsine Tables and other astronomical/astrological studies were compiled, and under his royal name they were elaborated.... This Alfonso, as the annals of the Spaniards relate, tried to judge and even to correct the divine works, which are most perfect and created with the highest wisdom in weight, number, and measure. For he used to state openly in blasphemous speech that if he had taken part in a council of God, Most High, at the beginning of the creation of Man, some things would have been arranged better and in a more orderly fashion (*ordinatius*). ... [10]

5. VOLTAIRE AND HIS SOURCES

The key text would seem to be that of Voltaire, for he apparently claims to be the first to understand Alfonso's words to refer to the Ptolemaic System. In a chapter on the history of Spain in the twelfth and thirteenth centuries in Voltaire's *Essai sur les moeurs et l'esprit des nations* (1756), we read:

It has been said of him [Alfonso X] that while studying the sky he lost the earth. This trivial thought would be just if Alfonso had neglected his affairs in favor of study; but he never did this. The same profound spirit that made him into a great philosopher [also] made him into a very good king. Yet many authors accuse him of atheism for having said, "If he had been in the council of God, he would have given Him good advice concerning the movement of the celestial bodies." These authors did not notice that this pleasantry of the wise prince fell uniquely on the Ptolemaic System, whose inadequacy and contradictions he felt.[11]

10. "Quare iste Alfonsus X. dictus est Astrologus, & quomodo arroganter opera Dei dixit melius fieri posse, & qualiter fuit diuinitus correptus, & de infortuniis & incommodis quae ex ea causa passus est, & de caeteris incidentiis tempore suo. Caput V.

Nec solum hic Alfonsus X. nominis sui gloriam in legibus condendis, in effundendis diuitiis, caeterisque magnificis gestis ampliare contentus est, sed aut arroganter, aut quia natura ad id eum impellebat, Astronomia delectabatur. Quare & Astrologus appellatus est. Cuius nomine, nescio an sapientia, tabulae Alfonsinae & aliae Astrologicae considerationes compilatae sunt, & sub eius regio nomine lustrantur. . . .[*lines 41–54 omitted*] . . . Hic Alfonsus (vt tradunt Hispanorum annalia) diuina opera quae perfectissima sunt, & cum summa sapientia, pondere, numero, & mensura [p. 377:1] creata, iudicare quippe & emendare conatus est. ore enim blasphemo dicebat pallam, Si a principio creationis humanae Dei altissimi consilio interfuisset, nonnulla melius ordinatiusque condita fuisse . . ." (Rodrigo Sanchez de Arévalo, "Roderici Santii Episcopi Palentini Historiae Hispanicae, partes quatuor," in R. Bell, *Rerum hispanicarum scriptores aliquot* [Frankfurt, 1579], pp. 376–77).

11. "On a dit de lui [i.e., Alfonso X] qu'en étudiant le ciel il avait perdu la terre. Cette pensée triviale serait juste si Alfonse avait négligé ses affaires pour l'étude; mais c'est ce qu'il ne fit jamais. Le même fonds d'esprit qui en avait fait un grand philosophe en fit un très bon roi. Plusieurs auteurs l'accusent encore d'athéisme, pour avoir dit "que s'il avait été du conseil de Dieu, il lui aurait donné de bons avis sur le mouvement

Notice that the blasphemy of Alfonso has been reduced to a "pleasantry"! Again no source is cited but, the divorce of this story from its context, the favorable attitude to Alfonso, and the allusion to astronomy, suggest that Voltaire depended on Rodrigo rather than on Zurita.

It can now be shown that Voltaire did in fact depend on Rodrigo and that P. Bayle's *Dictionnaire historique et critique* (first edition 1697) was probably his principal source. In a note to the article "Castille," Bayle cites the passage from Rodrigo in Latin and presents a French translation of it.[12] Bayle goes on to relate Alfonso's blasphemy to a criticism of Ptolemaic astronomy as follows:[13]

Although the silence of so wise an Historian[14] as to the System of Ptolemy ought to be of some weight, yet I believe that if Alphonsus exercised his audacious Censure on any part of the Universe, it was on the celestial Sphere. For, besides that he studied nothing more, it is certain that, at that time, Astronomers explained the Motion of the Heavens by such intricate (*embarrassées*) and confused Hypotheses, that they did no Honour to God, nor answered in any wise the Idea of an able Workman. So that it is likely it was from considering that Multitude of Spheres, of which Ptolemy's System is composed, so many eccentric Circles, so many Epicycles, so many Librations, so many Vehicles, that he happened to say, *That if GOD had asked his Advice, when he made the World, he would have given him good Counsel.*[15]

In a marginal note Bayle cites Fontenelle's *Entretiens sur la pluralité des mondes* (1686) where we read that the "King of Castile," who apparently

---

des astres." Ces auteurs ne font pas attention que cette plaisanterie de ce sage prince tombait uniquement sur le système de Ptolémée, dont il sentait l'insuffisance et les contrariétés" (Voltaire, *Essai sur les moeurs et l'esprit des nations* 1, ed. R. Pomeau [Paris: Classiques Garnier, 1963], p. 645).

12. "S'il avoit assisté au conseil de Dieu lors de la création de l'homme, il y avoit certaines choses qui seroient en meilleur order qu'elles ne sont" (Bayle, 1697), see note 15.

13. Des Maizeaux (trans.), *The Dictionary Historical and Critical of Peter Bayle*, 2d ed. (London, 1735), vol. 2: 379 (article "Castille", Note H).

14. A marginal note refers us to Juan de Mariana, *Historiae de rebus hispaniae* (Toledo, 1592), book 14, chap. 5, p. 668. In this chapter Mariana describes the reign of Alfonso and alludes to his blasphemy without citing it.

15. "Encore que le silence d'un si sage Historien [*mg. adds*: Mariana] par raport au Système de Ptolémée doive être de quelque poids, je ne laisse pas de croire que si Alfonse porta sa critique audacieuse sur quelque partie de l'Univers, ce fut sur les Spheres célestes. Car, outre qu'il n'étudia rien tant que cela, il est sûr que les Astronomes expliquoient alors le mouvement des cieux par des Hypotheses si embarrassées et si confuses, qu'elles ne faisoient point d'honneur à Dieu, et ne répondoient nullement à l'idée d'un habile Ouvrier. Il y a donc aparence que ce fut en considérant cette multitude de Spheres dont le Système de Ptolémée est composé, tant de cercles eccentriques, tant d'épicycles, tant de librations, tant de déférans, qu'il lui échapa de dire, que si Dieu l'eût appellé à son conseil, quand il fit le monde, il lui eût donné de bons avis" (P. Bayle, *Dictionnaire historique et critique* [Rotterdam 1697]: I have depended on the 3d ed. [Rotterdam: Michel Bohm, 1720], vol. 1:805 [article "Castille", Note H]).

was not very devout, said that if God had called him to His council when He made the world, he would have given Him good advice. Fontenelle adds that "[Alfonso's] thought is too impious (*libertine*), but it is quite amusing (*plaisant*) that this System [of astronomy] was then an occasion for sin, because it was too confused."[16] Fontenelle's account was translated into English shortly after it appeared in French:

> The Ancients imagin'd a strange Labyrinth of Circles to salve those extravagant Appearances. So great was the intricacy of those Circles, that then when men knew no better, it was said by a King of Arragon [sic], a great Mathematician, but something irreligious, That if he had been of God Almightys Council when he made the World, he would have advised him better. 'Twas the expression of a Libertine; but pleasant enough, that at that time the great confusion of that System was the occasion of sin.[17]

According to Shackleton,[18] in May 1686 P. Bayle wrote a review of the first edition of Fontenelle's *Entretiens* in which Bayle noted that "Aragon" should be corrected to "Castille". It seems that at least some of the subsequent French editions have been so corrected.

Note that Fontenelle does not use the word blasphemy, and calls Alfonso's critique of Ptolemy "amusing," and this seems to lie behind Voltaire's characterization of Alfonso's remark as a "pleasantry." Fontenelle cites no source, but, according to Shackleton,[19] it is likely to have been a historical work by Moreri (first edition, 1681)[20] in which Ro-

16. "Les Anciens avoient imaginé je ne sçay combien de Cercles differemment entrelassez les uns dans les autres, par lesquels ils sauvoient toutes ces bizarreries. L'embaras de tous ces Cercles estoit si grand, que dans un temps où l'on ne connoissoit encore rien de meilleur, un Roy de Castille, grand Mathematicien, mais apparemment peu devot, disoit que si Dieu l'eust appellé à son Conseil quand il fit le Monde, il luy eust donné de bons avis. La pensée est trop libertine, mais cela mesme est assez plaisant, que ce Sistème fust alors une occasion de peché, parce qu'il estoit trop confus" (B. de Fontenelle, *Entretiens sur la pluralité des mondes* [Paris, 1686]: I have depended on the nouv. éd. [Amsterdam: Pierre Mortier, 1687], pp. 32–33).

17. W. D. Knight (trans.), [Bernard de Fontenelle] *A Discourse on the Plurality of Worlds* (Dublin, 1687), pp. 9–10. Cf. Cotes's preface to the second edition of Newton's *Principia* dated 1713 (F. Cajori [trans.], *Sir Isaac Newton's Mathematical Principles* 1 [Berkeley, Los Angeles, London 1934], p. xxxii).

18. R. Shackleton (ed.), *Fontenelle: Entretiens sur la pluralité des mondes* (Oxford, 1955), p. 180.

19. Shackleton, 1955, p. 181.

20. "On dit qu'Alfonse lut quatorze fois toute la Bible aves ses Gloses, et que ses grandes occupations ne l'éloignoient point de l'étude et de ses observations Astronomiques. Il disoit ordinairement, qu'il auroit mieux aimé vivre en simple particulier, que de manquer de science et d'érudition. . . . On assure encore qu'il avoit de la piété; mais une réponse qu'on lui attribue, détruit ce sentiment: car considérant en Astronome les merveilles de la Création du Monde; il osa dire que si Dieu lui eût fait l'honneur de l'y appeller, il lui auroit donné de bons conseils" (L. Moreri, *Le Grand Dictionnaire historique* [Lyon, 1681]: I depended on the edition of Amsterdam: P. Brunel, 1740, vol. 1: 290b [article "Alfonse X"]). Mariana (1592) and Rodrigo Sanchez de Arévalo (1579) are included in the list of sources appended to the article.

drigo's work is cited. Since Moreri did not give an astronomical interpretation to Alfonso's remark, Fontenelle is the earliest source I have found in which this association is explicit. It is also noteworthy that only Bayle and Moreri cite their sources. There follows Moreri's account as it appeared in English in 1701:

> He had read the Bible 14 times, with several Commentaries upon it; he was a great Astrologer, and after he had deeply considered the Fabrick of the World, the following saying of his, reported by Lipsius, denotes him to have been none of the most Pious, *viz. That if God had advised with him in the creation, he would have given him good counsel.*[21]

As in Moreri's French text, Mariana (1592) and Rodrigo (1579) are included in the list of sources appended to the English translation of his article. On the other hand, Lipsius is not mentioned in the original French version; however, as Bayle indicated (in a note to the passage cited above), J. Lipsius mentioned Alfonso's blasphemy without any astronomical interpretation.[22]

6. THE HISTORY OF MATHEMATICS BY MONTUCLA

In Montucla (1799), we read that Alfonso was shocked by the [Ptolemaic] hypotheses, and this led him to utter a pleasantry that is not respectful. Montucla's use of the expression "better and simpler order" follows Bayle's account (taken in turn from Rodrigo), and not that of Voltaire; but the term "pleasantry" surely comes from Voltaire:

> Alfonso, shocked by the complicated hypotheses that he had to admit in order to reconcile all the celestial motions, could not restrain [himself from uttering] a pleasantry that is not very respectful. He said that if God had called him to His council when He created the Universe, things would have been in a better and simpler order. If we do not find in this expression much evidence for the religion of this prince, at least it teaches us that he regarded it as a blemish on the work of the universe.[23]

7. THE HISTORY OF ASTRONOMY BY DELAMBRE

Delambre's series of works on the history of astronomy composed at the beginning of the nineteenth century is still unsurpassed in many

21. J. Collier, *The Great Historical Dictionary . . . out of Lewis Morery,* 2d ed. (London, 1701), vol. 1, unpaginated (article "Alphonsus the Tenth").

22. J. Lipsius, *Monita et exempla politica,* libro duo (Antwerp, 1606), book 1, chap. iv, p. 23.

23. "Alphonse, choqué des hypothèses embarrassées qu'il falloit admettre pour concilier tous les mouvemens célestes, ne put retenir une plaisanterie peu respectueuse. Il dit que si Dieu l'eût appelé à on conseil, lorsqu'il créa l'Univers, les choses eussent été

respects. In a volume on medieval astronomy (1819), he indicates his awareness of this "pleasantry"; as usual, no source is cited and the "pleasantry" is different from the passages previously cited. In fact, it seems to be based on Voltaire's version, from which the reference to the pleasantry was taken, together with the passage in Bayle's *Dictionnaire*. This is Delambre's version of the story:

> Alfonso died at the age of 58 years in 1284, after a very unfortunate reign; he is often cited for his saying, that if God had consulted him at the moment of the creation, he would have given Him good advice. One has seen in this pleasantry a mark of impiety that concerns primarily the complication (*complication*) and the incoherence of the Ptolemaic System. However, it would seem that Alfonso accepted the reality of this system for, in regard to reforming [astronomical] tables, he did not give to his astronomers the advice he would have wished to give to God.[24]

Note that the expression found elsewhere (for example, in Montucla 1799, cited above), *hypothèses embarrassées,* is here called *complication,* which supports the rendering of *embarrassé* as "complicated" (compare Des Maizeaux's translation of Bayle, cited above).

8. TWENTIETH CENTURY REFERENCES TO THE MYTH OF ALFONSO

Dreyer's history of astronomy is often dependent on Delambre: here Dreyer has taken over Alfonso's statement while reversing Delambre's interpretation of it. According to Dreyer, Alfonso had his doubts about the physical truth of the Ptolemaic System, whereas Delambre claimed that he accepted the reality of that system. Following Dreyer, a number of recent scholars have continued to draw upon this legend for various purposes.

In T. S. Kuhn's now classic *The Structure of Scientific Revolutions* (1962), Alfonso's remark serves to show that in the thirteenth century some scholars were aware that the adjustments in the Ptolemaic System, made to account for the discrepancy between observations and calcula-

---

dans un ordre meilleur et plus simple. Si nous ne trouvons pas dans ce mot une preuve de la religion de ce prince, il nous apprend du moins qu'il le regardoit comme une tache à l'ouvrage de l'Univers" (J. F. Montucla, *Histoire des Mathématiques* [Paris Year 7 (1799)], vol. 1, p. 511).

24. "Alphonse mourut âgé de 58 ans, en 1284, après un règne très malheureux; on a cité souvent de lui ce mot, que si Dieu l'eût consulté au moment de la création, il lui eût donné de bons avis. On a cru voir une marque d'impiété dans cette plaisanterie, qui porte principalement sur la complication et l'incohérence du système de Ptolémée. Il paraîtrait cependant qu'Alphonse croyait à la réalité de ce système, puisqu'il n'a pas donné à ses astronomes, pour réformer les tables, les conseils qu'il aurait voulu donner à Dieu" (J. B. Delambre, *Histoire de l'astronomie du moyen âge* [Paris, 1819], p. 248).

tions, had led to complexity rather than increased accuracy.[25] Kuhn then cites Alfonso's remark in a form close to that of Dreyer. This is not the place to argue against Kuhn's view that adjustments in "Ptolemy's system of compounded circles" were made to account for observational discrepancies,[26] but we may note that there is no contemporary evidence to indicate that Alfonso expressed doubt about the Ptolemaic System, based on observations or for any other reason.

A recent allusion to Alfonso's remark in a context similar to that of Kuhn is to be found in Haugeland: "Unfortunately, as more careful observations accumulated, it became progressively harder and more complicated to square them with the accepted geocentric (Earth-centered) theory. The situation was so serious and exasperating, especially with regard to predicting the positions of the planets, that by the thirteenth century, Spain's King Alfonso X could exclaim: 'If God had consulted me when creating the universe, He would have received good advice!' "[27] In a note to this passage, Haugeland adds that "whether the story is true or not, the fact that it was told (and believed) suffices for the point in the text." Here, as in Kuhn, the theme is that the accumulation of observations and their discrepancies with theory led Alfonso to make his declaration. Haugeland cites Kuhn and notes that A. Koestler quoted Alfonso's remark in a similar context.[28]

In his account of Alfonso's blasphemy as reported in the *Crónica geral de 1344*, J. R. Craddock adds a note: "See . . . Dreyer . . . , and . . . Kuhn. . . . Bernard R. Goldstein, University of Pittsburgh, to whom I owe the references just cited, suspects that Alfonso X's alleged dissatisfaction with Ptolemy's epicycles is just another myth that has grown up around the figure of the learned King."[29] We can now put this myth aside as just another attempt to enliven the history of science.

25. "Given a particular discrepancy [between observations and calculations based on Ptolemy's models], astronomers were invariably able to eliminate it by making some particular adjustment in Ptolemy's system of compounded circles. But as time went on, a man looking at the net result of the normal research effort of many astronomers could observe that astronomy's complexity was increasing far more rapidly than its accuracy and that a discrepancy corrected in one place was likely to show up in another.

"Because the astronomical tradition was repeatedly interrupted from outside and because, in the absence of printing, communication between astronomers was restricted, these difficulties were only slowly recognized. But awareness did come. By the thirteenth century Alfonso X could proclaim that if God had consulted him when creating the universe, he would have received good advice" (T. S. Kuhn, *The Structure of Scientific Revolutions* [Chicago and London, 1962], pp. 68–69).

26. See P. Barker and B. R. Goldstein, "The Role of Comets in the Copernican Revolution," *Studies in the History and Philosophy of Science* 19 (1988): 299–319.

27. J. Haugeland, *Artificial Intelligence: The Very Idea* (Cambridge, MA, and London, 1985), p. 17.

28. A. Koestler, *The Sleepwalkers* (New York, 1959), p. 67.

29. Craddock, 1986, p. 205, n. 21.

PART IV

MATHEMATICS

8   Cavalieri's Indivisibles and Euclid's Canons

FRANÇOIS DE GANDT

Mathematicians of the seventeenth century were audacious, and their creativity was an essential element in the new achievements which are often characterized as the "Scientific Revolution."[1] It enlarged and irrigated mathematics itself, and it also made possible a richer study of nature. In particular the investigation of accelerated motion (curved trajectories of planets, trajectories of light, impact between bodies, and motion of fluids, for example) presupposed new tools, which involved infinitesimals.

These tools were eventually systematized under the form of the "calculus," that is, the Newtonian method of series and fluxions and the Leibnizian *algorithmus differentialis*. But before this systematized stage of the theory,[2] more informal sets of procedures were developed by the mathematicians of the years 1600–1680. Various versions of infinitesimals and "indivisibles," a sort of precalculus, became more and more

---

1. A first version of this paper was presented during the "Journées sur les Mathématiques Méditerrranéennes" organized in Marseille-Luminy by R. Rashed in April 1984. Unfortunately the planned Proceedings of that rich and original meeting remained in the realm of *potentia pura*. At that time there existed very little serious and detailed study of Cavalieri's theory: the two main contributions were the paper of A. Koyré and the excellent translation by L. Lombardo-Radice. I did not know of the existence of E. Giusti's study, which appeared *fuori commercio* in 1980. Then came the extensive and remarkable paper by K. Andersen. It rendered superfluous a good part of my original contribution, and therefore the present text concentrates upon a particular aspect of Cavalieri's theory: a confrontation with Greek geometrical requirements. The reader who wants more details and examples about the application of Cavalieri's method can now find them in the texts of K. Andersen and E. Giusti, in order to make his own way into the Latin text or the Italian translation (it must be noticed that the first book of the *Exercitationes*—of which there is a reprint—is an abridged and improved presentation of essential parts of the *Geometria*, for the most part excerpts from book II of the *Geometria*).

2. Even after the birth of the "calculus," indivisibles remained a useful tool among mathematicians and physicists. Cf. the textbooks mentioned by E. Giusti in "Dopo Cavalieri . . . " p. 87.

openly used, for instance in the works of Pascal and Wallis. The name of Cavalieri was often associated with the new method.

1. SEVENTEENTH-CENTURY ATTITUDES TOWARD THE CLASSICAL TRADITION

Were the creators of the new method audacious? Their own standard of reference was the classical tradition of Greek mathematics. How innovative and bold were they, by comparison with such canons?

To be perfectly rigorous, we should try to define exactly what we must understand by "the classical canons of Greek geometry." But it would lead us too far astray, and the task may indeed be insuperable, even if restricted to Euclid's *Elements*. There are strong obstacles and ambiguities that prevent a clear-cut definition of Euclid's axiomatic procedures (see Daniel Lacombe[3] and Ian Mueller[4]). We must be content with a case-by-case examination, a strategy that is more defensible. The particular features of the Greek models, the most specific prohibitions and stipulations of Greek mathematical discourse, will emerge through the comparison with seventeenth-century texts. Cavalieri will help us, in some sense, to read Euclid with a new precision. Reading Cavalieri offers answers to key questions: Does Euclid admit motion among the building blocks of his construction? How does he treat infinite quantities? What is the difference between a surface and an area?

The most important and obvious opposition between the seventeenth-century geometry and its traditional counterpart concerns the use of infinitesimal objects. Indivisibles had been forbidden and discarded during the classical era of Greek mathematics. The standard exposition of the dangers and contradictions in the use of such entities was given by Aristotle in book VI of his *Physics*;[5] his exposition was loosely glossed and expanded in the pseudo-aristotelian *De lineis insecabilibus*. If we except the very special case of Archimedes *Letter to Eratosthenes* (the "Method"), rediscovered in 1906, the major texts of ancient mathematics avoided any use of infinitesimal entities. This is a major

---

3. "Le champ [. . .] des évidences admises sans discussion par la géométrie grecque n'a jamais été explicité par elle" (Daniel Lacombe, "L'axiomatisation des mathématiques au IIIéme siècle avant J. C.," in *Thalès*, année 1949–50, [Paris: PUF, 1951], p. 46).

4. "Even the most extensive list of common notions in the manuscripts is inadequate to cover all of Euclid's inferences." "The deductive gaps in the *Elements* occur at a much more rudimentary level than the level of continuity or betweenness." Ian Mueller, Greek Mathematics and Greek Logic, in *Ancient Logic and Its Modern Interpretation*, J. Corcoran (ed.) (Reidel Dordrecht, 1974), p. 45.

5. The best exposition is given by Wieland, pp. 278–316. See also Waschkies, Sorabji, Kretzmann.

contrast, but our picture has to admit nuances and subtle gradations. There were many possible attitudes toward the classical tradition of mathematics. Among seventeenth-century pioneers in mathematics, some were very bold—or candid—and did not even attempt to justify their new freedom. Roberval, for instance, did not link his indivisibles, nor his kinematical method of tangents, to any ancestor.[6] Some other creators were convinced that the Ancients had similar tools, which were not transmitted in the official texts of the tradition. It was a frequent complaint addressed to the Ancients: Archimedes and other Greek mathematicians must have possessed a direct method of invention, although they disguised their results in the ritual dressing of rigorous demonstration (the double *reductio ad absurdum*). Among various expressions of this opinion, let us select two examples.

Torricelli writes in 1644, at the beginning of his "Quadrature of the parabola made through the new Geometry of the indivisibles" (a part of his *Opera Geometrica*):

I would not dare to affirm that this new Geometry of indivisibles is a totally new discovery. I could more easily believe that the ancient Geometers have made use of that method in the discovery of the most difficult theorems, though they preferred another way in the demonstrations, either to conceal the secret of the art, or not to give any occasion of contradiction to their jealous detractors.[7]

And Wallis, in a letter where he comments upon his own *Arithmetica Infinitorum* and Fermat's methods, writes:

The most serious and learned men regret, and almost consider as a defect, that Archimedes thus concealed the traces of his process of discovery, as if by jealousy he wanted to deprive the posterity of his means of discovery [. . .] But Archimedes was not the only one; most of the Ancients have hidden their Analytics (since there is no doubt that they possessed such a thing) to their posterity in such a way that it was easier for the Moderns to invent a new one.[8]

By claiming that the Ancients possessed analogous instruments, the creators felt justified in adopting new ways. Their invention was a restitution of the ancient skill, which had been veiled under the disguise of an inconvenient but rigorous scheme of demonstration. So far, Wallis and Torricelli show their respect for the Ancients. But in fact, in their work as creative mathematicians, they were closer to the Greek written tradition than Roberval was. A complete picture should also include

6. See notably the beginning of his "Traité des indivisibles" in *Mémoires de l'Académie Royale des Sciences*, vol. VI, 1693, pp. 247–48, and some decisive passages of his "Observations sur la composition des mouvements et le moyen de trouver les touchantes des lignes courbes," ibid., pp. 1–4 and 24–25.

7. *Opere di E. Torricelli* (Faenza, 1919), vol. 1, part 1, pp. 139–40.

8. *Wallis to Digby*, 21 Nov. 1657, in Fermat, *Oeuvres* 3, p. 439.

the role of the "classical" exercises, such as Torricelli's *On the Sphere and Spheroidal Solids*. These works reproduced and enlarged pieces of Greek geometry as a sort of training or display of virtuosity. Since modern readers are more interested by the obvious innovations, we run the risk of underestimating the role of this classical geometrical "practice," and of the step by step refinement of the inherited theories.

2. CAVALIERI

Cavalieri was not content with lip service to classical geometry. He tried, as very few did, to cast his new instruments in the mold of classical Euclidean exposition. It is striking to see how "Euclidean" Cavalieri is, compared to other creators in mathematics at the same time. He tried to give his new doctrine the classical solidity ensured by a set of demonstrative pillars, firmly delineated and rooted in the first principles of traditional geometry. He wanted to anchor the geometry of indivisibles in the Euclidean theory of proportions, and throughout the sequence of propositions in his book, he carefully made references to the *Elements* or other canonical texts.[9] He also maintained the appearance of the Euclidean procedure in the demonstration at the price of an extraordinary prolixity and a certain tediousness. Cavalieri's *Geometria Indivisibilibus* therefore proves to be a fruitful object for the study of continuity and change in mathematics. Since Cavalieri claims to be faithful to the Euclidean canons and tries to respect them in the course of his proofs, we can mark some of the crucial points where he departs from the classical tradition.

Cavalieri's strong respect for the Greek ideal of demonstration is not the only motive for a study of the *Geometria Indivisibilibus*. It can be considered as the starting point and the foundation for the seventeenth-century theory of indivisibles. During the following decades, the authors who expounded or used indivisibles appealed to Cavalieri's authority, so that his name became almost synonymous with the method of indivisibles. Other mathematicians (for instance Kepler and probably Roberval) had already forged similar tools. But only Cavalieri gave indivisibles the status of a genuine theory, clearly enunciated and founded on general theorems and preliminary definitions. Torricelli, Wallis, van Schooten, and Leibniz could then build on the foundations he had laid, although they did so with a basic misunderstanding, believing that Cavalieri considered a continuous magnitude as the sum of

9. As well as references to the preceding propositions in his book; unfortunately all these marginal references have disappeared in the excellent—and first—translation by L. Lombardo-Radice.

its indivisibles. Under the guarantee seemingly secured by the classical scaffolding of the *Geometria Indivisibilibus,* later mathematicians developed various new devices. In their view, the work of Cavalieri served to justify or authorize new freedoms. For us it is more important to try to check the solidity of the edifice's foundations, or at least to compare it to a well-established standard of rigor.

3. THE FUNDAMENTAL INSTRUMENT: A TRANSFER OF PROPORTIONS

In a letter to Galileo, Cavalieri displays the core of his method: "I did not dare to affirm that the continuum is composed of indivisibles, but I showed that between the continua there is the same proportion as between the collection of indivisibles."[10] Galileo asserted in his *Discorsi* that the line is made up of points, and that every continuous magnitude is composed of an infinite number of sizeless parts ("parti non quante"). Cavalieri contrasts himself to Galileo: "I do not compose the line with points, nor the surface with lines." His refusal to do so must be stressed heavily[11] against almost all of the second-hand versions of Cavalieri's theory. He never considers a surface as a sum, or a composition, or a totality of segments, nor a line as an aggregation of points.[12]

Concerning the continuum, Cavalieri does not put forward any such general thesis as Galileo's. He is not as bold as the Old Master ("I did not dare . . . "). Rather his strategy is indirect. He takes no definite position about the composition of the continuum, and does not pretend that his indivisibles are the elements which constitute magnitudes. He simply eludes the philosophical discussion. For Cavalieri, there is no need for the geometers to decide on the nature of the continuum. (We discuss this "neutrality" below.) Geometers want only to compare magnitudes, a comparison made possible by Cavalierian indivisibles. A Latin phrase from the preface of the *Geometria,* Book seven, clearly summarizes the

10. *Cavalieri to Galilei,* 21 June 1639, in *Opere di G. Galilei,* Ed. Naz., vol. 17, p. 67.
11. With A. Koyré and K. Andersen.
12. Except in a metaphorical and pedagogical passage in the *Exercitationes,* where he compares the surfaces in a solid with the sheets in a book, and the segments in a surface with the threads in a textile—unfortunately this passage was more widely known and quoted than his rigorous and technical formulations. *Exercitationes Geometricae* (Bologna, 1647), pp. 3–4. The analogy is expressed in the sort of preamble that precedes the formal theory: "Hinc manifestum est figuras planas nobis ad instar telae parallelis filis contextae concipiendas esse; solida vero ad instar Librorum, qui parallelis foliis coacervantur." It is followed by a sort of warning against a too gross interpretation: "Cum vero in telis sint semper fila, et in libris semper folia numero finita, habet enim aliquam crassitiem, nobis in figuris planis lineae, in solidis vero plana numero indefinita, ceu omnis crassitiei expertia, in utraque methodo supponenda sunt."

basic idea: "continua sequi indivisibilium proportionem" ("Continuous [magnitudes] follow the same ratio as indivisibles").[13]

Cavalieri's method is based on a transfer of proportions or ratios. Since there is the same ratio between two "continua" as between their indivisibles, it will suffice to establish proportions between indivisibles, and thence to draw statements concerning plane or solid figures. Following Cavalieri's own formulation:

> To discover which proportion there is between two plane figures, or two solids, it will be sufficient to find, in the plane figures, which is the proportion between all the lines of the figures, and, in solid figures, which is the proportion between all the planes of the figures, taken according to any rule.[14]

But we do not know what "all the lines" are or what a "rule" is until we have penetrated into the complex machinery of the *Geometria Indivisibilibus*.

4. THE DEFINITION OF ALL THE LINES AND THE ROLE OF MOVEMENT

What is to be called an indivisible in Cavalieri's doctrine? The word itself could be misleading. Cavalieri never uses the term indivisible in the course of demonstrations. The word belongs, so to speak, to the metalanguage; it is employed only in titles, prefaces, scholia, comments, and various glosses. In proper mathematical reasoning, Cavalieri uses another phrase: *omnes lineae* or *omnia plana* ("all the lines," "all the planes"). These *omnes lineae* are the result of a process of cutting that is governed by very strict conditions (we shall here restrict the discussion to the case of plane figures):

—the figure must be framed between two opposite and parallel tangents. Proposition I of book I has showed that it is always possible to insert a given plane figure between two parallel tangents, as a consequence of a postulate (postulate 1 of book I) which gives the right to move a straight line parallel to itself;

—through the two opposite tangents there pass two parallel planes;

—one of those planes slides or "flows" ("fluit") toward the other one which remains fixed, the moving plane remaining always parallel to another fixed plane;

—the interest is focussed on the successive positions of the moving plane: at each instant that plane cuts a particular straight line on the

---

13. *Geometria Indivisibilibus*, book VII (Praefatio, 1653), p. 483.
14. *Geometria Ind.*, book II, Cor. of prop. 3.

plane figure, and its mark or trace inside the boundaries of the figure is a particular segment (or several segments if the figure is concave); —if we consider together, collected as a whole, the segments successively cut off inside the figure by the moving plane, their aggregate forms what will be called "all the lines of a figure."

The stipulations are expressed in Definition 1 of book II:

Let there be given any plane figure with its two opposite tangents; let us draw two parallel planes passing through these tangents; the planes may be perpendicular or oblique to the plane of the figure, and they are supposed to be extended indefinitely on both sides. One of these planes moves toward the other, remaining parallel, until it coincides with the other plane (by the second postulate of book I); considering each of the right lines that are generated all along the movement as intersections of the moving plane and the given figure, let us call those lines, collected all together: all the lines of the figure, taken with one of those lines as their rule (by the E of definition 2 of book I) . . . [15]

Of that procedure let us offer the example that Cavalieri gives in the Appendix:

Let ABC be any plane figure, and EO and BC its two opposite tangents, drawn in any manner; let us understand also that two parallel and indefinitely extended planes are drawn through EO, BC, and that the one passing through EO for example moves toward the other which passes through BC, always parallel to it, until it becomes coincident with it; then I call the common sections of the moving or flowing plane and the figure ABC, which are generated all along the motion, when they are collected together, all the lines of the figure ABC, some of them being LH, PF, BC, taken with one of them as their rule, that is BC . . . [16]

Among the various conditions of the production of the *omnes lineae* we must stress the role of the "rule" (*regula*), which indicates the direction of the cutting or sweeping. One does not take all the lines of a figure ABC in general and absolutely speaking, but all the lines determined according to the rule AB. A later theorem will try to prove that the aggregate is invariant if the rule is changed.

The role of movement and time is a very interesting feature of the definition of the *omnes lineae*. With each instant is associated a position of the moving plane and a particular segment (or several segments) inside the plane figure. Continuity of movement and of time serves as a guarantee that "all" lines have been taken in the figure.[17] This is more explicit in Cavalieri's answer to Guldin, in the *Exercitationes:*

15. *Geometria Ind.,* 1653, p. 99.   16. Ibid., p. 104.
17. As Koyré notices. A. Koyré, "Bonaventura Cavalieri et la Géométrie des Continus," in *Etudes d'Histoire de la Pensée Scientifique,* p. 345.

Let us suppose a square, and through two of its opposite sides pass two indefinite planes, perpendicularly erected . . . and move one plane continuously toward the other. . . . At any moment the moving plane will cut the plane of the square. . . . Since in the aforesaid square no line can be assigned through which the moving plane does not pass once, that is at a certain moment (this is why I said that those lines are described by the plane) therefore all those lines, being so collected in mind that none is supposed to be excluded, I call "all the lines". . . . And if I say that those lines are produced in the totality of the movement, I imply: in all the instants of the time during which the total motion occurs. (198–99)

Time and motion therefore play a decisive role at that stage of the construction of the theory, both in the formulation of postulates 1 and 2 of book I (which give permission to let a line of a plane slide while remaining parallel to a given direction) and in the determination of the segments associated to the successive instants of the motion of the plane.

Cavalieri's procedures far exceed the frontiers of what was explicitly or implicitly allowed in traditional geometry. In Euclid's *Elements* special sorts of motion are required, for instance, tracing circles (postulate 3), tracing rectilinear segments (postulate 1), extending them in a straight line (postulate 2); and when superposition of figures is required, the act of superposing two figures and making them coincide ("epharmozein;" compare book I, axiom 7 and prop. 4, 8, etc.) is an almost fictitious motion, in which only the final position is considered.

There is indeed in the ancient texts some very interesting recourse to motion, which Cavalieri could have invoked, as Galileo did. Examples are the "mechanical" generation of certain special curves, like spirals (compare Archimedes *On spirals*) or the quadratrix (cf. Pappus, *Collection*, book IV, chapter 26). But these curves were excluded from the canon of geometry.

It is the continuity of motion that is important here. Because motion is continuous, it is certain that no lines have been forgotten, since every intersection of the moving plane and the figure has been realized once. But what exactly does this mean for the collection of lines? In which sense are "all" the lines determined and so to speak materialized? Let us notice, in the short text we were just quoting, the contrast between two formulations: "no line can be assigned through which the moving plane does not pass . . . at a certain moment"; "all those lines being so collected in mind." The cautious negative phrasing of the first statement is faithful to the spirit of the traditional Greek procedures, where the infinite is never actually given. For instance Euclid is perfectly able to state and to prove that there is an infinity of prime numbers; there is always a finite procedure to construct a new prime number, outside the collection (*Elements* IX, 20: "Prime numbers are more than any proposed multitude of prime numbers"). In the same vein, we shall later quote the Euclidean way to "exhaust" a given surface by successive removal of rectilinear areas, without any recourse to an actual infinity of terms.[18] The second phrasing is much different in spirit: if we want to speak of "all the lines," and to allow a manipulation with such aggregates, we have to enter a realm where the objects are much less determined. It is only "in mind" that we can consider the infinity of these entities; they cannot be at the disposal of the geometer, existing and particularized as one well-marked intersection between a plane and a figure, upon which one could build geometrical considerations and constructions. The adjective *omnes* is here affected by a sort of modalization[19]: "all the lines" are considered insofar as they become materialized in their uninterrupted succession during the motion, and insofar as they can be thought of as collected together. The unified treatment of "one" motion and "one" interval of time seems to allow the global manipulation of the infinite family of intersections materialized during the motion; but this disagrees with the standard Euclidean practice.

5. THE THREE FUNDAMENTAL THEOREMS

We now know what it means to consider "all the lines" of a figure. The next step is to learn how to handle these infinite aggregates of lines and to make them fit into demonstrations of proportionality, in order to draw consequences concerning the figures themselves.

Cavalieri's argumentation proceeds in three stages, corresponding to the first theorems of *Geometria*, book II:

1. There can be a proportion (a ratio) between the infinite aggregates of lines (the *omnes lineae*). In Cavalieri's words: "All the lines of right transit of arbitrary plane figures, and all the planes of arbitrary solids, are magnitudes which have a ratio to one another."

2. If plane figures have equal areas, the aggregates of their lines (their *omnes lineae*) are equal: "All the lines of equal plane figures, and all the planes of equal solids, are equal, taken with any rule." This second proposition could be split into two distinct and gradual statements: (i) the aggregates of lines are equal for a given figure, if one changes the direction of the "rule" (the mode of cutting); and (ii) the aggregates of lines are equal if the figures are equal in area.

18. Cf. our discussion in this paper, about Cavalieri's prop. II, 2, where we shall see that the word "exhaust" is particularly inadequate for Euclid.

19. Cf. Renaldini, quoted by E. Giusti in "Dopo Cavalieri . . . " p. 102: the *omnes lineae* are "possible lines."

3. The ratio between the aggregates of lines of two figures can be transferred to the figures themselves: "Plane figures have to one another the same ratio as all their lines taken according to any rule, and solid figures, as all their planes taken according to any rule."

With these three propositions Cavalieri forges a new tool. Its use is indicated in the following corollary:

From this it is clear that, in order to discover what is the ratio between two plane figures, or two solids, it will be sufficient for us to find, in plane figures, what ratio all the lines of the figures have to one another, and in solid figures, what ratio all the planes of the solids have to one another, taken according to any rule. This is what I lay as the supreme foundation of my new Geometry.[20]

6. THE TREATMENT OF INFINITE AGGREGATES IN PROPOSITION 1

Let us consider the first of those three propositions, namely theorem 1. Its aim is to give a meaning to the notion of ratio between two infinite aggregates. Cavalieri asked for Galileo's advice, precisely about that question, as early as 1621:

I would like to know if all the lines of a plane have a certain ratio to all the lines of another plane figure, because, since we can always draw more and more such lines, it seems that all the lines of a given figures are infinite, and hence excluded from the definition of the magnitudes which have a ratio to one another; but because on the other hand, if we enlarge the figure, the lines also become larger, since we then have the lines of the first figure and in addition the lines which are in the excess of the enlarged figures compared with the given figure, it seems that these lines are not excluded from the definition [of the magnitudes between which there is a ratio]; I wish you to deliver me from this incertitude.[21]

All the discussion is about the possibility of treating infinite aggregates as Euclidean "magnitudes": are we allowed to admit the *omnes lineae* or *omnia plana* among the objects between which there can be a ratio? They are entities that can be greater or less, and more precisely, that satisfy the Definition 4 of Euclid's *Elements*, book V: "Magnitudes are said to have a ratio to one another, when the less can be multiplied so as to exceed the other."

We do not have Galileo's answer. But we can guess that the text in the First Day of the *Discorsi*, although very late, contains Galileo's opinion on that point, arguing that the words "larger" and "smaller" lose their

20. *Geometria Indivisibilibus*, ed. 1653, p. 115.
21. *Cavalieri to Galileo*, 15 Dec. 1621, Ed. Naz., vol. 13, p. 81.

signification in the realm of the infinite. In that well-known text Salviati makes the following answer to an objection by Simplicio:

Such are the difficulties which derive from our making with our finite intellect a discourse concerning the infinites, and ascribing to them the attributes which we ascribe to finite and terminated things; which I think is inappropriate, since I think that these attributes of larger, smaller and equal are not appropriate to the infinites, of which it cannot be said that one is larger, or smaller or equal to another.[22]

Cavalieri's answer is different, and its essential point was already contained in the question posed to Galileo. Since all the lines cut off inside a given figure are enclosed within the boundary of the figure, and since there is a ratio between the figures themselves, there must also be a ratio between all the lines of one figure and all the lines of the other. This is what he seeks to prove in the theorem 1: "All the lines . . . of plane figures . . . are magnitudes which have a ratio to one another."[23]

How is it possible to demonstrate such a proposition? How do we treat those infinite bunches of lines?

Cavalieri's reasoning is supported by two figures in which "all the lines" are supposed to be cut off along the direction of the "rule" EQ, common to both (Figure 1). We first consider the case where the figures have equal height (if not, one cuts off and separates the excess of the higher, placing this part in the prolongation, thus restoring the case of two figures of equal height).

The lines that are as named "all the lines of EAG" and "all the lines of GOQ" are segments such as LM and NS, cut by successive positions of the moving plane and the corresponding horizontal line such as in LS, always parallel to the base or "rule" EQ.

The successive stages of the reasoning are the following:

1. Having placed the two figures along the common line EQ, one takes any of the positions of the moving horizontal, such as LS; it cuts several segments inside the boundary of the figures; if we suppose the figures to be convex (which Cavalieri implicitly does), there are just two segments, such as LM and NS.

2. These two segments (which are called "lines" according to the traditional vocabulary) satisfy the definition of the magnitudes that can have a ratio to one another, since they can, being multiplied a sufficient number of times, exceed one another. In this case, says Cavalieri, "if the line NS is smaller than the line LM, it can, being

22. *Discorsi,* Ed. Naz., vol. 8, 78–79.
23. *Geometria Indivisibilibus,* ed. 1653, pp. 108–9.

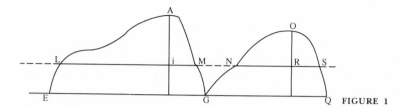

FIGURE 1

indefinitely extended, become at some time larger" ("potest indefinite producta aliquando fieri major").

3. The same thing is supposed to be done for all the "lines" ("si hoc intelligamus fieri de ceteris lineis"): all the lines virtually traced on GOQ are supposed to be extended enough to exceed the corresponding lines on EAG. Implicitly Cavalieri supposes that one chooses as common multiplier the one that is adequate for the largest of the lines in GOQ.

4. Then "all the lines of the figure EAG will be a part of all the lines of the figure GOQ that have been extended" ("patet ergo quod omnes lineae figurae EAG erunt pars omnium linearum figurae GOQ sic productarum").

5. But the whole is greater than its part ("totum autem est majus sua parte"); therefore all the lines of GOQ, being extended, are larger than all the lines of EAG. The same procedure can be applied in the other direction, EAG and GOQ exchanging their role.

6. As a consequence, these two infinite aggregates of lines are susceptible, if one multiplies them a sufficient number of times, to exceed one another. Therefore they satisfy the definition of magnitudes that can have a ratio to one another ("sunt magnitudines inter se rationem habentes").

In this way the aggregates comply with the criterion imposed by Euclid, to which Cavalieri refers explicitly in the margin of the demonstration ("Diffin 4.1.5 Elem."), and which has been restated at stage 6 of our enumeration ("magnitudines autem rationem habere inter se dicuntur, quae multiplicatae se invicem superare possunt"). Euclid's criterion was a very important piece in the Greek mathematical apparatus. It was restated, for instance, in Archimedes' *De Sphaera* (postulate 5). The Ancients also knew that there were objects that did not satisfy the requirement. Proclus mentions the case of rectilinear and curvilinear angles, which are not comparable and cannot exceed one another. They are then excluded from the class of magnitudes that can have a ratio to one another.[24]

But there is also a particular step in Cavalieri's reasoning that is not Euclidean. At stage 3, we have supposed the same operation of prolongation to be done an infinite number of times, for each of the particular positions of the moving horizontal line LS. As usual, an absence is difficult to prove; but, to the best of my knowledge, Euclid does not practice such an infinite iteration, and we should consider it as excluded by his "canons."

Let us also notice the crucial importance of one axiom: the whole is greater than its part. This axiom is essential, at stage 5, to prove that the *omnes lineae* can be declared larger than one another. Cavalieri does not refer to the *Elements* in this case, but it seems that this axiom[25] was accepted by the major editors or glossators (with the exception of authors like Hero and Boethius). The use of this particular axiom in the case of infinite quantities was rejected by Galileo in the *Discorsi*,[26] and it has become usual, since Dedekind,[27] to define infinite systems or sets by the fact that they do not satisfy this axiom: a set is infinite if it contains as many elements as one of its proper parts.

7. A FUNDAMENTAL PREMISE: THE COMPARABILITY OF BOUNDED FIGURES

The essential intuition in Cavalieri's demonstration is the same one expressed in his letter of 1621 to Galileo: if the infinity of the lines is enclosed within the frontiers of a given figure, then what is true concerning the size and comparability of figures must also hold, in some sense, for the aggregates of lines inside the figures. Cavalieri explains this idea in an answer to the objections of Guldin in book III of his *Exercitationes:* the aggregate of lines is not infinite absolutely speaking, but only infinite from a certain point of view (*secundum quid*), and it is finite or limited from another point of view. The infinity of lines is bounded by the sides of the figure, and since figures are comparable,

24. Proclus, *In Euclidem*, Friedlein 234; there is a vestige of this problem in Euclid's *Elements* III, 16.

25. Axiom 8 in Heiberg's edition, 9 in Clavius's version of 1603.

26. Galileo's assertion can be placed within a long tradition of discussions and speculations about the infinite and the continuum (see for instance Tony Lévy, *Figures de l'infini: les mathématiques au miroir des cultures* [Paris Le Seuil, 1987]). The application of the axiom concerning part and whole was denied in the case of the infinite, for instance by Duns Scotus ("Totum et pars inveniuntur solum in quantitatibus finitis, et de talibus ponitur illud principium: omne totum est majus sua parte," quoted in J. L. Gardies, *Pascal entre Eudoxe et Cantor* [Paris Vrin 1984], p. 46, note 31).

27. Cf. Dedekind, *Was sind und was sollen die Zahlen?*, 1887, sec. 64.

aggregates of lines must also be comparable. If, for example, a square is juxtaposed to an identical square, all the lines of the first square can also be found in the second square. Since all the lines of the first square can be found also in the second square, all the lines of the rectangle formed by the squares are double the lines of the square. No matter how many lines these are, the geometer can compare and manipulate them blindly, as the algebraist does his unknowns and radicals, without inquiring about what they are.[28]

The same conviction is expressed in the *Geometria Indivisibilibus,* in the scholium that immediately follows theorem 1, which we were just discussing. There the comparability of magnitudes is used as a trick to escape the philosophical question of the composition of the continuum. It is a well-known but difficult text:

SCHOLIUM

Someone could perhaps feel uncertain about this demonstration, failing to perceive how lines or planes indefinite in number, such as those which I call "all the lines" or "all the planes" of such and such a figure are thought to be, can be compared to one another. It seems therefore necessary to note that when I consider "all the lines" or "all the planes" of a figure, I do not compare their number, which we do not know, but only their magnitude [magnitudinem], which is equal [adaequatur] to the space occupied by the same lines, because it is congruent with it; and since that space is enclosed within limits, for that reason the magnitude of the lines is also enclosed within the same limits; therefore this magnitude can be subject to addition or subtraction, although we do not know the number of the lines. I say this is enough, in order for them to be comparable to one another: if not, the spaces themselves of the figures would not be comparable either.

For a continuum is either nothing else than the indivisibles themselves, or it is something else. If it is nothing else than the indivisibles, then of course if their assemblage [congeries] cannot be compared, the space or the continuum itself will not be comparable. And if a continuum is something else than the indivisibles themselves, it must be confessed that this something else lies between the indivisibles themselves. We then have a continuum which can be scattered [habemus ergo continuum disseparabile] into the things which compose the continuum, and which are indefinite in number. For between any two indivisibles there must lie something of the sort we have spoken of, when we said that there is something else in the continuum itself than the indivisibles; for if we had a reason to remove it from the interval between two indivisibles, then we would also remove it from the intervals between all the others. Things being so, we shall not be able to compare to one another the continua themselves or the spaces because then what is collected and compared after collection, namely the things which compose the continuum, would be indefinite in number. But it is absurd to say that continua enclosed within limits would not be comparable to one another; it is therefore absurd to say that the assemblages of all the lines

28. *Exercitationes Geometricae* 3, 8, pp. 202–203.

or planes of any two figures would not be comparable to one another, notwithstanding the fact that the things, which are collected and compose the assemblage, are indefinite in number, just as it is not an obstacle in a continuum. Therefore, whether a continuum is composed of indivisibles or not, the assemblages of indivisibles are comparable to one another and have a proportion.[29]

The argument is perplexing, but two basic statements are to be stressed. First, one does not compare the number of the lines, but only their "magnitude," which has some connection to the space they "occupy." This connection is expressed by the word *adaequatur*, which would deserve a long commentary by itself. It seems to denote something different from the usual congruence by superposition.[30] By contrast, *adaequatur* is used by Fermat in a central piece of argumentation in his method of maxima, to denote a quasi-equality between two expressions. Fermat, so to speak, "forces" two neighbouring points on a curve to be the same, in order to find the maximum. Fermat calls this *adaequatio* a "fictitious comparison."[31] Second, it would be absurd that bounded continuous surfaces or solids (*continua*) were not comparable, and from their comparability arises the comparability of the aggregates of lines or planes. For if the aggregates of indivisibles were not comparable, it would contradict the comparability of the continua.

Restricting the discussion to the two-dimensional case, two possibilities are open: Either the continuum is nothing but the aggregate of its lines (and then both must be equally comparable magnitudes), or the continuum is something else. Between any two lines there is some sort of stuff or intermediate material which contributes to the nature of the continuous space; but this intermediate something may also be split into an infinite number of elements, one piece of "stuff" between each pair of lines. The objection against comparability of "all the lines" should then also apply to the aggregates of these pieces of intermediate stuff. And this is obviously not the case, since we feel allowed to collect these pieces into a single bounded continuum, and to compare con-

29. *Geometria Indivisibilibus*, book II, scholium of prop. 1, ed. 1653, p. 111.

30. Clavius, when he transposes Euclid's axiom 7 of book I into his Notio Communis 8 [*Euclidis Elementorum* Libri XV, Romae 1603, pp. 67–68], renders the Greek verb "epharmozein" by "congruere," not by "adaequari." "Congruere" is also commonly used by Cavalieri himself to denote equality by superposition, for instance in postulate I and prop. II of book II of the *Geometria*.

31. Cf. Fermat, *Oeuvres*, vol. 1, p. 133ff., esp. p. 141; French trans. vol. 3, p. 127). E. Giusti mentions, in "Dopo Cavalieri . . . " p. 96, the assertion of Nardi, that all the lines "occupy a space similar and equal to the surface." Cf. also the biased presentation given by Torricelli in his *Opera Geometrica* of 1644, where the aggregates of indivisibles are identical to the figures themselves (cf. F. De Gandt, "Les indivisibles de Torricelli," in *L'Oeuvre de Torricelli, Science Galiléenne et Nouvelle Géométrie* [Nice, Presses de l'Université de Nice, 1988]).

tinua that are composed in this way. Therefore we should not be prevented from collecting and comparing aggregates of lines.

This argument presupposes an unusual or nonclassical conception of the continuum, in which a bounded figure is infinite in some sense, as an assemblage of infinitely many pieces of some sort of stuff. It is then no less infinite than the collection of lines that can be marked upon it. We have therefore an equal right to treat as a whole one figure and the corresponding collection of lines. Such a conception of a *continuum disseparabile*—if it is not advocated here only for the sake of this particular argument—does not correspond, for instance,[32] to the Aristotelian view of continuous magnitudes. Here the pieces, and a fortiori the indivisibles, do not exist until one has actualized them, by cutting a part or marking a point on a segment, or a line on a figure. It is always possible to cut new pieces, to draw new lines or points, but this does not confer on the magnitude the property of being infinite. This alleged similarity of status between the infinite collection of elements and the bounded magnitude is useful in Cavalieri's perspective. He has thus found a clever way to escape the traditional "labyrinth of the continuum," in order to build a theory which seems philosophically neutral but mathematically fruitful.

8. SOME APPLICATIONS

To show how Cavalieri uses the aggregates of lines in the course of the *Geometria Indivisibilibus*, let us follow a very elementary example, given in proposition 19 of book II (Figure 2). It concerns the area of the two triangles built on the diagonal (the "diameter") of a parallelogram:

THEOREM XIX, PROPOSITION XIX:
If one draws the diameter of a parallelogram, the parallelogram is double [in area] of any of the triangles constructed on the diameter.
Let AD be any parallelogram [ACDF, Fig. 2], and FC its diameter, which divides the parallelogram in the triangles FAC and CDF. I say that the parallelogram is double of any of the triangles FAC, CDF.
On FD and CA, let equal parts FE, CB be cut off toward the points F, C, and lines EH and BM be drawn parallel to the base CD through the points B and E,

---

32. It would be a gross simplification to suppose that the "implicit" philosophy of Euclid's *Elements* is Aristotelian. The system of Aristotle has many merits, among them its coherence, and the fact that it was transmitted to us, and it is certainly not without reason that Proclus, for instance, uses Aristotle's theory to formulate some philosophical questions and views concerning Euclid's practice, but we must refrain from tempting identifications. See the challenging views and warnings by W. R. Knorr in his "Infinity and Continuity: the Interaction of Mathematics and Philosophy in Antiquity," in Kretzmann, pp. 112–45.

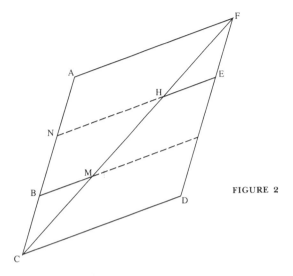

FIGURE 2

falling upon the diameter FC in the points H, M. Since in the triangles FHE and CBM the angle HFE is equal to its coalternate angle BCM, and the angle HEF is equal to its opposite angle FAC, which is itself equal to the angle MBC, the angle FEH is equal to the angle MBC. It follows that in the triangles FEH, MBC, there are two equal angles, with the adjacent sides equal, namely FE to BC; therefore (as a consequence of proposition 26 of book I of the *Elements*) the other sides will be equal, in particular HE to BM.

In the same way, about the other parallels to CD, namely those which cut off equal parts from the sides FD, CA, toward the points F, C, we show that they are also equal to one another, and the extremes AF, CD are equal as well.

Therefore, all the lines of the triangle CAF will be equal to all the lines of the triangle FDC, the lines being taken in both with CD as a rule; therefore (by proposition 3 of this book) the triangle ACF is equal to the triangle FDC; therefore the two triangles ACD, FDC, that is, the parallelogram AD, will be double of any of the triangles ACF, FCD. *Q. E. D.*[33]

This result, particularly simple and elementary, obtained through a sequence of rather trivial applications of *Elements*, book I, illustrates the cautious and slow pace of the demonstrations in the *Geometria Individibilibus* (and we are already around page 150 of the book).

In order to obtain more interesting results, it is necessary to enrich the theory with some new tools—for instance, the notion of "all the squares" built on all the lines of a given figure. An example will suffice to give an idea of the manner in which Cavalieri uses "all the squares."

33. *Geometria Indivisibilibus*, book II, prop. 19 (ed. 1653), p. 146.

It concerns the quadrature of the parabola, expounded in a proposition of the *Exercitationes*, which is a sort of abridgement of the corresponding arguments of the *Geometria* (Figure 3).

Let OAC be any parabola, with A its vertex, OC its base and AB its diameter. . . . I say that the parallelogram TC is to the parabola in the ratio of one and a half. Let BA be extended in F so that AF be equal to AD, and let the parallelogram FADE be completed, and in it the diameter AE be drawn; between CE and BF, let any line GR, parallel to CE and BF, serve as rule, cutting BC in G, the curve AC in H, and the lines AD in K, AE in I, FE in R; and through the point H let the line QS be drawn parallel to BC.

The square on CB is to the square on HS as BA is to AS (by proposition 20 of book I of the *Conics* [of Apollonius]). . . . Since BC is equal to AD, that is equal to AF or to RK, and since SH is equal to AK, that is to KI . . . therefore the square on RK is to the square on KI as GK is to KH. . . .

Thus all the squares of the parallelogram FD will be to all the squares of the triangle ADE, with CE as rule, as all the lines of the parallelogram BD to all the lines of the curvilinear triangle ADC, with the same CE as rule. But it has been shown in proposition 24 . . . that all the squares of FD are three times all the squares of the triangle ADE. Therefore, all the lines of the parallelogram DB will be three times all the lines of the curvilinear triangle ADC. Now, by proposition 3 of this book, all the lines of a figure are to all the lines of another as the figures themselves. Therefore, the parallelogram BD is three times the curvilinear triangle ADC, therefore it is to the semi-parabola in the ratio of one to one and a half. . . . Q. E. D.[34]

The evaluation of the area of the parabolic surface is achieved by comparing aggregates of lines and of squares, the infinite aggregates being inserted into relations of proportionality. The figures compared are of a different genus, and the reasoning consists in transmitting a ratio between two aggregates of squares to a ratio between two aggregates of lines, which itself is identical (via the fundamental proposition 3 of book II) to a ratio between two plane figures: the parabolic surface and the rectangle or parallelogram. At each stage, Cavalieri preserves homogeneity between terms that enter a ratio. He deals with a ratio between squares, then between lines, and finally between figures, never mixing heterogeneous entities into a single ratio.

9. REARRANGING INFINITE AGGREGATES AND PULVERIZING SURFACES

In these examples the crucial departure from traditional geometry consists in the manipulation of infinite aggregates or assemblages of lines or surfaces, in agreement with what was permitted by the funda-

---

34. *Exercitationes Geometricae*, 1647, book I, pp. 81–82.

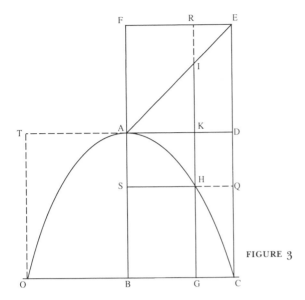

FIGURE 3

mental definitions and theorems we have presented previously. But the rest of the *Geometria* reserves some surprises, offering examples of a more radical departure from Greek canons, a departure of which Cavalieri himself seems not to be totally aware—at least in the *Geometria*— since it implicitly leads to a new and non-canonical treatment of the area of plane figures.

Our first surprise comes from the demonstration of an important and very general result in proposition 15 of book II: Similar figures are to one another as the squares of homologous sides (for instance, the areas of two hexagons are in the same ratio as the squares of their sides, the areas of disks are proportional to the squares of their radii, etc.). A general proof of this result was inaccessible to Greek methods. Euclid, for instance, demonstrates it successively for the case of similar triangles (*Elements* VI, proposition 19): "Similar triangles are to one another in the duplicate ratio of the homologous sides"; then for similar polygons, via a division into similar triangles (*Elements* VI, proposition 20); then he proves that similar polygons inscribed in circles are proportional to the squares of the diameters (*Elements* XII, proposition 1), and as a consequence, through a double reductio ad absurdum, that "circles are to one another as the squares of the diameters" (*Elements* XII, proposition 2).

Geometers of the seventeenth century felt the necessity of having the general assertion at their disposal. Newton makes it his lemma V of the

Principia (book I): "In similar figures all the sides which correspond to one another, curvilinear as well as rectilinear, are proportional, and the areas are in the duplicate ratio of the sides" (*Principia*, 3d ed., p. 30). And Leibniz considers the possibility of a systematic theory of figures in which the notion of similitude would allow a direct derivation of the general assertion that ". . . universally in similar [things], homologous lines, surfaces, solids will be respectively as the length, square and cube of homologous sides, which in its generality was until now more assumed than demonstrated."[35]

Cavalieri demonstrates this result—or tries to—with two similar figures weird enough to make the proof not too restricted (Figure 4). Both are framed between opposite tangents, according to the procedure we know, the tangents being in homologous positions. The inferior tangent serves as rule for the cutting of "all the lines."[36]

The vital spring of the demonstration consists in a reordering of "all the lines" of both figures along a given direction: each line is supposed to slide parallel to itself and to stop when it strikes a fixed vertical or oblique wall MK and M'K'. According to Cavalieri:

Between the opposite tangents [AK and FM, A'K' and F'M'], let us draw KM and K'M' . . . ; let us divide these incident lines KM and K'M' arbitrarily but similarly and on the same side at the points L and L'; let us draw through these points the lines BL and B'L', of which the portions intercepted by the figures will be BE and ID in the figure ABD, B'E' and I'D' in the figure A'B'D'. Let us then take on BL a line [= a segment] equal to BE and ID together and having its extremity on KM; this line is QL. In the same way, let us take Q'L' equal to B'E' and I'D' together, and having its extremity on K'M' at the point L'. Let us do the same for all the other lines that are parallel to the tangents and are enclosed inside the compass of the figures [sicque fiat de ceteris quae ipsis tangentibus aequidistant, et manent intra figurarum ambitus]. (*Geometria Indivisibilibus*, book II, proposition 15, ed. 1653, p. 128)

The benefit of that operation is to assemble together, along a rectilinear boundary, the successive segments cut by the sweeping plane during its motion. If the figure is concave, as in the illustration, segments such as BE and ID, corresponding to a single position of the moving plane, will correspond, after rearrangement, to one segment QL having its end L on the right line MK. Cavalieri's name for such an operation is "the translation of all the lines of the figure AFG into the figure KQM" (p. 128).

If this is done for all the lines of the figure AFG, another figure is obtained, containing segments all terminated on the right line MK at

35. Leibniz, "De Analysi Situs," *Mathematische Schriften* 5, p. 182.
36. Cf. a presentation of the demonstration in K. Andersen, pp. 327–29.

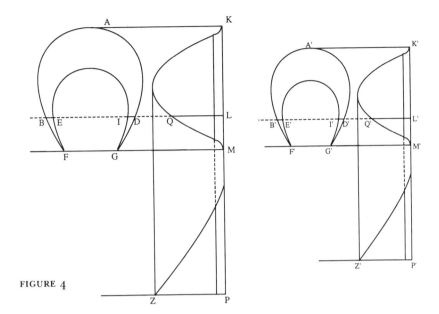

FIGURE 4

one extremity and forming with their other end a certain curve MQK. Cavalieri claims that the new figure MQKLM is equal in area to the former one, thanks to the fundamental theorem: all the lines of one figure are also found in the other. The same thing is done for the smaller figure F'A'G', which gives birth to a corresponding figure M'Q'K'L'M'. Then the same operation is performed in another direction, "all the lines" being cut along the direction of MK, which is chosen as a new "rule," and the lines being "translated" one by one in order to accumulate on the right lines ZP and Z'P'. After these two "translations," the figures have been metamorphosed into new figures, equal to the preceding ones. Both new figures have rectilinear sides along two directions, on which lengths of reference can be chosen, in order to prove that the ratio of the areas is the square of homologous lengths. The astonishing element of the proof resides in the permission to move one by one all the segments marked inside the boundaries of the figures, to place them against a given right line.

We have now to consider a reordering that is perhaps more radical and contradicts more directly the requirements of classical Greek geometry, wiping out any difference between two notions: a surface (which is a two-dimensional geometric object with a determined "shape") and an area (which is a number or an equivalence class of figures, attached to figures of various shapes as a measure of their "content") (Figure 5). Proposition 2 of book II tends to prove that "all the lines of equal plane figures are equal, and all the planes of equal solids are equal, any rule being taken" (*Geometria Indivisibilibus*, ed. 1653, p. 112).

Here is the essential passage of the demonstration:

I say that all the lines of the figure ADC, with the rule AC, are equal to all the lines of the figure AEB, with the rule AB. Let us suppose [intelligatur] that the figure AEB is superposed to the figure ADC, in such a way that their rules are superposed, as AB on AC, or that they are at least parallel. The result is that either the entire figure is congruent to the entire figure, or a part is congruent to a part.

Let us suppose that a part is congruent to a part. Therefore (by postulate 1 of book II [stating that all the lines of congruent plane figures are congruent]) all the lines of these congruent parts will be also congruent; that is to say, all the lines of the part ADB of the figure AEB will be congruent to all the lines of the part ADB of the figure ADC.

Let the remaining parts of the figures be again superposed, with the following requirement: that all their lines are always situated parallel to the rules AB and AC, or to the common rule AB or AC; and let this be performed again and again, until all the remaining parts are superposed to one another [et hoc semper fiat, donec omnes residuae partes ad invicem superpositae fuerint]. Then since the entire figures are equal, the superposed parts will be congruent. Therefore all their lines will also be congruent. But congruent figures are equal to one another [Euclid, *Elements* I, Axiom 7]. Therefore etc. . . . (*Geometria Indivisibilibus* book II, proposition 2, ed. 1653, p. 112).

As E. Giusti notes, the proof is "rather weak."[37] Possibly Cavalieri took inspiration from the procedure of alternate superposition, which Euclid uses in the case of numbers (for instance, *Elements* VII, proposition 1). But the manipulation of continuous magnitudes is subject to laws that are radically different. The iterated superposition of figures may in general fail to reach a final stage. In the Euclidean construction, the difference between congruence of figures and equality of areas is the mark of the specific status of continuous magnitudes. The path from equality by superposability to equality of areas is shown for instance in the cautious steps that lead Euclid from equality of rectangles to equality of parallelograms and triangles (*Elements* I, propositions 34–38).

Such refinements are made superfluous by the indefinite pulverization of figures allowed in Cavalieri's proof. Figures that are equal in area can be metamorphosed into congruent figures after an undetermined number of steps. Cavalieri was not totally unaware of that difficulty, at least some years after the final publication of the *Geometria*,

37. E. Giusti, *Bonaventura Cavalieri*, p. 37.

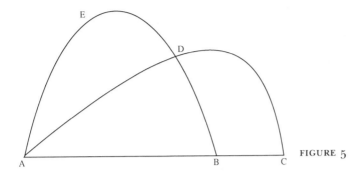

FIGURE 5

since in a letter to Torricelli he commented upon the shortcoming of his proof:

One could doubt whether it [the operation] has ever to reach its term, since through the mutual superposition one does not know if in the figures the parts which happen to be congruent are the half of the figure, or more than the half (because then it will be certain that at least one can arrive at a residue which is less than any proposed quantity) and therefore it could generate some doubts. (*Cavalieri to Torricelli*, 10 March 1643, *Opere di E. Torricelli* 3, p. 114)

He adds:

I believe also that I can give for this proposition a demonstration similar to those of Archimedes, through circumscription and inscription of solids, for these solids which have some regularity; but to find a demonstration different from the one which I have given for all solids, I consider to be of the highest difficulty. (Ibid., p. 115)

Cavalieri is here alluding to the procedure that is standard in Euclid's and Archimedes' demonstrations. In order to prove that a given figure has a certain area (or that the areas of two figures are in a certain ratio), they use an inscribed sequence of rectilinear polygonal figures that are less than the given one, then a sequence of figures that are greater, and they prove that the difference between the area of the figure and the area of the polygon can be made less than any proposed difference, however small, if the number of sides of the polygon is sufficiently large. The gist of the procedure, which has to be proved carefully, is that at each step the difference between the figure and the polygon diminishes by more than the half.

For instance, in the case of the circle, in order to prove that the area of the disk is proportional to the square of the diameter, Euclid (*Elements* XII, proposition 2) proves as a first step that "the inscribed square EFGH is greater than the half of the circle EFGH," then, doubling the number of sides and constructing an inscribed octagon, he proves that "each of the triangles EKF, FLG, GMH, HNE is also greater than the half of the segment of the circle about it" (Figure 6).

The result is true for every polygon in the sequence: square, octagon, 16-sided polygon, etc. . . . By increasing the number of sides by a factor of 2, one reduces the differing area by a factor larger than 2. The conclusion is: "Thus, by bisecting the remaining circumference and joining straight lines, and by doing this continually, we shall leave some segments of the circle that will be less than the excess by which the circle EFGH exceeds the area S [an area slightly inferior to the area of the circle]" (Euclid, *Elements* XII, proposition 2). To obtain that conclusion, Euclid uses a general theorem, proved as *Elements* X, proposition 1: "If two unequal magnitudes are proposed, and if from the greater there is subtracted a magnitude greater than the half, and from the remaining part a magnitude greater than its half, and if this is done continually, the remaining magnitude will be less than the lesser of the proposed quantities."

This explains why Cavalieri wanted to prove that by the successive superpositions he proposed, the part that is covered is "more than the half" of the given figures. But this was proved, in the traditional demonstrations, by relying on the particular properties of the given figures. The fact that in Euclid, *Elements* XII, proposition 2, the successive polygons cut away, at each stage, more than half of the remaining area, depends on the nature of the figure (a circle). For the general case, that is, for figures or solids that are deprived of "some regularity," such a method is inaccessible, or, as Cavalieri confesses, must at least be characterized as a result "of the highest difficulty."

10. CONCLUDING REMARKS

There is some ambivalence in Cavalieri's attitude toward the classical ideal. He recognizes the solidity of the Euclidean edifice and strives to maintain his *Geometria* within that frame. He regrets, in an afterthought, not being able to prove certain basic results rigorously. But it does not prevent him from building his theory upon "inadmissible" procedures.

In what sense are they inadmissible? It is not just a question of respect or admiration for the Ancients, a question of style or taste related to personal or collective sensitivity, as in the case of Petrarch, who, we may say, was more sensitive to the beauties of classical Latin prose than contemporary scholastic writers. In the appreciation of Euclid's subtleties, the very definition of mathematics is at stake. By discussing the

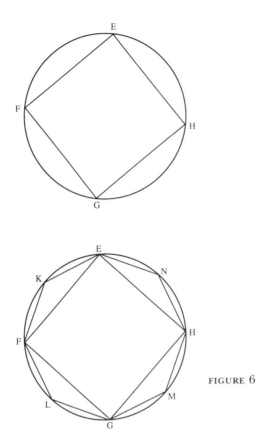

FIGURE 6

difference in status between surface and area, we come closer to what could be called the heart or the central enigma of Greek mathematics. It seems that the discovery of formally cogent demonstrations in Greece was essentially linked to the recognition of certain impossibilities and prohibitions. Chinese geometers, by contrast, were very skillful in manipulating truncated figures and solids to manifest unsuspected equalities by superposition,[38] but apparently they were blind to some questions of principle. They were not concerned whether the procedure of alternate superposition terminated or whether it is allowed to treat the relations between continuous magnitudes in the same manner

---

38. Cf. Donald B. Wagner, "An Early Chinese Derivation of the Volume of the Pyramid, Liu-Hui, 3d Century A.D.," *Historia Mathematica* 6 (1979), pp. 164–188. Cf. also Karine Chemla, "Démonstrations, Analogies, Résonances. Analyse des commentaires de Liu-Hui (3e siécle) aux 9 chapitres sur les procédures mathématiques," forthcoming.

as relations between numbers. Cavalieri holds a privileged position in the seventeenth century. He perceived, up to a certain point, the strength of the classical legacy in mathematics (as for instance Newton also did, from another point of view, and only at a certain stage of his evolution). But he seems, like his contemporaries, unable to reach the same level of rigor. It was probably not before the nineteenth century, among German-speaking mathematicians, such as Bolzano, Dedekind, and Weierstrass, that the deep significance of the Greek canons was fully understood.

9   Descartes' *Geometry* and the Classical Tradition

EMILY GROSHOLZ

In his *Geometry,* Descartes undertakes the rationalization of classical geometry, reorganizing, pruning and extending it by means of his new method. His method is designed to rearrange the subject matter of geometry according to the "order of reasons," so that one begins with starting points that are simple and easy to understand, and proceeds thence step by step to construct items of increasing complexity. Thus Descartes construes his project as both conservative and revolutionary. On the one hand, he is reordering and completing an already existing field, not founding a new one. On the other hand, the only starting points he acknowledges are those furnished by intuition, finite straight line segments standing in certain relations of proportionality. His method, presupposing the radically self-grounding powers of intuition, entails that his starting points have no notable relation to the tradition, nor to any inferential nexus of background knowledge in which they might be embedded.

In this essay, I will examine Descartes' attempt to classify curves according to genre, an attempt structured by his conception of method. In general, Descartes' use of method covers over his debt to the classical tradition. Thus, he often does not see how conservative his allegedly novel reconstructions are; specifically, he is not aware of how little his hierarchy of curves goes beyond the classical canon and how many classical prejudices about curves he shares. And he misses what is most innovative and promising among the ideas generated by his *Geometry,* that is, a way of understanding curves as hybrids, which are simultaneously spatially shaped configurations, algebraic equations in two unknowns, and an infinite array of number pairs. The central and important problem of how to classify curves will illustrate the continuities and discontinuities in Descartes' treatment of classical geometry, and his odd misunderstandings of his own accomplishment.

The three famous problems of classical antiquity were the squaring

of the circle, the duplication of the cube, and the trisection of the angle;
much of the most interesting mathematics in antiquity was generated
by the futile attempt to solve these problems with ruler and compass
and the elaboration of alternative but still unrespectable means of solution. The mathematician Hippias, who taught in Athens in the second
half of the fifth century B.C., is best remembered for his discovery of the
(transcendental) curve called the trisectrix or quadratrix. This curve is
generated by the composition of a rotating and a translational motion,
and provides easy solutions for the trisection of an angle and the squaring of a circle, when it is used as a constructing curve.[1] Menaechmus, a
student of Eudoxus who was in turn a student of Plato, discovered a
whole new family of curves, the conic sections, in an attempt to solve
the duplication of the cube. This problem is easily solved when two
parabolas, or a rectangular hyperbola and a parabola, are used as constructing curves. The great second century B.C. mathematician Archimedes developed the (transcendental) spiral named after him as a constructing curve for the trisection of the angle. Like the quadratrix, it
can also be used to square the circle.[2]

Thus, the range of curves with which the Greeks were familiar was
very limited; it included the straight line, the circle, the conic sections,
a few isolated transcendental curves and little else besides. In part, this
was because the Greeks did not regard curves so much as the source or
focus of problems, what problems are about, but rather as technical
means for solving problems. Curves for the Greeks were constructing
curves. And this habit was due in turn, I think, to the relative intractabil-
ity of curves for the Greeks. They had no flexible, powerful, multivalent
ways of representing curves that would give a purchase on their exploration, and therefore they had no motivation to make them the center
of a research program. The most sophisticated techniques of the age,
those of geometrical algebra (the application of areas and similar analytic tools) and of Archimedean quadrature, are suited for the representation and exploration not of curves, but of shaped areas.

In hindsight, we might think at first that Descartes' *Geometry* provides
precisely the means of representing curves needed to bring them to the
center of the mathematical stage and make them the subject of investigation. Fermat, Descartes' rival, and Leibniz, his respectfully dissenting
student, both recognized and exploited the powerful means furnished
by analytic geometry for representing curves as geometric-algebraic-
numerical hybrids and thus opening up their study. Descartes, how-

1. Carl B. Boyer, *A History of Mathematics* (Princeton University Press, 1985), pp.
140–42.
2. Ibid.

ever, never clearly understood this consequence of his work. His method left him unable to focus his attention on the study of curves as such; sharing this limitation of perspective with the Greeks, he was also unable to expand the collection of curves much beyond the classical canon.

Descartes opens the *Geometry* with the announcement, "Any problem in geometry can easily be reduced to such terms that a knowledge of the lengths of certain straight lines is sufficient for its construction."[3] My contention is that for Descartes the subject matter of geometry is primarily finite straight line segments and the relations of proportionality in which they stand. The demands of his method, which is supposed to be both ampliative and truth-preserving, are so strongly reductive that any other possible object of investigation is either excluded or disintegrated.

In particular, curves and their novel integrity as geometric-algebraic-numerical hybrids can never come into focus for Descartes. Throughout the *Geometry,* he rarely bothers to sketch a curve, and when he does so it is in fragmentary and sometimes erroneous fashion. To put this in slightly different terms, Descartes, only incidentally and not very clearly in the middle of Book II of the *Geometry,* mentions the fact that curves correspond to indeterminate algebraic equations in two unknowns.[4] He spends most of the *Geometry* studying determinate algebraic equations in one unknown, which represent problems like the trisection of the angle and generalizations of such problems. For him, this means constructing ordinates for given abscissas, that is, constructing further line segments on the basis of a given configuration of line segments.

My contention raises the question, what then is going on in Book II of the *Geometry,* where, according to the section headings, Descartes is investigating and classifying curves? In what follows, I will show that Descartes' primary interest in the *Geometry* is the classification of problems (articulated as determinate equations in one unknown), that he undertakes the classification of curves only in order to enhance his classification of problems, and that indeed sometimes he seems to confuse the two kinds of classification, as if classifying problems took care of the classification of curves.

3. D. E. Smith and M. L. Latham, *The Geometry of René Descartes* (Dover, 1954), p. 297. They follow the pagination of the original text, as I do; *La Géométrie* is one of the essays appended to the *Discours de la Methode* (Leiden, 1637). It is reprinted in vol. 6, pp. 367–485, of the *Oeuvres de Descartes,* C. Adam and P. Tannery (eds.) (Paris, 1897–1913).

4. Ibid., p. 335.

Descartes opens Book II by opposing his newly rationalized geometry to that of the ancients.[5]

> The ancients were familiar with the fact that the problems of geometry may be divided into three classes, namely, plane, solid and linear problems. This is equivalent to saying that some problems require only circles and straight lines for their construction, while others require a conic section and still others require more complex curves. I am surprised, however, that they did not go further, and distinguish between different levels (*degrés*) of these more complex curves, nor do I see why they called the latter mechanical rather than geometrical.

Descartes sees the central task he inherits from the Greeks as the classification of problems by means of the constructing curves needed to find a solution. His rationalization of the field consists in clarifying the level of problems occurring after those called solid. Transcendental curves are never appropriate means of construction for such geometric problems; appropriate, rather, are higher algebraic curves, in particular a certain cubic that has come to be known as the Cartesian parabola. Significantly, he does not say that the ancients missed out on many important curves because they weren't able to use the algebraic equation as a way of expressing and analyzing curves. For Descartes, algebra is not an originating source from which items, problems, and solutions might arise, but only a device for information storage, bookkeeping, and abbreviation. Rather, he points to their failure to generalize their means of construction and to submit those means to rational constraints.

An exposition of Descartes' classificatory scheme must begin with a description of Pappus's problem. Pappus of Alexandria flourished in the fourth century A.D.; in Book VII of his *Collection* he went beyond the two classical strategies for generating new curves, kinematic superposition and conic sections, in a significant fashion.[6] There he proposed a generalization of a problem that has been around since Euclid and that implied a whole new class of curves. Descartes had the great insight to take this problem up again and make it the centerpiece of the *Geometry*, but he turned it into a generous promissory note which he himself was not able truly to redeem.

In brief, the problem of Pappus asks for the determination of a locus whose points satisfy one of the following conditions illustrated in Fig-

5. Ibid., p. 315. Smith and Latham often mistranslate "*degré*" as "(algebraic) degree." I explain later on in the essay why it is extremely important to translate this term correctly. I have amended the Smith and Latham translation, and translate "degré" as "level," a somewhat vague appellation that Descartes tries to make precise in Book II with his notion of genre.

6. Boyer, *History of Mathematics*, p. 209.

ure 1. Let the d_i denote the length of the line segment from point P to L_i which makes an angle of ϕ_i with L_i. Choose α/β to be a given ratio and a to be a given line segment.

The problem is to find the points P that satisfy the following conditions. If an even number 2n of lines L_i are given in position, the ratio of the product of the first n of the d_i to the product of the remaining n d_i should be equal to the given ratio α/β. If an odd number (2n − 1) of lines L_i are given in position, the ratio of the product of the first n of the d_i to the product of the remaining (n − 1) d_i times a should be equal to the given ratio α/β. The case of three lines is exceptional, since it arises when two lines coincide in the four line problem; the condition there is

$$(d_1 \times d_2) / (d_3)^2 = \alpha/\beta.$$

The points that satisfy these conditions form loci on the plane.[7]

In the middle of Book I[8] Descartes describes his attack on the problem and then proudly announces, "I believe that I have in this way completely accomplished what Pappus tells us the ancients sought to do," as if he had solved the problem in a thoroughgoing way for any number of lines. While it is true that his combination of algebraic-arithmetical and geometrical results produced an important advance in the solution of the problem, it is not true that his treatment of the problem given in the *Geometry* was complete.

Descartes' explanation of how he proposes to solve this problem is accompanied by a diagram of a four-line instance of it (Figure 2).

He chooses y equal to BC (d_1) and x equal to AB, and then shows how all the other d_1 can be expressed linearly in x and y, by arguments that hinge on facts about similar triangles. Then the proportions defining the conditions for the cases of 3, 2n, and (2n − 1) lines given above can be rewritten as equations in x and y. For n lines, the equation will be of degree at most n; for (n − 1) lines, it will be of degree at most n, but the highest power of x will be at most (n − 1). [For 2n and (2n − 1) parallel lines, the result is an equation in one unknown, y.][9]

The pointwise construction of the locus is then undertaken as follows. One chooses a value for y and plugs it into the equation, thus producing an equation in one unknown, x. For the case of three lines, the equation in x is in general of degree 2; for 2n lines, it is of degree at most n; and for (2n − 1) lines, it is of degree at most (n − 1). [For

7. H. J. M. Bos gives a lucid exposition of Pappus's problem in his "On the Representation of Curves in Descartes' *Géométrie*," *Archive for History of Exact Sciences* 24, no. 4 (1981): 295–338.

8. *Geometry*, p. 309.   9. Ibid., p. 309.

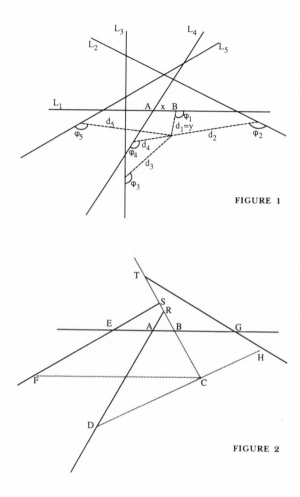

FIGURE 1

FIGURE 2

(2n − 1) parallel lines, there is only an equation in y, which is of degree n.] The roots of this equation can then be constructed by means of intersecting curves which one must decide upon. In the case of a quadratic equation, the choice is ruler and compass, that is, straight line and circle. This procedure generates the curve point by point and is thus potentially infinite, as Descartes notes: "If then we should take successively an infinite number of different values for the line y, we should obtain an infinite number of values for the line x, and therefore an infinity of different points, such as C, by means of which the required curve could be drawn."[10] Thus it seems that Pappus's problem has been

10. Ibid., p. 313.

reduced to the geometrical construction of roots of equations in one unknown: the construction of line segments on the basis of rational relations among other line segments.

The very first thing that Descartes says about his approach to Pappus's problem may seem odd if we expect him to be primarily interested in the loci, the curves, that the problem generates.[11]

First, I discovered that if the question be proposed for only three, four, or five lines, the required points can be found by elementary geometry, that is, by the use of the ruler and compasses only, and the application of those principles which I have already explained, except in the case of five parallel lines. In this case, and in the cases where there are six, seven, eight, or nine given lines, the required points can always be found by means of the geometry of solid loci, that is, by using some one of the three conic sections. Here, again, there is an exception in the case of nine parallel lines. For this and the cases of ten, eleven, twelve, or thirteen given lines, the required points may be found by means of a curve of level [*degré plus composé*] next higher than that of the conic sections. Again, the case of thirteen parallel lines must be excluded, for which as well as for the cases of fourteen, fifteen, sixteen, and seventeen lines, a curve of level [*degré plus composé*] next higher than the preceding must be used; and so on indefinitely.

For in this passage, he is classifying cases not by some feature of the locus generated but rather by what kind of curve can be chosen for the pointwise construction of the locus, that is, for the construction of the line segment x given the relevant equation in x and y and a definite value for y. This is a classification of problems representable by determinate equations in one unknown, not of curves.

In the passage that immediately follows the passage just quoted, Descartes finally has something to say about the loci themselves in this, rather different, classification of cases.[12]

Next, I have found that when only three or four lines are given, the required points lie not only all on one of the conic sections but sometimes on the circumference of a circle or even on a straight line. When there are five, six, seven, or eight lines, the required points lie on a curve of level [*degré plus composé*] next higher than the conic sections, and it is impossible to imagine such a curve that may not satisfy the conditions of the problem; but the required points may possibly lie on a conic section, a circle, or a straight line. If there are nine, ten, eleven, or twelve lines, the required curve can be of level [*degré plus composé*] higher than the preceding, but any such curve may meet the requirements, and so on to infinity.

The levels of complexity that Descartes invokes in this passage are the *genres* he introduces in Book II to classify curves, a classification whose peculiarities I will examine shortly. It is not clear from these remarks

11. Ibid., p. 308.   12. Ibid.

whether Descartes has any real acquaintance with any curves beyond those already familiar from the classical canon, the circle and the conic sections. What do the new curves look like? What are their properties? Do they fall into discernible kinds? What new problems do they suggest? What new problems might they help solve? In Books II and III it turns out that he does know something about a few of them, but this acquaintance is surprisingly minimal.

Before turning to Descartes' treatment of curves in Books II and III, I want to raise a question about the foregoing: Why did Descartes give two distinct classifications of the cases of Pappus's problem, and in that order? My first answer is that he was primarily interested in the classification of problems, and so he gives that classification first; the classification of curves is only subordinate to it. My second answer is that the classification he gives first is the most impressive, since it seems to classify cases for up to seventeen lines and beyond in a precise way. Yet Descartes' formulation obscures both the difficulties inherent in his notion of genre, and the fact that he had not the least inkling of what the loci themselves would look like for cases involving so many lines. Nor for that matter did he know what the constructing curves would be in the cases of fourteen lines or more.

The phrase we have noticed recurrently in the last two passages examined is *degré plus composé*. This phrase is emphatically not to be translated as the algebraic degree of an equation; Descartes never uses it in that way, and for the modern notion of degree he uses the word "dimension."[13] Rather, the phrase *degré plus composé* must be read as "next highest genre." But what is a "genre?" At the beginning of Book II, Descartes tries to make this notion precise by explaining it in terms of, but not identifying it with, algebraic degree. He introduces it after having described a couple of tracing machines, which he presents as rational generalizations of ruler and compass.[14]

I could give here several other ways of tracing and conceiving a series of curved lines, each curve more complex than any preceding one, but I think the best way to group together all such curves and then classify them in order, is by recognizing the fact that all points of those curves which we may call "geometric," that is, those which admit of precise and exact measurement, must bear a definite relation to all points of a straight line, and that this relation must be expressed by means of a single equation.

If this equation contains no term of higher degree than the rectangle of two unknown quantities, or the square of one, the curve belongs to the first and

13. This is an indication of the confusion that Smith and Latham's translation of "*degré*" by "degree" can lead to.
14. *Geometry*, p. 319.

simplest genre, which contains only the circle, the parabola, the hyperbola, and the ellipse; but when the equation contains one or more terms of the third or fourth degree in one or both of the two unknown quantities, the curve belongs to the second genre; and if the equation contains a term of the fifth and sixth degree in either or both of the unknown quantities, the curve belongs to the third genre, and so on indefinitely.

Though this classification of curves refers to the equation, it is not simply algebraic, for then Descartes would have no need to appeal to anything besides the degree of the equation as a classificatory mark and measure of complexity, which can thus order curves. Instead, he puts curves of first and second degree, the conic sections, together in one genre, curves of third and fourth degree together in the second genre, curves of fifth and sixth degree in the third genre, and so forth. Why did he do this?

There are at least three reasons (I will discuss two of them here) why Descartes sets up his classification this way, and each indicates that he was treating curves not as objects of interest in their own right, but as means of construction in the solution of problems. Each reason also involves an unfounded generalization, a rash extrapolation, which again unmasks Descartes' faith in the ability of his method to lead from simple and easy starting points to cases of greater and greater complexity with a kind of transparent, unproblematic reasonableness. He sums this up in the next to last line of the *Geometry:* "It is only necessary to follow the same general method to construct all problems, more and more complex, ad infinitum; for in the case of mathematical progressions, whenever the first two or three terms are given, it is easy to find the rest."[15] Descartes assumes that there are no hidden depths in the starting points, the network of rational relations raised upon them, nor the interaction between those two dimensions of the discovery process. All these assumptions turn out to be wrong.

Descartes gives one reason explicitly: "This classification is based upon the fact that there is a general rule for reducing to a cubic any problem of the fourth degree, and to a problem of the fifth degree any problem of the sixth degree, so that the latter in each case need not be considered any more complex than the former" ["Il y a reigle generale pour reduire au cube toutes les difficultés qui vont au quarré de quarré, et au sursolide toutes celles qui vont au quarré de cube, de façon qu'on ne les doit point estimer plus composées"].[16]

Here Descartes probably had in mind Ferrari's rule for reducing equations of fourth degree in one unknown to ones of third degree.

---

15. Ibid., pp. 412–13.   16. Ibid., p. 323.

But there is no such rule for equations of sixth and fifth degree, and so the extrapolation is unfounded. Moreover, Ferrari's rule is about equations in one unknown, which correspond to problems, not to curves; indeed, Descartes speaks here not of *lignes courbes* but of *difficultés*. He is thus justifying his way of classifying curves by an appeal to the reduction of one kind of problem to another. This is a very strange move and, I think, an indication of the extent to which, for Descartes, the study of curves is subordinate to the study of problems, that is, rational relations among straight line segments. Thus the classification of curves is an unstable project for Descartes; it can't quite come into focus, and it immediately transforms into a classification of problems.

The second reason for Descartes' way of classifying curves by genre reveals the significance of the first reason for him. It does not become explicit until Book III, entitled "On the construction of solid or supersolid problems." Descartes has shown in Book I that problems whose associated equation in one unknown is of degree at most 2 can be solved by ruler and compass. Descartes begins Book III by discussing a variety of problems, the equations associated with them, and the constructing curves needed to solve them, and summarizes his results with a "general rule." Make sure your equation (in one unknown) is reduced to its simplest form, he says, and then if it is of third or fourth degree, the problem is solid and the required constructing curve will be conic sections; indeed, one need employ only a circle and a parabola to solve all such problems.[17] And if the equation is of fifth or sixth degree, the problem belongs to the level next beyond solid. He goes on to solve a variety of solid problems by circle and parabola constructions, and then remarks that more complex problems require constructing curves that are themselves more complex than the conics.[18] At this point he reintroduces the Cartesian parabola, to serve as the pertinent constructing curve for problems at the next, supersolid level.

The Cartesian parabola is one of the very few cubics with which Descartes is really familiar. He derives it in Book II as the locus in a special case of the five-line problem of Pappus, where there are four equidistant parallel lines and a fifth line perpendicular to the others "given in position," and where the relevant ϕ_i are right angles (Figure 3).

In this diagram, he combines the nexus of lines for the Pappian case just described with, superimposed upon it, the second of the two tracing machines mentioned above, where a parabola has been plugged in. Thus, Descartes can furnish a pointwise construction of the curve and

17. Ibid., pp. 389–90.   18. Ibid., pp. 389–402.

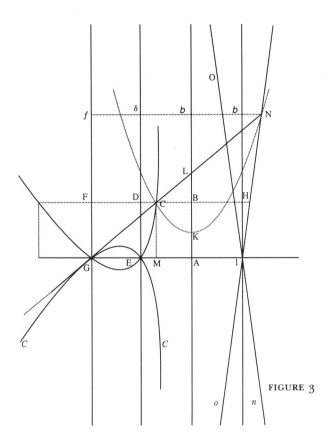

FIGURE 3

trace it continuously by one of his machines. And, as he shows,[19] both the pointwise construction and the tracing yield the same equation

$$y^3 - 2ay^2 - a^2y + 2a^3 = axy$$

The Cartesian parabola is then a respectable constructing curve, and is just what Descartes requires for the next, supersolid level of problems. He demonstrates this in the final pages of Book III,[20] where he uses it as the constructing curve to solve the class of problems associated with the sixth degree equation of general form

$$y^6 - py^5 + qy^4 - ry^3 + sy^2 - ty + u = 0.$$

Descartes then restates his "general rule" on the last page of Book III and of the *Geometry* as a whole. Significantly, in this passage he uses the

19. Ibid., pp. 335–38.   20. Ibid., pp. 403–12.

word "genre" to refer not to curves, but to problems, claiming that he has just reduced to a single construction all problems of the same genre: "reduit à une mesme construction tous les Problemes d'un mesme genre."[21]

... having constructed all plane problems by the cutting of a circle by a straight line, and all solid problems by the cutting of a circle by a parabola, and, finally, all that are but one level more complex [*d'un degré plus composé*] by cutting a circle by a curve only one level higher [*d'un degré plus composé*] than the parabola, it is only necessary to follow the same general method to construct all problems, more and more complex ad infinitum [*qui sont plus composés á l'infini*]; for in the case of mathematical progressions, whenever the first two or three terms are given, it is easy to find the rest.

The introduction of one higher curve as a means of construction apparently allows the construction of roots of equations in one unknown of two successive higher degrees. The introduction of the parabola covers problems whose associated equations are of degrees 3 and 4, and the Cartesian parabola covers equations of degrees 5 and 6. This strongly suggests to Descartes that equations (in one unknown) of degrees $(2n - 1)$ and $2n$ go together, and in this final passage he groups problems in this way, using the word "genre." Thus too he suggests that problems of genre n (where $n > 1$) should be solvable by some canonical curve of genre $(n - 1)$.[22]

Descartes' primary reason for classifying curves into genres in Book II is to provide constructing curves for higher genres of problems. His reasons for associating equations of degree $(2n - 1)$ and $2n$ pertain not to the equations in two unknowns (which correspond to curves), but to equations in one unknown (which correspond to problems). And yet his habit of transposing inquiry into curves back to inquiry into problems, of decomposing curves into indefinitely iterated cases of rational relations among straight line segments, leads him to associated curve-equations of degrees $(2n - 1)$ and $2n$ as well. This is my explanation of Descartes' peculiar notion of genre.

On the basis of this, I can also explain why Descartes knows so little about cubic (and quartic and higher) curves, and why this doesn't bother him, despite his own method's demand for completeness in expositions of this kind. My general claim is that Descartes is interested, not in the investigation of higher curves as such, but rather in "higher problems" and the constructing curves needed to solve them. Since one new constructing curve of a given genre n, like the Cartesian parabola,

21. Ibid., p. 413.
22. Ibid. Smith and Latham's mistranslation is again especially misleading here.

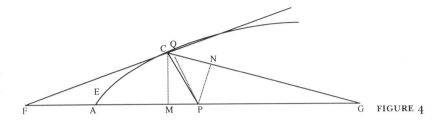

FIGURE 4

suffices for the solution of a whole class of problems of genre (n + 1), there is simply no need to catalogue the higher genres of curves and to discover all their subkinds. (Newton is the first to almost master the cubics; he catalogues seventy-two subkinds and even so omits six.) For Descartes, one good constructing curve as the representative of each higher genre will be sufficient.

Descartes does investigate one group of curves of higher genre: his optical ovals (Figure 4). But his treatment of them is characteristically off-center. He introduces one of them as an illustration of his method of constructing normals, with a diagram where the center of interest is not the curve but the problem, the construction of the line segment CP, the normal to the curve at the point C.[23]

When he discusses the optical ovals in order to exhibit their optical properties, he generates them by pointwise construction.[24] Nowhere does he give an equation for one of them, though he does use algebra to discuss their properties. Optics presents Descartes with a series of problems concerning the conditions under which reflected light rays (straight line segments) will converge at a given point. It's the construction of the point of convergence that interests Descartes, not the curve of the curved reflecting surface. He prefaces and closes his treatment of the optical ovals with an assertion of their usefulness to optics, as if one wouldn't be interested in the study of curves for their own sake. Ultimately, the only higher curve that Descartes understands in the thoroughgoing, multivalent fashion that characterized the important investigations of curves at the end of the century is the Cartesian parabola.

In sum, it is very easy for Descartes to slide from talking about genres as a classification for curves (loci) that refers to the degree of their equation (in two unknowns), to talking about genres as a classification for problems that refers to the degree of their representing equation (in

23. Ibid., pp. 342–45.   24. Ibid., pp. 352–68.

one unknown), especially since he wasn't centrally interested in curves to begin with. When he first uses the word "genre," in the passage from Book II that I cited earlier, he is explicitly classifying curves (albeit implicitly classifying problems). But at the very end of the *Geometry* and Book III, he uses the word explicitly to classify *problems* as he summarizes the fruits of his research. "But it is not my purpose here to write a large book. I am trying rather to include much in a few words, as will perhaps be inferred from what I have done, reducing to a single construction all the problems of one genre."[25] Descartes is really most concerned with and enthusiastic about the classification of problems, in which he has shown that problems of the first genre can be solved by circle and straight line, those of the second genre by circle and parabola, and those of the third genre by circle and Cartesian parabola.

And in this respect he shows himself to be a child of the classical tradition. Like the Greeks, Descartes was interested in curves primarily insofar as they offered means of construction for problems. His conception of method, far from providing radically new beginnings for geometry, obscured its generative links with Greek mathematics as well as its new promise of a multivalent way of understanding curves. Banishing transcendental curves and limiting the exploration of algebraic curves, Descartes allowed his method to close off part of the very mathematical domains it opened.

25. Ibid., pp. 412–13.

Bibliography

Adamson, I. "The Administration of Gresham College and Its Fluctuating Fortunes as a Scientific Institution in the Seventeenth Century." *History of Education* 9 (1980): 13–25.

———. "The Foundation and Early History of Gresham College, London, 1596–1704." Ph.D. dissertation, Cambridge University, 1975.

Aiton, E. J. "Peurbach's Theorica novae planetarum." *Osiris* 3 (1987): 5–44.

———. *Leibniz*. Bristol and Boston, 1985.

———. "Celestial Spheres and Circles." *History of Science* 19 (1981): 75–114.

———. *The Vortex Theory of Planetary Motions*. London and New York, 1972.

Agassi, J. *Towards an Historiography of Science. History and Theory*, Beiheft 2 (1963): 1–117.

Albertus de Saxonia. *Questiones subtillissimae Alberti de Saxonia in libros de caelo et mundo*. Venice, 1492.

Albertus Magnus. *Book of Minerals*. D. Wyckoff (trans.). Oxford: The Clarendon Press, 1967.

———. *Opera Omnia*. Aschendorff, 1951– . (This includes the *Meteororum, De mineralibus, De natura locorum* and *Liber de causis proprietatum elementorum*).

Allen, P. "Medical Education in Seventeenth-Century England." *Journal for History of Medicine* 1 (1946): 115–43.

Amici, G. B. *De motibus corporum coelestium juxta principia peripatetica sine eccentricis & epicyclis*. Paris, 1540.

Andersen, K. "Cavalieri's method of indivisibles," *Archives for History of Exact Sciences* 31 (1985): 291–367.

Aquinas, T. *Sancti Thomae Aquinatis . . . Opera Omnia*, Leonine (ed.). Rome and Turin, 1882– .

Archimedes. *Opera*, J. L. Heiberg (ed.). 3 vols. Leipzig, 1910–15.

Ariew, R. "The Phases of Venus Before 1610," *Studies in History and Philosophy of Science* 18 (1987): 81–92.

Aristotle. *Physica*, D. Ross (ed.). Oxford, 1960.

———. *On the Heavens*. W. K. C. Guthrie (ed. & tr.). London and Cambridge, MA: Loeb Classical Library, 1939.

———. *Physica*. R. P. Hardie & R. K. Gaye (tr.), in W. D. Ross (ed.) *The Works of Aristotle* 2. Oxford, 1930.

Aristotle (pseudo). *On indivisible lines*. In Aristotle, *Minor Works*. Cambridge: Harvard University Press, 1963.

Averroës. *Aristotelis Opera cum Averrois Commentariis*. 9 vols. + 3 supp. vols. Venice, 1562–74; Minerva reprint, Frankfurt-am-Main, 1962.

Avicenna. *Canon*. In Edward Grant (ed.) *A Source Book of Medieval Science*. Cambridge: Harvard University Press, 1974.

197

198   BIBLIOGRAPHY

———. *De congelatione et conglutinatione lapidum.* E. J. Holmyard and D. C. Mandeville (eds. & tr.). Paris: Librairie Orientaliste Paul Geuthner, 1927.
Bacon, Francis. *The Plan of the Great Instauration,* prefixed to his *New Organon.* London, 1620.
Bailly, J.-S. *Histoire de l'astronomie moderne depuis la fondation de l'Ecole d'Alexandrie, jusqu'a l'époque de MDCCXXX.* Paris, 1779–82.
Baker, K. *Condorcet: From Natural Philosophy to Social Mathematics.* Chicago: The University of Chicago Press, 1975.
Barker, P. "Jean Pena and Stoic Physics in the Sixteenth Century." In R. H. Epp (ed.) *Recovering the Stoics. Southern Journal of Philosophy* 23 (Supplement, 1985): 93–107.
Barker, P., and B. R. Goldstein. "The Role of Comets in the Copernican Revolution." *Studies in History and Philosophy of Science* 19 (1988): 299–319.
———. "Is Seventeenth Century Physics Indebted to the Stoics?" *Centaurus* 27 (1984): 148–64.
Bates, D. G. "Thomas Sydenham: The Development of His Thought, 1666–1676," Ph.D. dissertation, The Johns Hopkins University, 1975.
Bayle, P. *Dictionnaire historique et critique.* Rotterdam, 1697.
Becher, J. *Physica subterranea.* 1669.
Bellarmine, R. *The Louvain Lectures,* U. Baldini and G. V. Coyne (eds.). Vatican City: Vatican Observatory Publications. Studi Galileiani, vol. 1, no. 2, 1984.
Bennett, J. A. "The Mechanics' Philosophy and the Mechanical Philosophy." *History of Science* 24 (1986): 1–28.
Birch, T. *The History of the Royal Society of London.* London, 1756.
Blumenberg, H. *The Genesis of the Copernican World.* Cambridge, MA, 1987.
Box, H. J. M. "On the Representation of Curves in Descartes' *Géométrie.*" *Archive for History of Exact Sciences* 24, No. 4 (1981): 295–338.
Boyer, C. *A History of Mathematics.* Princeton: Princeton University Press, 1985.
Boyle, Robert. "The Origin of Forms and Qualities according to the Corpuscular Philosophy." In M. A. Stewart (ed.) *Selected Philosophical Papers of Robert Boyle.* Manchester, England, 1979.
———. *The Works of the Honourable Robert Boyle.* Thomas Birch (ed.). London, 1772.
———. *Of the Reconcileableness of Specific Medicines to the Corpuscular Philosophy.* London, 1685.
———. *Medicinal Experiments: Or, A Collection of Choice and Safe Remedies, for The most part Simple and easily prepared: Very useful in Families, and fitted for the Service of Country People.* 3d ed. London, 1669.
———. *Some Considerations touching the Usefulnesse of Experimental Naturall Philosophy, Propos'd in Familiar Discourses to a Friend, by way of Invitation to the Study of it.* Oxford, 1663.
Brian, Thomas. *The Pisse-Prophet.* London, 1637.
Brockliss, L. W. B. *French Higher Education in the Seventeenth and Eighteenth Centuries.* Oxford: Oxford University Press, 1987.
Broeckx, C. *Essai sur l'Histoire de la médecine Belge avant le XIX siècle.* Bruxelles: Société encyclographique des sciences médicales, 1838.
Brown, H. "L'Académie de Physique de Caen (1666–1675) d'après les lettres d'André de Graindorge." *Mémoires de l'Académie Nationale des Sciences, Arts et Belles-Lettres de Caen* 9 (1939): 117–208.

———. *Scientific Organizations in Seventeenth Century France*. Baltimore: The Johns Hopkins University Press, 1934.

Brown, T. M. "Medicine in the Shadow of the 'Principia'." *Journal of the History of Ideas* 48 (1987): 629–48.

———. "The College of Physicians and the Acceptance of Iatromechanism in England, 1665–1695." *Bulletin of the History of Medicine* 44 (1979): 12–30.

Browne, Richard. *Coral and Steel: A most Compendious Method of Preserving and Restoring Health. Or, a Rational Discourse, grounded upon Experience*. London, 1660.

Browne, Thomas. *Notes and Letters on the Natural History of Norfolk . . . from the MSS. of Sir Thomas Browne*, with notes by Thomas Southwell. London: Jarrold and Sons, 1902.

———. *Pseudodoxia Epidemica: or, Enquiries into Commonly Presumed Truths*. London, 1646.

Buridan, J. *Quaestiones super libri quattuor De Caelo et Mundo*, E. A. Moody (ed.). Cambridge, MA: Medieval Academy of America, 1942.

———. *Questiones super tres primos libros metheororum*. Bibliothèque Nationale, fonds Latin, ms. 14723, ca. 1350.

Burnet, T. *The Sacred Theory of the Earth*. London, 1684.

———. *Telluris theoria sacra*. London, 1680.

Campanus of Novara. *Campanus of Novara and Medieval Planetary Theory: Theorica planetarum*, F. S. Benjamin and G. J. Toomer (eds. & trs.). Madison: University of Wisconsin Publications in Medieval Science, 1971.

Cabeo, N. *Meteorologicorum Aristotelis commentarii*. Rome, 1646.

———. *Philosophica magnetica*. Ferrara, 1629.

Cassini, G. "De l'origine et du progres de l'astronomie, et de son usage dans la geographie et dans la navigation." *Mémoires de l'Academie Royale des Sciences 1666–1699* 8. Paris, 1729–1733.

———. *Recueil d'observations faites en plusiers voyages par ordre de Sa Majeste pour perfectionner l'astronomie et la geographie, avec divers traites astronomiques par Messrs de l'Academie Royale des Sciences*. Paris, 1693.

Cassini, J.-D. *Mémoires pour servir à l'histoire des sciences et à celle de l'Observatoire Royal de Paris, suivis de la vie de J.-D. Cassini, ecrite par lui-meme*. Paris, 1810.

Cassini, J. *Elemens d'Astronomie*. Paris, 1740.

Cavalieri, B. *Exercitationes Geometricae*. Bologna, 1647 (reprint Rome, 1980).

———. *Geometria degli indivisibili*. Turin, 1966.

———. *Geometria indivisibilibus continuorum nova quadam ratione promota*. Bologna, 1635.

Clément, P. *Lettres, Instructions et Mémoires de Colbert*. 8 vols. Paris, 1861–82.

Chalmers, A. "Planetary Distances in Copernican Theory." *British Journal for the Philosophy of Science* 32 (1981): 374–75.

Cochrane, E. W. *Tradition and Enlightenment in the Tuscan Academies 1690–1800*. Chicago, 1961.

Cohen, I. B. *The Newtonian Revolution, with Illustrations of the Transformation of Scientific Ideas*. Cambridge: Cambridge University Press, 1980.

———. "Newtonian Astronomy: The Steps Toward Universal Gravitation." *Vistas in Astronomy* 20 (1976): 85–98.

———. *Introduction to Newton's 'Principia'*. Cambridge: Cambridge University Press, 1971.

Collier, J. *The Great Historical Dictionary . . . out of Lewis Morery.* 2d ed. London, 1701.

Cook, H. "The New Philosophy and Medicine in Seventeenth-Century England." In D. C. Lindberg and R. S. Westman (eds.) *Reappraisals of the Scientific Revolution.* Cambridge: Cambridge University Press, 1990.

———. "Physicians and the New Philosophy: Henry Stubbe and the Virtuosi-Physicians." In R. French and A. Wear (eds.) *The Medical Revolution of the Seventeenth Century.* Cambridge: Cambridge University Press, 1989.

———. *The Decline of the Old Medical Regime in Stuart London.* Ithaca, NY: Cornell University Press, 1986.

Conimbricenses. *Commentarii in libros Meteorum.* Cologue, 1631.

Cornford, F. M. *Plato's Cosmology: The* Timaeus *of Plato* translated with a running commentary. London, 1937.

Craddock, J. R. "Dynasty in Dispute: Alfonso X el Sabio and the Succession to the Throne of Castile and Leon in History and Legend." *Viator* 17 (1986): 197–219.

Crombie, A. C. *Robert Grosseteste and the Origins of Experimental Science 1100–1700.* Oxford: Oxford University Press, 1953.

Crombie, A. C., and A. Carugo. "The Jesuits and Galileo's Ideas of Science and of Nature." *Annali dell'Istituto e Museo di Storia della Scienza di Firenze* 8, fasc. 2, 1983.

Dear, P. *Mersenne and the Learning of the Schools.* Ithaca, NY: Cornell University Press, 1988.

Debus, A. G. *The Chemical Philosophy: Paracelsian Science and Medicine in the Sixteenth and Seventeenth Centuries.* 2 vols. New York: Science History Publication, 1977.

De Gandt, F., ed. "L'oeuvre de Torricelli."*Science Galiléenne et Nouvelle Géométrie.* Nice: Presses de l'Université de Nice, 1988.

Delambre, J. B. *Histoire de l'astronomie moderne.* Paris, 1821. Johnson reprint, New York and London, 1969.

———. *Histoire de l'astronomie du moyen âge.* Paris, 1819.

Descartes, R. *Principles of Philosophy.* V. R. Miller and R. P. Miller (trans.). Collection des Travaux de l'Academie internationale d'histoire des sciences, No. 30. Dordrecht: Reidel, 1983.

———. *Oeuvres de Descartes.* C. Adams and P. Tannery (eds.). Paris: Vrin, 1964–74.

———. *La Géométrie,* appended to *Discours de la Methode.* Leiden, 1639.

Des Maizeaux, P. (trans.) *The Dictionary Historical and Critical of Peter Bayle.* 2d ed. London, 1735.

Dewhurst, K. *Dr. Thomas Sydenham (1624–1689): His Life and Original Writings.* Berkeley: University of California Press, 1966.

Dolaeus, John. *Systema Medicinale, A Compleat System of Physick, Theoretical and Practical.* William Salmon (trans.). London, 1696.

Donahue, W. H. *The Dissolution of the Celestial Spheres 1595–1650.* New York: Arno Press, 1981.

Drake, S. "Galileo's Pre-Paduan Writings: Years, Sources, Motivations." *Studies in History and Philosophy of Science* 17 (1986): 429–88.

———. *Galileo at Work.* Chicago: University of Chicago Press, 1978.

Draper, J. W. *The Conflict between Religion and Science.* 7th ed. London, 1876.

Dreyer, J. L. E. *A History of Astronomy from Thales to Kepler.* Cambridge, 1906. (2d ed. New York, 1953).

Duhem, P. *Uneasy Genius: The Life and Work of Pierre Duhem.* S. L. Jaki (ed.). Boston: Hingham, 1984.

———. *To Save the Phenomena.* E. Dolan and C. Maschler (trans.). Chicago: University of Chicago Press, 1969.

———. "Notice sur les titres et travaux scientifiques de Pierre Duhem, rédigée par lui-même lors de sa candidature à l'académie des sciences (mai 1913)." *Mémoires de la société des sciences physiques et naturelles de Bordeaux.* Series 7, vol. 1 (1917): 160.

———. *Le Système du Monde.* Paris: Hermann, 1914–59.

———. "History of Physics." In *The Catholic Encyclopedia* 12 (1911): 47–67.

———. *Etudes sur Léonard de Vinci.* Paris: Hermann, 1908–13.

———. *Les origines de la statique.* Paris: Hermann, 1906–8.

Eastwood, B. S. " 'The Chaster path of Venus' in the Astronomy of Martianus Capella." *Archive Internationale d'Histoire des Sciences* 32 (1982): 145–58.

Edelstein, L. "The Dietetics of Antiquity." Reprinted in O. Temkin and C. L. Temkin (eds.), C. L. Temkin (trans.), *Ancient Medicine: Selected Papers of Ludwig Edelstein,* pp. 303–16. Baltimore: Johns Hopkins University Press, 1967.

———. "The Relation of Ancient Philosophy to Medicine." Reprinted in O. Temkin and C. L. Temkin (eds.), C. L. Temkin (trans.), *Ancient Medicine: Selected Papers of Ludwig Edelstein,* pp. 349–66. Baltimore: Johns Hopkins University Press, 1967.

Essen, L. van der. *L'Université de Louvain (1425–1940).* Bruxelles: Editions Universitaires, 1945.

Euclid. *Elementa,* post Heiberg edidit Stamatis, 4 vols. Leipzig: Teubner, 1969–72.

———. *Elementorum libri XV quarto,* edited by C. Clavio. Rome, 1603.

———. *Elements,* trans. by T. L. Heath. New York: Dover, 1956.

Eustachius of Sancto Paulo. *Summa philosophiae quadripartita.* Cologne, 1629.

Evans, R. J. W. "Learned Societies in Germany in the Seventeenth Century." *European Studies Review* 7 (1977): 129–51.

———. *Rudolf II and His World: A Study in Intellectual History, 1576–1612.* Oxford: The Clarendon Press, 1973.

Favaro, A. "Galileo Galilei e i Doctores Parisienses." *Rencondita della Accademia dei Lincei* 27 (1918): 3–14.

———. "Léonard de Vinci a-t-il exercé une influence sur Galilée et son école?" *Scientia* 20 (1916): 257–65.

Feingold, M. *The Mathematicians' Apprenticeship: Science, Universities and Society in England, 1560–1640.* Cambridge, 1984.

———. "Universities and the Scientific Revolution: The Case of Oxford." In H. A. M. Snelders and R. P. W. Visser (eds.) *New Trends in the History of Science.* (in press).

Ferguson, W. K. *The Renaissance in Historical Thought: Five Centuries of Interpretation.* Boston: Houghton Mifflin Company, 1948.

Fermat, P. de. *Oeuvres.* P. Tannery and C. Henry (eds.). 4 vols. Paris, 1891–1912.

Fischer, D. H. *Historians' Fallacies: Toward a Logic of Historical Thought.* New York: Harper Torchbooks, 1970.

Flamsteed, J. *The Gresham Lectures of John Flamsteed.* Eric G. Forbes (ed.). London: Mansell, 1975.

Fontana, F. *Novae coelestium terrestriumque rerum observationes, et fortasse hactenus non vulgate, a Francisco Fontana, specillis a se inventis, et ad summam perfectionem perductis, editae.* Naples, 1646.

Fontenelle, B. de. *Histoire de l'Académie Royale des Sciences (1666–1699).* Paris, 1733.

——. *A Discourse of the Plurality of Worlds,* W. D. Knight (trans.). Dublin, 1687.

——. *Entretiens sur la pluralité des mondes.* Paris, 1686.

Forrest, Peter. *The Arraingment of Urines,* James Hart (trans.). London, 1623.

Frank, R. G. *Harvey and the Oxford Physiologists: Scientific Ideas and Social Interactions.* Berkeley and Los Angeles: University of California Press, 1980.

——. "Science, Medicine and the Universities of Early Modern England: Background and Sources." *History of Science* 11 (1973): 194–216, 239–69.

Freind, J. *The History of Physic.* London, 1725.

Galilei, G. *Dialogue on the Two Chief World Systems.* Berkeley: University of California Press, 1967.

——. *Le Opere.* A. Favaro (ed.) (Edizione Nazionale). Florence, 1929–39.

Gardies, J. L. *Pascal entre Eudoxe et Cantor.* Paris: Vrin, 1984.

Gascoigne, J. "The Universities and the Scientific Revolution: The Case of Newton and Restoration Cambridge." *History of Science* 23 (1985): 391–434.

Getz, F. M. "Medicine at Medieval Oxford University." In J. Catto (ed.) *The History of Oxford University* 2. Oxford: Clarendon Press, (forthcoming).

Gingerich, O., and R. S. Westman. *The Wittich Connection: Conflict and Priority in Late Sixteenth Century Cosmology,* Transactions of the American Philosophical Society, vol. 78, part 7, 1988.

Giusti, E. "Dopo Cavalieri. La Discussione sugli indivisibili." In *Atti del Convegno,* "La storia delle matematiche in Italia." Cagliari, 1982: 85–114.

——. "Bonaventura Cavalieri and the Theory of Indivisibles." Introductory essay to a reprint of *Exercitationes Geometricae,* a cura dell'U.M.I., Rome: Cremonese, 1980.

Godfrey, Robert. *Various injuries and abuses in chymical and Galenical physick; committed both by physicians and apothecaries, detected.* London, 1674.

Goldstein, B. R. "The Making of Astronomy in Early Islam." *Nuncius: Annali di Storia della Scienza* 1 (1986): 79–92.

——. *The Astronomy of Levi Ben Gerson (1288–1344).* New York: Springer-Verlag, 1985.

——. "The Arabic Version of Ptolemy's Planetary Hypotheses." *Transactions of the American Philosophical Society* 57, part 4 (1967): 3–55.

Gould, S. J. *Time's Arrow, Time's Cycle.* Cambridge: Harvard University Press, 1987.

Grant, R. *History of Physical Astronomy, from the Earliest Ages to the Middle of the Nineteenth Century.* London, 1852.

Gregory, D. *The Gregory Volume.* Royal Society Archives, MS 247, f. 61.

Groenevelt, Johannes. *The Rudiments of Physick Clearly and Accurately Describ'd and Explain'd, in the most easy and familiar Manner, by Way of Dialogue between a Physician and his Pupil . . . First collected from the Instructions of a celebrated Professor of Medicine in the Royal Academy of Paris: And since Improv'd from the best Authors, Ancient and Modern by John Groenvelt*. Sherborne and London, 1753.

———. *Fundamenta Medicinae Scriptoribus . . . editio noviss.* Venetiis, 1743.

———. *Fundamenta Medicinae Scriptoribus, Tam inter Antiquos quam Recentiores, Praestantioribus . . . Secundum Dictata D. Zypaei, M.D. et Medicinae Professoris Eruditissimi in Academia Lutetiana. Editio Secunda.* London, 1715.

———. *The Grounds of Physick, Containing so much of Philosophy, Anatomy, Chimistry, and the Mechanical Construction of the Humane Body, as is necessary to the Accomplishment of a Physitian: with the Method of Practice in Common Distempers.* London, 1715.

———. *Fundamenta Medicinae Scriptoribus, tam inter Antiquos quam Recentiores, Praestantioribus deprompta, Quorum Nomina Pagina sequens exhibet.* London, 1714.

Guldin, Paul. *Centrobaryca.* Vienna, 1635. (This includes the *Dissertatio physico-mathematica de motu Terrae* and an *Annotatio ad Dissertationem* responding to Cabeo's criticism.)

———. *Dissertatio physico-mathematica de motu Terrae.* Vienna, 1622.

Hahn, R. *The Anatomy of a Scientific Institution: The Paris Academy of Sciences, 1666–1803.* Berkeley: University of California Press, 1971.

Hannaway, O. *The Chemists and the Word: The Didactic Origins of Chemistry.* Baltimore: The Johns Hopkins University Press, 1975.

Hartner, W. "Ptolemaische Astronomie im Islam und zur Zeit des Regiomontanus." In G. Hamann (ed.), *Regiomontanus Studien*, pp. 109–24. Vienna: Österreicher Akademie der Wissenschaft (Phil-hist Kl. 364), 1980.

———. "Ptolemy, Azarquiel, Ibn Al-Shatir, and Copernicus on Mercury: A Study of Parameters." *Archives internationales d'histoire des sciences* 24 (1974): 5–25.

———. "Nasir Al-Din Al-Tusi's Lunar Theory." *Physis* 11 (1969): 287–304.

Harnack, A. *Geschichte der Königlichen preussischen Akademie der Wissenschaften zu Berlin.* 3 vols. Berlin, 1900.

Hart, James. *The Anatomie of Urines.* London, 1625.

Haugeland, J. *Artificial Intelligence: The Very Idea.* Cambridge, MA, and London: MIT Press, 1985.

Havighust, A. F. ed. *The Pirenne Thesis: Analysis, Criticism, and Revision.* 3d ed. Lexington, MA: D. C. Heath, 1976.

Heath, T. *Aristarchus of Samos, the Ancient Copernicus: A History of Greek Astronomy to Aristarchus together with Aristarchus's treatise on the sizes and distances of the Sun and the Moon.* Oxford, 1913.

Heersakkers, C. L. "Foundation and Early Development of the Athenaeum Illustre at Amsterdam." *Lias* 9 (1982): 4–55.

Herivel, J. *The Background to Newton's* Principia: *A Study of Newton's Dynamical Researches in the Years 1664–84.* Oxford: Clarendon Press, 1965.

Hessen, B. "The Social and Economic Roots of Newton's *Principia*." In *Science at the Cross Roads.* London: Frank Cass, 1971 [1931].

Hevelius, Johannes. *Epistolae II. prior: de motu lunae libratio, in certas tabulas*, G. Riccioli (ed.). Danzig, 1654.

———. *Selenographia: sive, lunae descriptio.* Danzig, 1647.

Higham, J. *History: The Development of Historical Studies in the United States.* Englewood Cliffs, NJ: Prentice Hall, 1965.

Hill, C. *Intellectual Origins of the English Revolution*. Oxford, 1965.

Hirschfield, J. M. *The Académie Royale des Sciences 1666–1683*. New York, 1981.

Historical Manuscripts Commission. *The Manuscripts of His Grace the Earl of Portland preserved at Welbeck Abbey*. London, 1893.

Hobbes, T. *De corpore*. 1655.

Hofstadter, R. *The Progressive Historians: Turner, Beard, Parrington*. New York: Alfred Knopf, 1968.

Hooke, R., and D. R. Oldroyd. "Some Writings of Robert Hooke on Procedures for the Prosecution of Scientific Inquiry, Including His 'Lectures of Things Requisite to a Natural History,' " *Notes and Records of the Royal Society* 41 (1987): 145–67.

Horrock, J. *Jeremiae Horroccii Liverpoliensis Opera posthuma*, John Wallis (ed.), 467–68. London, 1672, (2d ed. London, 1673).

Houzeau, J. C., and A. Lancaster. *Bibliographie générale de l'astronomie*. 2 vols. in 3. Brussels, 1882–89. (reprint London: Holland Press, 1964).

Hufbauer, K. *The Formation of the German Chemical Community (1720–1795)*. Berkeley: University of California Press, 1982.

Hunter, M. *The Royal Society and Its Fellows 1660–1700*, British Society for the History of Science, Monograph 4, corrected reprint. Chalfont St. Giles, 1985. First pub. 1982.

———. *Science and Society in Restoration England*. Cambridge: Cambridge University Press, 1981.

Jacob, M. C. *The Newtonians and the English Revolution, 1689–1720*. Hassocks: The Harvester Press, 1976.

Jardine, N. "The Significance of the Copernican Orbs." *Journal for the History of Astronomy* 13 (1982): 168–94.

Jervis, J. R. *Cometary Theory in Fifteenth-Century Europe: Studia Copernicana XXVI*. Warsaw: Polish Academy of Sciences, 1985.

Joannitius. *Isagoge*. In E. Grant (ed.), *A Source Book of Medieval Science*, pp. 705–15. Cambridge: Harvard University Press, 1974.

Joy, L. S. *Gassendi the Atomist: Advocate of History in an Age of Science*. New York: Cambridge University Press, 1988.

Kennedy, E. S., and I. Ghanem (eds.). *The Life and Work of Ibn al-Shatir*. Aleppo: Institute for the History of Arabic Science, 1976.

Kepler, J. *Kepler's Somnium: The Dream, or Posthumous Work on Lunar Astronomy*, E. Rosen (trans.). Madison: University of Wisconsin Press, 1967.

———. *Kepler's Conversation with Galileo's Sidereal Messenger*, E. Rosen (trans.), (The Sources of Science, No. 5). New York: Johnson Reprint Corp., 1965.

———. *Epitome of Copernican Astronomy, IV and V*, in Charles Glenn Wallis (trans.) *Great Books of the Western World* 16 (Ptolemy, Copernicus, Kepler). Encyclopaedia Britannica and University of Chicago, 1952.

———. *Opera Omnia*, C. Frisch (ed.). Frankfurt-Erlangen, 1858–71.

———. *Nova Sterometria Doliorum Vinariorum*. Linz, 1615.

———. *Dissertatio cum Nuncio Sidereo*. Prague, 1610.

Kircher, A. *Mundus subterraneus*. Amsterdam, 1644.

Klopp, O. "Leibniz' Plan der Gründung einer Societät der Wissenschaften in Wien." *Archiv für Osterreichische Geschichte* 40 (1869): 159–255.

Knowles Middleton, W. E. (ed. & trans.) *Lorenzo Magalotti at the Court of Charles II: His Relazione d'Inghilterra of 1668*. Waterloo, Ontario: Wilfrid Laurier University Press, 1980.

————. *The Experimenters: A Study of the Accademia del Cimento.* Baltimore: The Johns Hopkins University Press, 1971.

Koestler, A. *The Sleepwalkers.* New York, 1959.

Kopal, Z., and R. W. Carder. *Mapping of the Moon, Past and Present.* Dordrecht: Reidel, 1974.

Koyré, A. "Bonaventura Cavalieri et la géométrie des continus." In *Etudes d'histoire de la pensée scientifique.* Paris: Gallimard, 1973.

————. *The Astronomical Revolution.* Ithaca: Cornell University Press, 1973.

————. *Metaphysics and Measurement: Essays in Scientific Revolution.* Cambridge: Harvard University Press, 1968.

————. *From the Closed World to the Infinite Universe.* Harper Row, 1958.

————. *Etudes Galiléennes.* Paris, 1939.

Kretzmann, N. (ed.). *Infinity and Continuity in Ancient and Medieval Thought.* Ithaca: Cornell University Press, 1982.

Kuhn, T. *The Structure of Scientific Revolutions.* Chicago and London, 1962.

Lacombe, D. "L'axiomatisation des mathématiques au IIIème siècle avant J. C." In *Thales* (1949–50): pp. 37–58. Paris: PUF, 1951.

Lagrange. *Oeuvres de Lagrange,* J. A. Serret (ed.). Paris, 1867–1892.

Lakatos, I. *The Methodology of Scientific Research Programmes. Philosophical Papers.* Vol. I. Cambridge: Cambridge University Press, 1978.

Lakatos, I., and Zahar, E. "Why Did Copernicus' Research Programme Supercede Ptolemy's?" In R. Westman (ed.), *The Copernican Achievement,* pp. 354–83. Berkeley: University of California Press, 1975.

Latour, B. *Science in Action.* Cambridge: Harvard University Press, 1987.

Lefranc, A. *Histoire du Collège de France.* Paris, 1893 (reprint, Geneva, 1970).

Le Goff, J. *Time, Work, and Culture in the Middle Ages,* A. Goldhammer (trans.). Chicago: The University of Chicago Press, 1980.

Leibniz, G. W. *Mathematische Scriften.* Gerhardt (ed.) (reprint, Hildesheim: Olms, 1971).

————. *Leibnitii opera omnia.* L. Dutens (ed.). Geneva, 1768. (This includes *Epistola ad autorem dissertationes de figuris animalium quae in lapidibus observantur, & lithozoorum nomine venire possunt* and *Mémoire sur les pierres qui renferment des plantes & des poissons dessechés.*)

————. *Protogaea.* L. Sheidt (ed.). Göttingen, 1749.

Lerner, M.-P. " 'Sicut nodus in tabula': de la rotation propre du soleil au seizieme siecle." *Journal for the History of Astronomy* 11, 1980: 114–29.

Lévy, T. *Figures de l'infini, Les Mathématiques au miroir des cultures.* Paris: Le Seuil, 1987.

Lewis, C. S. *The Discarded Image: An Introduction to Medieval and Renaissance Literature.* Cambridge: Cambridge University Press, 1964.

Lipsius, J. *Monita et exempla politica.* Antwerp, 1606.

Lonie, I. M. "The 'Paris Hippocrats': Teaching and Research in Paris in the Second Half of the Sixteenth Century." In A. Wear, R. K. French, and I. M. Lonie (eds.), *The Medical Renaissance of the Sixteenth Century,* pp. 155–74, 318–26. Cambridge: Cambridge University Press, 1985.

Lux, D. S. *Patronage and Royal Science in Seventeenth-Century France: The Académie de Physique in Caen.* Ithaca: Cornell University Press, 1989.

————. "Colbert's plan for La Grande Académie: Royal Policy toward Science, 1663–1667." Paper delivered at the 51st annual meeting of the Southern Historical Association, Houston, 14–16 December 1985.

Lyons, H. *The Royal Society 1660–1940: A History of Its Administration under Its Charters.* Cambridge: Cambridge University Press, 1946.

Mackie, J. M. *Life of Godfrey William Leibnitz.* Boston, 1845.

Maier, A. *On the Threshold of Exact Science,* S. D. Sargent (ed. & trans.). Philadelphia: University of Pennsylvania Press, 1982.

———. *Zwei Grundprobleme der Scholastischen Naturphilosophie.* Rome: Edizioni di Storia e Letteratura, 1968.

Mairan, J. J. d'Ortous de. "Recherches sur l'equilibre de la lune dans son orbite. Premier Mémoire: De la rotation de la lune," *Histoire et Mémoires de l'Academie Royale des Sciences: Mémoires de mathèmatique et de physique,* 1747. Paris, 1752.

Mangeti, J. J. *Bibliotheca Scriptoribus Medicorum Veterum et Recentiorum.* Geneva: Perachon & Cramer, 1731.

Mariana, Juan de. *Historiae de rebus hispaniae.* Toledo, 1592.

Mercator, N. *Nicolai Mercatoris Holsati, e Soc. Reg. Institutionum Astronomicarum Libri Duo, De Motu Astrorum Communi & Proprio, Secundum Hypotheses Veterum & Recentiorum praecipuas; deque Hypotheseon ex observatis construtione: cum Tabulis Tychonianis Solaribus, Lunaribus, Lunae-Solaribus, Et Rodolphinis Solis, Fixarum, et Quinque Errantium; Earumque Usu Praeceptis & Exemplis commonstrato. Quibus accedit Appendix De iis, quae Novissimis temporibus Coeltius innotuerunt.* London: William Godbid, for Samuel Simpson in Cambridge, 1676.

Merrett, Christopher. *The Character of a Compleat Physician, or Naturalist.* London, 1680?

———. *Pinax Rerum Naturalium Britannicarum, Continens Vegetabilia, Animalia, et Fossilia, In haec Insula reperta inchoatus.* London, 1667.

———. "A Paper Concerning the Mineral Called Zaffora by Dr. Merrett found amongst Dr. Hook's papers by Mr. Waller" (Royal Society, RBO.RBC.9.360).

———. "The Art of Refining Lead" (RS, Cl.P.IX[i]1).

———. "Some Observations Concerning the Ordering of Wines" (RS, RBO,RBC.1.278; later published at the end of Walter Charleton's *Discourses on the Wits of Men* [1692]).

———. "An Account of the Tynn Mines and working of Tinn in the County of Cornewall" (RS, RBO.RBC.2.119).

———. "Observations concerning the Uniting of the Barks of Trees cut, to the tree itself" (RS, RBO.RBC.2.301).

Merton, R. *Science, Technology and Society in Seventeenth-Century England.* New York: Howard Fertig/Harper Torchbooks, 1970 (originally published in *Osiris* 4 (1938): 360–632).

Middlehurst, B. M., and G. P. Kuiper (eds.). *The Moon, Meteorites and Comets.* Chicago: The University of Chicago Press, 1963.

Montucla, J. F. *Histoire des Mathématiques.* Paris, 1799.

Moran, B. "German Prince-Practitioners: Aspects in the Development of Courtly Science, Technology, and Procedures in the Renaissance." *Technology and Culture* 22 (1981): 253–74.

———. *The Hermetic World of the German Court: Medicine and Alchemy in the Circle of Moritz of Hessen-Kassel* (forthcoming).

Moreri, L. *Le Grand Dictionnaire historique.* Lyon, 1681.

Muchembled, R. *Popular Culture and Elite Culture in France, 1400–1750.* Lydia Cochrane (trans.). Baton Rouge: Louisiana State University Press, 1985.

Mueller, J. "Greek Mathematics and Greek Logic." In J. Corcoran (ed.), *Ancient logic and its modern interpretation*, pp. 35–70. Dordrecht: Reidel, 1974.

Nedham, Marchamont. *Medela Medicina. A Plea for the free Profession, and a Renovation of the Art of Physick*. London, 1665.

Neugebauer, O. "On the Planetary Theory of Copernicus." *Vistas in Astronomy* 10 (1968): 89–103.

Newton, I. *Unpublished Scientific Papers of Isaac Newton*, A. R. Hall and M. B. Hall (eds.). Cambridge: Cambridge University Press, 1978.

———. *Isaac Newton's Theory of the Moon's Motion (1702)*. London: Dawson, 1975.

———. *Isaac Newton's Philosophiae Naturalis Principia Mathematica*, 3d ed. (1726) with variants. A. Koyré and I. B. Cohen (eds.). Cambridge, MA: Harvard University Press, 1972.

———. *The Correspondence of Isaac Newton*, H. W. Turnbull, J. F. Scott, A. R. Hall, and Laura Tilling (eds.). Cambridge: Cambridge University Press, 1959–77.

———. *Sir Isaac Newton's Mathematical Principles of Natural Philosophy and his System of the World*, A. Motte (tr.), F. Cajori (ed.). Berkeley: University of California Press, 1934.

———. Cambridge University Library, Add. MS 3965.

Neyman, J. (ed.). *Heritage of Copernicus: Theories Pleasing to the Mind*. Cambridge: MIT Press, 1974.

Oldenburg, O. *The Correspondence of Henry Oldenburg*, A. R. Hall and M. B. Hall (eds.). Madison and London, 1965–86.

Olivier-Martin, F. *l'Organization corporative de l'ancien régime*. Paris, 1938.

Oresme, N. *Livre du ciel et du monde*. A. D. Menu and A. J. Denomy (eds.), A. D. Menut (trans.). Madison: University of Wisconsin Press, 1968.

Ornstein, M. *The Rôle of Scientific Societies in the Seventeenth Century*. Chicago: The University of Chicago Press, 1928.

Pappus. *Collectio*. Hultsch (ed.). Berlin, 1876–1878.

Pechy, John. *The Store-house of physical practice*. London, 1695.

Pedersen, O. "Decline and Fall of the *Theorica Planetarum*." In *Science and History: Studia Copernicana XVI*, pp. 157–85. Warsaw: Polish Academy of Sciences, 1978.

Pedersen, O., and M. Pihl. *Early Physics and Astronomy*. New York: Elsevier, 1974.

Pingree, D. "Greek Influence On Early Islamic Mathematical Astronomy." *Journal of the American Oriental Society* 93 (1973): 32–43.

Porter, R. *Health for Sale: Quackery in England 1660–1850*. Manchester: Manchester University Press, 1989.

Potter, P. *Short Handbook of Hippocratic Medicine*. Quebec: Les Editions du Sphinx, 1988.

Poulle, E. "The Alfonsine Tables and Alfonso X of Castille." *Journal for the History of Astronomy* 19 (1988): 97–113.

Primrose, James. *Popular Errors, or the Errours of the People in Physick*, Robert Wittie (trans.). London, 1651.

———. *De vulgi in medicinâ Erroribus Libri quatuor*. London, 1638.

Proclus. *In Primum Euclidis Librum Elementorum Commentarii*. Friedlein (ed.). Repr. Hildesheim: Olms, 1967.

———. *Proclus, Commentaire sur le Timée*, A. J. Festugière (ed. & trans.). Paris, 1966–68.

————. *Procli Diadochi in Platonis Timaeum Commentaria,* E. Dieh (ed.). Leipzig, 1903–6.

Purver, M. *The Royal Society: Concept and Creation.* London: Routledge and Kegan Paul, 1967.

Rademaker, C. S. M. *Life and Work of Gerardus Joannes Vossius.* Assen, 1981.

Ragep, F. J. "Two Versions of the Tusi Couple." In D. A. King and G. Saliba (eds.), *From the Deferent to the Equant,* pp. 329–56. New York: New York Academy of Sciences, 1987.

————. *Cosmography in the "Tadhkira" of Nasir al-Din al-Tusi.* Ann Arbor, MI: University Microfilms, 1982. Harvard Ph.D. dissertation.

Rattansi, P. M. "Early Modern Art, 'Practical' Mathematics and Matter Theory." In Rom Harré (ed.), *The Physical Sciences since Antiquity,* pp. 63–77. London: Croom Helm, 1986.

————. "Paracelsus and the Puritan Revolution." *Ambix* 11 (1963): 24–32.

Raven, C. E. *English Naturalists from Neckam to Ray: A Study in the Making of the Modern World.* Cambridge: Cambridge University Press, 1947.

Redondi, P. *Galileo Heretic.* Princeton: Princeton University Press, 1987.

Reichenbach, E., and G. Uschmann (eds.). *Nunquam otiosus. Beiträg zur Geschichte der Präsidenten der Deutsche Akademie der Naturforscher Leopoldina.* Leipzig, 1970.

Reisman, D. *Thomas Sydenham Clinician.* New York: Paul B. Hoeber, 1926.

Riccioli, G. B. *Almagestum novum.* Bologna, 1651.

Robertson, D. M. *A History of the French Academy.* New York, 1910.

Roberval, G. P. de *Divers Ouvrages,* in *Mémoires de l'Académie Royale des Sciences* 6. Paris, 1693.

Rosen, E. "The Dissolution of the Solid Celestial Spheres," *Journal of the History of Ideas* 46 (1985): 13–21.

————. "Francesco Patrizi and the celestial spheres." *Physis* 26 (1984): 305–24.

Rosenberg, C. E. "Woods or Trees: Ideas and Actors in the History of Science." *Isis* 79 (1988): 565–70.

Rossi, P. *The Dark Abyss of Time.* Chicago: The University of Chicago Press, 1984.

————. *Philosophy, Technology and the Arts in the Early Modern Era.* S. Attanasio (trans.), Benjamin Nelson (ed.). New York: Harper and Row, 1970.

Royal Society. *The Record of the Royal Society of London.* 4th ed. London, 1940.

Rudwick, M. *The Meaning of Fossils.* 2d ed. Chicago: The University of Chicago Press, 1985.

Ruestow, E. G. *Physics at Seventeenth- and Eighteenth-Century Leiden.* The Hague, 1973.

Sabra, A. I. "The Andalusian Revolt against Ptolemaic Astronomy." In E. Mendelsohn (ed.), *Transformation and Tradition in the Sciences,* pp. 133–53. Cambridge: Cambridge University Press, 1984.

————. "An Eleventh Century Refutation of Ptolemy's Planetary Theory." In E. Hilfstein (ed.), *Science and History: Studia Copernicana XVI,* pp.117–31. Warsaw: Polish Academy of Sciences, 1978.

Saliba, G. "The Role of Maragha in the Development of Islamic Astronomy: A Scientific Revolution before the Renaissance." *Revue de Synthèse* 108 (1987): 361–73.

————. "Arabic Astronomy and Copernicus." *Zeitschrift fur Geschichte der Arabisch-Islamischen Wissenschaften* 1 (1984): 73–87.

————. "The Original Source of Qutb Al-Din Al-Shirazi's Planetary Model." *Journal for the History of Astronomy* 3 (1979): 4–18.

————. "The First Non-Ptolemaic Astronomy at the Maragha School." *Isis* 70 (1979): 571–76.

Sanchez de Arévalo, Rodrigo. *Roderici Santii Episcopi Palentini Historiae Hispanicae, partes quatuor.* In R. Bell, *Rerum hispanicarum scriptores aliquot.* Frankfurt, 1579.

Shackleton, R. (ed.). *Fontenelle: Entretiens sur la pluralité des mondes.* Oxford, 1955.

Shapin, S., and S. Schaffer. *Leviathan and the Air-Pump: Hobbes, Boyle, and the Experimental Life.* Princeton: Princeton University Press, 1985.

Shapiro, B. J. *Probability and Certainty in Seventeenth-Century England.* Princeton: Princeton University Press, 1983.

————. "The Universities and Science in Seventeenth Century England." *Journal of British Studies* 10 (1971): 47–82.

Scilla, A. *De corporibus marinis lapidescentibus.* Rome, 1747. (Translation of 1670).

————. *La vana speculazione disingannata dal senso.* Naples, 1670.

Siraisi, N. *Avicenna in Renaissance Italy: The Canon and Medical Teaching in Italian Universities after 1500.* Princeton: Princeton University Press, 1987.

Slack, P. "Mirrors of Health and Treasures of Poor Men: The Uses of the Vernacular Medical Literature of Tudor England." In C. Webster (ed.), *Health, Medicine and Mortality,* pp. 237–73. Cambridge: Cambridge University Press, 1979.

Smith, D. E., and M. L. Latham. *The Geometry of René Descartes.* Dover, 1954.

Smith, V. "Physical Puritanism and Sanitary Science: Material and Immaterial Beliefs in Popular Physiology, 1650–1840." In W. F. Bynum and R. Porter (eds.), *Medical Fringe and Medical Orthodoxy 1750–1850,* pp. 174–97. London: Croom Helm, 1987.

Smith, W. D. *The Hippocratic Tradition.* Ithaca, NY: Cornell University Press, 1979.

Sondervorst, F.-A. *Histoire de la médecine belge.* Zaventem: Sequoia, 1981.

Sorabji, R. *Time, Creation and the Continuum; Theories in Antiquity and the Early Middle Ages.* London: Duckworth, 1983.

Sprat, T. *History of the Royal Society.* Jackson I. Cope and Harold Whitmore Jones (eds.). St. Louis and London, 1959. Originally London: J. Martyn, 1667.

Spufford, M. *The Great Reclothing of Rural England: Petty Chapmen and Their Wares in the Seventeenth Century.* London: Hambleton Press, 1984.

Steneck, N. H. " 'The Ballad of Robert Crosse and Joseph Glanvill' and the Background to Plus Ultra." *The British Journal for the History of Science* 14 (1981): 59–74.

————. *Science and Creation in the Middle Ages.* Notre Dame, IN: Notre Dame University Press, 1976.

Steno, Nicholas *Prodromus.* 1916. (Translation of 1669; reprinted in 1968 by Haffner Press).

————. *The Prodromus to a dissertation concerning solids naturally contained within solids.* London, 1671. (Translation of 1669).

————. *De solido intra solidum naturaliter contento dissertationis prodromus.* Florence, 1669.

Stimson, D. *Ancients and Moderns.* London, 1949.

Stock, B. *Myth and Science in the Twelfth Century: A Study of Bernard Silvester.* Princeton: Princeton University Press, 1972.

Swerdlow, N. "Pseudodoxia Copernicana." *Archives Internationales d'Histoire des Sciences* 26 (1976): 108–58.

———. "The Derivation and First Draft of Copernicus's Planetary Theory: A Translation of the Commentariolus with Commentary." *Proceedings of the American Philosophical Society* 117 (1973): 423–512.

———. "Aristotelian Planetary Theory in the Renaissance: Giovanni Battista Amico's Homocentric Spheres." *Journal for the History of Astronomy* 3 (1972): 36–48.

Swerdlow, N., and O. Neugebauer. *Mathematical Astronomy in Copernicus's De Revolutionibus.* New York: Springer-Verlag, 1984.

Sydenham, Thomas. *The Whole Works of that Excellent Practical Physician Dr. Thomas Sydenham,* John Pechy (trans.). London, 1696.

Talbot, C. C. "Medicine." In David C. Lindberg (ed.), *Science in the Middle Ages,* pp. 391–428. Chicago: The University of Chicago Press, 1978.

Taylor, A. E. *A Commentary on Plato's* Timaeus. Oxford, 1928.

Temkin, O. "Greek Medicine as Science and Craft." *Isis* 44 (1953): 213–25.

Thirsk, J. *Economic Policy and Projects: The Development of a Consumer Society in Early Modern England.* Oxford: Clarendon Press, 1978.

Thomas, K. *Religion and the Decline of Magic: Studies in Popular Beliefs.* New York: Charles Scribner's Sons, 1971.

Thorndike, L. *History of Magic and Experimental Science.* New York: Columbia University Press, 1923–58.

Tilley, A. *Studies in the French Renaissance.* Cambridge: Cambridge University Press, 1922.

Tilmann, J. P. *An Appraisal of the Geographical Works of Albertus Magnus and His Contributions to Geographical Thought.* Michigan Geographical Publication No. 4. Ann Arbor: Department of Geography, University of Michigan, 1971.

Torricelli, E. *Opere.* G. Loria and G. Vassura (eds.). Faenza, 1919–44.

Trinkaus, C. *In Our Image and Likeness.* Chicago: The University of Chicago Press, 1970.

Tyacke, N. "Science and Religion at Oxford before the Civil War." In D. Pennington and K. Thomas (eds.), *Puritans and Revolutionaries.* Oxford, 1978.

Van Helden, A. *Measuring the Universe.* Chicago: The University of Chicago Press, 1985.

Van Schooten, F. *De organica conicarum sectionum in plano descriptione.* Leyden 1646.

Vasquez, G. *Commentariorum ac disputationum in primam partem Sancti Thomae.* Antwerp, 1596.

Voltaire, F. M. A. de. *Essai sur les moeurs et l'esprit des nations,* 1756. R. Pomeau (ed.). Paris: Classiques Garnier, 1963.

Wallace, W. *Galileo and His Sources: The Heritage of the Collegio Romano in Galileo's Science.* Princeton: Princeton University Press, 1984.

———. *Prelude to Galileo. Essays on Medieval and Sixteenth-Century Sources of Galileo's Thought.* Dordrecht: Reidel, 1981.

———. "Galileo Galilei and the *Doctores Parisienses.*" In R. E. Butts and J. C. Pitt (eds.), *New Perspectives on Galileo,* pp. 87–138. Dordrecht: Reidel, 1978.

———. *Galileo's Early Notebooks: The Physical Questions. A Translation from the*

Latin, with Historical and Paleographical Commentary. Notre Dame: Notre Dame University Press, 1977.

———. *Causality and Scientific Explanation* 1. Ann Arbor: Univeristy of Michigan Press, 1972.

Wallis, J. *Opera Mathematica.* 3 vols. Oxford, 1695–1699.

Walton, I. *The Compleat Angler,* London: Oxford University Press, 1935. Introduction by John Buchan.

Ward, J. *The Lives of the Professors of Gresham College.* London, 1740.

Waschkies, H. J. *Von Eudoxos zu Aristoteles, Das Fortwirken der Eudoxischen Proportionentheorie in der Aristotelischen Lehre vom Kontinuum.* Amsterdam: Grüner, 1977.

Webster, C. "Alchemical and Paracelsian Medicine." In C. Webster (ed.), *Health, Medicine and Mortality in the Sixteenth Century,* pp. 301–34. Cambridge: Cambridge University Press, 1979.

———. *The Great Instauration: Science, Medicine and Reform 1626–1660.* New York: Holmes and Meyer, 1975.

———. English Medical Reformers of the Puritan Revolution: A Background to the 'Society of Chymical Physitians,' " *Ambix* 14 (1967): 16–41.

Westfall, R. S. "Science and Patronage: Galileo and the Telescope." *Isis* 76 (1985): 11–30.

———. "Isaac Newton in Cambridge: The Restoration University and Scientific Creativity." In P. Zagorin (ed.), *Culture and Politics from Puritanism to the Enlightenment.* Berkeley and Los Angeles, 1980.

———. *Never at Rest: A Biography of Isaac Newton.* Cambridge: Cambridge University Press, 1980.

———. *The Construction of Modern Science.* New York, 1977.

Westman, R. S. "Proof, Poetics and Patronage: Copernicus's Preface to *De Revolutionibus.*" In D. C. Lindberg and R. S. Westman (eds.), *Reappraisals of the Scientific Revolution.* Cambridge: Cambridge University Press, 1990.

———. "The Astronomer's Role in the Sixteenth Century: A Preliminary Study." *History of Science* 18 (1980): 105–47.

———. "The Melanchton Circle, Rheticus and the Wittenberg Interpretation of the Copernican Theory." *Isis* 66 (1975): 165–93.

Whewell, W. *History of the Inductive Sciences.* 3 vols. London, 1837.

White, A. D. *A History of the Warfare of Science with Theology in Christendom.* 2 vols. New York, 1896.

White, Lynn. "Medical Astrologers and Medieval Technology." In L. White, *Medieval Religion and Technology,* pp. 297–315. Berkeley: University of California Press, 1978.

Whiteside, D. T. "Newton's Lunar Theory: From High Hope to Disenchantment." *Vistas in Astronomy* 19 (1976): 317–28.

———. "Before the *Principia:* The Maturing of Newton's Thoughts on Dynamical Astronomy, 1664–1684." *Journal of the History of Astronomy* 1 (1970): 5–19.

Wieland, W. *Die Aristotelische Physik.* Göttingen: Vandenhoek & Ruprecht, 1970.

Woodward, J. *An Essay toward a Natural History of the Earth.* London, 1695.

Zurita, Gerónimo de. *Anales de la Corona de Aragón.* Zaragoza, 1585.

Zypaeus, François. *Fundamenta medicinae physico-anatomica.* Brussels, 1683.

Notes on the Contributors

The editors of this collection are Roger Ariew and Peter Barker. Ariew is professor, Department of Philosophy and the Center for Programs in the Humanities, and adjunct professor in the Center for the Study of Science in Society, at Virginia Polytechnic Institute and State University (Virginia Tech). He is the author of articles on the history of philosophy, and the history and philosophy of science. He recently edited and translated *G. W. Leibniz's Philosophical Essays* (Indianapolis: Hackett, 1989) with Daniel Garber. He has also edited and translated Martial Gueroult's *Descartes' Philosophy Interpreted According to the Order of Reasons*, 2 vols. (Minneapolis: University of Minnesota Press, 1984–5); and a selection from P. Duhem's *Système du Monde—Medieval Cosmology: Theories of Infinity, Place, Time, Void, and the Plurality of Worlds* (Chicago: University of Chicago Press, 1985). He is currently working on the social and intellectual context of Cartesianism.

Peter Barker is associate professor, Center for the Study of Science in Society and Department of Philosophy, at Virginia Tech. He has published a number of papers on historical approaches to philosophy of science and on the philosophy of Wittgenstein. He is the editor, with C. G. Shugart, of *After Einstein* (Memphis: Memphis State Press, 1981), and the author of "Jean Pena and Stoic Physics in the Sixteenth Century," in R. H. Epp (ed.), *Recovering the Stoics: Spindel Conference 1984, Southern Journal of Philosophy* 23 (Supplement) (1985): 93–107, and, with B. R. Goldstein, of "The Role of Comets in the Copernican Revolution," *Studies in History and Philosophy of Science* 19 (1988): 299–319. He is currently studying the role of Stoic philosophy and physics in the sixteenth century.

Harold J. Cook is an associate professor in the Departments of the History of Medicine and History of Science at the University of Wisconsin—Madison. He is the author of: *The Decline of the Old Medical Regime in Stuart London* (Ithaca, NY: Cornell University Press, 1986), and a number of papers on the history of medicine, including "The

New Philosophy and Medicine in Seventeenth-Century England," in D. Lindberg and R. Westman (eds.), *Reappraisals of the Scientific Revolution* (Cambridge: Cambridge University Press, 1990), pp. 397–436 and "Practical Medicine and the British Armed Forces after the 'Glorious Revolution,' " *Medical History* 34 (1990): 1–26. He spent 1989–90 working in The Netherlands on studies of Dutch medicine and natural history in the latter seventeenth century.

Mordechai Feingold is Associate Professor of History and Science and Technology Studies in the Center for the Study of Science in Society, at Virginia Polytechnic Institute and State University. He is the author of *The Mathematicians' Apprenticeship* (Cambridge: Cambridge University Press, 1984) and the editor of *Before Newton: The Life and Work of Isaac Barrow* (Cambridge: Cambridge University Press, 1990). He is currently engaged in a study of Newtonianism.

Alan Gabbey is Reader Emeritus, Queen's University Belfast. He is an Effective Member of the Académie internationale d'histoire des sciences. His interests lie mainly in early modern science and philosophy, especially Cartesian and Newtonian natural philosophy, Henry More, and the seventeenth-century French scientific community (including archival resources). Among his publications are: "Force and inertia in the seventeenth century: Descartes and Newton," in S. Gaukroger (ed.), *Descartes: Philosophy, mathematics and physics*, 1980, pp. 230–320; "Philosophia cartesiana triumphata: Henry More 1646–1671," in T. M. Lennon et al. (eds.), *Problems of Cartesianism*, 1982, pp. 171–250); "The Mechanical Philosophy and Its Problems," in J. Pitt (ed.), *Change and progress in modern science*, 1985, pp. 9–84; "Newton and Natural Philosophy," in R. C. Olby et al. (eds.), *Companion to the History of Modern Science*, 1990, pp. 243–63. Currently he is investigating the nature-art distinction from medieval to modern times, and preparing a comparative study of Pascal and Roberval as philosophers.

François De Gandt is research fellow of the Centre National de la Recherche Scientifique, attached to the Centre Alexandre Koyré (Paris) and to the Département d'Astrophysique Rélativiste et de Cosmologie (Observatoire de Paris—Meudon). He has published a number of papers on Aristotelian science (especially the *Posterior Analytics, Physics*, and *Mechanics*), on seventeenth-century mathematics and natural philosophy, and on Newton (*Revue d'Histoire des Sciences*, 1986 and 1987, and *Graduate Faculty Philosophy Journal*, 1987). Most recently two chapters on percussion and on indivisibles appeared in F. De Gandt (ed.), *L'oeuvre de Torricelli: Science galiléenne et nouvelle géometrie* (Nice: Presses de l'Universite de Nice, Diffusion Soc. des Belles Lettres, 1989). He is also noted for his French translations of Kant, Hegel, Galileo, and

Torricelli, and serves as Secretary of the Committee for the Edition of the Works of D'Alembert.

Bernard R. Goldstein is professor of Religious Studies and History of Science at the University of Pittsburgh. He is a former Guggenheim fellow, a founding member of the Academy for Jewish Philosophy, a fellow of the Institute for Research in Classical Philosophy and Science, and an Effective Member of the Académie internationale d'histoire des sciences. Among his recent publications are: *The Astronomy of Levi ben Gerson (1288–1344)* (New York: Springer, 1985); a collection of twenty-four essays, *Theory and Observation in Ancient and Medieval Astronomy* (London: Variorum, 1985); and *From Ancient Omens to Statistical Mechanics: Essays on the Exact Sciences Presented to Asger Aaboe*, edited with J. L. Berggren (Copenhagen: University Library, 1987).

Emily Grosholz is an associate professor of philosophy at the Pennsylvania State University. Her book, *Cartesian Method and the Problem of Reduction*, was published by Oxford University Press in 1990. She has published articles on Descartes in *PSA 1986, PSA 1988*, the *Graduate Faculty Philosophy Journal* XII, 1 and 2; and as chapters in S. Gaukroger (ed.), *Descartes: Philosophy, Mathematics and Physics* (Harvester Press, 1980) and in G. Moyal (ed.) *Descartes: Critical Assessments* (Croom Helm, forthcoming). Her edition of "Two Leibnizian Manuscripts of 1690 Concerning Differential Equations" appeared in *Historia Mathematica* 14 (1987).

David Lux is associate professor, Department of History, Bryant College. He has published articles on historiography and the institutional history of French science. He is the author of *Patronage and Royal Science in Seventeenth-Century France* (Ithaca, NY: Cornell University Press, 1989). He is preparing a book on early modern scientific institutions and is also engaged in research on the historical processes of technological innovation.

Joseph Pitt is professor in the Department of Philosophy at Virginia Tech. He is the author of *Galileo, Human Knowledge and the Book of Nature* (Dordrecht: Kluwer, 1991) and *Pictures, Images and Conceptual Change: An Analysis of Wilfrid Sellar's Philosophy of Science* (Dordrecht: Reidel, 1981) and the editor of several anthologies, including *New Perspectives on Galileo* (with R. E. Butts, Dordrecht: Reidel, 1978); *Theories of Explanation* (Oxford: Oxford University Press, 1988), and *Rational Changes in Science* (with Marcello Pera, Dordrecht: Reidel, 1987). He has published numerous articles on the philosophy of science, including "The Character of Galilean Evidence" in *PSA 1986*, Vol. 2., and "Galileo, Rationality and Explanation" in *Philosophy of Science* 55 (1988).

Index

Index compiled with the assistance of Xiang Chen.

Revolution and Continuity
was composed in 10 on 12 Baskerville
by World Composition Services, Inc.,
and printed by McNaughton & Gunn Lithographers.